Fatigue of Structures and Materials

Jaap Schijve

Fatigue of Structures and Materials

Second Edition

Springer

Jaap Schijve
Professor of Aircraft Materials (Emeritus)
Delft University of Technology
Faculty of Aerospace Engineering
Delft, The Netherlands

Previously:
Structures and Materials Division of the National Aerospace Laboratory NLR
Amsterdam/North-East Polder, The Netherlands

The paper "Fatigue of structure and materials in the 20th century and the state of the art" has been reprinted with kind permission from *International Journal of Fatigue*, Elsevier, vol. 25, 2003, pp. 679–702.

ISBN-13: 978-94-007-8692-9   ISBN 978-1-4020-6808-9 (eBook)

Additional material to this book can be downloaded from http://extras.springer.com

Preface

Fatigue of structures and materials covers a wide scope of different topics. The purpose of the present book is to explain these topics, to indicate how they can be analyzed, and how this can contribute to the designing of fatigue resistant structures and to prevent structural fatigue problems in service.

Chapter 1 gives a general survey of the topic with brief comments on the significance of the aspects involved. This serves as a kind of a program for the following chapters. The central issues in this book are predictions of fatigue properties and designing against fatigue. These objectives cannot be realized without a physical and mechanical understanding of all relevant conditions. In Chapter 2 the book starts with basic concepts of what happens in the material of a structure under cyclic loads. It illustrates the large number of variables which can affect fatigue properties and it provides the essential background knowledge for subsequent chapters. Different subjects are presented in the following main parts:

- Basic chapters on fatigue properties and predictions (Chapters 2–8)
- Load spectra and fatigue under variable-amplitude loading (Chapters 9–11)
- Fatigue tests and scatter (Chapters 12 and 13)
- Special fatigue conditions (Chapters 14–17)
- Fatigue of joints and structures (Chapters 18–20)
- Fiber-metal laminates (Chapter 21)

Each chapter presents a discussion of a specific subject. The major aspects are recapitulated in the last section of a chapter, not as a summary, but just as the most important points to remember. In spite of a qualitatively good understanding of the fatigue phenomenon in structural elements, the quantitative accuracy of the prediction of fatigue properties can still be limited. As a consequence, it is important that all relevant issues are well recognized and understood. This was one of the major reasons for writing the present textbook. Knowledge of the fatigue mechanism in the material and how it can be affected by a large variety of practical conditions is of the utmost importance. The designer of a dynamically loaded structure should "design against fatigue". This approach includes not only the overall concept of the structure with related safety and economic aspects, but also questions on detail design, material surface quality, and joints. At the same time, the designer should try to predict the fatigue performance of a structure, fatigue limits, fatigue lives until crack initiation and the remaining life covered by crack growth until final failure. The approach requires a profound knowledge of the various influencing factors, also because predictions on fatigue properties have their limitations and shortcomings.

The first edition of the present book was published in 2001. It has been used for university courses and several courses with participants from industry, universities, research institutes, official agencies, and teachers of technical schools. Experience has shown that the book is generally appreciated as a textbook for basic knowledge about fatigue, but also for design applications and research programs. A new feature of the present edition is the CD-rom included in this book. The first part of the CD with exercises and summaries will be useful for students and teachers, and also for self-tuition. A number of instructive case histories on fatigue problems in service is covered in the second part of the CD. The third part includes aspects of designing against fatigue and planning experimental fatigue studies. In the last part personal reflections are presented on possible research of fatigue problems in the future.

References to the literature are added to each chapter, but the number of references is much smaller than usual for a monograph. Literature sources from which results or figures have been used in this book are listed at the end of each chapter. The lists are supplemented by a small number of general references, mainly books and conference proceedings. However, research workers who want to investigate specific problems in detail should access computerized retrieval systems.

After working for more than 40 years on fatigue problems, I finally had time to write the first edition of the present book, which was published in 2001. In the present edition, written seven years later, the text of all chapters has been carefully screened and corrected, but the fundamental information is still the same. However, the chapter on "Designing against Fatigue of Structures" (Chapter 19 in the first edition) has been replaced by a completely newly written text which is Chapter 20 in the present edition.

Another difference with the first edition is the CD-rom now attached to the book, which gives the reader more opportunity to work with the material offered and to guide and stimulate new work in this field. I trust that the present edition will be helpful for everybody engaged in fatigue of structures and materials, and also for a new generation of students.

Jaap Schijve
Delft, October 2008

Acknowledgments

For a period of 20 years I have been working in the Structures and Materials Division of the National Aerospace Laboratory NLR in Amsterdam and in the North-Eastpolder, and for another 20 years in the Structures and Materials Laboratory of the Faculty of Aerospace Engineering of the Delft University of Technology until my formal retirement. In both laboratories it was a cooperation with many people in a stimulating atmosphere. It is great to carry out investigations with colleagues and technicians who understand the essence of the fatigue experiments and microscopical examinations, too many to mention all their names. But I will make a single exception for Frans Oosterom for the indispensable assistance with microscopy and photography. It was also very stimulating to work together with many undergraduate and post-graduate students, eager to arrive at answers for unsolved questions. We shared various challenges, and at the same time a kind of a family relationship. Thanks to my successors, this continued after my retirement, first through professor Boud Vogelesang, afterwards the late professor Ad Vlot who most regretfully died young, and at the moment professor Rinze Benedictus. All three were fine colleagues always available for discussion. Informal brainstorming with many people was often the elixir of new ideas and progress.

With respect to preparing the manuscript for the first edition of this book, I want to acknowledge the unselfish help of Dr. Scott Fawaz, who did the work for his doctoral thesis in Delft. He read every chapter and suggested various improvements and corrections. The same was done by the late professor Hans Overbeeke of the Technical University of Eindhoven. Significant support was given by Harry van Lipzig, my first student in Delft, and now responsible for various post-academic courses on fatigue of structures.

For the second edition of this textbook I want to express my appreciation to Dr. René Alderliesten (Delft University) and Professor Malgorzata Skorupa (University of Mining and Metallurgy, Krakow) for useful comments on specific chapters of the book.

Publishing the book was a joint effort between the author and Springer Science+Business Media. I gratefully acknowledge the pleasant and effective cooperation with Mrs Nathalie Jacobs and Anneke Pot, who were concerned with getting the book printed, and Mrs. Jolanda Karada who arranged the typesetting of the book and figures, and suggested text improvements. It was really a great help.

Last but not least, without Janine, my love for uncountable years, this book would not exist. In addition of creating the circumstances which allowed me to write the book, her comments and questions about my work were meaningful.

Jaap Schijve
Delft, October 2008

Contents

Part IV Special Fatigue Conditions
(Chapters 14–17)

Part V Fatigue of Joints and Structures
(Chapters 18–20)

CD attached to this book

Contents:

I.   Exercises and Summaries

II. Case Histories

III. Special Topics

IV. Research on Fatigue Problems in the Future

Symbols, Acronyms and Units

Symbols

a	crack length, or semi-crack length, or depth of part through crack
a_0	initial crack length
a, b	semi-axes of ellipse
a_f	final crack length
c	(semi) crack length of surface crack
C	constant in Paris equation
D	diameter
	Also: damage parameter
da/dN	crack growth rate
dU/da	strain energy release rate
E	Young's modulus
G	shear modulus
k	slope factor in Basquin relation,
	or irregularity factor of random load
K	stress intensity factor
ΔK	$= K_{max} - K_{min}$
ΔK_{th}	threshold ΔK (Figure 8.6)
K_{op}	K at S_{op}
ΔK_{eff}	$= K_{max} - K_{op}$
K_f	fatigue strength reduction factor
K_{Ic}	fracture toughness
K_t	stress concentration factor
l	crack length from edge of notch
m	exponent in Paris equation
M	bending moment
M_t	torsion moment
n	number of cycles
N	fatigue life until failure
P	load
$p(x)$	probability density function
$P(x)$	distribution function
r	root radius of notch
	Also: polar coordinate
r_p	plastic zone size
R	stress ratio $= S_{min}/S_{max}$

S	nominally applied (gross) stress
S_1, S_2	biaxial stresses
S_a	stress amplitude
S_f	fatigue limit
S_{f1}	fatigue limit, unnotched specimen
S_{fk}	fatigue limit, notched specimen
S_m	mean stress
S_N	fatigue strength at fatigue life N
S_{op}	crack opening stress
S_{res}	residual stress
S_U	tensile strength of a material
$S_{0.2}$	yield strength of a material
ΔS	$= S_{max} - S_{min}$
ΔS_{eff}	$= S_{max} - S_{op}$
t	thickness
	Also: time
T	temperature
u, v, w	displacements in x, y, z direction
W	width
β	geometry factor in $K = \beta S \sqrt{\pi a}$
γ	surface roughness reduction factor
ε	strain
θ	polar coordinate
λ	biaxiality ratio (S_1/S_2)
ν	Poisson ratio
ρ	tip radius of notch
σ	local stress in material
$\sigma_{\log N}$	standard deviation of $\log N$
σ_a	stress amplitude
σ_m	mean stress
$\sigma_{nominal}$	nominal stress in net section
σ_{peak}	peak stress at notch
σ_{res}	residual stress
τ	shear stress
ϕ	location angle in ellipse (Figure 5.16)
ω	circular frequency in radian per second

Acronyms

CA	constant-amplitude
CCT	centered cracked tension (specimen)
COD	crack opening displacement
CT	compact tension (specimen)
M(T)	centered cracked tension (specimen)
OL	overload

PD   potential drop
VA   variable-amplitude
UL   underload

Units and conversion factors

1 meter (m) = 10^3 millimeters (mm) = 10^6 microns (μm)

1 inch = 25.4 mm,   1 mm = 0.04 inch

Stress: 1 MPa = 10^6 Pascal (1 Pascal = 1 Newton/m^2)

1 ksi = 6.90 MPa,   1 MPa = 0.145 ksi

Stress intensity factor: 1 MPa$\sqrt{m}$ = 0.910 ksi$\sqrt{in}$,   1 ksi$\sqrt{in}$ = 1.099 MPa$\sqrt{m}$

1 kc = 1000 cycles

Chapter 1
Introduction to Fatigue of Structures and Materials

Fatigue failures in metallic structures are a well-known technical problem. Already in the 19th century several serious fatigue failures were reported and the first laboratory investigations were carried out. Noteworthy research on fatigue was done by August Wöhler. He recognized that a single load application, far below the static strength of a structure, did not do any damage to the structure. But if the same load was repeated many times it could induce a complete failure. In the 19th century fatigue was thought to be a mysterious phenomenon in the material because fatigue damage could not be seen. Failure apparently occurred without any previous warning. In the 20th century, we have learned that repeated load applications can start a fatigue mechanism in the material leading to nucleation of a small crack, followed by crack growth, and ultimately to complete failure. The history of engineering structures until now has been marked by numerous fatigue failures of machinery, moving vehicles, welded structures, aircraft, etc. From time to time such failures have caused catastrophic accidents, such as an explosion of a pressure vessel, a collapse of a bridge, or another complete failure of a large structure. Many fatigue problems did not reach the headlines of the news papers but the economic impact of non-catastrophic fatigue failures has been tremendous. Fatigue of structures is now generally recognized as a significant problem.

The history of fatigue covering a time span from 1837 to 1994 was reviewed in an extensive paper by Walter Schütz [1]. Historical milestone papers were collected by Hanewinkel and Zenner [2] and Sanfor [3]. John Mann [4] compiled 21075 literature sources on fatigue problems covering the period from 1838 to 1969 in four books. Since that time the number of publications on fatigue has still considerably increased and it may be estimated to be around 100,000 in the year 2000. Fortunately, consulting the literature on specific topics can now be done with computerized literature retrieval systems.

As a result of extensive research and practical experience, much knowledge has been gained about fatigue of structures and the fatigue

(a) Front wheel, broken spikes, axle part with drum   (b) Fatigue fractures
brakes

Fig. 1.1 Collapse of the front wheel of a motorcycle by fatigue of the spokes.

mechanism in the material. Qualitatively our understanding of fatigue problems is fairly well developed in the 20th century as discussed in a survey paper by the author [5] (this paper is copied on the CD attached to this book). Much has been learned from laboratory research. However, accident investigations has also highly contributed to the present state of the art. Fatigue failures in service can be most instructive and provide convincing evidence that fatigue may be a serious problem. The analysis of failures often reveals various weaknesses contributing to an insufficient fatigue resistance of a structure. This will be illustrated here by a case history. The front wheel of a heavy motorcycle completely collapsed, see Figure 1.1a. Ten spokes of the light alloy casting were broken. Examination of the failure surfaces indicated that fatigue cracks occurred in all spokes, see Figure 1.1b. Why was the fatigue life of this wheel insufficient? A first question of a failure analysis must be: Was the failure a symptomatic failure or was it an incidental case? If it is a symptomatic failure, all motorcycles of the same type are in danger and immediate action is required. However, the failure may be an incidental case for some special reason applicable to that single motorcycle only: for instance, unusual and severe damage of the material surface. In the case of this motorcycle, the same failure had occurred in several wheels in different countries, although predominantly in motorcycles of the police. The wheel shown in Figure 1 collapsed when a policeman suddenly had to use the brakes to stop before a railway crossing.[1] He survived after some heavy shocks. The two most common questions usually put forward after a fatigue

---

[1] A policeman from Delft, and a railway crossing in Pijnacker where the author is living.

failure are: (i) was the fatigue resistance of the material too low, or (ii) was the stress level at the failure location too high? However, the list of questions is larger. For instance, (iii) What is so special about the load spectrum of the police motorcycle, and was the fatigue load spectrum known and taken into account by the motorcycle industry? (iv) How was the material surface quality at the location where the fatigue cracks started? This is a production question. Answers to each of these questions could lead to different clues for improvements. Another good question is whether a fatigue test had been done on this wheel before starting large scale production? Actually, the test was done after the failure, and the fatigue failure was reproduced in the test.

A structure should be designed and produced in such a way that undesirable fatigue failures do not occur during the design life of the structure. Apparently there is a challenge which will be referred to as *"designing against fatigue"*.[2] It will be discussed in later chapters that various design options can be adopted to ensure satisfactory fatigue properties with respect to sufficient life, safety and economy. They are related to different structural concepts such as more careful detail design, less fatigue sensitive materials, improved material surface treatments, alternative types of joints, and lower design stress levels. Also, less obvious approaches can be considered, e.g. design for damage tolerance (fail safe), damage prevention (e.g. corrosion protection), alleviation of the dynamic loads in service. The spectrum of possibilities is extensive due to the large number of variables which can affect the fatigue behavior of a structure. Scenarios of designing against fatigue are also influenced by questions about the cost-effectivity of design efforts to improve the fatigue quality of a structure.

People working in the design office of an industry usually adopt standardized calculation procedures for predictions on fatigue strength, fatigue life, crack growth and residual strength. Standardized procedures can be useful, but it must be realized that such procedures may be unconservative or overconservative. Such calculation procedures start from some generalized conditions, which need not be similar to the conditions of the structure in service. It requires understanding, experience and engineering judgement to evaluate the significance of calculated results. The predictions may have a limited accuracy and reliability. In cases of doubt about calculated predictions, it is useful to perform supporting fatigue tests. Some people feel that an experiment is highly superior to theoretical calculations. Statements like "Experiments never lie" are well known. Unfortunately, an experiment gives results applicable to the conditions of the

---

[2] This is the title of a book by R.B. Heywood, published in 1962 [6].

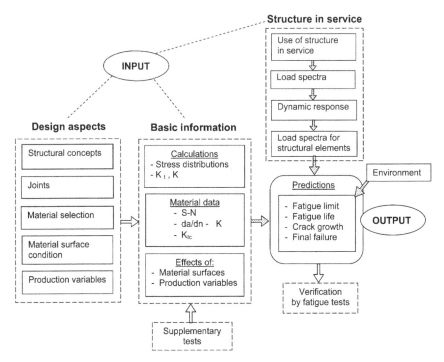

Fig. 1.2 Survey of the various aspects of fatigue of structures, a multidisciplinary problem setting [6].

experiment. The question is, are the test conditions a realistic representation of the conditions in service. Also this question asks for understanding, experience and judgement. In other words, whether designing against fatigue is done by analysis, calculations or experiments, it requires a profound knowledge of the fatigue phenomenon in structures and materials and the large variety of conditions that can affect fatigue. The aim of the present textbook is to present basic knowledge about this multi disciplinary problem setting.

A summary of aspects of fatigue design procedures is given in Figure 1.2. The first column contains major topics of the design work. Various aspects of basic information are listed in the second column. This information should be used for selections of materials, material surface treatments and production variables, but also for detail design issues, noteworthy for joints. In order to arrive at an evaluation of the fatigue quality of a structure, predictions have to be made. It then is a prerequisite to have relevant information on the fatigue loads. This includes a number of steps listed in the third column, starting with considerations about how the structure is used in service. This should lead to load spectra and subsequently to stress spectra for the fatigue critical locations in the structure. As also indicated in Figure 1.2, it may be desirable to do supplementary tests on specific issues or verification tests to cover uncertainties of predictions.

A special issue is how to account for environmental effects. Experimental data used in the predictions are generally obtained under laboratory conditions and relatively high testing frequencies. However, in service corrosive environments may be present and the load frequency can be much lower. As an example, think of a welded structure for a drilling platform in the sea. The environment is salt water, and the loading rate of water waves is relatively low.

About the contents of the book

Topics of Figure 1.2 are discussed in separate chapters, but a number of questions somewhat hidden in this diagram are also covered, such as notch effect, size effect, residual stress, statistics of fatigue. The book starts with chapters on basic issues:

- Fatigue as a phenomenon occurring in the material (Chapter 2). Some knowledge of basic aspects of the fatigue mechanism is essential to understand the various influencing factors of fatigue.
- Because stress concentrations in a structure are of paramount importance for fatigue, stress distributions around notches and stress concentration factors (K_t) are discussed in Chapter 3.
- Residual stresses can have a significant effect on fatigue. A special chapter is presented on these stresses (Chapter 4).
- After fatigue cracks have been nucleated, the stress intensity factor (K) is a controlling parameter for the severity of the stress distribution around the crack tip and fatigue crack growth. The concept of the stress intensity factor is discussed in Chapter 5.
- Fatigue properties, including the fatigue limit, S-N curves, fatigue diagrams, and also fatigue crack growth are discussed in Chapters 6 to 8. Some prediction problems are discussed in these chapters.

Chapters 2 to 8 cover the basic aspects of fatigue of materials and this is Part I of the book (see the list of chapters in the Contents).

The subjects of Part II are load spectra and fatigue under variable-amplitude loading. Fatigue load spectra are discussed in Chapter 9, starting with a description of various types of loads in service and the statistical analysis of load histories obtained by counting methods. Fatigue predictions without a load spectrum are impossible. If the load spectrum contains loads of different amplitudes, prediction problems require a fatigue damage accumulation model. This topic is discussed in Chapter 10 for fatigue lives and in Chapter 11 for fatigue crack growth.

Part III covers fatigue testing and scatter of fatigue properties. In view of uncertainties about predictions, it is desirable to verify the results by experiments. Fatigue tests are also important for obtaining material data and for exploring various effects on fatigue properties by comparative tests. Aspects of fatigue tests are discussed in Chapter 13. Unfortunately, fatigue properties can show significant scatter. It then must be recognized that scatter of the fatigue behavior of a structure in service is usually caused by different conditions than scatter in fatigue tests. Problems about scatter are discussed in Chapter 12.

Different conditions which are important for fatigue are treated in Part IV. Surface treatments can have a large effect on the fatigue properties as discussed in Chapter 14. Fretting corrosion is usually unfavorable for the fatigue life. This is the subject of Chapter 15. Environmental effects including the influence of corrosion on the fatigue performance are considered in Chapter 16. The effect of temperature, both high and low temperatures are briefly reviewed in Chapter 17. Actually, high-temperature fatigue is somewhat outside the scope of this book because it is more a problem of material selection and material development rather than primarily a problem of designing against fatigue.

Fatigue of joints and structures is discussed in Part V. Joints are often the most fatigue critical part of a structure. Moreover, the variety of joints is large. Chapter 18 covers different types of joints except welded joints which are covered in Chapter 19. The discussion in Chapter 20 offers general reflections on fatigue of structures with comments on how to deal with fatigue of structures as a design problem, also in view of uncertainties, safety and economic issues.

Finally, the last chapter, Chapter 21, covers fiber-metal laminates. The fiber-metal laminates (Arall and Glare) developed in Delft are a new class of hybrid materials with a high fatigue resistance. A new addition to this family of fiber-metal laminates is CentrAl. With these laminated materials, the designer does not only design the geometry of the structure, but also the lay-up parameters of the laminate in order to achieve optimal properties for specific fatigue critical components.

The major aspects of each chapter are summarized in the last section (except for this Chapter 1). These summaries are useful to reconsider the contents of a chapter. References used in the text are added to each chapter. Some general references are added for further study of the subject of a chapter. They are listed because the author has consulted these references, but the list cannot be expected to give a full coverage of all relevant publications. Moreover, each year numerous publications appear. For further in depth research more information should be retrieved.

About using the book

As pointed out in the Preface, the present book is written as a textbook for engineers, designers, researchers, students and teachers, and also for self-tuition. It is not a material data handbook. The main emphasis is on understanding the analysis of fatigue problems of structures. It requires that the fatigue mechanism in terms of fatigue crack nucleation and fatigue crack growth must be understood, as well as the influence of relevant variables on the fatigue mechanism. Fortunately, fatigue is no longer a mysterious phenomenon in the material. In qualitative terms, the fatigue mechanism is reasonably well understood. It is because of this understanding that we must accept that quantitative predictions in many cases can only be an approximation without being absolutely precise and accurate.

The prerequisites for the book are elementary knowledge of materials (material structure and material properties) and linear-elastic structural analysis (stress and strain distributions, tension, bending and torsion). The first edition of the book has been used in courses for students and in workshop courses for people in the industry, research institutes and other agencies involved in fatigue of structures, and safety and durability issues. In courses a teacher can use specific chapters and omit other ones depending on the background of the participants. If participants are already acquainted with linear-elastic stress analysis and fracture mechanics, Chapters 3 and 5 can be omitted or briefly summarized as a refreshment. Chapter summaries compiled in the last section of Chapters 2 to 20 should be useful for the evaluation of personal understanding. Questions on the topics of these chapters are compiled on the CD attached to this book. They can be useful for examinations, refreshing of knowledge and self-tuition.

In any course, case-histories about fatigue failures in service are most important. Several case-histories are discussed in the book. More case histories are presented on the CD with comments on design aspects, failure analysis and typical fatigue issues. In a course a teacher should also bring in his own case-histories together with the hardware of broken parts. Participants can then hold these parts in their own hands and observe fractures with their own eyes, a magnifying glass and a microscope. Attention should be focused on the initiation point of the fatigue failure.

Fatigue tests to be carried out as part of a course can be instructive, but they are time consuming. Demonstrations of fatigue experiments and explaining the purpose of the test can be much more rewarding.

About the CD attached to this book

Additional information is provided on a separate CD to keep the size of the book manageable. The information on the CD should not be considered to be an essential part of the various chapters in the book itself. It is supplementary to the book.

The CD encloses four parts. The first part contains exercises. It covers questions on various chapters and hints for answers in a separate section. The objective of the questions is to verify whether the important concepts are well understood. It can be useful for course work, both for students and teachers, and also for self-tuition. A summary of chapters is included for refreshing purposes.

In the second part selected case histories are summarized. The purpose is to indicate how accident investigations can be evaluated in order to be instructive for remedial activities and future fatigue issues.

The third part is covering two topics associated with designing against fatigue and planning experimental fatigue programs respectively. The discussion is addressing people engaged in designing against fatigue as well as others involved in research on fatigue problems.

In the fourth part, questions about the objectives of research topics are approached from a more philosophical point of view. Which research should be done, and why should it be done. Is it worthwhile? And is it a challenge? These questions must be considered by research institutes and university in view of research strategies. As a kind of an addendum to the fourth part, the text of a paper published in 2003 [5] has been included. The title is "Fatigue of structures and materials in the 20th century and the state of the art". The future is in the 21st century.

References

1. Schütz, W., *A history of fatigue*. Engrg. Fracture Mech., Vol. 54 (1996), pp. 263–300.
2. Hanewinkel, D. and Zenner, H., *Fatigue strength. A facsimile collection of historical papers until 1950*. Technical University of Clausthal (1989) [historical papers in English and German].
3. Sanfor, R.J. (Ed.), *Selected Papers on Foundations of Linear Elastic Fracture Mechanics*. SEM Classic papers, Vol. CP1, SPIE Milestone Series, Vol. MS 137 (1997).
4. Mann, J.Y., *Bibliography on the Fatigue of Materials, Components and Structures*, Vols. 1 to 4. Pergamon Press, Oxford (1970, 1978, 1983 and 1990).
5. Schijve, J., *Fatigue of structures and materials in the 20th century and the state of the art*. Int. J. Fatigue, Vol. 25, No. 8 (2003), pp. 679–702.
6. Heywood, R.B., *Designing against Fatigue*. Chapman and Hall, London (1962).
7. Schijve, J., *Predictions on fatigue life and crack growth as an engineering problem. A state of the art survey*. Fatigue 96, Proc. 6th International Fatigue Congress, Berlin, Vol. II, G. Lütjering and H. Nowack (Eds.). Pergamon (1996), pp. 1149–1164.

Part I
Fatigue under Constant-Amplitude Loading

Chapter 2
Fatigue as a Phenomenon in the Material

2.1 Introduction

In a specimen subjected to a cyclic load, a fatigue crack nucleus can be initiated on a microscopically small scale, followed by crack grows to a macroscopic size, and finally to specimen failure in the last cycle of the fatigue life. In the present chapter the fatigue phenomenon will be discussed as a mechanism occurring in metallic materials, first on a microscale and later on a macroscale.

Understanding of the fatigue mechanism is essential for considering various technical conditions which affect fatigue life and fatigue crack growth, such as the material surface quality, residual stress, and environmental influence. This knowledge is essential for the analysis of fatigue properties of an engineering structure. Fatigue prediction methods can only be evaluated if fatigue is understood as a crack initiation process followed by a crack growth period. For that reason, the present chapter is a prerequisite for most chapters of this book.

The fatigue life is usually split into a *crack initiation period* and a *crack growth period*. The initiation period is supposed to include some microcrack growth, but the fatigue cracks are still too small to be visible. In the second period, the crack is growing until complete failure. It is technically significant to consider the crack initiation and crack growth periods separately because several practical conditions have a large influence on the crack initiation period, but a limited influence or no influence at all on the crack growth period. This chapter starts with a general definition of the fatigue life as consisting of different phases (Section 2.2). Crack initiation and growth are then described as different phenomena (Sections 2.3 and 2.4). In order to understand the effects of the variables, the fatigue mechanism is considered in more detail in Section 2.5. It also covers the basics of several variables which are discussed as technical problems in later chapters. Characteristic features of fatigue failures are summarized in Section 2.6. The major concepts developed in this chapter are summarized in the final section, Section 2.7.

2.2 Different phases of the fatigue life

Microscopic investigations in the beginning of the 20th century [1] have shown that fatigue crack nuclei start as invisible microcracks in slip bands. After more microscopic information on the growth of small cracks became available, it turned out that nucleation of microcracks generally occurs very early in the fatigue life. Indications were obtained that it may take place almost immediately if a cyclic stress above the fatigue limit is applied. The fatigue limit is the cyclic stress level below which a fatigue failure does not occur.[3] In spite of early crack nucleation, microcracks remain invisible for a considerable part of the total fatigue life. Once cracks become visible, the remaining fatigue life of a laboratory specimen is usually a small percentage

---

[3] The fatigue limit depends on the mean stress of the cyclic load, but for clarity of the text, this will be disregarded if it does not affect the discussion.

of the total life. The latter percentage may be much larger for real structures such as ships, aircraft, etc.

After a microcrack has been nucleated, crack growth can still be a slow and erratic process, due to effects of the microstructures, e.g. grain boundaries. However, after some microcrack growth has occurred away from the nucleation site, a more regular growth is observed. This is the beginning of the real crack growth period. Various steps in the fatigue life are indicated in Figure 2.1. The important point is that the fatigue life until failure consists of two periods: the *crack initiation period* and the *crack growth period*. Differentiating between the two periods is of great importance because several surface conditions do affect the initiation period, but have a negligible influence on the crack growth period. Surface roughness is just one of those conditions as discussed in Section 2.5. Corrosive environments can affect initiation and crack growth, but in a different way for the two periods. Differences between the crack initiation period and crack growth period are discussed in Sections 2.3 and 2.4. It should already be noted here that fatigue prediction methods are different for the two periods. The stress concentration factor K_t is the important parameter for predictions on crack initiation. The stress intensity factor K is used for predictions on crack growth. These two parameters are discussed in Chapters 3 and 5 respectively.

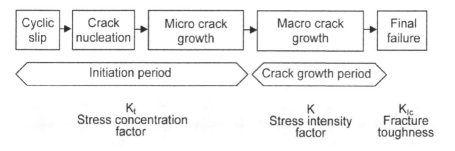

Fig. 2.1 Different phases of the fatigue life and relevant factors.

2.3 Crack initiation

Fatigue crack initiation and crack growth are a consequence of cyclic slip. It implies cyclic plastic deformation, or in other words dislocation activities. Fatigue occurs at stress amplitudes below the yield stress. At such a low

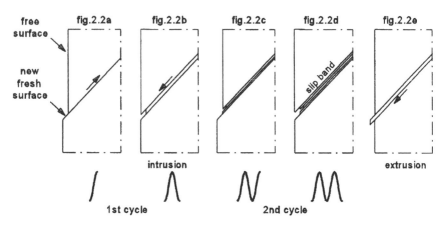

Fig. 2.2 Cycle slip leads to crack nucleation.

stress level, plastic deformation is limited to a small number of grains of the material. This microplasticity preferably occurs in grains at the material surface because of the lower constraint on slip. At the free surface of a material, the surrounding material is present at one side only. The other side is the environment, usually a gaseous environment (e.g. air) or a liquid (e.g. sea water). As a consequence, plastic deformation in surface grains is less constrained by neighbouring grains than in subsurface grains; it can occur at a lower stress level.

Cyclic slip requires a cyclic shear stress. On a microscale the shear stress is not homogeneously distributed through the material. The shear stress on crystallographic slip planes differs from grain to grain, depending on the size and shape of the grains, crystallographic orientation of the grains, and elastic anisotropy of the material. In some grains at the material surface, these conditions are more favorable for cyclic slip than in other surface grains. If slip occurs in a grain, a slip step will be created at the material surface, see Figure 2.2a. A slip step implies that a rim of new material will be exposed to the environment. The fresh surface material will be immediately covered by an oxide layer in most environments, at least for most structural materials. Such monolayers strongly adhere to the material surface and are not easily removed. Another significant aspect is that slip during the increase of the load also implies some strain hardening in the slip band. As a consequence, upon unloading (Figure 2.2b) a larger shear stress will be present on the same slip band, but now in the reversed direction. Reversed slip will thus preferably occur in the same slip band. If fatigue would be a fully reversible process, this book would not have been written. However,

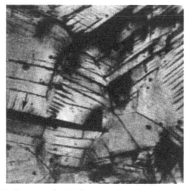

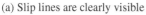

(a) Slip lines are clearly visible   (b) Same as in (a) but plastically strained
                                       (5%) which opens a microcrack, see arrow

Fig. 2.3 Development of cyclic slip bands and a microcrack in a pure copper specimen. $S_m = 0$, $S_a = 77.5$ MPa, $N = 2 \times 10^6$ [2].

we have already mentioned two reasons why it cannot be fully reversible. First, the oxide monolayer cannot simply be removed from the slip step. Secondly, strain hardening in the slip band is also not fully reversible. As a consequence, reversed slip, although occurring in the same slip band, will occur on adjacent parallel slip planes. This is schematically indicated in Figure 2.2b. The same sequence of events can occur in the second cycle, see Figures 2.2c and d.

Of course Figure 2.2 offers a simplified picture, but there are still some important lessons to be learned:

(i)   A single cycle is sufficient to create a microscopical intrusion into the material, which in fact is a microcrack.

(ii) The mechanism occurring in the first cycle can be repeated in the second cycle, and in subsequent cycles and cause crack extension in each cycle.

(iii) The first initiation of a microcrack may well be expected to occur along a slip band. This has been confirmed by several microscopic investigations, see Figure 2.3. A slip band seen in Figure 2.3a is actually a microcrack as confirmed in Figure 2.3b after the band is opened by applying a 5% plastic strain to the material. A part of this slip band was already visible after no more than 0.5% of the fatigue life.

(iv) In Figure 2.2 the small shift of the slip planes during loading and unloading is leading to an intrusion. However, if the reversed slip would occur at the lower side of the slip band, an extrusion is obtained, see Figure 2.2e. Such extrusions have been reported in the literature, noteworthy by Forsyth [3]. However, from a potential strain energy point of view, the intrusion is the more probable consequence of cyclic slip in a slip band.[4]

(v)   The simple mechanism of Figure 2.2, and even if it would be different or more complicated, implies disruption of bonds between atoms, i.e. *decohesion* occurs, either by tensile decohesion, shear decohesion, or both. It occurs if a slip step penetrates through a free surface. It can also occur at the tip of a growing fatigue crack. The disruption of bonds at the crack tip might also be caused by a generation of dislocations from the crack tip. It should be expected that the decohesion can be accelerated by an aggressive environment.

The lower restraint on cyclic slip at the material surface has been mentioned as a favorable condition for crack initiation at the free surface. However, more arguments for crack initiation at the material surface are present. A very practical reason is the inhomogeneous stress distribution due to a notch effect of a hole or some other geometric discontinuity. Because of an inhomogeneous stress distribution, a peak stress occurs at the surface (stress concentration). Furthermore, surface roughness also promotes crack initiation at the material surface. Other surface conditions with a similar effect are corrosion pits and fretting fatigue damage both occurring at the material surface. These technical conditions are discussed later. The most important conclusion to be drawn here is:

In the crack initiation period, fatigue is a material surface phenomenon.

2.4 Crack growth

As long as the size of the microcrack is still in the order of a single grain, the microcrack is obviously present in an elastically anisotropic material with a crystalline structure and a number of different slip systems. The microcrack contributes to an inhomogeneous stress distribution on a microlevel, with

---

[4] The author has found extrusions in slip bands of the pure aluminium cladding layer of 2024-T3 aluminium alloy sheet material. Under cyclic loading plastic shake down will occur in this soft layer. It implies that it will also see compressive stresses even if the nominal stress on the sheet material is cyclic tension. Under compression the extrusion mechanism should be favored.

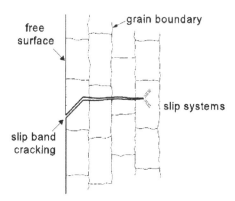

Fig. 2.4 Cross section of microcrack.

a stress concentration at the tip of the microcrack. As a result, more than one slip system may be activated. Moreover, if the crack is growing into the material in some adjacent grains, the constraint on slip displacements will increase due to the presence of the neighbouring grains. Similarly, it will become increasingly difficult to accommodate the slip displacements by slip on one slip plane only. It should occur on more slip planes. The microcrack growth direction will then deviate from the initial slip band orientation. In general, there is a tendency to grow perpendicular to the loading direction, see Figure 2.4.

Because microcrack growth is depending on cyclic plasticity, barriers to slip can imply a threshold for crack growth. This has been observed indeed. Illustrative results are presented in Figure 2.5. The crack growth rate measured as the crack length increment per cycle decreased when the crack tip approached the first grain boundary. After penetrating through the grain boundary the crack growth rate increased during growth into the next grain, but it decreased again when approaching the second grain boundary. After passing that grain boundary, the microcrack continued to grow with a steadily increasing rate.

In the literature, several observations are reported on initially inhomogeneous microcrack growth, which starts with a relatively high crack growth rate and then slows down or even stops due to material structural barriers. However, the picture becomes different if the crack front after some crack growth passes through a substantial number of grains, as schematically indicated in Figure 2.6. Because the crack front must remain a coherent crack front, the crack cannot grow in each grain in an arbitrary direction and at any growth rate independent of crack growth in the adjacent grains. This continuity prevents large gradients of the crack growth rate along the

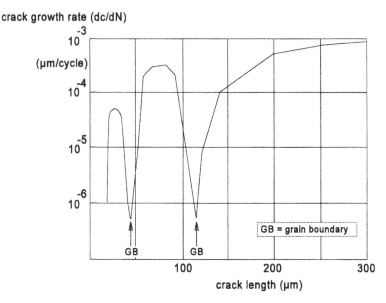

Fig. 2.5 Grain boundary efect on crack growth in an Al-alloy [4]. The crack length was measured along the material surface.

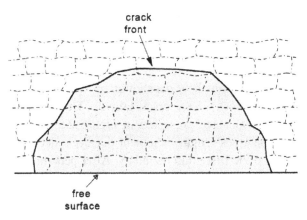

Fig. 2.6 Top view of crack with crack front passing through many grains.

crack front. As soon as the number of grains along the crack front becomes sufficiently large, crack growth occurs as a more or less continuous process along the entire crack front. The crack front can be approximated by a continuous line, which could have a semi-elliptical shape. How fast the crack will grow depends on the crack growth resistance of the material. Two important surface aspects are no longer relevant. The lower restraint on cyclic slip at the surface is not applicable at the interior of the material. Secondly, surface roughness and other surface conditions do not affect crack growth. This leads to the second important conclusion:

Crack growth resistance when the crack penetrates into the material depends on the material as a bulk property. Crack growth is no longer a surface phenomenon.

2.5 The fatigue mechanism in more detail

In the previous section, the fatigue life was discussed as consisting of a crack initiation period and crack growth period. The transition from the initiation period to the crack growth period has not yet been defined. The definition cannot really be given in quantitative terms, but in a qualitative way the following definition will be used:

The initiation period is supposed to be completed when microcrack growth is no longer depending on the material surface conditions.

It implies that the crack growth period starts if the crack growth resistance of the material per se is controlling the crack growth rate. The size of the microcrack at the transition from the initiation period to the crack growth period can be significantly different for different types of materials. The transition depends on microstructural barriers to be overcome by a growing microcrack, and these barriers are not the same in all materials.

The crack initiation period includes the initial microcrack growth. Because the growth rate is still low, the initiation period may cover a significant part of the fatigue life. This is illustrated by the generalized picture of crack growth curves presented in Figure 2.7 which schematically shows the crack growth development as a function of the percentage of the fatigue life consumed ($= n/N$), with n as the number of fatigue cycles and N as the fatigue life until failure. Complete failure corresponds to $n/N = 1 = 100\%$. There are three curves in Figure 2.7, all of them in agreement with crack initiation in the very beginning of the fatigue life, however, with different values of the initial crack length. The lower curve

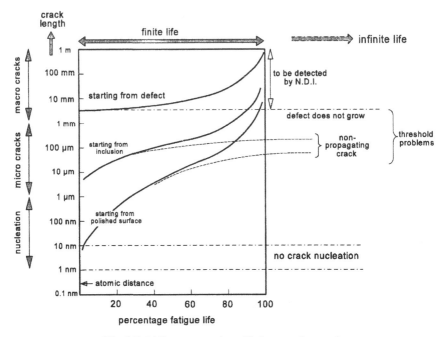

Fig. 2.7 Different scenarios of fatigue crack growth.

corresponds to microcrack initiation at a "perfect" surface of the material. Here, the mechanism of Figure 2.2 could be applicable. The middle curve represents crack initiation from an inclusion, which is briefly discussed later. The upper curve is associated with a crack starting from a material defect which should not have been present, such as defects in a welded joint. Figure 2.7 illustrates some interesting aspects:

(i)   The vertical crack length scale is a logarithmic scale, ranging from 0.1 nanometer (nm) to 1 meter (1 nanometer $= 10^{-9}$ m $= 4 \cdot 10^{-8}$ inch). Microcracks starting from a perfect free surface can have a sub-micron crack length (<1 μm $= 10^{-6}$ m). However, cracks nucleated at an inclusion will start with a size similar to the size of the inclusion. The size can still be in the sub-millimeter range. Only cracks starting from macrodefects can have a detectable macrocrack length immediately.

(ii) The two lower crack growth curves illustrate that the major part of the fatigue life is spent with a crack size below 1 mm, i.e. with a practically invisible crack size.

(iii) Dotted lines in Figure 2.7 indicate the possibility that cracks do not always grow until failure. It implies that there must have been barriers in the material which stopped crack growth.

Figure 2.7 gives generalized scenarios about possible crack growth developments. In order to understand more about fatigue under various practical conditions, several aspects of the fatigue mechanism are discussed in more detail. The aspects covered in this section are:

1. Crystallographic nature of the material;
2. Crack initiation at inclusions;
3. Small cracks, crack growth barriers, crack growth thresholds;
4. Number of crack nuclei;
5. Surface effects;
6. Macrocrack growth and striations;
7. Environmental effects;
8. Cyclic tension and cyclic torsion.

2.5.1 Crystallographic aspects

As pointed out before, the initial growth of a microcrack shows a tendency to grow along a slip band. It thus must be expected that the crystallography of a material has some influence on the mechanistic behavior during the initiation period. The crystallographic properties vary from one material to another. As a consequence, the initial microcracking depends on the material. Aspects to be mentioned here are:

• Type of crystal lattice, elastic anisotropy, allotropy;
• Slip systems, ease of cross slip;
• Grain size and shape;
• Variation of the crystal orientation from grain to grain (texture).

These subjects will not be discussed extensively, but some significant features are reviewed for understanding differences between the fatigue mechanisms in different materials, especially so for the crack initiation period, i.e. for small microcracks.

The three well-known crystal lattices are face centered cubic (f.c.c.) for Al, Cu, Ni and γ-Fe, body centered cubic (b.c.c.) for α-Fe and β-Ti, and hexagonal close packed (h.c.p.) for α-Ti and Mg. The elastic and plastic behavior of a material depends on the crystal structure, but even for the same crystal lattice large differences can occur. The elastic anisotropy can vary considerably as illustrated by the E-moduli in Table 2.1.

The anisotropy is quite large for copper and fairly small for Al, with α-Fe (ferrite) at an intermediate position. Fatigue generally occurs at low stress

Table 2.1 Some data on elastic anisotropy.

Material	$E_{\max}$ [111] (MPa)	$E_{\min}$ [100] (MPa)	Ratio max/min
α-Fe	284500	132400	2.15
Al	75500	62800	1.20
Cu	190300	66700	2.85

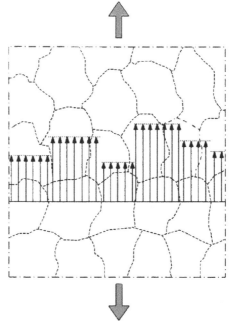

Fig. 2.8 Simplified picture of the inhomogeneous stress distribution from grain to grain due to elastic anisotropy.

levels without macroplastic deformation. As a result of the elastic anisotropy, the stress distribution from grain to grain should be inhomogeneous as schematically indicated in Figure 2.8. Such a picture suggests a homogeneous stress in each single grain, which is an approximation. However, one conclusion remains valid. The inhomogeneity of the stress distribution from grain to grain are small for Al-alloys and much larger for steel. In other words, in Al-alloys most grains will be subjected to similar stress levels, whereas for steel and other materials the stress level can vary significantly from grain to grain.

Slip systems are characterized by crystallographic planes on which slip occurs and by crystallographic slip directions. Slip systems have been extensively studied and are well documented in textbooks on material science. The possibility of cross slip is important for dislocation movements in order to circumvent obstacles and to continue slip on adjacent parallel slip planes. Cross slip is easier if the stacking fault energy of a material is high, and difficult if the stacking fault energy is low. Aluminium is a noteworthy example for easy cross slip and nickel for difficult cross slip. As a result, slip lines in Al are wavy and cyclic slip can lead to slip bands with a measurable thickness. In Ni- and Cu-alloys, slip lines are more sharply defined straight lines, see as an example Figure 2.3. Moreover, if the number of easily activated slip systems is limited, microcracks persist much longer in growing along crystallographic directions. This behavior may continue until a crack length in the order of 1 mm (0.04 inch), while for Al-alloys cracks as small as 0.1 mm already grow more or less perpendicular to the main principal stress, usually the tensile stress along the material surface. Such observations show that the microcrack growth behavior can be essentially different for different types of material.

2.5.2 Crack initiation at inclusions

In most technical materials, a variety of inclusions can be present, such as impurities introduced during the melting production process of the alloys. In the present section, inclusions of a microscopic size are considered. Larger macroscopic inclusions are generally regarded as material defects which should not be present, for example slag streaks, weld defects, major porosities. Large defects have occasionally caused disastrous failures in service, but they are not considered in this section.

Non-metallic inclusions of a microscopic size (10 to 100 μm) have been observed in low-alloy high-strength steels. Fatigue crack nucleation occurred at these inclusions located at the material surface or slightly below the surface. Crack nucleation can occur in aluminium alloys at intermetallic inclusions which partially contain alloying elements. These inclusions are not considered to be harmful for the static strength, but the inclusions can reduce the ductility of a material due to generating internal voids at large plastic strains. However, significant plasticity does not occur under fatigue at a relatively low stress level. But the inclusions are still foreign constituents which can interact with cyclic slip. The inclusions affect the stress distribution on a microlevel and thus can contribute to crack

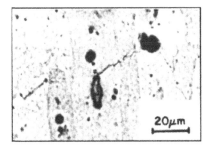

Fig. 2.9 Slip band microcrack nucleated at the tip of an intermetallic inclusion in the polished material surface of an aluminium alloy (2024-T3) specimen [7].

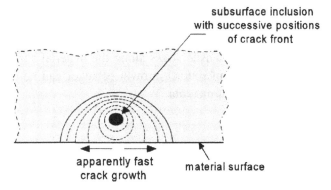

Fig. 2.10 Subsurface nucleation of fatigue crack at inclusion which suggests initial fast crack growth [8].

nucleation. Nucleation starting from inclusions has been shown in several publications, see e.g. [5, 6]. An example is shown in Figure 2.9.

The nucleation of a microcrack at an inclusion can occur slightly subsurface and not necessarily at the material surface. However, the free surface argument (lower restraint on slip) remains valid. It is for this reason that crack nucleation far below the material surface is rarely observed, although it can occur if the inclusion is large, or if a residual tensile stress is present away from the material surface.

An interesting situation arises if a microcrack initiated at a subsurface inclusion penetrates through the ligament between the inclusion and the free surface, see Figure 2.10. Crack growth observed on the material surface corresponds to a kind of a *break-through*. As a result, an initially high growth rate is observed which slows down later when the crack becomes a true surface crack. This is another reason for an apparently unsystematic development of the crack growth rate in the crack initiation period.

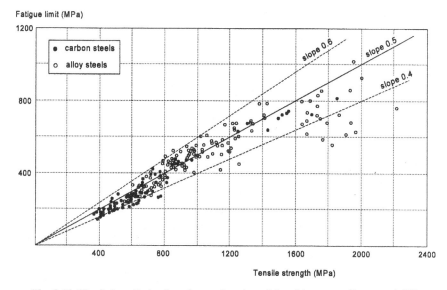

Fig. 2.11 The fatigue limit of steels as a function of the ultimate tensile strength [9].

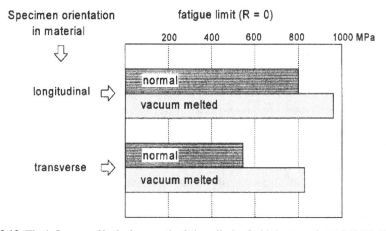

Fig. 2.12 The influence of inclusions on the fatigue limit of a high-strength steel (SAE 4340) [10].

The significance of inclusions for fatigue in steel has received considerable interest in the past. It was observed that the fatigue limit of different types of steel increased approximately proportionally to the tensile strength, see Figure 2.11. However, at very high values of the tensile strength, this trend was not continued and lower fatigue limits were obtained. The explanation was found by considering inclusions in high-strength steel. The inclusions act as micronotches in the material which can generate

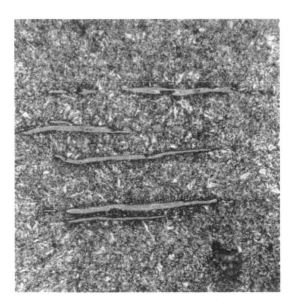

Fig. 2.13 Non-metallic slag inclusions in a martensitic matrix of a high-strength steel. (Courtesy H.P. van Leeuwen, NLR, Amsterdam)

fatigue crack nuclei, especially in high-strength material. Because of the high-yield stress of these materials they have a high notch sensitivity, even for micronotches. In low-strength steels, inclusions are less harmful because of the lower notch-sensitivity. Improvements of the fatigue limit of high-strength steels could be obtained by purifying the composition to eliminate inclusions. The improvement is illustrated by the results in Figure 2.12 for a high-strength NiCrMo steel. It is noteworthy that the fatigue limit of the "normal" material shows a significant anisotropy. Specimens in the transverse direction have a 32% lower fatigue limit than in the longitudinal direction. This is due to the directionality of the elongated inclusions with the long dimension in the longitudinal direction. An example of such inclusions from another investigation is shown in Figure 2.13. After a vacuum melting process, the number of inclusions is drastically reduced, which increases the fatigue limit considerably and at the same time the directionality effect is also smaller as illustrated by the data in Figure 2.12.

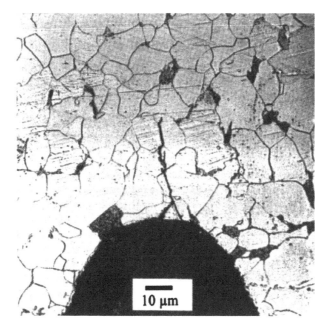

Fig. 2.14 Non-propagating fatigue crack at the root of a rotating beam specimen of mild steel after 24×10^6 cycles at $S_a \pm 39$ MPa. Circumferential V-notch, depth 1.3 mm, root radius 0.07 mm [11].

2.5.3 Small cracks, crack growth barriers, thresholds

It has been observed in laboratory experiments as well as in service that minute cracks were nucleated which stopped growing at a small crack length (Figure 2.7). Apparently, the cracks encountered a type of a crack growth barrier and could not grow any further. The barrier was a threshold for crack growth.

In the fifties, Frost et al. [11] studied so-called non-propagating fatigue cracks. He observed small fatigue cracks in notched specimens tested under a low cyclic load. Frost found small cracks which stopped growing at a crack length of some grain diameters. An example is shown in Figure 2.14. Frost did not observe a similar behavior on unnotched specimens, but for sharply notched specimens he concluded that small fatigue cracks could be initiated at stress amplitudes below the fatigue limit. His results for specimens with different stress concentrations are illustrated by the graph in Figure 2.15. For severe stress concentrations with large K_t-values,[5] crack initiation at low

---

[5] K_t is the stress concentration factor, an elastic concept discussed in Chapter 3. By definition K_t is the ratio of the peak stress in the notch and the nominal stress ($K_t = S_{\text{peak}}/S_{\text{nom}}$).

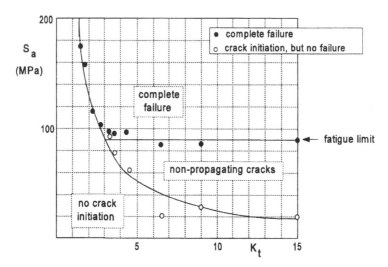

Fig. 2.15 Observations of Frost [12] on non-propagating cracks as a function of K_t. Material: mild steel.

stress amplitudes could not be avoided because of the high peak stress at the root of the notch. However, the stress amplitudes were insufficient to maintain continuous crack growth and failure did not occur. As discussed before, initiation occurs at the material surface, where the restraint on cyclic slip is minimal. After some crack growth the crack tip stress field changes from plane stress at the free surface to plane strain deeper in the material. It implies an increased restraint on cyclic slip, and apparently crack arrest could occur as a consequence of insufficient cyclic slip. In this case, it is not really a material barrier that stops crack growth, but a change in the crack tip stress field. Such small cracks should be referred to as *mechanically short cracks*.

In the seventies and afterwards, it was recognized that non-growing microcracks can also occur in unnotched specimens. Again it means that crack nucleation occurs below the fatigue limit. The reasons for crack arrest, however, are different from those for the experiments of Frost. Several types of barriers to the growth of these cracks are possible. An example of a microstructural effect was already given in the discussion on Figure 2.5 with the grain boundaries as apparent barriers to a continuous growth of the microcrack. Although the grain boundaries did not stop microcrack growth in this case, it is possible that microcracks are nucleated in a grain, but cannot penetrate into the neighbouring grains, and thus remain arrested within that grain. Examples of other microstructural barriers are two-phase

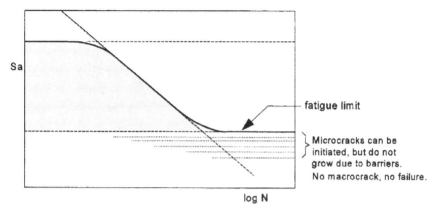

Fig. 2.16 Microstructurally small cracks below the fatigue limit of unnotched specimens.

boundaries, such as pearlite islands in low-carbon steel and α/β interfaces in Ti-alloys. In general, the size of such non-growing cracks is in the order of the spacing between the microstructural barriers. Such cracks are referred to as *microstructurally small cracks*. The significance of material structural barriers is associated with their effect on cyclic slip at the tip of the microcrack.

Because of non-growing cracks below the fatigue limit, see Figure 2.16, the definition of the fatigue limit should be reconsidered. A frequently used definition says that the fatigue limit is the stress amplitude for which the fatigue life goes to infinity. Mathematically, the definition implies that the fatigue limit corresponds to the lower horizontal asymptote of the S-N curve. Another definition can now be postulated: the fatigue limit is the lowest stress amplitude for which crack nucleation is followed by crack growth until failure. Of course, this definition could also be formulated as the fatigue limit is the largest stress amplitude which does not lead to continuous crack growth until failure. The fatigue limit is thus recognized as a threshold for the growth of small cracks, and not as a threshold for crack nucleation. This observation is relevant for notch effects on fatigue discussed in Chapter 7.

It may be questioned if microscopically small barriers to slip can still have a significant effect on continuous crack growth, for instance, inclusions in the order of 1 μm (or in the sub-micron range). They can even be small compared to the size of microcracks. Although such inclusions will have a local effect on cyclic slip and local crack growth, the microcrack can grow around the inclusion. It thus will not be a barrier as a permanent threshold for further

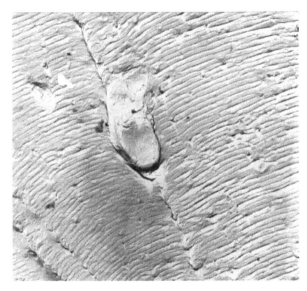

Fig. 2.17 Fatigue crack growth around an inclusion. Striation spacing about 0.5 μm.

crack growth. Evidence for growing around inclusions has been shown for macrocracks in Al-alloys, see the example in Figure 2.17. Similar evidence for other materials is not abundant, but overcoming local barriers should also be expected to occur.

2.5.4 Number of crack nuclei

It is an old and good question to ask why fatigue apparently occurs as a localized failure. Many times, a fatigue failure in service seems to be the result of the growth of a single fatigue crack only. An example of such a failure is shown in Figure 2.18. The fracture surface of a broken car axle suggests that the fatigue failure started at a single nucleation point. However, also for unnotched specimens tested in the laboratory the trend appears often to be valid. As said before, fatigue crack initiation is a surface phenomenon, but thousands of grains are found at the material surface of unnotched specimens. The question is whether the conditions for nucleation of a fatigue crack are similar for all surface grains? That is not true for several reasons. Nucleation depends on the occurrence of cyclic slip. However, the stress is not equal in all surface grains due to the anisotropy of the material (recall the discussion in Section 2.5.1). Moreover, the variation of the shape

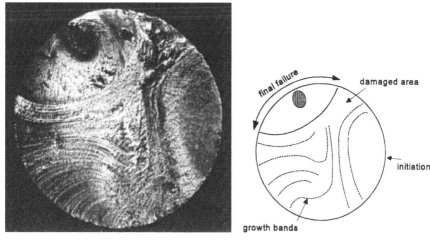

Fig. 2.18 Fatigue fracture surface of a rear axle of a motor car, diameter 20 mm. The fatigue failure started as a single dominant fatigue crack.

and size of the grains also contributes to the inhomogeneity of the stress distribution on a microlevel. Furthermore, the orientation of the crystal lattice differs from grain to grain, which amplifies different possibilities for cyclic slip. If crack nucleation occurs at small inclusions near the free surface of the material, the size, shape and orientation of the inclusions is another source for differences between grains. Surface roughness can also contribute to some preferred locations for crack initiation. Nucleation of fatigue cracks in an unnotched specimen will therefore occur at a location where all combined conditions are most favorable for cyclic slip and crack nucleation. It seems to be reasonable that in a fatigue test only one dominant fatigue crack nucleus is detected on the fracture surface. This is particularly true if the cyclic stress level is close to the fatigue limit. The location of crack nucleation site may then be called the "weakest link" of the specimen.

Fatigue fracture surfaces as shown in Figure 2.18 suggest a single fatigue crack nucleus, but it is possible that more small fatigue cracks are present which are still too small to be seen on the fracture surface. They can also occur in different cross sections of a specimen without being linked up by the main crack. Furthermore, it should be recalled that the crack initiation period covers a relatively large part of the fatigue life, see Figure 2.7. It then is possible that a single crack has grown into the macrocrack growth period until complete failure while other cracks are still in the crack initiation period as a consequence of scatter of the local conditions for microcrack nucleation.

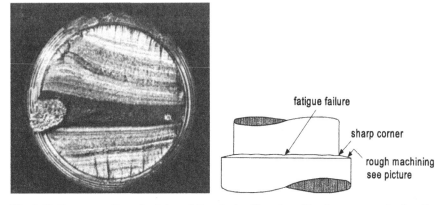

Fig. 2.19 Fracture surface of a fatigue failure in the tiller of a sailing boat as a result of cyclic bending. Crack nucleation at several locations in the sharp corner edge. Overlap of crack nuclei produces lines in the crack growth direction.

The situation is different if the stress amplitude is high. The significance of microstructural thresholds becomes less important because they are more easily overcome at higher stress levels. As a result, more microcracks can develop successfully to a larger size. This has been observed, especially in Al-alloys. Stress levels between different grains in these alloys are relatively low because of the low elastic anisotropy (Section 2.5.1). Microscopic investigations on specimens of Al-alloys at relatively high stress levels have shown fairly large numbers of microcracks in different surface grains, e.g. [13, 14]. At a later stage linking up of these cracks occurred to form one single larger crack.

A high stress at the material surface is also obtained if sharp notches are present due to the high stress concentration. Crack initiation then is relatively easy and can occur at many places more or less simultaneously. Figure 2.19 shows the fatigue failure of a tiller of a sailing-vessel, which was primarily loaded by plane bending. Cracks initiated at the sharp corner of a diameter reduction, both at the top side and the lower side in Figure 2.19. The bending stress level was not high because a large amount of fatigue crack growth occurred as shown by the area of fatigue bands in Figure 2.19. As a consequence, the final failure in the last cycle of the fatigue life covers a relatively small area which is the horizontal dark band in the middle of Figure 2.19. However, the K_t-value at the root of the notch was extremely high due to the sharp corner, and thus crack nucleation occurred at many neighbouring locations. Overlapping of these cracks leads to the vertical lines in the crack growth direction. The lines are very small steps

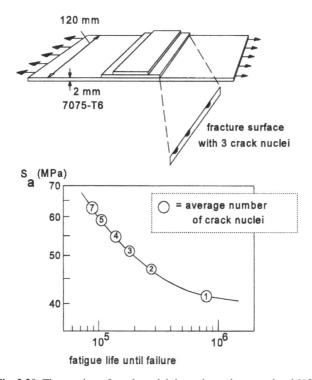

Fig. 2.20 The number of crack nuclei depends on the stress level [15].

in the fracture surface because the cracks initially started to grow in slightly staggered planes, see also Figure 2.33 to be discussed later.

Another illustrative example of multi-site crack nucleation is shown in Figure 2.20. The specimen consists of a sheet with adhesively bonded doublers. Fatigue cracks occur at the edge of the doubler because the doubler implies a local stress concentration which is further enhanced by additional bending of the specimen due to the doublers being present at one side of the sheet only. Tests were carried out at several stress levels to determine an S-N curve. All fracture surfaces were examined, and the number of visible crack nuclei were counted for each specimen. Average numbers of crack nuclei are indicated in the graph for different stress levels. Obviously the number of visible fatigue crack nuclei increases at higher stress levels, whereas only one nucleus was evident at a low stress level (the weakest link).

The above results confirm that crack nucleation at a low stress level is more problematic. It is indeed a threshold problem. At high stress levels the situation is entirely different because nucleation can occur anyway, and it can occur at several locations. A technically significant conclusion to be

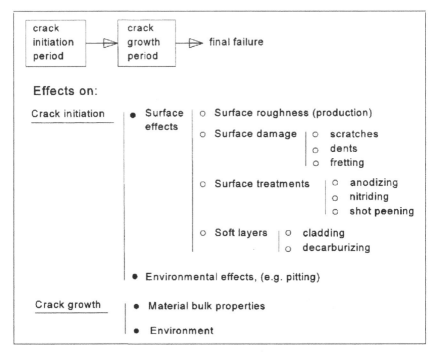

Fig. 2.21 Effects of the crack initiation and crack growth period.

drawn now is: If crack nucleation is problematic, scatter of fatigue properties must be expected. If crack nucleation is not problematic, scatter will be less sensitive to microstructural characteristics of the material, and thus scatter will be lower. This topic is addressed later in Chapter 12 on scatter of fatigue.

2.5.5 Surface effects

The discussion in Section 2.3 has indicated that fatigue in the crack initiation period is a surface phenomenon. It was also pointed out that the initiation period may cover the major part of the fatigue life until failure (Figure 2.7). As a consequence, various kinds of surface effects can be of great importance for the fatigue life. Surface effects include all conditions which can reduce the crack initiation period. In other words, they cover the phenomena which enhance the crack initiation mechanism. Several effects are mentioned in Figure 2.21. The list is not necessarily complete, but various well-known effects are listed. They are briefly discussed below while most effects are covered more extensively in later chapters. The purpose of the discussion

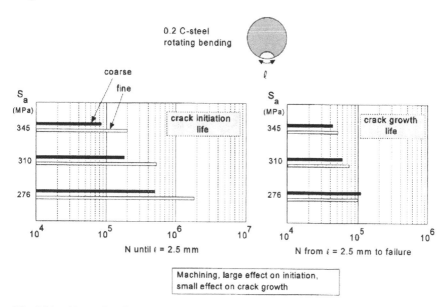

Fig. 2.22 Effects of surface roughness on the crack initiation and crack growth period [16].

here is to reveal general aspects related to the fatigue process in terms of the crack initiation period and crack growth period.

Surface roughness and surface damage imply that the free surface is no longer perfectly flat. As a consequence, small sized stress concentrations along the material surface occur. Although the stress concentration will rapidly fade away from the surface, it is still significant for promoting cyclic slip and crack nucleation at the material surface. As an example, results of de Forest [16] are shown in Figure 2.22. He carried out rotating bending fatigue tests on specimens with two different surface roughnesses, coarse machining and fine machining. Rough machining causes deeper circumferential grooves than fine machining. De Forest periodically interrupted his tests to observe possible crack growth. He then defined the crack initiation period as the fatigue life until a circumferential crack length $l = 2.5$ mm (0.1 inch) was reached, while the crack growth period covered the life from $l = 2.5$ mm until failure. As shown by the test results at three different stress amplitudes, the crack initiation life is significantly shorter for rough machining if compared to fine machining, see the left-hand graph in Figure 2.22. However, according to the right-hand graph of Figure 2.22 the crack growth period is hardly affected by the surface roughness. A comparison of the two graphs in Figure 2.22 also reveals that the crack growth period is much shorter than the crack initiation period.

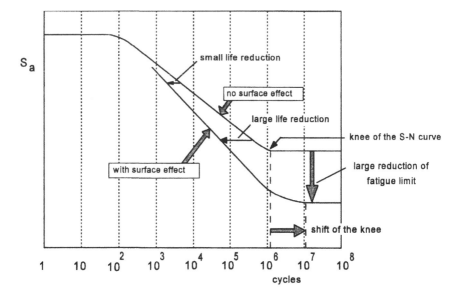

Fig. 2.23 Surface effects on the S-N curve. Both S_a and N are plotted on a logarithmic scale.

There is still another message in Figure 2.22. The crack initiation period significantly increases for a lower stress level, as should be expected. If the stress amplitude is reduced further to the fatigue limit, the crack initiation life becomes very large, confirming the threshold idea about the fatigue limit as discussed before. The crack growth period in Figure 2.22 also increases for a lower stress amplitude, but the effect is much smaller. This trend is generally observed for surface effects. The amplitude-dependent sensitivity is illustrated in Figure 2.23. The most detrimental consequence of an unfavorable surface effect is the large reduction of the fatigue limit. This is especially important for structural components designed for an infinite life, i.e. with all amplitudes in service below the fatigue limit. Unintentional surface damage, such as nicks and dents, can then be very harmful. The same is true for damage due to fretting, a phenomenon described in more detail in Chapter 15. The large reduction of the fatigue limit indicates that there is a range of stress amplitudes between the original S_f and the reduced S_f which can be harmful if surface damage is present. Without surface damage such fatigue cracks are not initiated, but with the assistance of surface damage cracks can be started and cause failure. Due to the relatively low stress amplitude, the crack growth life can be large. As a consequence, the inflection point of the S-N curve to the horizontal part (the so-called knee of the S-N curve) occurs at a higher fatigue life as for the original S-N curve, see the shift of the knee in Figure 2.23.

If a design is made for a finite life, detrimental surface effects may be less important, specifically if the design life is short. Although surface damage can accelerate crack initiation, the high stress amplitude cycles can also generate crack nuclei early in the fatigue life, and the assistance of surface damage is less important for the initiation process. However, if the design life is large in numbers of cycles, the significance of adverse surface effects should be recognized. The high sensitivity for surface effects at low stress amplitudes and the relatively low sensitivity for surface effects at high stress amplitudes can lead to more scatter of the fatigue life at low amplitudes and less scatter at high amplitudes. This trend is generally observed in fatigue experiments. The trend was already mentioned in the previous section for similar reasons.

Shot peening is used as a remedy if fatigue problems are anticipated. Shot peening introduces plastic deformation in the surface layer of the material. As a result of the plastic deformation, residual compressive stresses are left in a thin surface layer. Because residual stresses do not affect cyclic shear stresses, cyclic slip may still occur. It even can lead to small microcracks, but crack growth is difficult. The residual compressive stresses reduce or prevent crack opening of the microcracks. As a result, the stress concentration at the crack tip is much lower and crack growth will be impeded. It may well be stopped completely. The residual compressive stress zone serves as a barrier for microcrack growth. There are many other ways to introduce residual stresses in the material of structural components. Residual stresses are very important for practical reasons. Possibilities for introducing residual stresses are discussed in Chapter 4.

The list in Figure 2.21 also refers to environmental effects. A brief discussion on this topic is presented in Section 2.5.7 and a more extensive treatment in a later chapter (Chapter 16).

2.5.6 Crack growth and striations

As discussed before, fatigue crack growth in the crack growth period is no longer affected by the material surface conditions. The crack growth is then a bulk material phenomenon. Usually the crack is growing perpendicular to the main principal stress. For uniaxial loading conditions in symmetric specimens, it implies that the crack growth direction is macroscopically

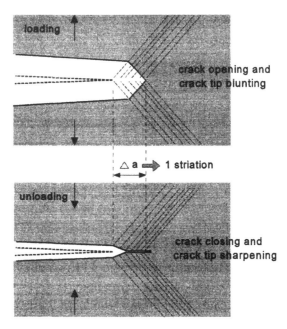

Fig. 2.24 Crack extension in a single load cycle.

perpendicular to the loading direction. After the crack has grown away from the free surface, slip deformations will occur on more than one slip plane. Figure 2.24 gives a schematic visualization of a possible mechanism for crack extension in one load cycle. During loading the crack will be opened by crack tip plastic deformations, which in Figure 2.24 are supposed to occur on two symmetric slip systems. Stress analysis of a solid with a crack indicates that the slip zones in Figure 2.24 are indeed the zones with the maximum shear stress, both during loading and unloading. During loading the slip deformation will cause some crack extension. Also for these larger cracks, the crack extension implies decohesion, which should be associated with dislocations flowing into the crack tip, or being emitted by the crack tip. It also seems plausible that crack extension occurs in every successive load cycle in a similar way as sketched in Figure 2.24.

The slip deformations are not fully reversible due to strain hardening and other possible mechanisms. As illustrated by Figure 2.24, a ridge of microplastic deformation is left on the new upper and lower crack tip surfaces created in that particular cycle. These ridges are called "striations", which can be observed on the fatigue fracture surface in the electron microscope. Although they were already observed long ago [17, 18] under the optical microscope, much better pictures were obtained in the electron

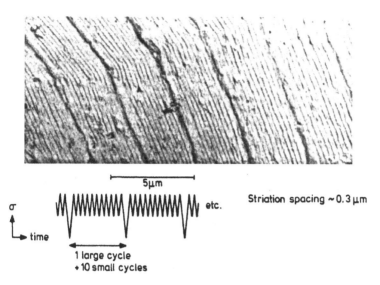

Striation spacing ~ 0.3 μm

Fig. 2.25 Correspondence between striations and load cycles during macrocrack growth in an Al-alloy sheet. (Picture National Aerospace Laboratory, NLR, Amsterdam)

microscope, originally in the Transmission Electron Microscope (TEM) and later abundantly in the Scanning Electron Microscope (SEM). An illustrative picture is shown in Figure 2.25. An aluminium alloy sheet specimen was loaded by a repetition of 10 small cycles and a single larger load cycle. There is a striking correspondence between the load history and the striation pattern. The 10 small cycles correspond to the smaller striation spacings, while the single larger cycle is responsible for the wider and more dark striation. Such pictures prove that crack extension did occur in each cycle of the load history. It also allows measurements of the crack growth rate (i.e. the crack extension per cycle) from such fractographs. The striation spacing for the small cycles in Figure 2.25 is about 0.3 μm which thus corresponds to a crack growth rate of 0.3 μm/cycle.

The occurrence of striations can give essential information to the analysis of failures occurring in service. First, if striations are observed, it shows that at least part of the failure occurred due to cyclic loads. Moreover, striations can give information about the crack growth direction and crack growth rate. Unfortunately, striations are not equally visible for all materials. Most useful observations have been obtained on aluminium alloys. However, striations were also observed on various types of steels, titanium alloys and some other alloys, but usually less abundantly and with less well-defined striations. If striations cannot be observed it should not be concluded immediately that fatigue did not occur.

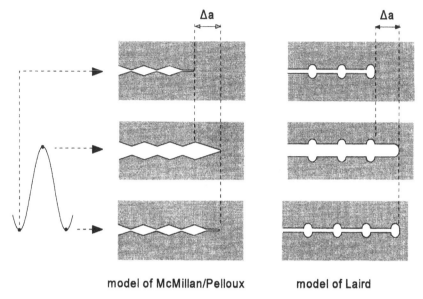

model of McMillan/Pelloux   **model of Laird**

Fig. 2.26 Two models for striaton forming during fatigue crack growht [19, 20].

The geometry of the blunted crack tip in Figure 2.24 is a rather simple one. In reality it might be much more complex. The mechanism of fatigue crack growth at the crack tip has been a subject of much speculation in the literature. Two well-known models are presented in Figure 2.26. Both models create one striation in each cycle, but the crack tip blunting and crack tip resharpening is different for the two models. The first model by McMillan and Pelloux [19] is similar to the sketch in Figure 2.24. It implies that the crack tip upon unloading is closed, starting at the crack tip. However, in the model of Laird [20] closing of the crack starts behind the crack tip, which creates a kind of an ear marking at the end of the crack extension in a cycle. Because the formation of striations is a result of cyclic plasticity, it is very well possible that the striation geometry depends on the type of material. The McMillan–Pelloux model was developed for Al-alloys, while Laird supported his model with observations for pure metals.

Microscopic research studies on the mechanism of crack extension under cyclic loading have received much attention for years, but some major problems were encountered. Fatigue cracks observed at the free surface of the material do not necessarily show the same behavior as the crack front inside the material. At the free surface, the deformation restraint on cyclic slip is different from the restraint inside the material. Secondly, at the interior of the material, observations on striations can be made only after opening of

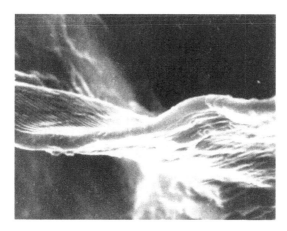

Fig. 2.27 Casting of a fatigue crack (2024-T3 alluminium alloy, width of picture 16 μm). Technique developed by Quinton Bowles [21]. Curved crack front and striations at both sides of the casting.

the fatigue crack. It gives the final topography, but not the situation during the dynamic process of crack extension and crack tip closure. Moreover, the dimensions of details of striations are in the sub-micron range where observations are difficult. An interesting technique was developed by Bowles [21]. He used a vacuum infiltration method, which implies that the crack at any selected load can be filled with a plastic. The crack serves as a mould. After hardening of the plastic, the material of the aluminium specimen is removed by a chemical solvent. A casting of the crack is then obtained, which can be studied in the SEM. A picture is shown in Figure 2.27. It shows striations of the upper and lower side of the fatigue crack. More important, it shows that the tip of the fatigue crack is not sharp but rounded (estimated radius 0.4 μm), and the crack front is not a straight line. The latter observation was already confirmed by Figure 2.25. The infiltration technique until now has been applied to Al-alloys only. The method cannot be used for materials where chemical removal of the specimen material is difficult which applies to steel. However, knowledge about the geometry of the crack front is significant for the justification of applying fracture mechanics to predictions on fatigue crack propagation, the subject of Chapters 8 and 11.

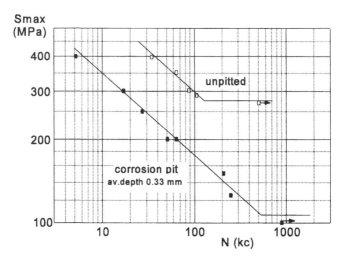

Fig. 2.28 The effect of a corrosion pit on the S-N curve of unnotched specimens of an Al-alloy [22].

2.5.7 Environmental effects

As indicated in Figure 2.21, the environment can affect both crack initiation and crack growth. It should be pointed out that there is a difference between fatigue of corroded specimens in a non-aggressive environment, and fatigue of initially undamaged specimens in a corrosive environment. The first case is illustrated by Figure 2.28. Unnotched specimens of a notch sensitive Al-alloy (7075-T6) were provided with a corrosion pit to a depth of approximately 0.3 mm. These specimens were then used to obtain an S-N curve. For the material used, such a corrosion pit is severe surface damage, also because the shape of a corrosion pit represents a significant stress raising geometry. As shown by the S-N curves in Figure 2.28 the fatigue life at high stress amplitudes is reduced about six times. But the most dramatic effect is the reduction of the fatigue limit S_f. Without the corrosion pits S_f is 275 MPa, whereas it is only 110 MPa if there is a corrosion pit. It confirms a basic idea of Figure 2.23. Stress amplitudes below the original fatigue limit are unable to create a fatigue crack. However, with the corrosion pit, cracks can be initiated and cause failure. Figure 2.28 also illustrates the shift of the knee already defined in Figure 2.23. A similar effect of corrosion pits on the S-N curve of a martensitic 12% Cr-steel for turbine blades was found by Zhou and Turnbull [23].

In general, corrosion fatigue refers to acceleration of crack initiation and crack growth under the combined action of fatigue and corrosion. The

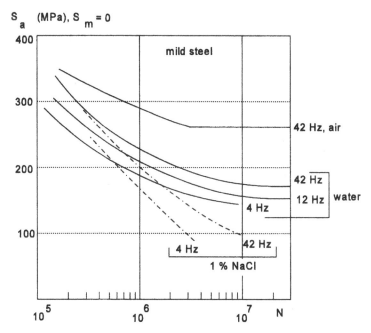

Fig. 2.29 The effect of environment and load frequency on the S-N curve of unnotched mild steel specimens [24].

acceleration should be associated with some contribution of a corrosive environment to the fatigue fracture mechanism. The S-N curves in Figure 2.29 illustrate a large effect. Endo and Miyao [24] carried out rotating bending tests on unnotched mild steel specimens in three environments; air, tap water and salt water. A dramatic effect on the S-N curve is observed. Again, a large reduction of the fatigue limit is evident. Although the effect at a high stress amplitude is still significant, it is much smaller, again in agreement with the trends shown in Figure 2.23. Unfortunately, the occurrence of a fatigue limit in salt water is not indicated by the curves in Figure 2.29. Apparently the salt water corrosion is successful in assisting the crack initiation and subsequent crack growth at very low stress amplitudes.

The results in Figure 2.29 also show that the frequency of the fatigue load has a systematic effect both in tap water and in salt water, compare the results for 42 Hz and 4 Hz. Because corrosion is a time dependent phenomenon, a frequency effect has to be expected. Actually, this effect is disturbing if structures are operating in a marine environment (ships, offshore structures) where load frequencies can be very low.

The effect of corrosion on crack nucleation and crack growth is a difficult phenomenon to be described in physical and electro-chemical terms. It may

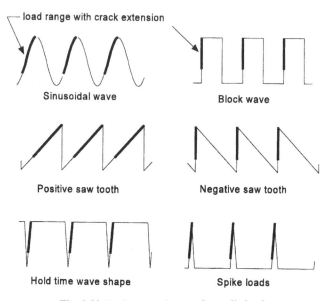

Fig. 2.30 Basic wave shapes of a cyclic load.

well be possible that crack extension is promoted by an aggressive agent at the crack tip. Because crack extension is decohesion of the material, foreign ions of the environment can weaken the cohesive strength of the material in some way. It has also been thought that some resolving of the material at the crack tip can occur. The mechanism will depend on the specific combination of the material and the environment. Aspects of cyclic crack tip plasticity, crack extension and environmental contributions are a difficult problem to study in physical details. One important aspect has to be mentioned here. If the load frequency has a significant effect, because of some time dependent corrosion mechanism, it should be expected that the wave shape of the load cycle can also have an effect on crack growth. In fatigue tests, the wave shape usually is sinusoidal, but in service it can be highly different. Several basic wave shapes are shown in Figure 2.30. Even if the load frequency of these wave shapes is the same in numbers of cycles per minute, corrosion fatigue does not necessarily occur at the same rate. It should be recognized that crack extension as a result of crack tip plasticity occurs during loading to S_{max} as a progressive process. It is not just a crack length jump Δa at the moment that S_{max} is reached. Crack extension occurs in the period before S_{max} is reached. In this period a corrosive environment can enhance the decohesion, and thus amplify the crack extension. The short period of the crack extension process becomes the significant variable. The period is indicated as a black line in the

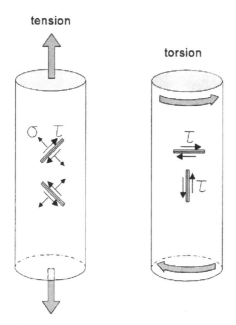

Fig. 2.31 Slip planes with maximum shear stress.

wave shapes shown in Figure 2.30. It then is remarkable to see a significantly different period for the positive and the negative saw tooth wave shape. The loading rate of the positive saw tooth wave is relatively low. Crack extension occurs during a large part of the load cycle period. By contrast, the loading rate is high for the negative saw tooth wave shape. Crack extension by crack tip plasticity occurs in a very small fraction of the cyclic load period. As a consequence, not much time is available for a corrosive contribution to the crack extension process. A much lower effect on the crack growth rate should be expected. Relevant experiments were carried out by Barsom [25] on a high-alloy steel tested in salt water at 0.1 Hz. It turned out that crack growth for the positive saw tooth wave was about three times faster than for the negative saw tooth wave (see also Chapter 16).

2.5.8 *Cyclic tension and cyclic torsion*

As discussed before, cyclic slip is essential for microcrack nucleation and early microcrack growth. Crack nucleation in an unnotched specimen will now be considered for two loading cases: (1) cyclic tension, and (2) cyclic torsion, see Figure 2.31. Under cyclic tension the maximum shear stress

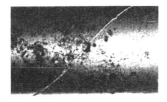 

(a) Spiral crack due to cyclic torsion. No macroplastic deformation

(b) Same fatigue failure of (a). Arrow indicates starting point of fatigue crack at a surface pit

Fig. 2.32 Fatigue failure of drive shaft of a scotter starting at surface damage.

occurs on planes at an angle of 45° with respect to the longitudinal axis. Under cyclic torsion, planes with a maximum shear stress are perpendicular and parallel to the longitudinal axis. An important difference between the two loading systems is that the plane of maximum shear stress in the tension case also carries a normal stress component ($\sigma = \tau$). For cyclic torsion, however, this normal stress component on that slip plane is zero. As long as the initiation is still a matter of cyclic slip in a single surface grain, the two cases are essentially different. In the cyclic tension case the normal stress tries to open the microcrack and that will enhance the transition from cyclic slip into microcrack growth along the slip band. However, under cyclic torsion this crack opening effect is absent. Microscopical investigations have shown that nucleation in a slip band under cyclic torsion is problematic if the load amplitude is low, i.e. close to the fatigue limit. But for higher amplitudes above the fatigue limit, microcracks under cyclic torsion are generated which then grow further in a direction perpendicular to the main principal stress. In the cylindrical bar of Figure 2.31 this direction occurs at an angle of 45° with the axis of the bar. As a consequence, cracks in a round axle under cyclic torsion grow as a spiral around the surface of the axis. An example is shown in Figure 2.32, a drive shaft of a scooter, broken by torsional fatigue. The fatigue failure started at a surface damaging pit.

2.6 Characteristic features of fatigue failures

Several characteristic features of fatigue fracture surfaces have been discussed in previous sections. They are repeated here together with some more typical aspects. The characteristic features are significant in

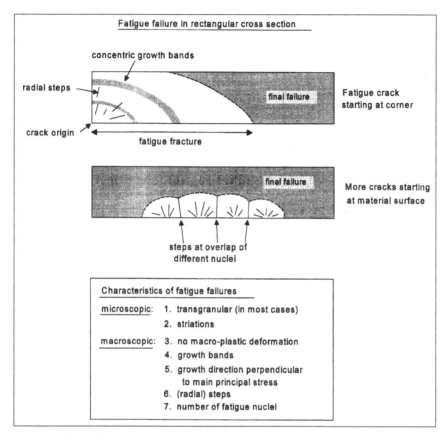

Fig. 2.33 Survey of characteristic features of fatigue fracture surfaces.

failure analysis in order to distinguish between fatigue failures, static failure, including brittle failures, stress corrosion failures and creep failures. Moreover, if a fracture surface indicates that a failure is due to fatigue, it may well be possible to arrive at more details about the service fatigue load history. Examinations of fatigue failures of laboratory specimens can also give worthwhile information to validate fatigue prediction models.

The characteristics of a fatigue fracture are divided into two groups, microscopic features and macroscopic features, see Figure 2.33. It may be noted here that papers and test reports on fatigue experiments should always include a description of the fracture surfaces. Without these observations the discussion of the test results is inadmissibly incomplete.

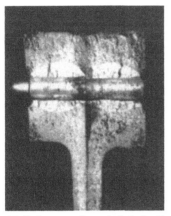
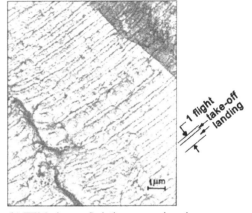

(a) Fatigue cracks at bolt hole. Both   (b) TEM picture. Striations occur in pairs
sides of the fracture surface are shown

Fig. 2.34 Striations in pairs on a fatigue fracture in a flap beam of an aircraft [8].

2.6.1 Microscopic characteristics

(1) *Transgranular crack growth*
Fatigue cracks in almost all materials are growing along a transgranular path, i.e. through the grains. They do not follow the grain boundary, contrary to stress corrosion cracks and creep failures. Because fatigue crack growth is a consequence of cyclic slip, it is not surprising that fatigue crack prefers to grow through the grains. Restraint on slip exerted by the grain boundaries is minimal inside the grains. The transgranular character can easily be observed on microscopic samples in the optical microscope.

(2) *Striations*
As discussed before, striations are remnants of microplastic deformations of individual load cycles, see Figures 2.24 to 2.27. The striations indicate the cyclic nature of the load history. The visibility of striations depends on the type of material and the load history as well. If striations cannot be observed, it need not imply that the failure is not due to fatigue.

An example of striations on a failure occurring in service is given in Figure 2.34. A flap beam of an aircraft failed during landing due to fatigue at a bolt hole (Figure 2.34a). The electron microscope revealed that striations occurred in pairs with a larger and a smaller striation spacing respectively (Figure 2.34b). The flap beam was loaded significantly twice in each flight, viz. by a large load during landing (flaps fully out) causing the wider striation, and a lower load during take off (flaps only partly out) causing the smaller striation. It implies that each pair of striations corresponds to a single flight. The number of flights could thus be counted on the larger part of the fatigue fracture surface. It corresponded to a crack growth life of approximately 5000 flights. This example of striations also confirms that crack propagation occurs cycle-by-cycle. It also indicates that the information on crack growth in the component can be useful in order to assess periodic inspections for fatigue cracks and thus ensure safe flights.

2.6.2 *Macroscopic characteristics*

Macroscopic characteristics of a fatigue failure can be observed with the naked eye. However, it always should be advised to look also with a small magnifying glass with a magnification of 6 to 8×. It is often surprising how many details then can be observed, e.g. small crack nuclei, sites of crack nuclei, surface damage causing a crack nucleus, etc., details which escape visual observations by eye but which may be significant for the evaluation of the fatigue problem. Moreover, some details are sometimes overlooked at larger magnification in the electron microscope.

(3) *No macroplastic deformation and a flat fatigue fracture surface*
The fracture surface of a fatigue failure usually shows two different parts:

(i)   The real fatigue failure caused by fatigue crack growth (see Figure 2.33) is characterized by practically no macroplasticity. Because fatigue is a result of microplastic deformations, which are largely reversed in each cycle, it is not surprising that macroscopic deformations appear to be absent. For the major part of the fatigue life, the crack can hardly be seen on the material surface. Crack detection during service inspections may become problematic. Various NDI techniques have been developed for that purpose.

(ii)   The second part of the fracture surface is caused by the final failure in the last load cycle. It occurs if the remaining uncracked cross section of the material can no longer carry the maximum load of the load cycle. In general, the final failure can be considered to be a quasi-static failure. It will exhibit macroplastic deformation, depending on the ductility of the material. The difference between a fatigue failure without visible plastic deformation and a static failure with visible plastic deformation is illustrated by two pictures of a failure in a light alloy helicopter blade

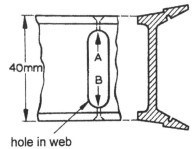

hole in web
(a) Cross section with hole in the web of the spar of a helicopter blade

(b) View of fatigue failure according to arrow A. No macroplastic deformation. Both parts fit well together

(c) View of static failure according to arrow B. Macroplastic deformation, ovalized hole, necking at right edge

Fig. 2.35 Difference between a fatigue failure and a static failure, both occurring in the same spar of a helicopter blade. The fatigue failure of (b) caused the static failure of (c). Hole diameter 3 mm.

in Figure 2.35. The blade separated form the helicopter during starting procedures before takeoff; the helicopter capsized, but no fire and no fatalities. Blade failure occurred in a section with a lightening hole in the spar of the blade with a rivet hole at the top and the bottom side, see Figure 2.35a. The failure in section A did not show any macroplastic deformation as can be seen in Figure 2.35b where the two sides of the failure still fit nicely together. However, such a nice fit was not possible for the other rivet hole, see Figure 2.35c, because of macroplastic deformation visible by hole ovality and necking at the edges.

(4) *Growth bands*
Fatigue failures obtained in service often show growth bands which are visible with the naked eye, see Figure 2.36, and also Figures 2.18 and 2.19. The bands are also referred to as oyster-shell markings or beach markings. The bands indicate how the crack has been growing. The different colours

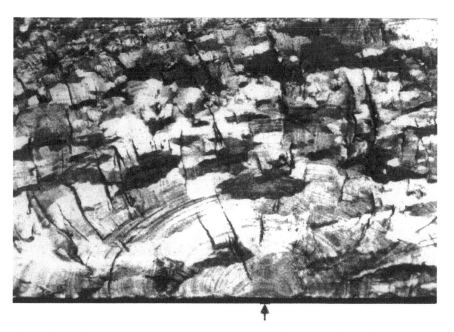

Fig. 2.36 Fracture surface of a light metal compressor blade. The fatigue failure started at the lower surface (arrow). Top to bottom of picture: 3 mm.

are associated with variations of the stress level of the cyclic load. This is illustrated by a picture obtained in some elementary experiments, see Figure 2.37. Blocks of cycles with two different S_{max}-values were applied alternately to specimens with edge notches. Approximately quarter circular cracks have grown from both edge notches. The dark bands were caused by the cycles with the higher S_{max}. Different degrees of corrosive attack can also cause bands, especially if cracks are dormant for certain periods.

(5) The growing direction, perpendicular to the main principal stress
As discussed before, fatigue cracks are growing in a direction perpendicular to the main principle stress (provided the crack growth rate is not very high). Depending on the geometry of the component, it implies for a cyclic tension load that crack growth will be perpendicular to the loading direction. For cyclic torsion the crack in a circular bar will occur at 45° with the longitudinal axis. An example of the spiral crack growth was given before, see Figure 2.32.

A noteworthy exception to the "perpendicular" crack growth is observed in some materials, e.g. Al-alloys and some steels. The growing fatigue crack at the material surface exhibits so-called shear lips, see Figure 2.38, at an

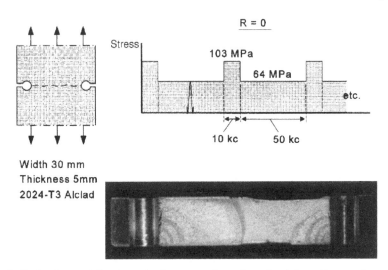

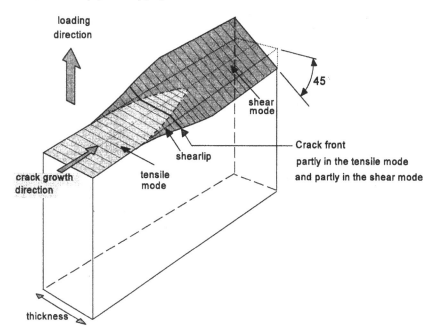

Fig. 2.37 Crack growth bands on the fracture surface of an alluminium alloy specimen with two side notches ($K_t = 2.85$) [27].

Fig. 2.38 Transition from crack growth in the tensile mode to the shear mode.

angle of approximately 45° with the central part of the crack front. The shear lip width increases during faster crack growth until they cover the full thickness. The shear lips start at the material surface, and in fact it should be considered to be a free surface phenomenon. The possibilities for plastic deformation at the location where the crack front meets the free surface are less restrained than at mid-thickness. It apparently allows plastic shear deformations which promote the formation of shear lips. These shear lips are of a similar nature as shear lips observed on the fracture surface of static failures along the free surface edges. For crack growth predictions, the shear lips could be a problem. This question is considered in Chapter 11. Shear lips also imply that microscopic observations on the material surface are not necessarily characteristic for the fatigue mechanism below the material surface.

(6) *Radial steps and the number of fatigue crack nuclei*
These two characteristics have already been discussed in Section 2.5.4. The radial steps as schematically indicated in Figure 2.33, are also visible in Figure 2.36 where the deformation texture of the forged alloy promotes crack growth on planes in slightly different directions. This is also confirmed by the dark and light areas, which are reflections of crack growth in different grains. The radial steps in Figure 2.19 were already mentioned previously.

2.7 Main topics of the present chapter

This section should not be considered to be a summary of the previous sections. The purpose is to present a list of the major findings, which are significant for applications in subsequent chapters.

1. The fatigue mechanism in metallic materials should basically be associated with cyclic slip and the conversion into crack initiation and crack extension. Details of the mechanism are dependent on the type of material.
2. The fatigue life until failure comprises two periods, the crack initiation period and the crack growth period. The crack initiation period includes crack nucleation at the material surface and crack growth of microstructurally small cracks. The crack growth period covers crack growth away from the material surface.
3. In many cases the crack initiation period covers a relatively large percentage of the total fatigue life.
4. Fatigue in the crack initiation period is a surface phenomenon, which is very sensitive to various surface conditions, such as surface roughness, fretting, corrosion pits, etc.

5. In the crack growth period, fatigue is depending on the crack growth resistance of the material and not on the material surface conditions.

6. Microstructurally small cracks can be nucleated at stress amplitudes below the fatigue limit. Crack growth is then arrested by microstructural barriers. The fatigue limit as a threshold property is highly sensitive to various surface conditions. At high stress amplitudes, and thus relatively low fatigue lives, the effect of the surface conditions is much smaller.

7. In view of possible effects during the crack initiation period, it can be understood that scatter of the fatigue limit and large fatigue lives at low stress amplitudes can be large, whereas scatter of lower fatigue lives at high amplitudes will be relatively small.

8. Aggressive environments can affect both crack initiation and crack growth. The load frequency and the wave shape are then important variables.

9. Predictions on fatigue properties are basically different for the crack initiation life and for the crack growth period.

10. The various characteristics of fatigue fractures can be understood in terms of crack initiation and crack growth mechanisms. These characteristics are essential in failure analysis, but they are also relevant to understand the significance of technically important variables of fatigue properties.

References

1. Ewing, J.A. and Humfrey, J.C.W., *The fracture of metals under repeated alternations of stress*. Phil. Trans. Roy. Soc., Vol. A200 (1903), pp. 241–250.
2. Bullen, W.P., Head, A.K. and Wood, W.A., *Structural changes during the fatigue of metals*. Proc. Roy. Soc., Vol. A216 (1953), p. 332.
3. Forsyth, P.J.E., *The Physical Basis of Metal Fatigue*. Blackie and Son, London (1969).
4. Blom, A.E., Hedlund, A., Zhao, W., Fathalla, A., Weiss, B. and Stickler, R., *Short fatigue crack growth in Al 2024 and Al 7475*. Behaviour of Short Fatigue Cracks, Symp., September 1985, Sheffield. EGF 1, MEP (1986), pp. 37–66.
5. Cummings, H.N., Stulen, F.B. and Schulte, W.C., *Tentative fatigue strength reduction factors for silicate-type inclusions in high-strength steels*. Proc. ASTM, Vol. 58 (1958) pp. 505-514.
6. Murakami, Y., Takada, M. and Toriyama, T., *Super-long life tension-compression fatigue properties of quenched and tempered 0.46% carbon steel*. Int. J. Fatigue, Vol. 20 (1998), pp. 661–667.
7. Kung, C.Y. and Fine, M.E., *Fatigue crack initiation and microcrack growth in 2024T4 and 2124-T4 aluminum alloys*. Metall. Trans. A, Vol. 10A (1979), pp. 603–610.

8. Schijve, J., *The practical and theoretical significance of small cracks. An evaluation.* Fatigue 84. Proc. Int. Conf. on Fatigue Thresholds, Birmingham. EMAS (1984), pp. 751–771.

9. Forrest, P.G., *Fatigue of Metals.* Pergamon Press, Oxford (1962).

10. Ransom, J.T., *The effect of inclusions on the fatigue strength of SAE 4340 steels.* Trans. Am. Soc. Metals, Vol. 46 (1954), pp. 1254–1269.

11. Frost, N.E. and Phillips, C.E., *Studies in the formation and propagation of cracks in fatigue specimens.* Proc. Int. Conference on Fatigue of Metals, London, September 1956. The Institution of Mechanical Engineers (1956), pp. 520–526.

12. Frost, N.E., Marsh, K.J. and Pook, L.P., *Metal Fatigue.* Clarendon, Oxford (1974).

13. Kung, C.Y. and Fine, M.E., *Fatigue crack initiation and microcrack growth in 2024T4 and 2124-T4 aluminum alloys.* Metall. Trans. A, Vol. 10A (1979), pp. 603–609.

14. Sigler, D., Montpetit, M.C. and Haworth, W.L., *Metallography of fatigue crack initiation in an overaged highstrength aluminium alloy.* Metall. Trans. A, Vol. 14A (1983), pp. 931–938.

15. Schijve, J., *Fatigue predictions and scatter.* Fatigue Fract. Engng. Mater. Struct., Vol. 17 (1994), pp. 381–396.

16. de Forest, A.V., *The rate of growth of fatigue cracks.* J. Appl. Mech., Vol. 3 (March 1936), pp. A-23 to A-25.

17. Zappfe, C.A. and Worden, C.O., *Fractographic registrations of fatigue.* Trans. Am. Soc. Metals, Vol. 43 (1951), pp. 958–969.

18. Forsyth, P.J.E. and Ryder, D.A., *Some results of the examination of aluminium alloy specimen fracture surfaces.* Metallurgia, Vol. 63 (1961), pp. 117–124.

19. McMillan, J.C. and Pelloux, R.M.N., *Fatigue crack propagation under program and random loading.* Fatigue Crack Propagation, ASTM STP 415 (1967), pp. 505–535.

20. Laird, C., *The influence of metallurgical structure on the mechanisms of fatigue crack propagation.* Fatigue Crack Propagation, ASTM STP 415 (1967), pp. 131–180.

21. Bowles, C.Q. and Schijve, J., *Crack tip geometry for fatigue cracks grown in air and vacuum.* Advances in Quantitative Measurement of Physical Damage. ASTM STP 811 (1983), pp. 400–426.

22. Scheerder, C., *The danger of single corrosion pits with respect to fatigue.* Master Thesis, Faculty of Aerospace Engineering, Delft University of Technology (1992).

23. Zhou, S. and Turnbull, A., *Influence of pitting on the fatigue life of a turbine blade steel.* Fatigue Fract. Engng. Mater. Struct., Vol. 22 (1999), pp. 1083–1093.

24. Endo, K. and Miyao, Y., *Effects of cycle frequency on the corrosion fatigue strength.* Bull. Japan Soc. Mech. Engrs., Vol. 1 (1958), pp. 374–380.

25. Barsom, J.M., *Effect of cyclic stress form on corrosion fatigue crack propagation below K_{Iscc} in a high yield strength steel.* Corrosion Fatigue: Chemistry, mechanics and Microstructure, O.F. Devereux, A.J. McEvily and R.W. Staehle (Eds.), Vol. NACE-2. National Association of Corrosion Engineers, Houston (1972), pp. 424–436.

26. Broek, D., *Accident investigation.* Report of the National Aerospace Laboratory NLR, Amsterdam.

27. Schijve, J., *Fatigue crack propagation in light alloys.* National Aerospace Laboratory NLR, Amsterdam, Report M.2010 (1956).

Some general references

28. Murakami, Y., *Metal Fatigue: Effects of Small Defects and Nonmetallic Inclusions.* Elsevier (2002).

29. Suresh, S., *Fatigue of Materials*, 2nd edn. Cambridge University Press, Cambridge (1998).
30. Miller, K.J., *The three thresholds for fatigue crack propagation.* ASTM STP 1296, R.S. Piascik, J.C. Newman, and N.E. Dowling (Eds.). ASTM (1997), pp. 267–286.
31. *Fatigue and Fracture.* American Society for Materials, Handbook Vol. 19, ASM International (1996).
32. Carpinteri, A., *Handbook of Fatigue Cracking – Propagation in Metallic Structures.* Elsevier, Amsterdam (1994).
33. *Fractography.* American Society for Materials, Handbook Vol. 12. ASM International (1987).
34. Ritchie, R.O. and Lankford, J. (Eds.), *Small Fatigue Cracks.* Proc. 2nd Engineering Foundation Int. Conf., 1986. The Metallurgical Society (1986).
35. Miller, K.J. and de los Rios, E.R. (Eds.), *The Behaviour of Short Fatigue Cracks.* EGF Publication 1, Mechanical Engineering Publications, London (1986).
36. Fuchs, H.O. and Stephens, R.I., *Metal Fatigue in Engineering.* John Wiley & Sons (1980).
37. Klesnil, M. and Lukás, P., *Fatigue of Metallic Materials*, 2nd edn. Elsevier, Amsterdam (1992).
38. Fong, J.T. (Ed.), *Fatigue Mechanisms.* ASTM STP 675 (1979).
39. Kocanda, S. *Fatigue Failure of Metals.* Sijthoff & Noordhoff (1978).

Chapter 3
Stress Concentration at Notches

3.1 Introduction

Calculations on the strength of structures are primarily based on the theory of elasticity. If the yield stress is exceeded plastic deformation occurs and the more complex theory of plasticity has to be used. Fatigue, however, and also stress corrosion, are phenomena which usually occur at relatively low stress levels, and elastic behavior may well be assumed to be applicable. The macroscopic elastic behavior of an isotropic material is characterized by three elastic constants, the elastic modulus or Young's modulus (E), shear modulus (G) and Poisson's ratio (v). The well-known relation between the constants is $E = 2G(1 + v)$.

In a structure, geometrical notches such as holes cannot be avoided. The notches are causing an inhomogeneous stress distribution, see Figure 3.1, with a stress concentration at the "root of the notch". The (theoretical) stress concentration factor, K_t,[6] is defined as the ratio between the peak stress at the root of the notch and the nominal stress which would be present if a stress concentration did not occur.

---

[6] K_t is often referred to as the theoretical stress concentration factor. However, the factor is not a theoretical one. It is based on the assumption of linear elastic material behavior.

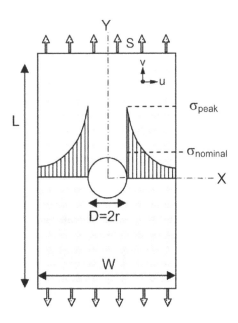

Fig. 3.1 Strip with central hole as a prototype of a notched part.

$$K_t = \frac{\sigma_{\text{peak}}}{\sigma_{\text{nominal}}} \qquad (3.1)$$

The severity of the stress concentration is depending on the geometry of the notch configuration, generally referred to as the shape of the notch. Designers should always try to reduce stress concentrations as much as possible in order to avoid fatigue problems. The present chapter deals with various aspects of stress concentrations and the effect of the geometry (the shape) on K_t. This is one of the fundamental issues of designing a fatigue resistant structure, i.e. designing against fatigue. Problems discussed in the present chapter cover definitions of stress concentration factors, calculations and estimations of K_t-values, stress gradients, aspects related to size and shape effects, superposition of notches and methods to determine K_t-values.

3.2 Definition of K_t

The strip with a central hole shown in Figure 3.1 is a prototype of a notched element. It is frequently used in fatigue experiments to study notch effects on fatigue. If the strip is loaded by a homogeneous stress distribution, the hole will cause an inhomogeneous stress distribution in the critical section, which

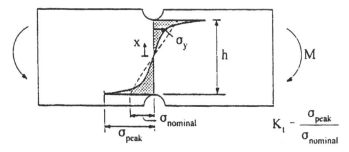

Fig. 3.2 Stress distribution in a beam with two grooves loaded in bending.

is the minimum section at the hole. This stress distribution is characterized by a peak stress σ_{peak} at the root of the notch, and a nominal net section stress $\sigma_{nominal}$. The ratio of the peak stress and the nominal stress in the net section leads to the commonly used definition of the stress concentration factor K_t given in Equation (3.1). It should be emphasized that all deformations are supposed to be elastic. K_t is essentially an elastic concept. It gives a direct indication of the severity of the stress concentration, because it is an amplification factor on the stress level which is nominally present in the net section of the notch.

$$\sigma_{peak} = K_t\, \sigma_{nominal} \qquad (3.2)$$

Sometimes it is informative to see the ratio between the peak stress and the gross stress S, applied to the element. This ratio with the symbol K_{tg} is:

$$K_{tg} = \frac{\sigma_{peak}}{S} \qquad (3.3a)$$

The two factors are obviously interrelated. With the dimensions W (specimen width) and D (hole diameter):

$$K_{tg} = \frac{\sigma_{nominal}}{S} K_t = \frac{W}{W - D} K_t \quad \text{and thus} \quad K_{tg} > K_t \qquad (3.3b)$$

K_t and K_{tg} are the symbols used by R.E. Peterson in his book *Stress Concentration Factors* [1], which is a standard book on stress concentrations. In general, K_t is the preferred factor to indicate the stress concentration.

For bending and torsion the definition of K_t is the same as given in Equation (3.1); K_t is the ratio between the peak stress at the root of the notch and the nominal stress in the critical net section. The nominal stress for the strip with the side grooves in Figure 3.2 is the bending stress which would be present if the stress concentration did not occur:

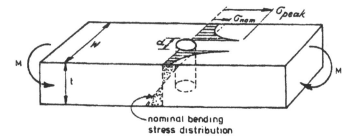

Fig. 3.3 Stress distribution in a beam with a transverse hole loading in bending.

$$\sigma_{\text{nominal}} = \frac{6M}{th^2} \tag{3.4}$$

(t = thickness of the beam; h = height of minimum section, see Figure 3.2).

The nominal stress for the strip with a transverse hole also loaded in bending shown in Figure 3.3 is:

$$\sigma_{\text{nominal}} = \frac{6M}{(W - D)t^2} \tag{3.5}$$

K_t-values can be obtained with different methods:

- by calculations: analytical methods, finite-element methods (FEM),
- by measurements: strain gage measurements, photo-elastic measurements.

Some comments on measurements versus calculations are given in Section 3.7.

3.3 Analytical calculations on stress concentrations

Analytical solutions based on the theory of elasticity are not treated in detail here. The analysis can be found in various textbooks (e.g. [2, 3]). Basically, the following procedure is used. For a two-dimensional problem, as shown in Figure 3.1, the displacement functions $u(x, y)$ and $v(x, y)$ have to be found. If these functions are obtained, the strains follow from these functions, and the stresses are linked to the strains by Hooke's law. The problem then is apparently solved. As part of finding the solution, the tensile strains, $\varepsilon_x(x,y)$ and $\varepsilon_y(x,y)$, and the shear strain $\gamma_{xy}(x, y)$ must satisfy the compatibility equation. Furthermore, there are equilibrium equations for σ_x, σ_y and τ_{xy}. These stresses are linked to the strains by three equations representing

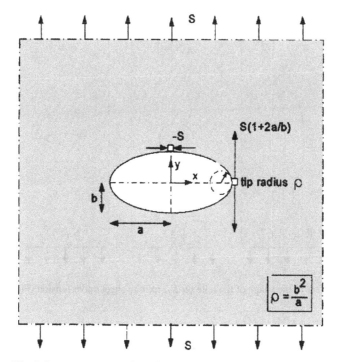

Fig. 3.4 Stress concentration of an elliptical hole in an infinite sheet.

Hooke's law, including the elastic constants of the material. The equations obtained are rewritten by introducing the Airy stress function ϕ which leads to a biharmonic equation. The problem then is to find a function ϕ that satisfies this equation. The solution will still contain unknown constants, which should follow from the boundary conditions. These conditions are essential for solving a particular problem. For the tensile strip with a central hole in Figure 3.1, the boundary conditions are:

1. At the upper and lower edge: $\sigma_y = S$, $\sigma_x = 0$, $\tau_{xy} = 0$.
2. At the side edges ($x = \pm W/2$): $\sigma_x = 0$, $\tau_{xy} = 0$.
3. At the edge of the hole: the stress perpendicular to the hole edge and the shear stress are zero.

An exact analytical solution for the apparently simple case of Figure 3.1, a strip with a hole, is not available, but accurate numerical approximations were obtained. However, for an infinite sheet with an elliptical hole the exact solution was obtained [2, 3]. This problem is known as a classical problem in the theory of elasticity. It is not really a simple problem. Elliptical coordinates and complex functions are used to arrive at the solution, which

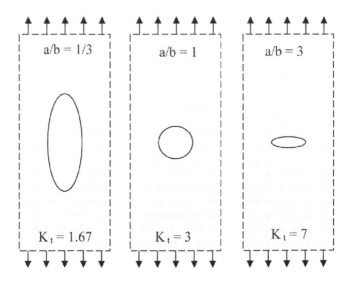

Fig. 3.5 Effect of the shape of the hole on K_t, infinite sheet under tension (Figure 3.4).

then provides the stress distribution in the entire plate. The results illustrate several interesting features of stress distributions around the hole. The tangential stresses along the edge of the hole are of great interest. The maximum stress, σ_{peak}, occurs at the end of the main axis ($x = a$, $y = 0$), see Figure 3.4. The semi-axes of the elliptical hole are a and b respectively. The tip radius at the end of the major axis follows from $\rho = b^2/a$. The equations for the peak stress and K_t are simple:

$$\sigma_{peak} = S\left(1 + 2\frac{a}{b}\right) = S\left(1 + 2\sqrt{\frac{a}{\rho}}\right) \tag{3.6a}$$

$$K_t = 1 + 2\frac{a}{b} = 1 + 2\sqrt{\frac{a}{\rho}} \tag{3.6b}$$

The last equation indicates that a small notch root radius ρ will give a high K_t, whereas a large radius will give a low K_t. This is illustrated in Figure 3.5. Although a structure is not directly comparable to a sheet with an elliptical hole, it is always profitable to use large radii in notched components to reduce the stress concentration.

A circular hole is a special case, obtained from an ellipse with equal axes; $a = b$. The K_t-value according to Equation (3.6b) is equal to 3. This is a classical value. For an open hole in a structural part, the K_t-value will be somewhat lower because the component has a finite width. In practice, fatigue cracks have indeed frequently occurred in structures at open holes.

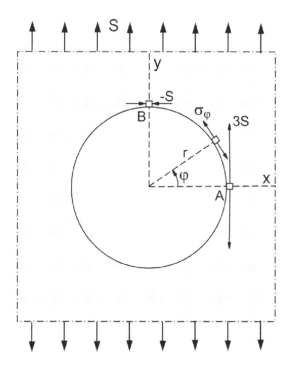

Fig. 3.6 Tangential stress around a circular hole in a sheet loaded by a remote tensile stress S.

It is noteworthy that the tangential stress at the end of the vertical axis ($y = b$, $x = 0$ in Figure 3.4) is a compressive stress, which is equal to the tensile stress applied to the infinite plate. This result is valid for all ellipses, and thus also for a circular hole, see Figure 3.6. Along the edge of the hole, starting from A to the top of the hole B the tangential stress changes from $+3S$ to $-S$, following the equation:

$$\sigma_\varphi = S(1 + 2\cos 2\varphi) \tag{3.7}$$

The value of the tangential stress must go through zero ($\sigma_\varphi = 0$) which occurs at $\varphi = 60°$.

Stress gradients

Although the peak stress is of great importance, it is also interesting to know how fast the stress decreases away from the root of the notch, see Figure 3.7. The stress gradient of σ_y along the X-axis is used in some prediction models

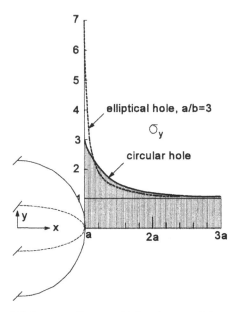

Fig. 3.7 Stress gradients are high at the root of the notch.

to account for the size effect on the fatigue limit of notched parts. For an elliptical hole in an infinite sheet, the exact solution for σ_y along the X-axis is given by:

$$\frac{(\sigma_y)_{y=0}}{S} = 1 + \frac{a(a-2b)(x-\sqrt{x^2-c^2})(x^2-c^2) + ab^2(a-b)x}{(a-b)^2(x^2-c^2)\sqrt{x^2-c^2}} \quad (3.8)$$

where $c^2 = a^2 - b^2$. For $x = a$, the equation reduces to Equation (3.6b). For x very large, the asymptotic result is $(\sigma_y)_{y=0} = S$ as should be expected far away from the hole.

For a circular hole, Equation (3.8) becomes more simple:

$$\frac{(\sigma_y)_{y=0}}{S} = 1 + \frac{1}{2}\left(\frac{a}{x}\right)^2 + \frac{3}{2}\left(\frac{a}{x}\right)^4 \quad (3.9)$$

The distributions of $(\sigma_y)_{y=0}$ for a flat elliptical hole ($a/b = 3$, $K_t = 7$) and for a circular hole are given in Figure 3.7. Obviously, the peak stress drops off much faster for the higher peak stress of the elliptical hole. The stress gradient of $(\sigma_y)_{y=0}$ at the root of the notch ($x = a$) is obtained by differentiation of Equation (3.8). With $K_t = 1 + 2a/b$ (Equation 3.6b), the gradient can be written as

$$\left(\frac{d\sigma_y}{dx}\right)_{x=a} = -\left(2 + \frac{1}{K_t}\right)\frac{\sigma_{\text{peak}}}{\rho} = -\alpha\frac{\sigma_{\text{peak}}}{\rho} \quad (3.10)$$

The negative gradient is proportional to the peak stress (expected for linear elastic behavior) and inversely proportional to the root radius. The proportionality constant is:

$$\alpha = 2 + \frac{1}{K_t} \tag{3.11a}$$

which implies

$$2 < \alpha < 3 \tag{3.11b}$$

Apparently, K_t does not have a large effect on the stress gradient coefficient α. For notches in a structure with K_t in the range of 2 to 5, the value of α is about 2.2 to 2.5 [5].

The stress gradient at the root of a notch should give an indication of the volume of the highly stressed material. As a numerical example, an estimate is made of the distance δ along the X-axis for drop of σ_y from σ_{peak} to $0.9\sigma_{peak}$, a drop with 10%. A circular hole ($K_t = 3.0$) with a diameter of 5 mm ($\rho = 2.5$ mm) is considered. Assuming a linear stress gradient, the value of δ can be derived with Equation (3.10):

$$\left(\frac{d\sigma_y}{dx}\right)_{x=a} \approx -\frac{\sigma_{peak} - 0.9\sigma_{peak}}{\delta} = -\left(2 + \frac{1}{3}\right) \cdot \frac{\sigma_{peak}}{2.5}$$

which gives $\delta \approx 0.1$ mm $= 100$ μm. For an average grain size of 50 μm the depth δ corresponds to just a few grains. Conclusion: Especially the grains at the notch root surface are the highly loaded grains. This is important for fatigue.

The stress gradient along the edge of the notch

In the previous paragraphs, it was discussed how σ_y is dropping off away from the edge of the hole. However, in Chapter 2 it was pointed out that fatigue crack nucleation is a surface phenomenon. It then appears to be of interest to know how fast the tangential stress along the notch edge is decreasing. This is illustrated by Figure 3.8, again for a circular hole. Lines of a constant principal stress were calculated for stress levels of 95, 90, 80 and 50% of the peak stress. The 90% line corresponds to a 10% reduction of the peak stress. An interesting result should be noted. The tangential stress along the edge of the hole decreases relatively slowly in comparison to the stress away from the edge (along the X-axis). It then should be recalled that crack nucleation starts at the material surface. Apparently, the highly stressed

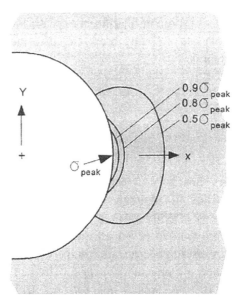

Fig. 3.8 Relatively slow decrease of the stress along the edge of a notch, illustrated by lines of constant principal stress for a circular hole loaded in tension [5].

surface layer is stretched along the edge of the hole. It implies that the stress gradient along the edge of the hole is of greater importance than the stress gradient perpendicular to the hole edge.

Larger notches have a larger material surface along the root of the notch, which is significant for the notch size effect on fatigue to be discussed in Chapter 7. Furthermore, the extent of the highly stressed material along the wall of a hole emphasizes the significance of the surface quality obtained in the production. This topic is also discussed in Chapter 7.

The calculated results discussed above were obtained for an infinite sheet with an elliptical hole, and with a circular hole as a special case. More results presented in [5] indicate that the trends with respect to the stress gradients are more or less similar for all notches in the engineering range of relevant K_t-values and notch root radii. Similar peak stresses and notch root radii give comparable stress distributions around the root of the notch, and the trends discussed in this section are thus valid for engineering design considerations.

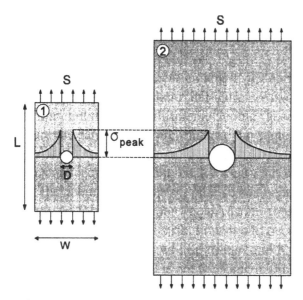

Fig. 3.9 Geometrically similar specimens have the same K_t, but different stress gradients.

3.4 Effect of the notch geometry on K_t

For a circular hole in an infinite sheet, the only dimension is the diameter D. However, for the simple tension specimen with a central hole, see Figure 3.1, there are already three dimensions: the specimen width (W), the specimen length (L) and the hole diameter (D). The specimen thickness is not yet considered here. In Figure 3.9 two specimens are shown, which are geometrically similar, but the size is different. Geometric similarity implies that all ratios of the dimensions are the same, in the present case the same D/W and L/W.

Because K_t is a dimensionless ratio, it can depend on dimensionless geometrical ratios only. Assume that all dimensions of specimen 2 in Figure 3.9 are two times larger than the dimension of specimen 1. As a result of the geometric similarity, all displacements are also two times larger, but the relative displacements will be the same. As a result, the strains are the same. Consequently, a geometrically similar stress distribution should occur in both specimens as depicted in Figure 3.9. The same peak stress will be found, and K_t is the same. However, due to the difference in size, the stress gradient is not the same in the two specimens because the gradient is not dimensionless. According to Equation (3.10), the gradient is inversely proportional to the root radius ρ. The consequence is that larger

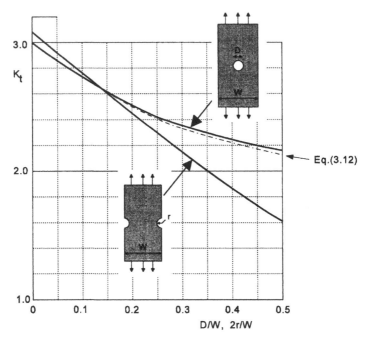

Fig. 3.10 K_t for a specimen with a central hole and a specimen with two edge notches [1].

specimens have larger volumes and larger notch surface areas of highly stressed material, which is significant for the size effect on fatigue.

Many K_t graphs for various shapes and different types of loading can be found in the book by Peterson [1]. Collections of K_t-values are also presented in other sources, e.g. in the ESDU Data Sheets [4]. Furthermore, software packages also containing a K_t database are now commercially available. Some simple examples of K_t graphs will be shown here to illustrate the effect of the shape on the stress concentration. Figure 3.10 shows K_t for a central hole and a double edge notch, geometries frequently adopted for fatigue investigations in laboratories. For an increasing notch radius (r) the value of K_t decreases, although much more for the edge notched specimen than for the central hole specimen. For the edge notched specimen $K_t \rightarrow 1$ for $2r/W \rightarrow 1$ (zero ligament between the notches), while for the central hole specimen $K_t \rightarrow 2$ for $D/W \rightarrow 1$ (also zero ligaments).

The $K_t(D/W)$ curve for the central hole specimen, based on calculations of Howland [6], was approximated by Heywood [7] by Equation (3.12) to:

$$K_t = 2 + \left(1 - \frac{D}{W}\right)^3 \qquad (3.12)$$

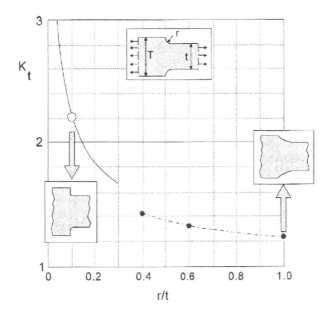

Fig. 3.11 K_t for a fillet, $t/T = 2/3$. Data from two graphs in [1].

The value of K_t should also depend on the length of the specimen, but this influence is negligible if the length is more than twice the specimen width.

Another illustration is given in Figure 3.11 for a fillet where the plate thickness (T) is reduced to a lower thickness (t) with a transition radius r. In this case two geometric ratios have to be considered, e.g. t/T and r/t. Figure 3.11 applies to a single value of the thickness reduction ratio t/T $(= 2/3)$. The graph shows K_t as a function of r/t as obtained from two different sources. The two curves suggest some disagreement because the $K_t(r/t)$ function appears to be discontinuous which cannot be correct. In spite of some inaccuracies of the curves, Figure 3.11 clearly illustrates that a larger radius leads to a significantly smaller K_t, see the two inset figures for $r/t = 0.1$ and 1.0 respectively. The corresponding K_t-values are 2.24 and 1.24, which means a 45% lower K_t for the larger root radius.

It is also instructive to see how the K_t-values compare for different shapes with the same geometry ratios, see Figure 3.12. The highest K_t-value in this figure applies to the edge notches. The fillet geometry is obtained by removing material from the edge notch geometry. K_t is then reduced by 25%. The fillet geometry is less disturbing for the "stress flow". The stress flow can be visualized by the main principal stress trajectories, which have to bend around the notch, see Figure 3.13. Thinking in terms of the stress

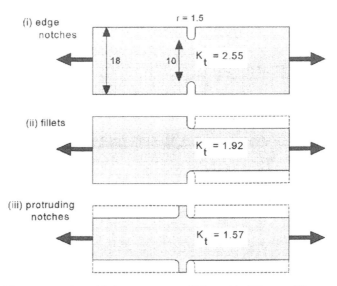

Fig. 3.12 Three geometries with the same root radii but with different stiffness transitions. K_t-values derived from graphs in [1].

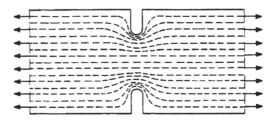

Fig. 3.13 Main principal stress trajectories bending around the notch.

flow, it could be qualitatively expected that the fillet in Figure 3.12, case (ii), does not obstruct the stress flow as much as the edge notch. The third case in Figure 3.12 is derived from the fillet case by a further omission of material, leaving two external ledges. Such a protruding notch has a limited effect on the stress flow, which reduces K_t still further. The highest K_t-value in Figure 3.12 is obtained for the intruding notch, which applies also to intruding damage to the surface material, such as dents, corrosion pits or imprinted letter codes of the manufacturer. An illustration of a corrosion pit with an estimated K_t-value is given in Figure 3.14. The effect of corrosion pits on the S-N curve was discussed in Chapter 2, see Figure 2.28.

The effect of the root radius on K_t is again illustrated by Figure 3.15 for a shaft with a shoulder fillet. Values are given for a diameter reduction $d/D = 2/3$, for two loading cases; tension and bending. The K_t-values

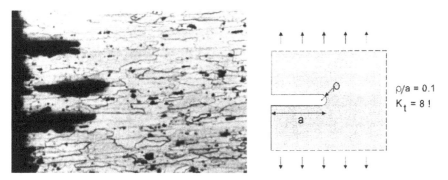

Fig. 3.14 Relative deep corrosion pits in a so-called end grain structure at the material surface of an aluminium alloy part. Pit depth = 0.15 mm. The geometrically equivalent shape leads to a very high K_t.

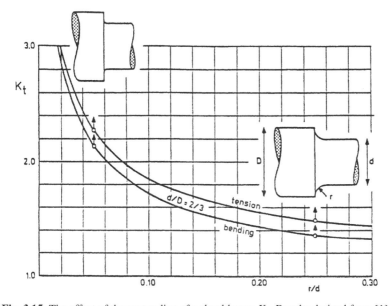

Fig. 3.15 The effect of the root radius of a shoulder on K_t. Results derived from [1].

for bending are slightly smaller than for tension. Similar to the graph for fillets in Figure 3.11, it illustrates that fairly low K_t-values can be obtained provided that a "generous" root radius is adopted. However, this is not always possible in practice, for instance if a ball bearing has to support a shaft as shown in Figure 3.16. The radius at the fillet can then be increased by a stress relieving groove in the shoulder of the thicker part, which reduces the stress concentration.

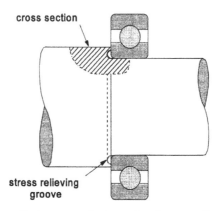

Fig. 3.16 Stress relieving groove by increasing the root radius at a shoulder.

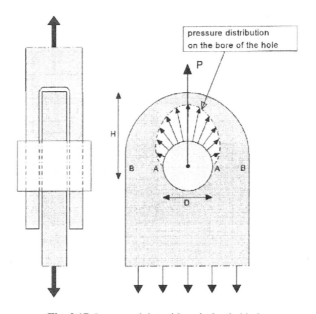

Fig. 3.17 Lug type joint with a pin-loaded hole.

Pin-loaded hole

The most elementary case of a pin-loaded hole is the connection between a lug (or lug head) and a clevis, see Figure 3.17. Load transmission between the lug and fork occurs by a single pin or a bolt. The pin applies a distributed pressure load to the upper half of the bolt hole in the lug. In many practical cases it is essential that some rotation in the joint is possible, which requires a clearance fit between the bolt and the hole and no clamping between the fork

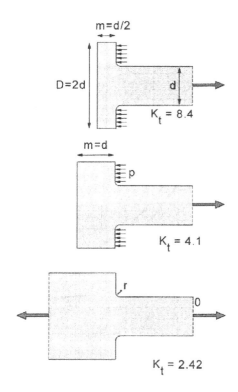

Fig. 3.18 High K_t-values of the flat T-heads with pressure loads close to the root on the notch ($r/d = 0.1$). Comparison to fillet.

and lug. For the critical net section of the lug (B-B in Figure 3.17), it implies pressure loads on the hole in the near vicinity of the root of the notch (points A in Figure 3.17), which is the location where crack nucleation should be expected under cyclic loading. This proximity of pressure on the hole surface and the critical section usually leads to relatively high stress concentrations. An elementary illustration of this observation is given in Figure 3.18 for a so-called T-head. In this case, the load P is balanced by a surface pressure p on the head edges, which is close to the root of the notch. For the two T-heads, K_t is significantly higher than for the fillet where no pressure loads in the vicinity of the notch root are present. Note also the difference between the two T-heads with the same radius ($K_t = 8.4$ and $K_t = 4.1$ respectively). The high K_t-value in the upper case of Figure 3.18 is associated with the lower bending stiffness of the T-head.

Values of K_t for lugs are the wellknown results obtained by Frocht and Hill from photo-elastic measurements [8]. These results are presented in

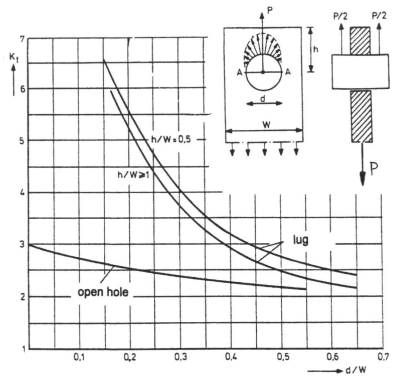

Fig. 3.19 K_t-values for a lug. Results of Frocht and Hill [8]. Comparison to an open unloaded hole.

Figure 3.19 in comparison to K_t-values for an open unloaded hole. The pin-loaded hole obviously causes the more severe stress concentration. This should be associated with the pressure load distribution applied closely to the root of the notch as mentioned earlier. In view of the high K_t-values, lugs are fatigue critical parts, but that is also due to fretting corrosion occurring inside the hole to be discussed in Chapter 15. Values of D/W below 1/3 are generally avoided to keep K_t-values below about 3.5.

3.5 Some additional aspects of stress concentrations

Pure shear

The stress concentration factor for a circular hole in a plate under pure shear loading can be obtained by superposition of two uniaxial loading cases, see

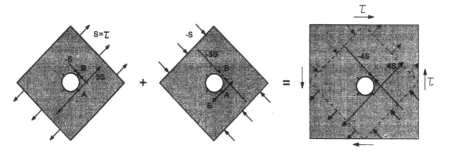

Fig. 3.20 K_t for pure shear obtained by superposition of two cases.

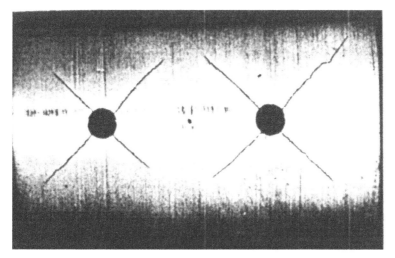

Fig. 3.21 Fatigue craks growing from holes in a shaft subjected to cyclic torsion. Material low-C steel [9].

Figure 3.20. Pure shear can be split into a pure tension case S at 45° and a pure compression case $-S$ at $-45°$. Summing the stresses at points A and B lead to tangential stresses of $4S$ and $-4S$. Because $S = \tau$, it leads to $K_t = 4$, quite a high value. Fatigue cracks have indeed been observed under cyclic torsion at the critical points at $\pm 45°$, see Figure 3.21.

Biaxial loading

Another simple case is the stress concentration of an elliptical hole under biaxial loading, as shown in Figure 3.22. The tangential stresses at the ends of the two axes of the elliptical hole are obtained by superposition of the two mutually perpendicular load cases, S and βS, with β as the biaxiality ratio.

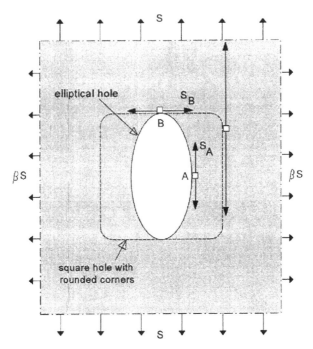

Fig. 3.22 Elliptical hole in biaxial stress field.

With the stress values presented earlier in Figure 3.4, the following equations are obtained:

$$\text{Point A:} \quad S_A = S(1 + 2a/b) - \beta S$$

$$\text{Point B:} \quad S_B = \beta S(1 + 2b/a) - S \tag{3.13}$$

The biaxiality in the shell of a pressure vessel is due to the circumferential hoop stress and the longitudinal tension stress. The ratio is $\beta = 0.5$ if stiffeners are not present. For this ratio and a circular hole ($a = b$), substitution in Equation (3.13) gives:

$$S_A = 2.5S \quad \text{and} \quad S_B = 0.5S$$

The value of S_A is lower than $3S$ applicable to uniaxial loading. The biaxiality relieves the stress concentration. However, a further reduction is possible for an elliptical hole. If the b/a ratio is chosen to be 2, and again $\beta = 0.5$, see Figure 3.22, the result with Equation (3.13) is:

$$S_A = 1.5S \quad \text{and} \quad S_B = 1.5S$$

or $S_A = S_B$. Actually, the tangential stress in the latter case is equal to $1.5S$ along the full edge of the hole.

A completely different type of hole is also indicated in Figure 3.22. The dashed line represents a square hole with rounded corners (radius is 10% of hole width). For the same biaxiality ($\beta = 0.5$) the K_t-value for this hole is 4.04 [1], a large difference as compared to the elliptical hole. Of course it should be realized that these results are theoretical results, because open holes in a pressure vessel cannot exist. However, they illustrate that shapes can have a large effect on the stress concentrations. Designers nowadays can avoid undesirable shapes more easily than in the past by using computer controlled machining techniques.

Reinforcements of open holes

In various structures, openings cannot be avoided for several reasons associated with the usage of the structure. Approximately rectangular openings can be desirable in large shell structures such as ships and aircraft in view of cargo transportation or other reasons. These openings have caused significant fatigue problems. Carefully designed reinforcements of the edges of an opening can alleviate the local stress level around the opening, but a stress analysis by FE calculations should then be made.

On a much smaller scale, a hole in a plate element of a structure can be necessary for various arguments, usually related to passing of something through the hole. However, the stress concentration factor can easily be in the order of three or even higher for shear loading. Designers can apply a collar to the edge of the hole to reduce the stress level around the hole. Values of K_t for integral collars are presented in [1] and [4]. However, the variety of reinforcements around a hole is large in view of the method of joining the reinforcing material to the plate. Ring elements can be attached around the hole by fasteners (bolts, rivets) or adhesive bonding (in aircraft). It should be realized that the advantage of such reinforcements is not always obvious because fatigue critical locations can now occur at the edge of the reinforcement or fastener holes. Moreover, a high-stiffness reinforcement attracts load to the hole area and can also introduce bending due to the eccentricity of the reinforcement. Again, stress analysis as well as engineering judgement is required to deal with such problems.

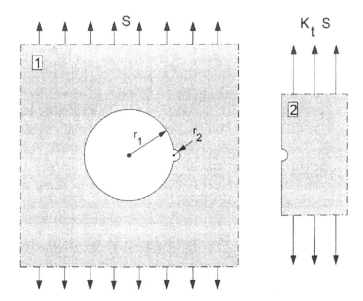

Fig. 3.23 Superposition of a very small notch on the root of a larger notch.

3.6 Superposition of notches

If a relatively small notch is added to the root of the main notch, there is a superposition of notches. A simple example is shown in Figure 3.23. A small semi-circular notch with $r_2/r_1 \ll 1$ occurs at the critical section of an open hole. This small additional notch could be mechanical damage inside the hole. The small notch occurs at a location where the peak stress would have been $K_{t1} \cdot S$. If the stress reduction away from the hole is still moderate, it can be assumed that the stress condition for the small notch is comparable to the second case shown in Figure 3.23. The K_t-value for this case is $K_{t2} = 3.07 \approx 3$, and the peak stress at the small notch added to the edge of the hole can be approximated by

$$\sigma_{\text{peak}} = K_{t1} K_{t2} S \quad \text{or} \quad K_t = K_{t1} K_{t2} \tag{3.14}$$

This is a reasonable first estimate if r_2 is much smaller than r_1. The K_t-value thus could be in the order of 9, which is very high. In reality, it will be lower because the small notch does not occur in a homogeneous stress field. Equation (3.14) will overestimate the real K_t. However, an amplification effect of the superposition will apply. This means that mechanical damage at the root of a notch can have a rather detrimental effect on the fatigue resistance.

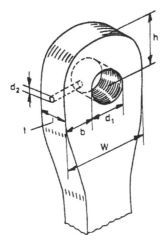

Fig. 3.24 Lug with a small lubrication hole to the lug hole. An example of superposition of two notches.

The above example refers to the superposition of an unintentional notch on another notch. This does not apply to the second example, shown in Figure 3.24. As discussed before, K_t for a lug is relatively high. In Figure 3.24, a superposition of notches occurs because of the small lubrication hole drilled to reach the main hole of the lug. Again Equation (3.14) will be a first approximation, and although it will be an overestimate, the total K_t-value will be very high. From a design point of view, the lubrication hole is entering the lug hole at the most unfavorable location. A much better solution would be to locate the lubrication hole at the top of the lug head. The two examples illustrate that a designer should try to avoid superposition of notches. If functional holes are necessary, this superposition effect can be limited by selecting appropriate positions in low-stress areas.

An other example of a superposition of notches is illustrated by Figure 3.25. It shows a fatigue crack in a bracket. A generous radius was applied between the vertical flange and the base plate. Unfortunately, the favorable radius was fully destroyed by machining a flat surface into the base plate to accommodate the positioning of an attachment bolt of the bracket. This caused a local superposition with a very sharp notch and fatigue cracking occurred in service. Such a mistake of detail design should be observed in the design office.

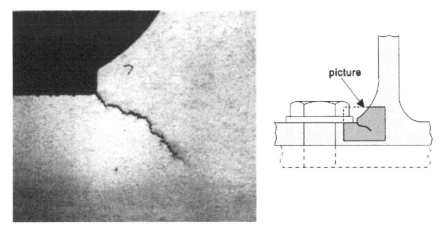

Fig. 3.25 Cross section of fatigue cracks at sharp corner.

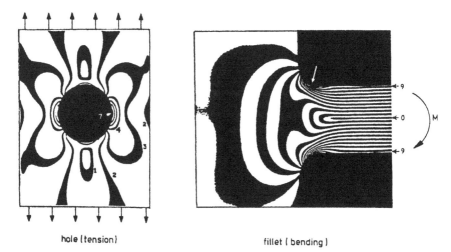

hole (tension)   fillet (bending)

Fig. 3.26 Two examples of photo-elastic pictures [10].

3.7 Methods for the determination of stress concentrations

Before FE techniques were generally available, say before 1960, many K_t-values were obtained by measurements, e.g. by photo-elasticity [10, 11]. Various graphs in Peterson's book [1] are based on such measurements. Two pictures obtained by photo-elasticity are shown in Figure 3.26. The stress field can be derived from the black and white interference bands, called fringes. Local stresses can be obtained from the fringes as explained in [10, 11]. The parallel and equally spaced fringes in the thin section of the fillet bending case illustrate the linear bending stress distribution across the thickness of the horizontal beam. Notice in this figure that the peak stress occurs at the transition of the beam to the very beginning of the fillet (see the arrow). In the past, the advantage of the photo-elastic method was that an impression of the entire stress field was obtained. Moreover, the photo-elastic model allows modifications of the shape to see how improved stress distributions can be obtained. However, the accuracy of the method is problematic.

An alternative measuring technique is to use strain gages. Strains can be measured fairly accurately. Strain gages with a small filament length should be used because of the large stress gradients at the notch. Such gages are available (gage length $\leq$ 1 mm). Unfortunately, the root of a notch is not always easily accessible for applying the strain gage. Accurate measurements of the peak stress are difficult. The strain gage technique is still used to measure nominal stress levels in full-scale structures or components.

It was pointed out before that similarity of notches (especially similar root radii) and superposition may be helpful to estimate K_t-values. Furthermore, interpolation between data for existing geometries is possible. Available data (e.g. [1]) should always be consulted to see whether information for similar geometries is available. However, the accuracy of several K_t graphs in the book by Peterson may be limited. More reliable K_t-values require a thorough elasto-mechanic analysis which in most cases will be FE calculations. An illustration of a simple FE model is given in Figure 3.27. The geometry of the component has to be modeled by a large number of small interconnected elements. Many more elements are required at places where stress gradients are high which generally applies to the area around a notch.

Nisitani and Noda [12] carried out calculations with the boundaryelement technique to obtain K_t-values for cylindrical bars with circumferential notches under tension, bending and torsion. They found K_t-values which were about 10% higher than data reported in the book by Peterson [1] which were based on a Neuber analysis. The authors found similar trends for strips with double and single edge notches [13]. Gooyer and Overbeeke [14] carried out FE calculations for a shaft with a shoulder fillet loaded under tension and torsion. They found K_t-values even more than 10% higher than reported in the book by Peterson. Similar discrepancies for shoulder fillets were obtained by Noda et al. [15].

An example of photo-elastic results compared to FE results is given in Figure 3.28 for four specimens with an increasing notch severity. Such specimens are used in laboratory experiments to study the fatigue notch

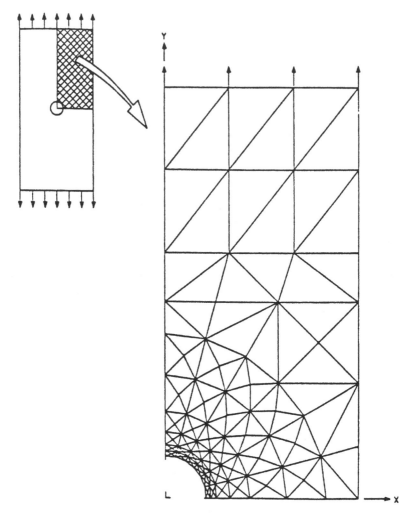

Fig. 3.27 Example of a FE model. In view of the symmetry of the load and the specimen, only one quarter needs to be modeled.

sensitivity of a material. The notch of specimen 3 consists of a circular hole with two semi-circular edge notches with a smaller radius (hole with ears). The notch of specimen 4 is obtained by drilling two small holes connected by a saw cut. This notch is the most severe one simulating a slit with rounded ends. The photo-elastic results confirm the increasing notch severity for specimens 1 to 4 in agreement with the trend of the FE results. Differences between the measured and calculated K_t-values are 10 to 16%. The calculated values should be expected to be more accurate.

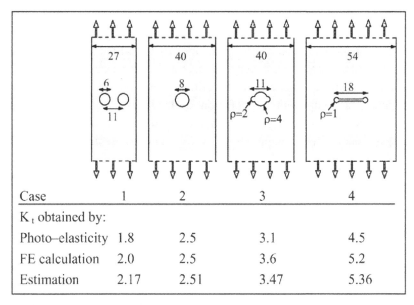

Fig. 3.28 Comparison of K_t-values of notched specimens as obtained by photo-elastic measurements and FE calculations [16]. Estimated values are included. Dimensions in mm.

The bottom line of Figure 3.28 also gives estimated K_t-values obtained with simple approximations. Specimen 1 with two holes is replaced by a strip with a single hole and half the width of specimen 1. It leads to $D/W = 6/13.5$, and with the Heywood equation (Equation 3.12) the result is $K_t = 2.17$. The approximation for specimens 3 and 4 is obtained with the K_t equation for an elliptical hole (Equation 3.6b) with the semi-axis "a" equal to half the total width of the notch, and the same root radius at the end of the notch. This leads to K_t-values of 4.41 and 7.0 respectively for such holes in an infinite sheet. The width correction is assumed to be the same as for a circular hole according to the Heywood equation, implying a correction factor $\{2 + (1 - D/W)^3\}/3$. The K_t-values then become 3.47 and 5.36 respectively. A comparison with the FE results in Figure 3.28 shows that the estimates are reasonably accurate. The predominant significance of the notch root radius is thus emphasized again. It may also be pointed out that such K_t estimates are useful as a preliminary validation of FE calculations. It is desirable to have some idea about the magnitude of the value to be calculated.

For complex shapes, analytical calculation and estimates become impracticable. This certainly applies to components with a significant three-dimensional (3D) character, but also for 2D components with a complex geometry, a limited symmetry and no comparable simple shapes. If stress concentration factors or stress distributions are needed, calculations should be made with finite element techniques for which several computer codes have been developed. Very satisfactory results can be achieved. FE calculations require experience and a critical view on the FE model adopted to simulate a component. Application of loads to the model, boundary conditions and mesh distributions should be given careful attention. It is useful to check computer programs by calculations on simple models for which solutions are already available.

The obvious criteria for a comparison between FE calculations and experimental techniques are: accuracy, cost-effectivity and time efficiency. If sufficient experience on FE calculations is available, the criteria are all in favor of the FE calculations. The computer also can provide colorful pictures of the stress distribution. Each color then corresponds to a certain interval of the stress level. Locations of peak stresses are easily recognized in such pictures. Complex shapes can be handled. Of course, it requires a computer program and a computer suitable for complex 3D problems. It is desirable to have experience with such calculations in order to understand what the computer program is doing.[7] Some strain gage measurements for the validation of the calculation may be useful if the full-scale structure is available for that purpose.

In practical problems, it should be questioned whether it is justified to spend substantial efforts on obtaining accurate K_t-values. Graphs in the book by Peterson with a limited accuracy can still give good indications of the reduction of K_t by increasing root radii. Moreover, it should be recognized that other input data required for fatigue life predictions may also be afflicted by uncertainties, e.g. the load spectrum. An other variable is scatter of root radii in production which can have a significant effect on K_t if small radii were specified on the production drawings.

3.8 Main topics of the present chapter

This section is not a summary of the present chapter, but the major results of the present chapter are recollected.

---

[7] It may be noted that the von Mises stress distribution is often calculated in FE programs. However, this is incorrect for a fatigue analysis for which the largest principal tensile stress is the most indicative stress component.

1. The most important variable for the stress concentration factor K_t is the root radius ρ. Sharp notches can give unnecessarily high K_t-values.
2. Loads applied close to the notch root usually give higher K_t-values than remotely applied loads.
3. Pin-loaded holes are a more severe stress raiser than open holes.
4. The gradient of the peak stress at the notch root to lower values away from the root surface ($d\sigma_y/dy$) is inversely proportional to the root radius ρ, and linearly proportional to the peak stress at the root of the notch, σ_{peak}. This gradient is relatively high and σ_y drops off rapidly away from the material surface. Contrary to this large gradient , the tangential stress along the material surface at the notch root decreases relatively slowly. This observation is significant for considering notch size effects on fatigue and notch surface qualities.
5. Superposition of notches can lead to a multiplication effect on K_t ($= K_{t1} \times K_{t2}$).
6. Accurate K_t-values can be calculated with FE techniques. With the current computers, these calculations are more accurate and cost-effective in comparison to measurements of stress distributions around notches.

References

1. Peterson, R.E., *Stress Concentration Factors.* John Wiley & Sons, New York (1974). New edition by Pilkey, W.D. and Pilkey, D.F., *Peterson's Stress Concentration Factors*, 3rd rev. edn., John Wiley & Sons (2008).
2. Timoshenko, S. and Goodier, J.N., *Theory of Elasticity*, 3rd edn.). McGraw Hill (1987).
3. Muskhelisvili, N.I., *Some Basic Problems of the Mathematical Theory of Elasticity.* Noordhoff (1963).
4. *Stress Concentration Factors.* ESDU Engineering Data, Structures, Vol. 7, Issued 1964 to 1995. Engineering Sciences Data Unit, London.
5. Schijve, J., *Stress gradients around notches.* Fatigue Engrg. Mater. Struct., Vol. 3 (1981), pp. 325–338.
6. Howland, R.C.J., *On the stress in the neighbourhood of a circular hole in a strip under tension.* Phil. Trans. Roy. Soc., Vol. 229 (1930), pp. 49–86.
7. Heywood, R.B., *Designing against Fatigue.* Chapman and Hall, London (1962).
8. Frocht, M.M. and Hill, H.N., *Stress concentration factors around a central circular hole in a plate loaded through pin in the hole.* J. Appl. Mech., Vol. 7 (1940), pp. A5–A9.
9. Frost, N.E., Marsh, K.J. and Pook, L.P., *Metal Fatigue.* Clarendon, Oxford (1974).
10. Heywood, R.B., *Designing by Photo-Elasticity.* Chapman and Hall, London (1952), and *Photo-Elasticity for Designers.* Pergamon Press (1969).
11. Durelli, A.J., Phillips, E.A. and Tsao, C.H., *Theoretical and Experimental Analysis of Stress and Strain.* McGraw-Hill (1958).

12. Nisitani, H. and Noda, N.-A., *Stress concentration of a cylindrical bar with a V-shaped circumferential groove under torsion tension or bending.* Engrg. Fracture Mech., Vol. 20 (1984), pp. 743–766.

13. Nisitani, H. and Noda, N.-A., *Stress concentration of a strip with double edge notches under tension or in-plane bending.* Engrg. Fracture Mech., Vol. 23 (1986), pp. 1051–1065. See also: *Engrg. Fracture Mech., Vol. 28* (1987), pp. 223–238 for a single edge notch.

14. Gooyer, L.E. and Overbeeke, J.L., *The stress distributions in shouldered shafts under axisymmetric loading.* J. Strain Anal., Vol. 26 (1991), pp. 181–184.

15. Noda, N.-A., Takase, Y. and Monda, K., *Stress concentration factors for shoulder fillets in round and flat bars under various loads.* Int. J. Fatigue, Vol. 19 (1997), pp. 75–84.

16. Ostermann, H., *Stress concentration factors of plate specimens for fatigue tests.* Laboratorium für Betriebsfestigkeit, LBF, Darmstadt, Report TM Nr. 61/71 (1971) [in German].

Some general references

17. Sih, G.C., *Stress Analysis of Notch Problems. Mechanics of Fracture, Vol. 5.* Noordhoff (1978).

18. Nisitani, H. (Ed.), *Computational and Experimental Fracture Mechanics.* Computational Mechanics Publications, Southampton (1994).

19. Leven, M.M., *Stress gradients in grooved bars and shafts.* Proc. SESA, Vol. 13 (1955), pp. 207–213.

20. Savin, G.N., *Stress distribution around holes.* NASA TT F-607 (1970) [translated from Russian].

21. Neuber, H., *Kerbspannungslehre.* Springer, Berlin (1937). Translation: *Theory of Notch Stresses.* J.W. Edwards, Ann Arbor, Michigan (1946).

Chapter 4
Residual Stress

4.1 Introduction

The significance of residual stresses for fatigue is important in various practical problems. Unintentional tensile residual stress can have an adverse effect on the fatigue resistance, while compressive residual stress can significantly improve the fatigue behavior. The existence of residual stress and the introduction of such stresses in components are the subjects of the present chapter. It is restricted to basic aspects, while some specific topics will return in later chapters.

By definition, residual stress refers to a stress distribution, which is present in a structure, component, plate or sheet, while there is no external load applied. In view of the absence of an external load, the residual stresses are sometimes labeled as internal stresses. The background of the terminology "residual stress" is that a residual stress distribution in a material is often left as a residue of inhomogeneous plastic deformation.

Residual tensile stress and residual compressive stress always occur together. A possible residual stress distribution is presented in Figure 4.1. If there is no external load, residual tensile stresses must be balanced by residual compressive stresses. More precisely, in view of the absence of an external load, the residual stress distribution must satisfy the equilibrium equation:

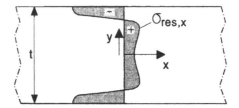

Fig. 4.1 A residual stress distribution is an equibibrium distribution.

$$\int_{-t/2}^{t/2} \sigma_x \, \mathrm{d}y = 0 \tag{4.1}$$

Similar, because of the absence of an external moment the following equation must also be satisfied:

$$\int_{-t/2}^{t/2} y \cdot \sigma_x \, \mathrm{d}y = 0 \tag{4.2}$$

An external load applied to a component will introduce a stress distribution in agreement with the external load and the geometry of the component. If the behavior is still elastic, the material will respond to the sum of the stress distribution of the external load and the residual stress distribution.

$$\sigma = \sigma_{\text{external load}} + \sigma_{\text{residual}} \tag{4.3}$$

If a cyclic fatigue load is applied, $\sigma_{\text{external load}}$ in the material is a cyclic stress with a certain stress amplitude (σ_a) and mean stress (σ_m). However, σ_{residual} is permanently present. It does not affect the stress amplitude, but it gives a shift to the mean stress:

$$\sigma_a = \sigma_{a,\text{external load}} \tag{4.4a}$$

$$\sigma_m = \sigma_{m,\text{external load}} + \sigma_{\text{residual}} \tag{4.4b}$$

If the local residual stress is positive, it increases σ_m (unfavorable for fatigue), and if it is negative, it reduces σ_m (favorable for fatigue). Residual stresses can be quite high. As a result of a high compressive residual stress it is possible that σ_{peak} is low, or even negative. In the latter case a microcrack can hardly grow. Because residual stresses do not affect the stress amplitude, cyclic slip at the material surface is still possible, and some microcrack nucleation can occur. However, if such microcracks are not opened at σ_{peak}, microcrack growth will not occur. If σ_{peak} including the compressive residual stress is positive, microcrack growth is possible, but the growth rate is reduced in view of the lower σ_{peak}.

Residual stresses have more consequences than for fatigue only. It is well known that tensile residual stresses can be most harmful if the material is sensitive to stress corrosion. Secondly, machining of the material with a residual stress distribution, can lead to distortions of the material (warpage). For instance, if a surface layer is removed from the plate in Figure 4.1 at one side only, the residual stress distribution does no longer satisfy the equilibrium equations (4.1) and (4.2) if warpage did not occur. As a consequence, warpage must occur (in this case plate bending) which changes the residual stress distribution until these equations are satisfied again.

It should be pointed out that the residual stresses discussed in this chapter occur on a macroscale. They have the same meaning as the stresses induced by an external load. On a much smaller scale, another type of residual stress can be present. Plastic deformation on a microscale is not a homogeneous process. It will be different from grain to grain, and even inside a single grain it may be concentrated into a few slip bands. Also in this case, equilibrium requires that the sum of the residual microstresses is zero. The microresidual stresses are significant for explaining the fatigue mechanism on a microlevel (Chapter 2) and also for the Bauschinger effect. These aspects are not considered in this chapter.

4.2 Different sources of residual stresses

Residual stresses can be present in a material as a result of different processes. In this section attention is paid to:

1. Inhomogeneous plastic deformation, in many cases at notches.
2. Production processes.
3. Shot peening.
4. Plastic hole expansion.
5. Heat treatment.
6. Assembling components

Inhomogeneous plastic deformation

A simple theoretical model will be discussed first. In Figure 4.2 two tension bars of different lengths are connected to the same infinitely stiff clampings at the ends. If a load is applied to this two-bar system, the elongation $\Delta \ell$ is the same for the two bars. As a consequence, the strain (ε) in the shorter

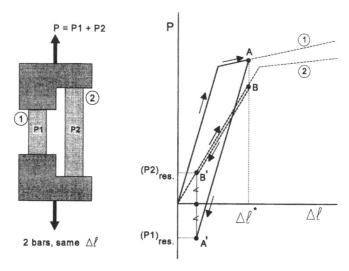

Fig. 4.2 A two-bar system to illustrate residual stress as a result of local plastic deformation.

bar will be larger because its length is shorter. In view of the larger ε, the stress will be higher and the bar thus will carry more load than the other bar. This is shown in the load-displacement ($P - \Delta\ell$) diagram in Figure 4.2. There is a load concentration in bar 1. Assuming that both bars are similar, it implies that permanent plastic deformation can occur in bar 1, while bar 2 is still fully elastic. This occurs at points A and B where $\Delta\ell = \Delta\ell^*$. During reversion of the loading direction elastic unloading occurs in both bars. After full unloading $P = 0$, which means that the sum of the residual loads in the two bars is zero; and thus $(P1)_{\mathrm{res}} = -(P2)_{\mathrm{res}}$, see Figure 4.2. Because of the plastic elongation of bar 1, this bar is longer than it was before. As a result, it will be in compression at $P = 0$ while bar 2 will be in tension. Residual stresses have been introduced as a result of plastic deformation in one part of the two-bar system.

A similarly inhomogeneous plastic deformation occurs in a strip with a hole loaded in tension, see Figure 4.3. If a high load is applied to the specimen, σ_{peak} at the edge of the hole exceeds the yield limit, and a small plastic zone is created at the root of the notch. As a consequence of the plastic deformation, σ_{peak} is smaller than $K_t\sigma_{\mathrm{nom}}$. The peak of the stress distribution is flattened by local plastic yielding. In the plastic zone permanent plastic deformation has occurred. The plastic zone is elongated; it is larger than it was before. After removing the tensile load on the strip, i.e. in the unloaded condition, the elongated plastic zone will be under compression. It does no longer fit stress-free in the elastic surrounding which tries to constrain the

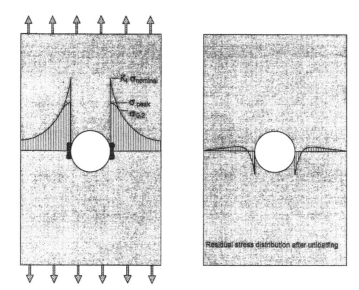

Fig. 4.3 Residual compressive stress at the root of a notch.

permanent plastic deformation. A residual stress distribution is introduced with a residual compressive stress at the root of the notch, which is the fatigue critical location under cyclic loading. The residual compressive stress is balanced by residual tensile stresses away from the notch. The residual compressive stress at the root of the notch can be most favorable for fatigue. In general, local plastic deformation causes an inhomogeneous residual stress distribution as illustrated by the right-hand picture in Figure 4.3. High residual stresses can be introduced up to a level approaching the compressive yield stress.

Production processes

Two common production processes are cold working and machining. Cold working implies that the material is plastically deformed, which should leave a residual stress distribution in the product. An elementary example is plastic bending. As illustrated by Figure 4.4, a bending moment will induce plastic deformation in the outer fibers of the material. After unloading, elastic spring back occurs, and a residual stress distribution as schematically presented in Figure 4.4 will remain in the material. The stress distribution should satisfy both equations (4.1) and (4.2).

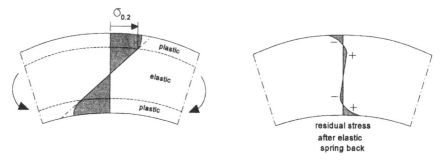

Fig. 4.4 Residual stresses as a result of a plastic bending operation.

In a similar way, residual stresses can exist after a variety of cold working processes. Forging in many cases is a hot-working process, which still can leave residual stresses. The same is true for rolling of plate and sheet material. Afterwards straightening is usually done at room temperature. A residual stress system may be present in the final product.

It is not always realized that machining operations can also introduce residual stresses. Metal cutting implies removal of a layer of material, which includes a failure process near the tip of the cutting tool. But the failure process is preceded by plastic deformation. Depending on machining conditions (sharpness of the cutter, feed rate, depth of cut, etc.) and also on the material, residual stresses can be significant in a thin surface layer.

Shot peening

Shot peening is a well-known process to introduce favorable residual stresses at the material surface of a component. In various practical cases it is applied to prevent fatigue or stress corrosion problems. The peening operation is plastically stretching the surface layer of a material. Because this layer must remain coherent with the elastic substrate material, residual compressive stresses are introduced at the surface. It can lead to warpage of the component, but dimensional distortions can sometimes be prevented by a symmetric peening operation.

The intensity of the peening operation can be checked by peening a so-called Almen strip, which is a steel strip (76 × 19 mm, 3 × 0.75 inch). The strip is fixed by bolts to a stiff foundation, and peened under well defined conditions from one side only, see Figure 4.5. After removing the bolts, the strip is curved. The arc height is measured, which gives a direct indication of the shot peening intensity [1]. An example of a residual stress distribution

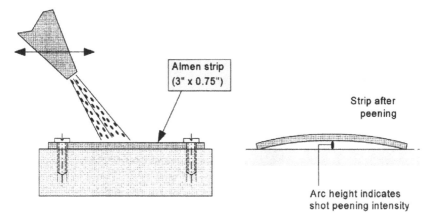

Fig. 4.5 Measurement of shot peening intensity with an Almen strip [1].

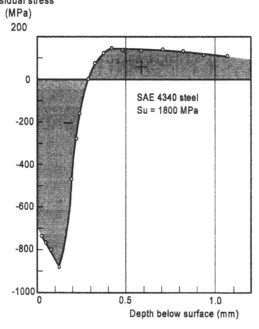

Fig. 4.6 Residual stress introduced by shot peening of SAE 4340 steel, heat treated to $S_U = 1800$ MPa [2].

obtained by shot peening is shown in Figure 4.6 for a high-strength steel, which is fatigue sensitive.

Surface rolling is another process to plastically deform the material surface, see the discussion in Chapter 14 (Figure 14.10). It can be applied

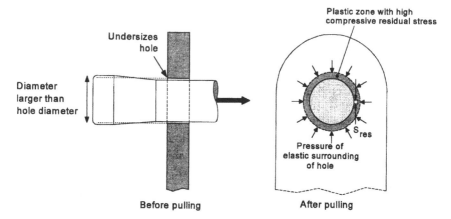

Fig. 4.7 Plastic hole expansion of a lug.

locally to the notch root area, e.g. to the root of fillet radii frequently present in axles. Rolling does not produce a rough surface.

Plastic hole expansion

Plastic hole expansion has been developed to improve the fatigue resistance of holes, also for bolted and riveted joints. The hole is drilled with a slightly undersized hole, e.g. with a diameter a few per cent smaller than the design value. A tapered pin is then pulled through the hole to expand the hole, see Figure 4.7. As a result, plastic deformation does occur around the hole. The plastic zone has been stretched tangentially because it was pushed outwards in the radial direction. The plastic zone has a larger diameter than before. It implies that the elastically strained material around this plastic zone will exert a pressure on the zone, see Figure 4.7. As a result tangential compressive stresses around the hole are introduced. The method is very effective for improving the fatigue resistance because the residual stresses can be high, i.e. almost in the order of the compressive yield stress. Moreover, the depth of the plastic zone can be a few millimeters (compare to the small depth in Figure 4.6). Small distortions of the cylindrical shape of the hole can be corrected afterwards by reaming, which hardly reduces the residual stress. Commercial apparatus has been developed for hole expansion, and a large favorable effect on fatigue can be obtained, see the discussion in Chapter 18.

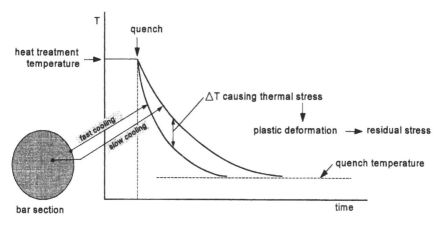

Fig. 4.8 Different cooling rates during quenching cause thermal stresses which can lead to a residual stress distribution.

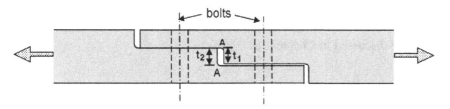

Fig. 4.9 Bolted joint with built-in stresses if $t1 \neq t2$.

Heat treatment

Quenching is an abrupt step of many heat treatments applied to alloys of various materials. Cooling usually occurs very fast at the outside of a component, see Figure 4.8, and significantly slower inside the material. The inhomogeneous cooling introduces thermal stresses. The faster thermal contraction at the outside causes local tensile stresses balanced by compressive stresses inside. At the still elevated temperature, the yield stress is low and plastic deformation can easily occur. Residual stresses are then introduced. In the rotational symmetric case of Figure 4.8 it should lead to the favorable situation of compressive residual stress at the outside balanced by tensile residual stress inside the material. Unfortunately, many components have complex shapes which makes it difficult to know the residual stress distribution obtained after quenching. Tensile residual stresses, also at the outside are possible. They can be reduced, or even reversed, by shot peening.

Assembling stresses

The previous examples of sources of residual stresses were associated with inhomogeneous plastic deformation. A completely different category of residual stresses in a structure is due to mounting of components to form a single structure. In many cases, bolted connections are involved. The residual stresses in the structure depend on the dimensional tolerances of the components. A simple example is shown in Figure 4.9. If $t1$ and $t2$ of this joint are not exactly equal, and the bolts are fastened, the misfit will introduce bending. Maximum internal stresses occur at the root of the fillet notches A in Figure 4.9 in the still unloaded joint. In this case the term "internal stresses" appears to be more correct. These stresses due to assembling a structure are also referred to as built-in stresses. The occurrence of the stresses can be avoided by a strict tolerance system.

In special cases, built-in stresses are desirable. This applies to bushes pushed with an interference fit into a hole, and to pretensioned bolts. These cases are discussed in Chapter 18.

4.3 Measurements or calculations of residual stresses

Residual stresses cannot directly be observed which is rather unfortunate because there are no simple techniques for measuring these stresses. A non-destructive measurement can be done by X-ray diffraction techniques, but it is a fairly elaborate method, which is not easily adopted on a routine basis. It becomes even more problematic if the residual stress has to be measured at the root of a notch, e.g. at the bore of a hole, where high strain gradients are present.

Destructive measurements are possible, but again it is not a simple routine procedure. Small strain gages are bonded to the locations of interest. Cuts in the material around the gages relax the residual strains. The strain variations indicated by the gages during this operation must be measured, and the residual stresses can then be calculated.

Calculations on residual stress distributions in notched elements can be made by FE analysis. A nonlinear elasto-plastic stress-strain relation must then be assumed. In the past simplification of the analysis was obtained by assuming that strain hardening did not occur. Later a linear strain hardening was adopted, and nowadays it is possible to employ a more realistic

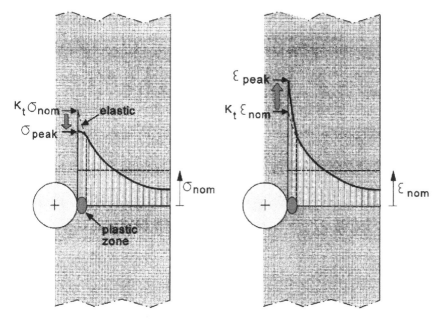

Fig. 4.10 Differences between the σ- and ε-distribution caused by a plastic zone at the root of a notch.

nonlinear strain hardening, e.g. the Ramberg–Osgood equation. However, elasto-plastic FE analysis requires expert experience.

Designers are not particularly fond of relying on good fatigue properties obtained by introducing favorable residual compressive stresses at fatigue critical locations. The argument is indeed that residual stresses cannot easily be measured. If residual compressive stresses at the material surface are still desirable, it is essential to have a closely controlled production process to introduce these stresses (e.g. by shot peening or hole expansion).

4.4 Estimation of the residual stress at a notch after a high load

In view of stress concentrations, it is possible that loads in service introduce a residual stress distribution at notches in the structure. These stresses are significant for crack nucleation, fatigue life, and fatigue damage accumulation in general. An analytical calculation of residual stresses around notches is practically impossible if plastic deformation occurs. Such calculations can be made with FE techniques, although they cannot be classified as simple calculations. However, a reasonable estimate of σ_{peak}

can be made with a simple procedure, based on a postulate of Neuber [3].[8] As long as plastic deformation does not occur, all strains are proportional to the applied load, which is Hooke's law. The shapes of the stress and strain distribution are not depending on the load. However, as soon as a plastic zone is created at the notch root, the shape of the two distributions will be changed. The stress at the notch root (σ_{peak}) is lower than the elastic prediction (Figure 4.10a) and the strain at the same location (ε_{peak}) is larger (Figure 4.10b). In other words:

$$\sigma_{peak} < K_t \sigma_{nom}$$

$$\varepsilon_{peak} > K_t \varepsilon_{nom} \tag{4.5}$$

The fact that σ_{peak} is smaller than the elastic prediction is related to the other fact that ε_{peak} is larger than the elastic prediction. According to the postulate of Neuber the product $\sigma_{peak}\varepsilon_{peak}$ still agrees with the elastic prediction:

$$\sigma_{peak}\varepsilon_{peak} = K_t^2 \sigma_{nom}\varepsilon_{nom} \tag{4.6}$$

It implies that in the product, σ_{peak} being smaller than predicted is compensated by ε_{peak} being larger than predicted. Defining plastic concentration factors K_σ and K_ε as:

$$K_\sigma = \sigma_{peak}/\sigma_{nom} \quad (< K_t)$$

$$K_\varepsilon = \varepsilon_{peak}/\varepsilon_{nom} \quad (> K_t) \tag{4.7}$$

the postulate of Neuber becomes:

$$K_\sigma K_\varepsilon = K_t^2 \tag{4.8}$$

Neuber has proven that the postulate is correct for a hyperbolic notch under shear loading. He then assumed that it will be approximately correct for other types of notches and loading. This was more or less confirmed empirically, provided the plastic zone is small.

Substitution of $\varepsilon_{nom} = \sigma_{nom}/E$ into Equation (4.6) leads to:

$$\sigma_{peak}\varepsilon_{peak} = (K_t \sigma_{nom})^2/E \tag{4.9}$$

For a given load and K_t, the right-hand side of Equation (4.9) has a known constant value. The equation thus gives one (hyperbolic) relation between

---

[8] A different proposal for the same problem was made by Glinka [4]. His analysis is based on energy density considerations.

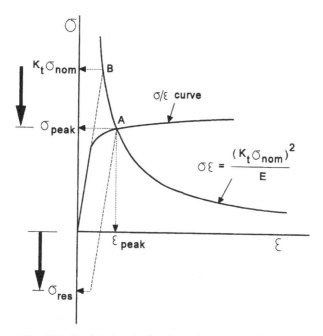

Fig. 4.11 Graphical method to determine σ_{peak} and σ_{residual}.

the unknown σ_{peak} and $\varepsilon_{\text{peak}}$. A second relation is necessary for which the stress-strain curve of the material obtained in a tensile test is used. The graphical solution then becomes as simple as shown in Figure 4.11. At the intersection point A of the two curves, the values of σ_{peak} and $\varepsilon_{\text{peak}}$ satisfy both relations. If no plasticity had occurred, the peak stress would have been found at point B. The difference between the two points B and A gives the reduction of the peak stress. After elastic unloading the residual stress is found as:

$$\sigma_{\text{residual}} = \sigma_A - \sigma_B = \sigma_A - K_t\sigma_{\text{nom}}$$

As a numerical example: $K_t = 2.5$, $\sigma_{\text{nom}} = 200$ MPa, $E = 210000$ (steel). Assume a bilinear σ-ε curve with a yield stress of 300 MPa and a plastic modulus $E_{\text{pl}} = E/20$.[9] According to the above equations the residual stress at the notch root becomes -176 MPa, which is a considerable compressive residual stress.

---

[9] It implies that the plastic σ-ε relation is: $\sigma - 300 = E/20 * (\varepsilon - 300/E)$. Substitution in Equation (4.9) leads to a second-order equation form which the solution is obtained as $\sigma_A = 323.6$ MPa. If the more realistic Ramberg–Osgood stress-strain relation is adopted the problem can also be solved numerically.

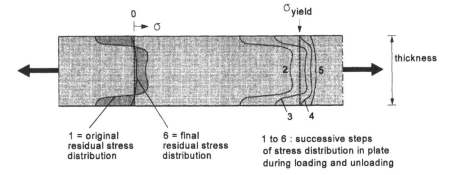

Fig. 4.12 Elimination of residual stress in a plate by plastic stretching.

4.5 How to remove residual stresses

It can be desirable to remove residual stresses. Two arguments are:

1. unfavorable residual tensile stresses can be harmful for fatigue and stress corrosion, and
2. dimensional distortions (warpage) can occur after machining.

A heat treatment may remove all residual stresses, especially if recrystallization occurs. The original permanent plastic deformation causing the residual stress system implies a high dislocation density. During a high temperature anneal, recrystallization removes the initial dislocation structure and a much lower dislocation density is obtained. The initial permanent plastic deformation (strain hardening) is removed, and as a consequence the residual stresses are eliminated. Annealing at a lower temperature will not always induce recrystallization, but certain rearrangements of the dislocation structure can reduce the residual stresses to some extent. Unfortunately, an annealing treatment is not always a feasible option because it can soften the material. The original heat treatment hardening may vanish.

As said before, unfavorable residual tensile stresses at the surface of a material can be reduced or reversed into favorable residual compressive stresses by shot peening. The surface quality obtained after peening can be rather rough, but usually the improvement by residual compressive stresses will prevail.

Another simple mechanical method to eliminate residual stresses from sheets and plates is to stretch the material to a small plastic strain. Figure 4.12 shows an originally inhomogeneous residual stress distribution, No. 1 in this figure. During stretching of the material to a stress level exceeding the yield stress plastic deformation leads to a more homogeneous stress distribution. Finally, when the yield stress is exceeded through the full thickness, the fairly homogeneous stress distribution No.5 is obtained. This distribution is retained during elastic unloading. After full unloading, the original residual stresses are practically eliminated. In the aluminum industry, such prestrained plates are produced. It is more expensive, but warpage problems in the workshop are prevented.

There is another interesting conclusion to be derived from Figure 4.12. If the stretching operation is continued until failure, the stress distribution before failure is approximately homogeneous. As a consequence, the strength of the material will not be affected by the original residual stresses. The same conclusion is also true for the static strength of a notched part. Actually pulling the central hole specimen in Figure 4.10 to failure is just a continuation of the high load that caused the residual stress and strain distribution of Figure 4.10 if unloading is done before failure.

4.6 Main topics of the present chapter

1. Residual stresses are usually caused by inhomogeneous plastic deformation. Due to permanent plastic deformation, the plastic zone no longer fits stress-free in the elastic surrounding, which introduces a residual stress system.
2. Residual stresses can be introduced on purpose (shotpeening, plastic hole expansion). They can also occur unintentionally (production processes, heat treatment). Another important source is assembling of components, which can cause significant built-in stresses.
3. A residual stress system is an equilibrium system. There is never a favorable compressive residual stress without an unfavorable tension residual stress at an other location.
4. Residual stresses can have a significant effect on fatigue and stress corrosion. During machining residual stresses can cause warpage.
5. Measuring of residual stresses at the root of a notch introduced by a high load is not a simple technique. An estimate can be obtained with a simple calculation technique. FE calculations of a residual stress distribution with an FE analysis is possible but it requires experience.

References

1. Marsh, K.J. (Ed.), *Shot Peening: Techniques and Applications.* EMAS, Warley, UK (1993).
2. Lessels, J.M. and Broderick, A.G., *Shot-peening as protection of surface damaged propeller-blade materials.* Proc. Int. Conf. on Fatigue of Metals, London 1956. The Institute of Mechanical Engineers, London (1956), pp. 617–627.
3. Neuber, H., *Theory of stress concentration for shear strained prismatical bodies with arbitrary nonlinear stress-strain law.* Trans. ASME. J. Appl. Mech., Vol. 28 (1961), pp. 544–550.
4. Glinka, G., *Relations between the Strain Energy Density Distribution and Elastic-Plastic Stress-Strain Fields near Cracks and Notches and Fatigue Life Calculations.* ASTM STP 942 (1988), pp. 1022–1047.

Some general references

5. Rice, C.R. (Ed.), *SAE Fatigue Design Handbook*, 3rd edn. AE-22, Society of Automotive Engineers, Warrendale (1997).
6. Champoux, R.L., Underwood, J.H. and Kapp, J.A. (Eds.), *Analytical and Experimental Methods for Residual Stress Effects in Fatigue.* ASTM STP 1004 (1989).
7. Niku-Lari, A. (Ed.), *Advances in Treatments: Technology – Applications – Effects, Vol. 4: Residual Stresses.* Pergamon Press, Oxford (1987).
8. Frost, N.E., Marsh, K.J. and Pook, L.P., *Metal Fatigue.* Clarendon, Oxford (1974).
9. Forrest, P.G., *Fatigue of Metals.* Pergamon Press, Oxford (1962).

Chapter 5
Stress Intensity Factors of Cracks

5.1 Introduction

Stress concentrations around notches were considered in Chapter 3 with the stress concentration factor K_t as an important parameter for characterizing the severity of the stress distribution around the notch. The crack initiation life is highly dependent on the K_t-value. The crack initiation period is followed by the fatigue crack growth period, recall Figure 2.1. For a crack, the K_t-value is no longer a meaningful concept to indicate the severity of the stress distribution around the crack tip. Because a crack is a notch with a zero tip radius, K_t would become infinite, and this would be true for any crack length. A new concept to describe the severity of the stress distribution around the crack tip is the so-called stress intensity factor K. This concept was originally developed through the work of Irwin [1]. The application of

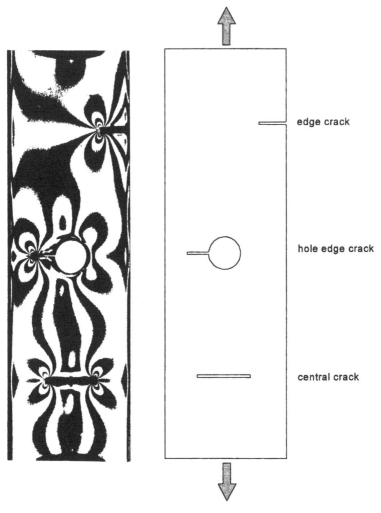

Fig. 5.1 Photo-elastic picture of a specimen with three different types of cracks. Note the similar butterfly pattern at each crack tip.

the stress intensity factor to present fatigue crack growth data and to predict fatigue crack growth is referred to as "linear elastic fracture mechanics".

The difference between a notch and a crack can be illustrated by considering an elliptical hole (semi-axes a and b, tip radius ρ). In an infinite sheet loaded in tension, as previously shown in Figure 3.4, the stress concentration factor K_t is given by

$$K_t = 1 + 2\frac{a}{b} = 1 + 2\sqrt{\frac{a}{\rho}} \qquad (5.1)$$

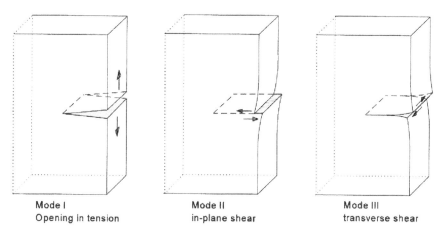

Mode I   Mode II   Mode III
Opening in tension   in-plane shear   transverse shear

Fig. 5.2 Three different crack opening modes.

with the tip radius $\rho = b^2/a$. The elliptical hole becomes a crack by decreasing the minor axis b to zero. If $b = 0$ the hole is a crack with a tip radius $\rho = 0$, and according to Equation (5.1) $K_t = \infty$, regardless of the semi-crack length. This result is not useful. However, the stress distribution around the tip of a crack shows a characteristic picture. This is illustrated in Figure 5.1 by photo-elastic results of a specimen with three cracks loaded in tension. Apparently, similar isochromatic pictures occur at the tips of the three cracks, which suggests similar stress distributions at the crack tips. The "intensity" of the crack tip stress distribution is depending on the stress intensity factor K which can be written as

$$K = \beta S \sqrt{\pi a} \qquad (5.2)$$

In this equation, S is the remote loading stress, a is the crack length, and β is a dimensionless factor depending on the geometry of the specimen or structural component. The important feature is that stress distribution around the crack tip can be fully described as a linear function of the stress intensity factor K. The concept of the stress intensity factor is presented in this chapter. First, different types of cracks are listed, followed by more details about stress intensity factors for several geometries. Some basic aspects of the stress analysis of cracked configurations are addressed including differences between plane stress and plane strain situations, crack tip plasticity and determination of K factors. The basic principle of the application of K factors to fatigue crack growth is considered. Finally, the main topics of the present chapter are listed.

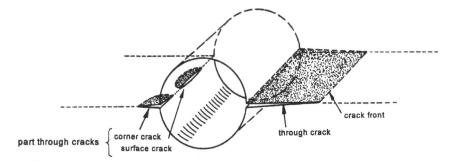

Fig. 5.3 Different types of cracks starting from a hole.

5.2 Different types of cracks

As discussed in Chapter 2, fatigue cracks in service usually grow in a direction which is macroscopically perpendicular to the main principle stress. In many cases that is perpendicular to the tensile stress that tries to open the crack. This kind of crack opening is called "mode I", see Figure 5.2. In principle, other modes of crack opening are possible, see modes II and III in the same figure. It could be thought that these modes will occur under cyclic shear stress. However, experience has shown that small cracks, nucleated under pure shear loading, quickly exhibit a transition to fatigue crack growth in the tensile mode, i.e. mode I. This appears to be reasonable from a physical point of view, because the tensile stress component, which opens the crack, will certainly promote the conversion of cyclic plastic deformation into crack extension (see Chapter 2).

If shear lips are created at the surface of the material, see Figure 2.38, a mixed mode occurs, modes I and III. This aspect is briefly addressed in Section 5.11, but the present chapter is dealing mainly with mode I cracks, the most relevant mode for fatigue cracks.

Fatigue cracks growing through the entire thickness are referred to as through the thickness cracks, or simply *through cracks*, see Figure 5.3. If the fatigue load is cyclic tension only, without bending, the crack front is perpendicular to the material surface. However, in thick material, and in general for smaller cracks, the crack front does not pass through the full thickness. The crack front then is generally curved. Figure 5.3 shows two such cracks, a *corner crack* and a *surface crack*. Corner cracks and surface cracks are also labeled as *part through cracks*.

A through crack with the crack front perpendicular to the material surface is usually treated as a two-dimensional (2D) problem. For part through

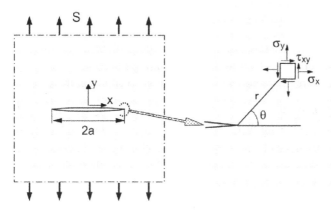

Fig. 5.4 Infinite sheet with a mode I crack. Definition of coordinates.

cracks this is no longer allowed. The stress analysis then is a 3D problem, and thus more complicated. Sections 5.3 to 5.5 deal with 2D cases. In Section 5.6 cracks with a curved crack front are considered.

5.3 Definition of the stress intensity factor

In Chapter 3 on the stress concentration factor, it was discussed that an exact solution exists for an infinite sheet with an elliptical hole. Because a crack can be considered to be an elliptical hole with zero height, an exact solution is also available for an infinite sheet with a crack. Unfortunately, the stress components cannot be written as simple functions of the (x, y) coordinates. However, if the attention is restricted to the area around the crack tip, explicit functions are obtained by considering analytic limits for a short distance of a point (x, y) to the crack tip, i.e. relatively short as compared to the crack length. In Figure 5.4 it implies that

$$r \ll a \tag{5.3}$$

With polar coordinates the following equations are then obtained for a crack in an infinite sheet loaded in tension, see Figure 5.4:

$$\sigma_x = \frac{S\sqrt{\pi a}}{\sqrt{2\pi r}} \cos\frac{\theta}{2}\left(1 - \sin\frac{\theta}{2}\sin\frac{3\theta}{2}\right) - S \tag{5.4a}$$

$$\sigma_y = \frac{S\sqrt{\pi a}}{\sqrt{2\pi r}} \cos\frac{\theta}{2}\left(1 + \sin\frac{\theta}{2}\sin\frac{3\theta}{2}\right) \tag{5.4b}$$

$$\tau_{xy} = \frac{S\sqrt{\pi a}}{\sqrt{2\pi r}} \cos\frac{\theta}{2} \sin\frac{\theta}{2} \cos\frac{3\theta}{2} \qquad (5.4c)$$

The equations are asymptotically correct for small values of r. It implies that they are a good approximation in the crack tip region.

The equations are characterized by some noteworthy features:

(i)   The stress distribution at the crack tip shows a singularity, because all stress components go to infinity for $r \rightarrow 0$ for each value of θ. They increase with $1/\sqrt{r}$. Of course an infinite stress cannot be present at the crack tip. Some crack tip plasticity must occur. This topic is addressed in Section 5.8.

(ii) A non-singular finite term (S) occurs in Equation (5.4a) for σ_x. In general, this term is negligible as compared to the first singular term of the equation. For this reason, the term -S is usually omitted. But an interesting result applies to the upper and lower edge of the crack, i.e. for $\theta = \pm 180°$. For these θ-values the singular term is zero, and thus $\sigma_x = S$, a compressive stress equal to the remote tensile stress. Although Equations (5.4) apply to the crack tip area only, the exact solution for the full sheet indicates that $\sigma_x = S$ along the entire upper and lower edges of the crack.[10]

(iii) It is noteworthy that π occurs in the square root terms of both numerator and denominator in Equations (5.4). Consequently, π could have been omitted, and that would change the definition of the K factor by a factor $\sqrt{\pi}$. Although this was done in some older publications, it was not done later on, and historically the π survived.

After neglecting the non-singular term $(-S)$ in Equation (5.4a), the three equations can be written as

$$\sigma_{i,j} = \frac{K}{\sqrt{2\pi r}} f_{i,j}(\theta) \qquad (5.5)$$

with

$$K = S\sqrt{\pi a} \qquad (5.6)$$

The above equations apply to a mode I crack in an infinite sheet loaded in tension. However, it can be shown that the equations are also applicable to elements with finite dimensions provided that the geometry is symmetric

---

[10] An empirical confirmation of a compression stress along the crack edges is obtained in a static tensile test on a thin sheet specimen with a central crack. An increasing load on the specimen causes crack edge buckling with crack edge displacements out of the plane of the sheet. This is a complication for fracture toughness tests.

with respect of the X-axis in order to have still mode I cracks. The finite dimensions are accounted for by a geometry correction factor β in the equation for the stress intensity factor:

$$K = \beta S \sqrt{\pi a} \qquad (5.7)$$

for an infinite sheet $\beta = 1$.

The function $f_{i,j}(\theta)$ in Equation (5.5) determines the picture of the stress distribution around the crack tip (e.g. the butterfly isochromatics in Figure 5.1), while K represents the severity of the stress intensity with β as the dimensionless geometry factor. The term $1/\sqrt{(\pi r)}$ indicates how fast stresses decrease away form the crack tip. In summary, Equation (5.7) defines K as the characteristic parameter for the stress intensity around the crack tip.

According to Equation (5.7), the unit of the stress intensity factor is $(\text{N/m}^2)\sqrt{\text{m}} = \text{Pa}\sqrt{\text{m}}$ (N = Newton, Pa = Pascal = N/m^2). Because Pa$\sqrt{\text{m}}$ is a rather small unit, the 10^6 times larger unit, MPa$\sqrt{\text{m}}$, is usually adopted. Another unit is ksi$\sqrt{\text{in}}$ with 1 ksi$\sqrt{\text{in}} = 1.10$ MPa$\sqrt{\text{m}}$ which implies a 10% difference.

It is not correct to think that the stress concentration factor K_t and the stress intensity factor K are somewhat similar concepts for a notch and a crack respectively. Both concepts are based on the theory of elasticity. However, K_t is a dimensionless shape factor which accounts for the geometry only, and not for the load applied. The stress intensity factor K accounts for both, and thus can no longer be dimensionless. It may be said that K_t and β both serve a similar purpose by accounting for the shape only.

5.4 Examples of stress intensity factors

Many results of calculations on stress intensity factors for various geometries and loading cases have been published. The results of the calculations are often referred to as K solutions which are presented as values of the geometry correction factor β in the equation $K = \beta S \sqrt{(\pi a)}$. Results are usually presented in graphs which show how β depends on geometry ratios. Compilations of K solutions can be found in some handbooks. Frequently cited books are [3–5]:

1. *Stress Intensity Factors*, 1976, by D.P. Rooke and D.J. Cartwright.
2. *The Stress Analysis of Cracks Handbook*, 1985, by H. Tada, P.C. Paris and G.R. Irwin.

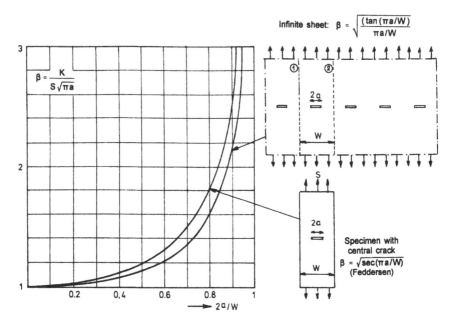

Fig. 5.5 Sheet with infinite row of collinear cracks. Comparison to finite width specimen with a single central crack.

3. *Stress Intensity Factors Handbook*, 1987, three volumes, edited by Y. Murakami.

More compilations are available, also as software packages.

In the present section, some examples of K-values for simple geometries are discussed. The prime purpose is to show certain trends of geometry effects and to see how they can be understood, at least qualitatively. In spite of the extensive literature on K-values, it should be pointed out that in many practical cases K solutions for cracks are not available. Sometimes the values can be approximated by available solutions for less complicated geometries. Otherwise, FE calculations are necessary, see Section 5.10.

Infinite row of collinear cracks

An exact analytical solution was derived by Westergaard [2] for the stress distribution in an infinite sheet with an infinite row of collinear cracks, see Figure 5.5. The solution for the stress intensity factor is:

$$K = \beta S \sqrt{\pi a} \qquad (5.7)$$

with

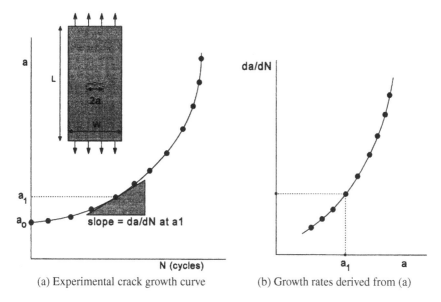

(a) Experimental crack growth curve   (b) Growth rates derived from (a)

Fig. 5.6 Results of crack growth fatigue test.

$$\beta = \sqrt{\frac{\tan(\pi a / W)}{\pi a / W}} \tag{5.8}$$

The geometry factor β is plotted in Figure 5.5 as a function of $2a/W$, i.e. the relative crack length. If W goes to infinite ($W \to \infty$) this case is reduced to an infinite sheet with a single crack. According to Equation (5.8) the result then is $\beta = 1$, which is in agreement with the value mentioned in the previous section.

Central crack tension specimen

A specimen with a central crack (see Figure 5.6) is frequently used in fatigue tests to study various effects on fatigue crack propagation, and for obtaining the crack growth properties of a material. The specimen is referred to as a CCT specimen (Center Cracked Tension). In an ASTM Standard [6] it is labeled as an $M(T)$ specimen (Middle Cracked Tension). The specimen is provided with a central notch, quite often a simple saw cut, or a small hole with two saw cuts or spark eroded slits. Under cyclic loading, fatigue cracks will start from the starter notch. If constant-amplitude loading is applied, the stress intensity factor will increase while the crack is growing. The crack growth curve (a-N) is determined, and the crack growth rate (da/dN) can be derived as a function of the crack length a, see Figure 5.6. The crack rate is increasing because the stress intensity at the crack tip is increasing. In Section 5.11 it will be explained how such results can be described by calculating da/dN as a function of the range of the cyclic stress intensity factor ($\Delta K = K_{max} - K_{min}$) in order to generalize the meaning of the data. For this purpose, K must be known as a function of the crack length which implies that the geometry factor β must be available. An exact solution for the CCT specimen does not exist. Feddersen [7] has proposed the following equation:

$$\beta = \sqrt{\sec \frac{\pi a}{W}} \tag{5.9}$$

Substitution in $K = \beta S \sqrt{\pi a}$ gives:

$$K = S \sqrt{\frac{\pi a}{\cos \dfrac{\pi a}{W}}} \tag{5.10}$$

Although this is not an exact solution, it agrees very well with K-values obtained by FE calculations. Because the geometry factor must be dimensionless, it can depend on ratios of dimensions only. In this case, it is the ratio a/W. A dimension which does not occur in Equation (5.9) is the specimen length L. However, if L/W is larger than 2 the effect of the length L on β is negligible.

It should be noted that K for a crack in a center cracked specimen increases during crack growth because of the $\sqrt{\pi a}$ in the K equation. However, an additional increase is contributed by the geometry factor β which also increases for an increasing crack length (Figure 5.5 and Equation 5.9). In general K is increasing for a growing fatigue crack, but there is a noteworthy exception to be discussed later.

It is of some interest to compare the Feddersen equation with the Westergaard solution given as Equation (5.8). This is done in Figure 5.5. The difference between the two geometry factors β is small, although it increases for larger cracks. If a cut is made along the two lines ① and ② in the infinite plate with collinear cracks, a strip with a single crack is obtained, which is a similar configuration as the CCT specimen. However, along these lines stresses are present contrary to the edges of the CCT specimen. Some differences in the β-values must occur.

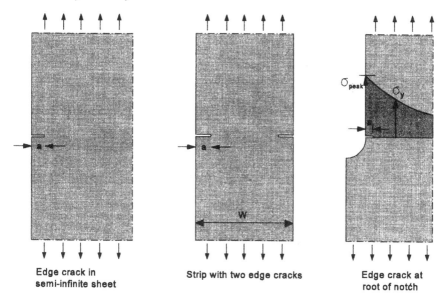

Fig. 5.7 Three cases with edge cracks.

Edge crack

An exact solution exists for an edge crack in a semi-infinite plate, the first case in Figure 5.7. The solution is:

$$K = 1.1215 \ S\sqrt{\pi a} \tag{5.11}$$

Apparently, the geometry factor is constant, $\beta = 1.1215$, which in the literature is often rounded off to 1.12. As said before, the geometry factor is dimensionless and will thus depend on ratios of the dimensions of the geometry. Because in an infinite sheet with an edge crack, only one dimension is present, i.e. the crack length a. If only one dimension is available, ratios of dimensions cannot be defined, and as a consequence β must be a constant, similar to $\beta = 1$ for an infinite sheet with a single crack. Another case of a constant β-value will turn up later.

For a long strip with two symmetric edge cracks, the second case in Figure 5.7, an exact solution does not exist, but an accurate approximation is presented in [4]:

$$\beta = \frac{1.122 - 1.122\left(\dfrac{a}{W}\right) - 0.060\left(\dfrac{a}{W}\right)^2 + 0.728\left(\dfrac{a}{W}\right)^3}{\sqrt{1 - 2\dfrac{a}{W}}} \tag{5.12}$$

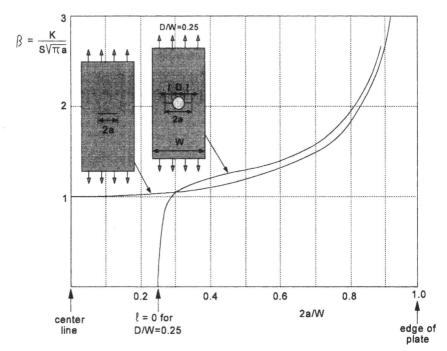

Fig. 5.8 Comparison of the stress intensity of a hole with two edge cracks [8] and a central crack in a finite width strip with the same tip to tip distance.

Note that this equation goes to $\beta = 1.122$ for $W \to \infty$ in agreement with the β-value for the first case in Figure 5.7.

The third case is of more practical interest. It is an edge crack at the root of a notch, the usual location for fatigue crack initiation. As long as the crack is small, the crack remains in the high-stress area of the notch root, where the stress is σ_{peak}. The stress intensity can then be approximated by adopting Equation (5.11) with the rounded constant:

$$K = 1.12\sigma_{\text{peak}}\sqrt{\pi a} = 1.12K_t S\sqrt{\pi a} \qquad (5.13)$$

As long as the crack length is much smaller than the root radius, this approximation is useful for estimating crack growth of a small fatigue crack.

Cracks at the edge of a hole

The values of K for a strip with a circular hole and two equal hole edge cracks was determined by Newman with finite-element calculations [8], see

Figure 5.8. It is interesting to compare the results with K-values for a strip with a central crack only, but the same tip to tip distance, i.e.:

$$2a = D + 2\ell \qquad (5.14)$$

where D is the hole diameter and ℓ is the crack length measured from the edge of the hole. The comparison in Figure 5.8 applies to a hole diameter to specimen width ratio $D/W = 0.25$. The geometry factor $\beta = K/(S\sqrt{\pi a})$ for the CCT specimen follows the Feddersen equation (Equation 5.10), beginning with $\beta = 1$ for a zero crack length. For the hole edge cracks, the curve of the β-value starts at $\ell = 0$, which corresponds to $2a/W = 0.25$. The β-value is then zero because $\ell = 0$, but it rapidly increases, initially in agreement with Equation (5.13) for small cracks. Intersection of the two curves occurs at $2a/W = 0.30$, corresponding to $\ell/D = 0.10$, i.e. an edge crack length of 10% of the hole diameter. For larger cracks, the difference between the two curves remains relatively small. It implies that edge cracks larger than 10% of the hole diameter have a K-value of a comparable magnitude as the single crack with the same tip to tip length. In other words; for relatively small cracks at a hole, the stress intensity factor is approximately the same as for a much larger crack with a length that includes the hole diameter. This is generally true for cracks originating at notches. If the width or the depth of the notch is added to the real crack length, see Figure 5.9, an effective crack length (a_{eff}) is obtained. This effective length should be used for an approximate indication of the stress intensity at the crack tip. In Figure 5.9a, a crack at the edge of a window can thus introduce a rather severe situation. Actually, this is what happened during the Comet accidents in the early fifties when two aircraft exploded at cruising altitude as a result of the cabin pressure, while only relatively short window cracks were present [9]. Another illustration is given in Figure 5.9b which shows that a crack at the root of a notch in the wall of a pressurized cylinder can have an effective crack length which is the sum of the notch depth and crack length. Such a crack once caused a complete failure of a thick-walled steel cylinder.

The geometry of a hole (radius ρ) with two edge cracks is considered again in Figure 5.10, but now including biaxial loading. The geometry factors are presented for three biaxiality ratios λ ($= S_1/S_2$). Some remarkable trends can be observed:

(i) A comparison can be made between the uni-axial tension case ($\lambda = 0$) and the biaxial case ($\lambda = 1$). The biaxiality gives a reduction of the stress concentration factor if there are no cracks, see the discussion in

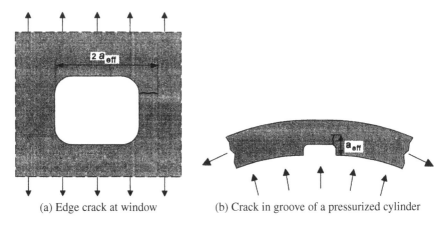

(a) Edge crack at window   (b) Crack in groove of a pressurized cylinder

Fig. 5.9 Large effective crack length by a contribution of a notch.

Chapter 3 (Section 3.5). The peak stress, σ_{peak}, for a circular hole and $\lambda = 1$ is $2S$ instead of $3S$ for $\lambda = 0$ (Equation 3.13). In other words, it is reduced with a factor of 2/3. This is significant for small cracks. With Equation (5.11), the K_t-value for small edge cracks should also be decreased with a reduction factor of 2/3. The exact reduction factors for three small crack lengths are given in the table below:

ℓ/ρ	0.05	0.10	0.20	0.30
reduction factor	0.70	0.72	0.75	0.79

Because the factors for low ℓ/ρ-values are still close to the theoretical value of 0.67, they confirm that Equation (5.11) is useful for small cracks.

(ii) For large cracks, the biaxiality effect in Figure 5.8 is vanishing. For long edge cracks the hole is relatively small, and according to the SaintVenant principle, it should no longer affect the stress distribution at the remote crack tips. As a consequence, the situation becomes very much similar to a single crack. In that case, the lateral stress λS does not cause any disturbance of the stress distribution at the crack tip. Only the vertical stress S leads to the characteristic singular stress distribution at the crack tip. As confirmed by Figure 5.10 the geometry factor $\beta \rightarrow 1$ for both $\lambda = 0$ and $\lambda = 1$.

(iii) As shown in Figure 5,10, the difference between β-values for a single hole edge crack and two symmetric hole edge cracks is rather small. For small cracks this should be expected according to Equation (5.11), while for large crack lengths β should go to 1 for both cases.

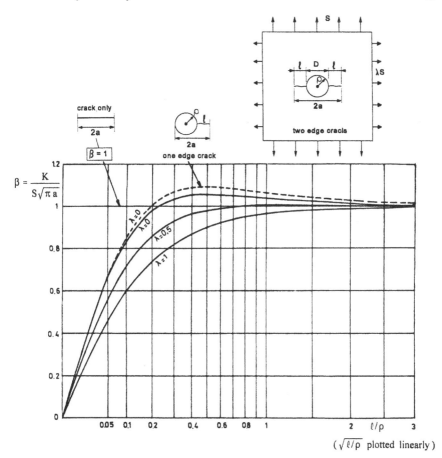

Fig. 5.10 The geometry factor for hole edge cracks in an infinite sheet under biaxial loading [8, 10].

Crack edge loading

In Figure 5.11, a few cases of crack edge loading are indicated, with P as a point load per unit thickness. It appears to be unrealistic to have loads which act on a single point of the edge of a crack. However, cracks starting from rivet holes or bolt holes (loaded holes) come fairly close to this configuration. The load applied by a bolt to a hole can be considered as a concentrated load at a single point if the crack length is sufficiently large compared to the hole diameter, see case 3 in Figure 5.11. For an infinite sheet with two loads on the crack edges, case 1 and case 2, an exact solution is available. For case 2 in Figure 5.11 the K solution for the right-hand crack tip is:

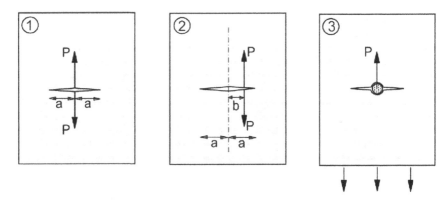

Fig. 5.11 Three cases of crack edge loading.

$$K = \frac{P}{\sqrt{\pi a}} \sqrt{\frac{a+b}{a-b}} \tag{5.15}$$

For case 1 in Figure 5.11, $b = 0$, and the equation reduces to the simple equation:

$$K = \frac{P}{\sqrt{\pi a}} \tag{5.16}$$

Because the dimension of P is N/m, the dimension of K is still N/m$^{3/2}$. Equation (5.16) seems to deviate from the general equation $K = \beta S \sqrt{\pi a}$, but it can be rewritten as

$$K = \frac{2}{\pi} \left(\frac{P}{2a} \right) \sqrt{\pi a} \tag{5.17}$$

The load $(P/2a)$ is now expressed as a bearing pressure, and the factor $2/\pi$ is the geometry factor β. The equation thus is again similar to $K = \beta S \sqrt{\pi a}$. Note that β is a constant $(2/\pi)$ because there are no dimensionless geometric ratios to affect β as discussed earlier.

A second noteworthy aspect of Equation (5.16) is that K decreases for an increasing value of a. That might appear to be in conflict with intuitive expectations, but the intuition in this case is not right. Fatigue crack growth tests on specimens loaded under crack edge loading conditions have confirmed a decreasing crack growth rate for a growing crack. This will be discussed in Chapter 8 (Figure 8.8). The third case in Figure 5.11 is addressed in Section 5.5 on superposition.

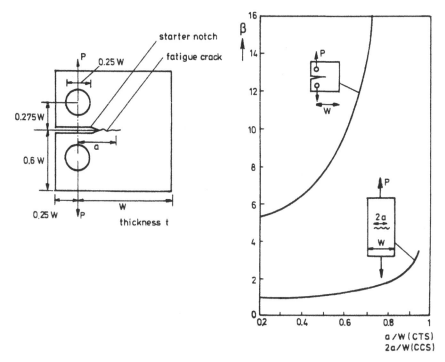

Fig. 5.12 Compact tension specimen (CTS). Comparison to center cracked specimen (CCT).

Compact tension specimen

The compact tension specimen (CT specimen) shown in Figure 5.12 is generally adopted to determine the so-called fracture toughness of a material. This material property, usually indicated by the symbol K_{Ic} (I for opening mode, c for critical) is obtained in a static test on the CT specimen provided with a fatigue crack. At the moment of failure, the K-value calculated for the failure load is the critical value; K_{Ic}. The fracture toughness K_{Ic} is considered to be a material property which indicates the sensitivity of the material for cracks under static loading. This property is used for residual strength calculations of a cracked structure under static loading, which is an important issue for a premature failure of a structure with cracks (risk analysis).

The CT specimen is also used in fatigue crack growth experiments. The equation of the K-value of the CT specimen is a curve fit to FE analysis results:

$$K = \beta S \sqrt{\pi a} \quad \text{with} \quad S = \frac{P}{Wt} \quad \text{and}$$

$$\beta = \frac{\left(2 + \dfrac{a}{W}\right)}{\left(1 - \dfrac{a}{W}\right)^{1.5} \left(\dfrac{\pi a}{W}\right)^{0.5}} \tag{5.18}$$

$$\times \left[0.886 + 4.64 \left(\frac{a}{W}\right) - 13.32 \left(\frac{a}{W}\right)^2 + 14.72 \left(\frac{a}{W}\right)^3 - 5.6 \left(\frac{a}{W}\right)^4\right]$$

Two advantages of the CT specimen are: (i) it is small and does not require much material, and (ii) a relatively low load is sufficient for a high K-value, which can be useful if the load capacity of a testing machine is limited. A comparison between K-values of the CT specimen and a CCT specimen is made in the graph of Figure 5.12. It shows the β-values of the two specimens for the same load P. There are also some disadvantages of the CT specimen [11]. (iii) Production of a CT specimen is not as simple as for a CCT specimen, and even more important, (iv) the CT specimen crack is also opened by a significant bending moment on the specimen. Cracks in a real structure usually do not show a similar type of loading, see the discussion in [11].

5.5 *K* factors obtained by superposition

If a loading system 1 is leading to stresses $(\sigma_{i,j})_1$, and a second loading system 2 to stresses $(\sigma_{i,j})_2$, these stresses have to be summed to obtain $\sigma_{i,j}$ for the case of both loading systems applied simultaneously: $\sigma_{i,j} = (\sigma_{i,j})_1 + (\sigma_{i,j})_2$. For the crack tip it implies:

$$\frac{K}{\sqrt{2\pi r}} f_{i,j}(\theta) = \frac{K_1}{\sqrt{2\pi r}} f_{i,j}(\theta) + \frac{K_2}{\sqrt{2\pi r}} f_{i,j}(\theta) \tag{5.19}$$

Because $f_{i,j}(\theta)$ is the same for all mode I cracks, the equation gives:

$$K = K_1 + K_2 \tag{5.20}$$

In other words, superposition implies that the corresponding K factors should be added. A simple, but interesting example is shown in Figure 5.13. Case 1 is the sum of the two loading cases 2 and 3. Case 2 is equivalent to a plate without a crack and a homogeneous stress $\sigma_y = S$ through the entire plate. If a cut is made to arrive at the same geometry of case 1, nothing will change if the crack edges are loaded by the same stress S to keep the cut just closed. As a consequence, a singular stress distribution at the crack tip

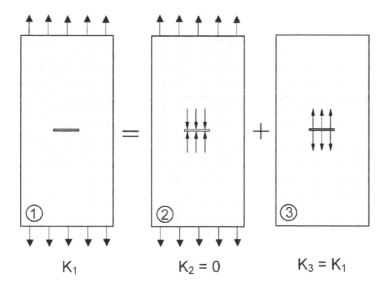

$$K_1 \qquad\qquad K_2 = 0 \qquad\qquad K_3 = K_1$$

Fig. 5.13 The stress intensity factor for a crack under internal pressure (case 3).

is not present in this case, or $K_2 = 0$. However, to make case 1 equal to the sum of cases 2 and 3, the opposite stress on the crack edges of case 2 must be applied to the crack edges of case 3. It thus becomes a crack under an internal pressure $p = S$. Because $K_2 = 0$, the stress intensity factor for the crack under internal pressure is K_3 is equal to K_1. The superposition here shows that $K_3 = K_1$.

It should be recognized that this example of superposition also indicates that internal pressures on crack edges are causing a singular stress field at the crack tip. For thin sheet material in a pressure vessel, the K contribution of the pressure on the crack edges should be negligible if compared to K of the external load. However, for a thick-walled cylinder the contribution can be more important and should be considered.

A second example of superposition is given in Figure 5.14, which is the problem of case 3 in Figure 5.11. Case 1 in Figure 5.14 is representative for load transmission to a plate by a fastener which connects an other part to the plate, e.g. a stiffener. It is a non-symmetric loading case. Figure 5.14 shows how the K-value can be estimated by summing and splitting loading cases. Case 2 is obtained by rotation of case 1 by $180°$, which does not affect the stress intensity, and thus $K_2 = K_1$. Case 3 is the sum of cases 1 and 2, which implies that $K_1 = \frac{1}{2}K_3$. Case 3 can be split into cases 4 and 5. If the cracks are not very small as compared to the hole diameter, case 4 is similar to the

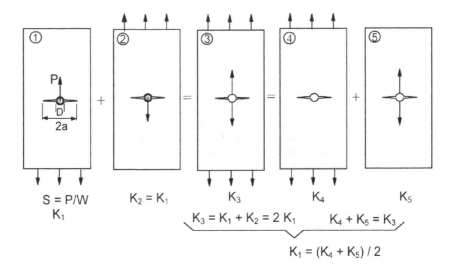

Fig. 5.14 Summing and splitting loading cases.

center cracked strip, and case 5 is similar to the case of crack edge loading (case 1 in Figure 5.11). The final result is: $K_1 = \frac{1}{2} K_3 = \frac{1}{2} (K_4 + K_5)$. The value of K_4 is given by Equation (5.10). The value of K_5 should be estimated from (5.16) for an infinite sheet by applying a width correction. As a first estimate, the Feddersen width correction (Equation 5.10) can also be used for this purpose. Although the final result is not exactly correct, it can be considered as a fair estimate. If more accurate K-values are required, FE calculations must be made.

5.6 Cracks with curved crack fronts

In Figure 5.3, the through crack with a straight crack front perpendicular to the plate surface is a simple configuration. This type of a 2D crack configuration was considered in the previous sections. In the same figure, two examples are shown with curved crack fronts. For a corner crack, the shape of the crack front is usually approximated by a quarter ellipse. For a surface crack it will be close to a semi-ellipse. If the crack front is curved, the problem has a 3D character. The K factor varies along the crack front; it is no longer a constant K-value.

Sneddon [12] analytically solved the problem of the stress distribution around a circular crack in an infinite solid in tension, see Figure 5.15. For

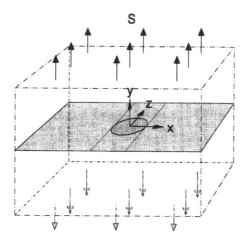

Fig. 5.15 Penny-shaped crack in an infinite solid under tension.

this penny-shaped crack, the K factor is still constant in view of the rotational symmetry:

$$K = \frac{2}{\pi} S \sqrt{\pi a} \tag{5.21}$$

Note that the geometry factor $\beta = 2/\pi$ is smaller than $\beta = 1$ for the through crack in an infinite plate.

If the circular crack in Figure 5.15 is replaced by an elliptical crack, see Figure 5.16, the K factor varies along the crack front. Irwin [13] derived the following equation:

$$K(\varphi) = \frac{S\sqrt{\pi b}}{\phi} \left[\sin^2 \varphi + \left(\frac{b}{a}\right)^2 \cos^2 \varphi \right]^{1/4} \tag{5.22}$$

The parametric angle φ determines the location C at the crack front in the way as shown in Figure 5.16. The symbol ϕ represents a so-called complete elliptical integral of the second kind, which depends on the aspect ratio a/b of the ellipse:

$$\phi(a/b) = \int_0^{\pi/2} \sqrt{1 - k^2 \sin^2 \theta}\, d\theta \quad \text{with} \quad k^2 = 1 - (b/a)^2 \tag{5.23}$$

Elliptical integrals cannot be solved analytically, but numerical values can be found in tables in the literature. A very good approximation for Equation (5.23) is [14]:

$$\phi(a/b) = [1 + 1.464(b/a)^{1/65}]^{1/2} \quad (\text{for } a \geq b) \tag{5.24}$$

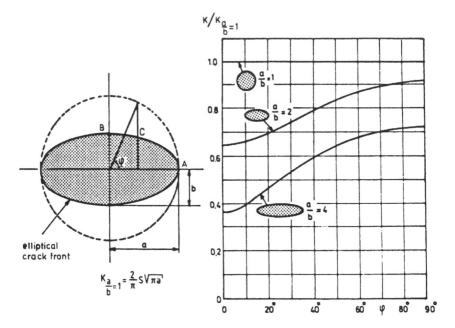

Fig. 5.16 Variation of K along the crack front of an elliptical crack in an infinite solid loaded in tension.

An illustration of the variation of K along elliptical crack fronts is given in Figure 5.16. The K-value is normalized by dividing the value by K for a circular crack. The graph shows values for two ellipse aspect ratios, $a/b = 2$ and 4 respectively. Apparently, K has a minimum at the end of the long axis (in A) and a maximum at the end of the short axis (in B). For a fatigue crack it implies that there will be a tendency to grow to a circular shape.

For quarter- or semi-elliptical cracks in finite dimension components, the K formula deviate from Equation (5.22). Solutions in the literature are usually presented as a correction on that equation:

$$K(\varphi) = K_{\text{Eq.}(5.22)} \cdot F\left(\frac{a}{b}, \frac{a}{r}, \frac{a}{W}, \frac{b}{t}, \varphi\right) \tag{5.25}$$

The corrections to be made for a finite width (W), thickness (t), the presence of a hole (radius r) if present, are contained in the correction function F. This correction also depends on the aspect ratio a/b and the location angle φ. Exact solutions are not available, but results of FE calculations can be found in the literature [14]. A practical case is the occurrence of semi-elliptical surface cracks initiated by surface damage (dents, fretting, corrosion pits, poor machining). Some K results are presented in Figure 5.17 for a flat

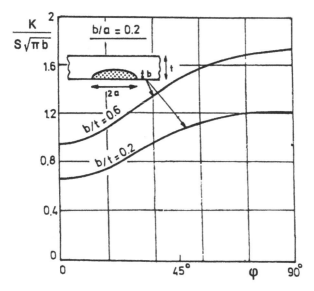

Fig. 5.17 *K* along the crack front of a semi-elliptical surface crack. Effects of the crack depth and the location angle φ.

semi-elliptical surface crack ($b/a = 0.2$) for two values of the relative crack depth b/t. The K-value at the deepest point ($\varphi = 90°$) is significantly larger than at the material surface. The fatigue crack growth in the depth direction will then be faster than along the material surface.

5.7 Crack opening and the state of stress

If a cracked component is loaded in tension, the crack will be opened. For an infinite sheet with a central crack and a remote tensile stress S, see Figure 5.18, the displacements $u(x, y)$ and $v(x, y)$ are exactly known. The vertical crack edge displacements, $v_{(y=0)}$ are of special interest because they indicate the opening of the crack. At the center of the crack, the equation is simple. The full crack opening displacement (COD) for $x = 0$ (plane stress) is:

$$\text{COD}_{(x=0)} = 2v_{(x=0, y=0)} = 4\frac{S}{E}a = 4a\varepsilon_\infty \tag{5.26}$$

with $\varepsilon_\infty = S/E$ as the nominal strain in the sheet. According to the equation, a crack of 25 mm (tip to tip) in steel ($E = 210000$ MPa), loaded by a stress of 200 MPa, the COD in the center of the crack is 0.05 mm. Such a crack

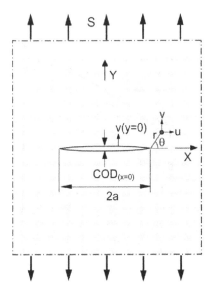

Fig. 5.18 Crack edge displacements.

opening is visible for the unaided eye, provided a smooth material surface is present.

The displacements close to the crack tip, $r \ll a$, can again be written in relatively simple explicit equations, usually presented as

$$u_i = \frac{K}{G}\sqrt{\frac{r}{2\pi}} f_i(\theta) \qquad (5.27)$$

with u_i representing u, v and w in the x, y and z direction respectively. As shown by the equation, it includes an elastic constant, for which the shear modulus $G\ (= E/2(1+v))$ is chosen. Secondly, instead of $(r)^{-0.5}$ as applicable for the singular stresses $\sigma_{i,j}$ (Equation 5.5), the singular displacement behavior is represented by $r^{+0.5}$. The displacements u_i go to zero for $r \rightarrow 0$ at the crack tip, as should be expected. For large values of r, the displacements u_i increase, but it should be recalled here that Equation (5.27) is asymptotically correct for r-values much smaller than the crack length, and not for large r-values. The function $f_i(\theta)$ contains the Poisson's ratio, v. More important, the function $f_i(\theta)$ is different for plane stress and plane strain. At the crack tip, an interesting case is the vertical displacement of the crack edge which is obtained for $\theta = 180°$. Substitution in the relevant $f_i(\theta)$ of Equation (5.27) gives

$$v = \frac{4K}{E}\sqrt{\frac{r}{2\pi}} \quad \text{(plane stress)} \qquad (5.28a)$$

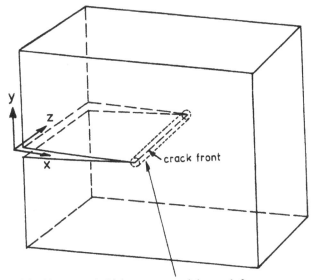

Material with extremely high stress around the crack front.
Contraction in the thickness direction (Z) is prevented by the
surrounding material, and thus ϵ_z is small ($\approx$ plane strain).

Fig. 5.19 Prevention of lateral contraction along crack front.

$$v = \frac{4K(1 - v^2)}{E} \sqrt{\frac{r}{2\pi}} \quad \text{(plane strain)} \qquad (5.28b)$$

The value of K is the same for both states of stress but displacements are
smaller for plane strain by a factor of $(1 - v^2)$, compare Equations (5.28a)
and (5.28b). The displacement equations (5.27) and (5.28) are also valid for
finite dimensions (mode I cracks). The finite dimensions are accounted for
by the geometry factor β in the stress intensity factor K.

It should be recalled that plane stress is characterized by $\sigma_z = 0$; no
stresses in the z-direction (= thickness direction), while plane strain refers
to $\varepsilon_z = 0$; zero strain in the z-direction. Because $\varepsilon_z = 0$ implies a constraint
on deformations in the xy plane, the material behaves as if it has an increased
elastic stiffness. The crack tip opening is lower by about 10% ($= v^2$).

In Section 5.8, it will be discussed that the effect of the state of stress on
crack tip plasticity is much larger. In view of that issue, the stress distribution
around a crack tip is considered again. For a through crack in a plate, see
Figure 5.19, σ_x and σ_y close to the crack front (very small r) are very high.
As a result, the material wants to contract in the lateral direction, which is the
z-direction or thickness direction. If the contraction could occur, the lateral
strain, ε_z, would theoretically be very large because of the very high stress

components and

$$\varepsilon_z = \frac{\sigma_z}{E} - \frac{\nu(\sigma_x + \sigma_y)}{E} \tag{5.29}$$

For plane stress $\sigma_z = 0$ and thus $\varepsilon_z = \nu(\sigma_x + \sigma_y)/E$. Close to the crack tip, ε_z becomes extremely large, causing an extreme contraction (singular behavior). However, the highly stressed material at a small distance of the crack front is surrounded by material that carries a much lower stress due to the larger r-value. The contraction in the surrounding material is relatively low. In view of the continuity of the material, the extremely high contraction for $r \to 0$ cannot occur. It will be restrained by the surrounding material. As a result, the material at the crack front can hardly contract, which implies that ε_z remains relatively small instead of becoming very large. The state of stress will thus be close to $\varepsilon_z = 0$, i.e. plane strain. A very high σ_z will then be present along the crack front, because $\varepsilon_z = 0$ implies:

$$\sigma_z = \nu(\sigma_x + \sigma_y) \tag{5.30}$$

At the plate surface (ends of the crack front) σ_z must go to zero (plane stress at the material surface). There is a thin surface layer where plane stress prevails. As a consequence, the singular behavior leads to a 3D stress problem. For a thick plate with a through crack, a fairly large part of the crack front is approximately in plane strain; whereas a relatively small part of the crack front at the material surface is approximately in plane stress. Only in a very thin sheet, the major part of the crack front can be considered to be in plane stress because lateral contraction cannot be successfully prevented. As will be pointed out below, crack tip plasticity promotes the tendency for plane stress situations.

5.8 Crack tip plasticity

According to the equation

$$\sigma_{i,j} = \frac{K}{\sqrt{2\pi r}} f_{i,j}(\theta) = \beta S \frac{\sqrt{\pi a}}{\sqrt{2\pi r}} f_{i,j}(\theta) \tag{5.31}$$

the stress becomes infinite for $r \to 0$. This would be a disaster if the material is fully brittle. High-strength structural materials very often have a low ductility, but they are not really brittle. As a result of the ductility, a small plastic zone will be created. The infinite peak stress is leveled off, see Figure 5.20b. It implies that the above equation can no longer be valid in the

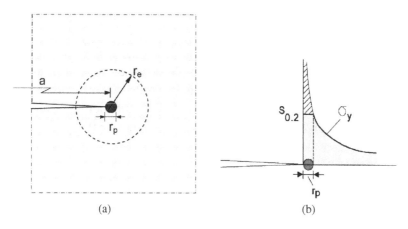

(a)   (b)

Fig. 5.20 K-dominated zone around plastic zone at crack tip.

crack tip plastic zone because the equation was based on an assumed elastic behavior. The question then is whether the small plastic zone completely destroys the meaning of K? Fortunately this is not the case. In Figure 5.20a a circle is drawn around the crack tip with radius r_e. The value of r_e is chosen small enough to ensure that Equation (5.31) is still approximately correct for $r = r_e$. But r_e is significantly larger than the size of the plastic zone. It implies that small plastic zones are considered here (small-scale yielding). If a small plastic zone (size r_p) is formed inside the r_e-zone, a certain redistribution of the stress will occur. However, if r_p is significantly smaller than r_e, then the stress redistribution on the periphery of the r_e-zone will not be significant. In other words; the stresses working on this zone are still well represented by Equation (5.31) if $r = r_e$ is substituted. As a consequence, the value of K should still give a meaningful indication of the severity of the stresses acting on the crack tip zone. The r_e-zone will be called the *K-dominated zone*.

Now compare two specimens, or a specimen and a component. Loads are applied to both parts. If the same K-values are applicable to both parts, then the same K-dominated zones will be present. This implies that the same stresses are applied on the crack tip region. As a consequence, the same cyclic plasticity and crack extension should occur in both parts. In view of the K-dominated zone concept a comparison can thus be made between fatigue crack growth in a laboratory specimen and a in component of a structure. Actually, crack growth in the laboratory specimen is used as a calibration result to predict the crack growth in a component. This similarity approach is addressed again in Section 5.11.

From the requirements $r_e < r_p$ and $r_p < a$, it will be clear that large plastic zones will invalidate the usefulness of K. Large plastic zones will be formed if the material has a low yield stress and if the load applied is relatively high. However, the size of the plastic zone is also dependent on the state of stress (plane stress or plane strain). A first estimate of the plastic zone size r_p can be made as follows. The stress distribution of σ_y along the x-axis is given by Equation (5.4b) with $f(\theta) = 0$ for $\theta = 0$:

$$\sigma_y = \frac{K}{\sqrt{2\pi r}} \tag{5.32}$$

This hyperbolic relation is shown in Figure 5.20b. A first estimate of r_p for a plane stress situation is obtained from $\sigma_y = S_{0.2}$. Substitution in Equation (5.32) gives

$$S_{0.2} = \frac{K}{\sqrt{2\pi r_p}} \rightarrow r_p = \frac{1}{2\pi} \left(\frac{K}{S_{0.2}} \right)^2 \tag{5.33}$$

This r_p estimate is generally supposed to be an underestimate, because it ignores the leveled off part of the stress distribution (shaded area in Figure 5.20b). A better estimate has been proposed in the literature to be twice the value of Equation (5.33):

$$r_p = \frac{1}{\pi} \left(\frac{K}{S_{0.2}} \right)^2 \quad \text{(plane stress)} \tag{5.34}$$

For plane strain, an estimate can be made by adopting a yield criterion, e.g. the von Mises criterion. Because plane strain implies a constraint on lateral contraction, the effective yield stress is substantially higher (substitute $\sigma_z = \nu(\sigma_x + \sigma_y)$ in the von Mises criterion). As a result, the plastic zone is significantly smaller. The relation frequently quoted in the literature is

$$r_p = \frac{1}{3\pi} \left(\frac{K}{S_{0.2}} \right)^2 \quad \text{(plane strain)} \tag{5.35}$$

This r_p is three times smaller than for plane stress. There are several critical comments to be made on the two r_p equations. The equations suggest that the plastic zone has a single dimension (r_p) only. The shape of the zone is not considered. Lines of a constant von Mises stress suggest a butterfly shape, but that is again a result of an elastic analysis. Another approach used in the literature is that the shape is a thin strip along the x-axis (Dugdale yield strip, [15]). As shown by surface deformations in Figure 5.21, the real

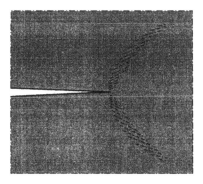

Fig. 5.21 Plastic zones at tip of a crack observed on the surface of an aluminium specimen with a polished surface.

shape is that plastic deformation is fanning outwards from the tip of the crack along two symmetric zones. A second aspect about r_p estimates is that the yield stress adopted is usually $S_{0.2}$, but of course the $S_{0.1}$ yield stress could also be used which would lead to larger plastic zone estimates. In spite of these shortcomings, the proportionality in Equtions (5.34) and (5.35)

$$r_p \propto \left(\frac{K}{S_{0.2}} \right)^2 \tag{5.36}$$

is approximately correct, and the equations can be used for an estimate of the plastic zone size. A numerical example: Assume $S = 100$ MPA, $S_{0.2} = 400$ MPA, and $a = 10$ mm. It gives $r_p = 0.6$ mm and $r_p = 0.2$ mm for plane stress and plane strain respectively. If r_p is much smaller than the material thickness (e.g. $t = 10$ mm), lateral contraction at the crack front will be largely restrained, and plane strain is applicable at the above stress level. However, in a thin sheet (e.g. $t = 1$ mm), lateral contraction in the plastic zone cannot effectively be restrained, and the crack tip will be largely in plane stress.

5.9 Some energy considerations

If a plate is loaded by an increasing stress S, see Figure 5.22, it becomes longer. The work done by the load is stored in the plate as potential energy, also referred to as elastic strain energy. The load is $P = S(Wt)$ and the elongation of the plate $\delta = (S/E)H$. The elastic energy, U, thus becomes

$$U = \frac{1}{2} P\delta = \frac{1}{2} \cdot SWt \cdot \frac{S}{E} H = \frac{1}{2} \frac{S^2}{E} \cdot HWt \tag{5.37}$$

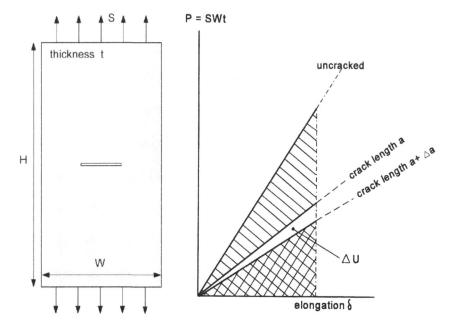

Fig. 5.22 Elastic response of plate without crack and with crack.

HWt is the volume of the plate, and $\frac{1}{2}\,S^2/E$ the strain energy per unit volume (i.e. the strain energy density). The stiffness of a plate with a crack is lower, and less energy will be stored in the plate at the same elongation, see Figure 5.22. In other words, if a crack is made in the plate and the elongation is not changed, relaxation of the elastic strain energy occurs.

If a crack in the same specimen is extended with a small increment Δa, again without a change of the length of the specimen, a small relaxation of the strain energy, ΔU, will occur. The calculation of ΔU in terms of the stress intensity factor is possible [4], because the crack edge displacements in the crack tip area (Equation 5.28) are known, as well as the relaxed stresses in the same area (Equation 5.4). The calculation leads to

$$\Delta U = \frac{K^2}{E^*} \cdot \Delta a \tag{5.38}$$

with $E^* = E$ for plane stress and $E^* = E/(1 - \nu^2)$ for plane strain. The so-called *strain energy release rate* becomes

$$\frac{\mathrm{d}U}{\mathrm{d}a} = \frac{K^2}{E^*} \tag{5.39}$$

The incremental strain energy tries to extend the crack length at the crack tip. For this reason, the strain energy release rate is also referred to as the *crack driving force*. Sometimes this term is even used for K which seems to be less correct. The strain energy release rate is used in FE calculations on K-values, see next section.

It is of historical interest that Griffith in 1924 used the concept of strain energy release to predict the static strength of glass [16]. He assumed that glass was a brittle material with crack-like defects. The potential energy increment (dU) had to provide the energy increment for the new crack surfaces of the incremental crack extension. The strength was thus associated with surface energy. However, for fatigue crack growth during cyclic loading the energy required for crack tip plastic deformation is a multitude of the surface energy. The Griffith approach cannot be applicable.

5.10 Determination of stress intensity factors

The first approach to obtain K-values should be to consult the literature, e.g. the handbooks mentioned earlier [3–5]. Sometimes it is possible to make clever estimates by evaluating available data for "similar" geometries. Comments on edge cracks and equivalent crack depth made in Section 5.4 can be helpful. In other cases, superposition arguments can be used, see Section 5.5. However, when accurate K-values are required, calculations or measurements have to be made. There is a large variety of FE techniques available which are outside the scope here. If calculations are made with the FE technique a very fine mesh is required around the crack tip in view of the large gradients of stress and strain. Close to the crack tip, calculated stresses or displacements should be compared to the results obtained with the equations $\sigma_{i,j} = K/\sqrt{(2\pi r)} \cdot f_{i,j}(\theta)$ or $u_i = (K/G)\sqrt{(r/2\pi)} \cdot f_i(\theta)$ (Equations 5.4 and 5.28 respectively). The value of K follows from such comparisons. The accuracy can be limited because in the equations the behavior is singular for $r \rightarrow 0$. Special crack tip elements have been developed which have the singular behavior as a property of the element. Boundary elements techniques can also be used.

A different calculation method is based on the strain energy release rate $dU/da = K^2/E$ (Equation 5.39). It implies that U has to be calculated for some values of "a" in order to determine dU/da. In FE models the crack length is increased by uncoupling of nodes along the crack line. Experience has shown that an acceptable accuracy can be obtained without the necessity to go to an extremely fine mesh.

The above comments are concerned with plane 2D problems. The calculation techniques are significantly more complex for 3D cases with curved crack fronts. Calculations are possible, but extensive computer capacity may be required. Experts are needed to program the computer codes. Validations of the computer programs is not always easy.

If computations offer problems, the alternative is measurements. However, again in view of the large stress and strain gradients, measurements are also not a simple alternative. Similar to comments made on obtaining stress concentrations factors (K_t), calculations are probably the most expedient solution. Special strain gages have been developed [17] to measure the K-value at the point where the crack front is coming to the free surface. The K gages must be bonded on the crack tip.

A fundamentally different method to obtain K-values is based on fatigue crack growth experiments. The crack growth rate is measured, and the K-value can then be deduced from an empirical da/dN-ΔK relation obtained on simple specimens of the same material. An example is discussed in Chapter 8.

5.11 The similarity concept and the application of the stress intensity factor K

The stress intensity factor is a measure of the severity of the stress distribution around the tip of a crack. A fatigue load on a cracked specimen introduces a cyclic stress intensity at the crack tip varying between $K_{\max}$ and $K_{\min}$. The range between the two extremes is

$$\Delta K = K_{\max} - K_{\min} = \beta S_{\max}\sqrt{\pi a} - \beta S_{\min}\sqrt{\pi a}$$

$$\text{or:} \quad \Delta K = \beta \Delta S \sqrt{\pi a} \tag{5.40}$$

The so-called stress ratio R, defined as $S_{\min}/S_{\max}$, also applies to the corresponding K-values:

$$R = \frac{S_{\min}}{S_{\max}} = \frac{K_{\min}}{K_{\max}} \tag{5.41}$$

A cyclic stress intensity defined by $K_{\max}$ and $K_{\min}$ is also defined by ΔK and R. As discussed before, the results of a fatigue crack growth experiment under constant-amplitude (CA) loading can be expressed as a crack growth curve (*a*-*N*), or as the crack growth rate da/dN as a function of the crack length, see Figure 5.6. For each crack length, the stress intensity range ΔK can be calculated (Equation 5.40), and the results can then be presented as da/dN-ΔK curves. Such results are shown in Chapter 8 on crack propagation data.

It appears to be fully correct to assume that the crack growth rate, which is the crack extension in one cycle, is a function of K_{max} and K_{min}, or, which is the same statement, a function of ΔK and R:

$$\frac{da}{dN} = f(\Delta K, R) \tag{5.42}$$

This function should be characteristic for the fatigue crack growth properties of a material. It represents the fatigue crack growth resistance of a material under specified ΔK and R conditions. If the crack growth is considered in another specimen or in a structural component, crack growth can then be predicted, based on the similarity concept, which says:

Similar cyclic conditions (ΔK and R) applied to fatigue cracks in different specimens or structures of the same material should have similar consequences, i.e. similar crack extensions per cycle, thus the same da/dN.

In the literature, the similarity principle is also referred to as the similitude approach. The similarity principle is a physically sound concept which is abundantly used in many prediction problems. However, a basic problem to be considered is the question, whether the conditions for crack growth in the experiments and in a structure are physically similar, a requirement which should be satisfied for predictions.

In Section 5.8 it was already discussed that crack tip plasticity can affect the similarity depending on the state of stress. Plane strain conditions occurring in thick plates are causing relatively small plastic zones, whereas plane stress conditions in thin sheet material are causing larger plastic zones. In other words, similar K conditions in thick and thin plates will produce different plastic zone sizes. Different crack growth rates should then be expected. Such a thickness effect on fatigue crack growth has been observed indeed. Apparently, the similarity condition "the same material" is not sufficient. The state of stress should also be comparable. A related aspect is the occurrence of shear lips as mentioned in Section 5.2 (see also Chapter 2, Figure 2.38). Shear lips affect the shape of the crack front, which can invalidate the similarity. In later chapters on fatigue crack growth (Chapters 8 and 11), it will turn out that the similarity has to be considered again in order to evaluate the reliability of prediction methods. The question is: Are fatigue crack growth data obtained with simple specimens sufficiently representative for fatigue cracks in structures under service load spectra?

5.12 Main topics of the present chapter

1. The stress intensity factor $K = \beta S \sqrt{\pi a}$ gives an indication of the stress severity around the tip of a crack. S accounts for the stress level, a for the crack length, and the geometry factor β for the shape of the specimen or structure. The equations presented for stresses and displacements in the crack tip area are valid only at a relatively short distance from the crack tip. The stress intensity factor K is essentially an elastic concept. The singular character of the equations $(r^{-1/2})$ leads to infinite stresses at the crack tip, which causes plastic deformation at the crack tip. As long as the plastic zone is relatively small, the stress intensity factor still gives a good indication of the stress system acting on the K-dominated zone around the crack tip. K can thus still be used for two purposes: (i) to describe the fatigue crack growth resistance properties of a material, usually as $da/dN = f(\Delta K, R)$; and (ii) to use such data for predictions on fatigue crack growth in other specimens and structures.

2. A large amount of data on K-values (actually β-values) is available in the literature and in software packages, but it is generally related to well defined shapes, and not to more complex geometries. First estimates can be obtained by considering more simple geometries, in particular for small edge cracks. For large cracks at notches an effective crack length including the notch size can also yield good K estimates. More accurate K-values can be obtained by calculations. Part-through cracks and curved crack fronts offer 3D problems with K-values varying along the crack front. Expertise on such calculations and substantial computer capacity are then required.

3. The state of stress conditions at the tip of a crack can vary from plane strain $(\varepsilon_z = 0)$ to plane stress $(\sigma_z = 0)$. If lateral contraction along the crack front is difficult, plane strain will prevail. At the material surface plane stress applies. The state of stress has a significant influence on the

size of the plastic zone. Smaller plastic zones occur under plane strain. Crack tip plasticity allows more lateral contraction and thus promotes plane stress.

4. The application of the K factor to fatigue cracks is based on the similarity concept. An essential question to be considered is whether the crack tip conditions of the laboratory specimen and the structural part are sufficiently similar to allow predictions based on relevant K-values.

References

1. Irwin, G.R., *Analysis of stresses and strains near the end of a crack traversing a plate*. Trans. ASME, J. Appl. Mech., Vol. 24 (1957), pp. 361–364.
2. Westergaard, H.M., *Bearing pressures and cracks*. J. Appl. Mech., Vol. 6 (1939), pp. A-49 to A-53.
3. Rooke, D.P. and Cartwright, D.J., *Stress Intensity Factors*. Her Majesty's Stationary Office, London (1976).
4. Tada, H., Paris, P.C. and Irwin, G.R., *The Stress Analysis Handbook*, 2nd edn. Paris Productions Inc., St. Louis (1985).
5. Murakami, Y. (Ed.), *Stress Intensity Factors Handbook*. Pergamon Press, Oxford (1987).
6. *Standard Test Method for Measurement of Fatigue Crack Growth Rates*. ASTM Standard E647-91a (1991).
7. Feddersen, C.E., *Discussion in ASTM STP 410* (1966), pp. 77–79.
8. Newman Jr., J.C., *An Improved Method of Collocation for the Stress Analysis of Cracked Plates with Various Shaped Boundaries*, NASA Report TN D-6376 (1971).
9. Swift, T., *The effects of stress level, geometry, and material on fatigue damage tolerance of pressurized fuselage structure*. Plantema Memorial Lecture, Proc. 14th ICAF Symp., Ottawa, EMAS (1987), pp. 177.
10. Tweed, J. and Rooke, D.P., *The distribution of the stress near the tip of a radial crack at the edge of a circular hole*. Int. J. Engrg. Sci., Vol. 11 (1973), pp. 1185–1195.
11. Schijve, J., *Fatigue specimens for sheet and plate material*. Fatigue Fract. Engng. Mater. Struct., Vol. 21 (1998), pp. 347–357.
12. Sneddon, I.N., *The distribution of stress in the neighbourhood of a crack in an elastic solid*. Proc. Roy. Soc. London, A, Vol. 187 (1946), pp. 229–260.
13. Irwin, G.R., *The crack extension force for a part-through crack in a plate*. Trans. ASME, J. Appl. Mech., Vol. 29 (1962), pp. 651–654.
14. Raju, I.S. and Newman, J.C., *Stress-intensity factors for a wide range of semi-elliptical surface cracks in finite-thickness plates*. Engrg. Fracture Mech., Vol. 11 (1979), pp. 817–829.
15. Dugdale, D.S., *Yielding of steel sheets containing slits*. J. Mech. Phys. Solids, Vol. 8 (1960), pp. 100–104.
16. Griffith, A.A., *The theory of rupture*. Proc. First Int. Conf. for Applied Mechanics, Delft 1924. C.B. Biezeno and J.M. Burgers (Eds.), Waltman, Delft (1925).
17. Miyake, S., Nawa, Y., Kondo, Y. and Endo, T., *Application of the "K-gage" to aircraft structural testing*. Proc. 15th ICAF Symposium, Jerusalem, June 1989, A. Berkovits (Ed.). EMAS Warley (1989), pp. 369–394.

Some general references (see also [3–5])

18. Sanfor, R.J. (Ed.), *Selected papers on Foundations of linear elastic fracture mechanics.* SEM Classic papers, Vol. CP1, SPIE Milestone Series, Vol. MS 137 (1997).
19. Anderson, T.L., *Fracture Mechanics: Fundamentals and Applications*, 2nd edn. CRC Press (1995).
20. Nisitani, H. (Ed.), *Computational and experimental fracture mechanics.* Computational Mechanics Publications, Southampton (1994).
21. Carpinteri, A., *Handbook of Fatigue Crack Propagation in Metallic Sstructures.* Elsevier, Amsterdam (1994).
22. Broek, D., *Elementary Engineering Fracture Mechanics*, 4th edn.). Martinus Nijhoff Publishers, the Hague (1985).
23. Ewalds, H.L. and Wanhill, R.J.H., *Fracture Mechanics.* Edward Arnold (1983).
24. Sih, C.M., *Handbook of Stress Intensity Factors.* Lehigh University (1973).
25. Paris, P.C. and Sih, C.M., *Stress analysis of cracks.* ASTM STP 381 (1965), pp. 30–83.

Chapter 6
Fatigue Properties

6.1 Introduction

In the present chapter, fatigue properties of materials are described in terms
of the fatigue limit, fatigue curves (S-N curves) and a fatigue diagram.
The properties are restricted to results of constant-amplitude (CA) tests on
unnotched specimens ($K_t = 1.0$) It is generally thought that the results
of these tests reflect the basic fatigue behavior of a material. Mechanical
properties of a material should include fatigue properties, but quite often
reporting of fatigue properties is restricted to the fatigue limit on unnotched
specimens obtained in rotating beam experiments ($S_m = 0$).

   If fatigue has to be considered as part of the design analysis of a structure,
it is well recognized that a stress cycle is characterized by a stress amplitude
(S_a) and a mean stress (S_m), see Figure 6.1. Instead of S_a and S_m, a second
equivalent definition is given by S_{max} and S_{min}, while a third one uses the
stress range $\Delta S \, (= 2S_a)$ together with the stress ratio R, defined as

$$R = \frac{S_{min}}{S_{max}} \qquad (6.1)$$

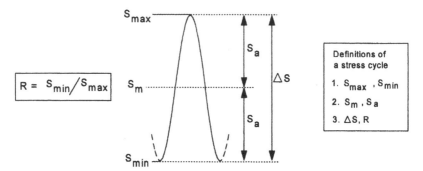

Fig. 6.1 Characteristic stress levels of load cycle.

Which definition of a stress cycle should be preferred? From a fatigue mechanistic point of view the obvious choice is S_{max} and S_{min}. Those are the stress levels at which the loading direction is reversed, and thus cyclic slip is reversed. Crack extension in a cycle stops at S_{max}. However, in service, a structure quite often is carrying a stationary load with a superimposed cyclic load. The stationary load can be a result of the weight of the structure, cargo, etc. The stationary load accounts for the mean stress, whereas loads in service induce cycles with certain stress amplitudes. If the severity of the cyclic load spectrum can be reduced, the S_a-values becomes smaller but S_m remains the same. Another situation occurs if a designer wants to increase the fatigue life by reducing the design stress level, e.g. by increasing the cross section of the fatigue critical area. All stress levels are then reduced with the same ratio. In other words, the stress ratio R remains constant, but ΔS is reduced. All three definitions of a stress cycle are used in the literature on fatigue.

In addition to the stress levels, another characteristic involved in the definition of a stress cycle is the *wave shape of the cycle*. The influence of the wave shape was briefly touched upon in Section 2.5.7. Effects of the wave shape and cyclic loading rate (i.e. the frequency in cycles per minute) can be important if a time dependent phenomenon is affecting the fatigue mechanism. It can be corrosion or creep, and also diffusion mechanisms in the material. Such effects are not yet considered in the present chapter.

The various ways to describe fatigue properties of a material are discussed in this chapter. It includes the fatigue limit, S-N curves and fatigue diagrams, all related to unnotched specimens. This is done in Section 6.2. General aspects of the fatigue strength are discussed in Section 6.3 which includes the correlation of the fatigue limit for $S_m = 0$ with the material ultimate tensile stress, the effect of the mean stress, different types of loading (tension,

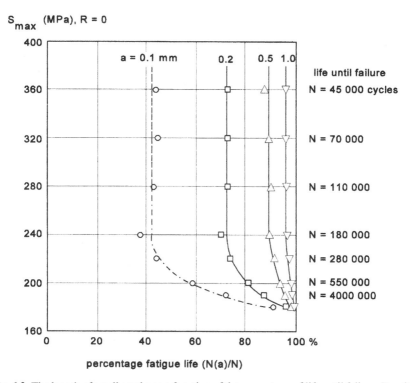

Fig. 6.2 The length of small cracks as a function of the percentage of life until failure. Results of unnotched specimens of an Al-alloy, 2024-T3 [1].

bending, torsion) and also combined loading. Low-cycle fatigue is addressed in Section 6.4. The main items of the present chapter are summarized in Section 6.5. The fatigue strength of notched specimens and predictions on the fatigue strength of notched elements are discussed in Chapter 7.

6.2 Description of fatigue properties of unnotched material

Fatigue properties of unnotched specimens are generally supposed to be material properties, such as S-N curves (fatigue lives until failure N), or the fatigue limit defined as the horizontal asymptote of an S-N curve. This information is coming from fatigue tests carried out until failure, or until a very high number of cycles if failure does not occur, e.g. 10^7 cycles. Observations on crack growth are not included. In Chapter 2, it was discussed that microcracks are nucleated early in the fatigue life which implies that the fatigue life covers two phases: (i) an initiation period including microcrack growth, and (ii) a crack growth period with macro crack growth, see Figure 2.1. The second period is relatively short for unnotched specimens. Illustrative data are given in Figure 6.2 for unnotched specimens of an Al-alloy. Cracks as small as 0.1 mm were detected during continuous observations with binocular microscopes. For the higher fatigue stress levels, a crack of 0.1 mm occurred at about 40% of the fatigue life until failure. However, at 95% of the life the crack was still small, in the order of 1.0 mm. Such cracks cannot be seen with the unaided eye. In other words; the life until failure is only slightly larger than the crack initiation life, and practically almost the same.

The results in Figure 6.2 show another noteworthy trend. For low fatigue stress levels with fatigue lives in the order of 10^6 cycles and more, the curves for a constant crack length values go to 100% fatigue life. This illustrates that it becomes more difficult for microcracks to grow until failure if the stress level goes down to the fatigue limit. It confirms the threshold character of the fatigue limit.

A similar trend was already observed in the discussion on Figure 2.22 (Section 2.5.5) with the results of rotating beam specimens of mild steel. It was shown that the fatigue life until a crack of 2.5 mm was large if compared to the remaining life from 2.5 mm until failure. This was more obvious for the lower stress amplitude and the longer fatigue life. Although 2.5 mm is too large to be a microcrack, the trends agree with those of Figure 6.2 for the Al-alloy. It then seems reasonable to use S-N data of unnotched specimens for predictions on the crack initiation life of notched elements.

The S-N curve

An S-N curve, also called a Wöhler curve, is obtained as a result of a number of fatigue tests at different stress levels. An example of such results is given in Figure 6.3 for unnotched specimens of a CrMo steel (SAE 4130). In this tests $S_m = 0$, and thus the stress ratio is $R = S_{min}/S_{max} = -1$. The variable in Figure 6.3 is the stress amplitude S_a. The fatigue life N is usually plotted on a logarithmic scale. In the literature, the stress amplitude is presented both on a linear scale and on a logarithmic scale. Here, it is preferred to adopt the log scale because it frequently leads to an approximately linear relation between $\log S_a$ and $\log N$ for a substantial range of N-values. Mathematically, this linear relation can be written as

$$S_a^k N = \text{constant} \qquad (6.2)$$

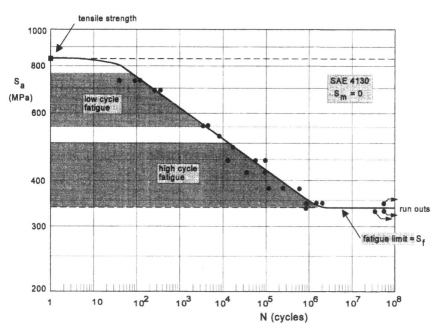

Fig. 6.3 Fatigue test results of unnotched specimens of a low-alloy steel (SAE 4130) [2].

The equation is known as the Basquin relation. The slope of the linear part is equal to $-1/k$. Some comments on Figure 6.3 should be made:

(i) Experiments carried out at the same S_a do not give the same fatigue life. Scatter between results of similar tests occurs. Three experiments at low stress amplitudes were stopped at 2.5 or 5×10^7 cycles without failure (so-called run outs). Scatter is the subject of Chapter 12.

(ii) The number of test results in Figure 6.3 is 25. The tests were run at a loading frequency of 30 Hz. Testing time for the experiments in the fatigue machine is in the order of 60 full days. Obviously, the simple S-N data in Figure 6.3 require an expensive test program.

(iii) In Figure 6.3 the lower horizontal asymptote is the fatigue limit S_f. However, a second horizontal asymptote occurs at the upper side of the S-N curve. If $S_{max} = S_U$ (the tensile strength of the material), the specimen will fail in the first cycle as in a tensile test. For $S_m = 0$, this occurs if $S_a = S_U$, and for $S_m > 0$ if $S_a + S_m = S_U$. However, if S_a is slightly smaller, the specimen does not fail in the first cycle. Apparently, the specimen can then survive many cycles in the order of 100 or even more which is a result of strain hardening, see Figure 6.4. In the first uploading half cycle from O to A, a large plastic strain

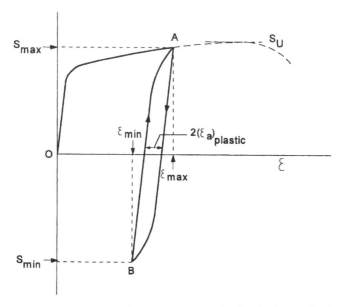

Fig. 6.4 Stress-strain loop of a high-stress amplitude cycle after the first application of S_{max}.

deformation occurs because S_{max} is exceeding the yield stress. But the following unloading to B and subsequent reloading to A cause a much smaller plastic strain amplitude due to plastic strain hardening of the material, see the hysteresis loop in Figure 6.4. The hysteresis loop can be sustained a substantial number of cycles before microcracking leads to failure. As a result, an upper horizontal asymptote is found. The situation is different if the specimen is not loaded under a constant stress amplitude, but under a constant strain amplitude as discussed in Section 6.4.

Fatigue at high amplitudes and fatigue lives up to some 10^4 cycles is called *low-cycle fatigue* (or high-level fatigue), see Figure 6.3. If fatigue covers a large number of cycles, say 10^5 cycles or more, it is called *high-cycle fatigue* (or low-level fatigue). The boundary between low and high-cycle fatigue is not exactly defined by a specific number of cycles. The more relevant difference between the two conditions is that low-cycle fatigue is associated with macroplastic deformation in every cycle. High-cycle fatigue is more related to an elastic behavior on a macro scale of the material. Actually, high-cycle fatigue is the more common case in practice, whereas low-cycle fatigue is associated with specific structures and load spectra. In Section 6.4, attention is paid to low-cycle fatigue as a material phenomenon. The topic returns in some later chapters.

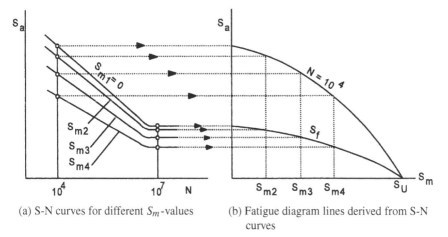

(a) S-N curves for different S_m-values

(b) Fatigue diagram lines derived from S-N curves

Fig. 6.5 Fatigue diagram as a cross plot of S-N curves.

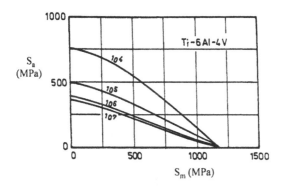

Fig. 6.6 Fatigue diagram of a Ti-alloy with lines for a constant fatigue life.

Fatigue diagrams

The mean stress for the S-N curve in Figure 6.3 is $S_m = 0$. Different S-N curves are obtained if fatigue tests are carried out at other S_m-values, see Figure 6.5a. A higher mean stress will give a lower S-N curve. Cross plots can now be made to arrive at a fatigue diagram with lines for a constant fatigue life as illustrated in Figure 6.5b for $N = 10^4$ and for the fatigue limit. Lines for other fatigue lives can be drawn in a similar way in the same diagram. An example of such a fatigue diagram is given in Figure 6.6 for a Ti-alloy. All lines for a constant N are converging to the same point on the S_m axis, $S_m = S_U$ for $S_a = 0$ (i.e., no cyclic stress) which theoretically should be expected. Fatigue diagrams generally suggest that the effect of S_m is not

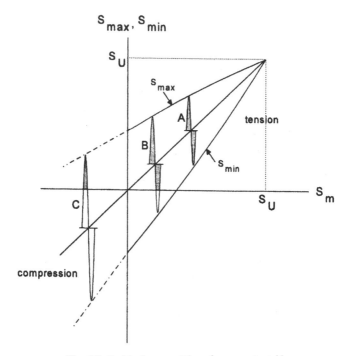

Fig. 6.7 Smith diagram. Lines for a constant N.

large, especially if N is high. It implies the general experience that *the stress amplitude S_a has a much larger effect on fatigue than the mean stress.* The trend reflects that fatigue is primarily a consequence of cyclic loads.

The fatigue diagram in Figure 6.6 is usually referred to as a Goodman diagram. Another way to present such data is offered by the Smith diagram where both S_{max} and S_{min} are plotted as a function of the mean stress S_m, see Figure 6.7. The two lines apply to one specific fatigue life, in many cases to a high fatigue life, e.g. 10^7 cycles, in order to represent the fatigue limit. Three load cycles for different S_m-values are indicated, A, B and C. In case A for a positive mean stress both S_{max} and S_{min} are positive. However, in case B for a lower S_m the minimum stress is negative (compression) although the larger part of the load cycle occurs still in tension. In case C with a negative mean stress, the larger part of the load cycle occurs in compression. A fatigue crack is closed under compression, and the negative part of the cycle should thus be expected to be non-damaging. Reversed loading is significant in view of reversed (crack tip) plasticity, but crack opening is necessary for crack extension and for reversed plasticity in the crack tip plastic zone. If a negative S_m is present, crack opening requires a larger S_a. In practice, it implies that fatigue is rarely a problem for a negative mean stress. As a consequence, fatigue diagrams are usually given for positive mean stresses only.

Perhaps it should be recalled from Chapter 2 that compressive cyclic stresses can produce microcracks at the material surface as a result of cyclic slip. Cyclic slip primarily depends on the shear stress amplitude, and not on the tensile stress along the surface of the material. Some microcracking in slip bands at the material surface can occur. Under a cyclic compressive fatigue load, these cracks are not effectively opened at S_{max}. As a result, microcracks will be non-propagating.

6.3 Some general aspects of the fatigue strength of unnotched specimens

Extensive literature is available on the fatigue strength of a material and how it is affected by various characteristics of the material and testing conditions. Some general aspects covered in this section are primarily discussed in relation to the effects on the fatigue limit. These effects are significant for the nucleation period and less important for the crack growth period. With reference to Figure 6.3, it can also be said that these effects are larger for *high-cycle fatigue* and relatively small for *low-cycle fatigue*. This is especially true for surface effects as discussed already in Section 2.5.5, see Figure 2.23.

Some classical topics associated with fatigue properties are:

- Relation between the fatigue limit, usually S_f for $S_m = 0$, and the strength of the material, in general the ultimate tensile strength, S_U.
- Mean stress effect.
- Size of the unnotched specimen.
- Type of loading; tension, bending, torsion.
- Combined loading, e.g. tension and torsion, or bending and torsion.
- Low-cycle fatigue.

6.3.1 Relation between S_f and S_U

An old idea is that the fatigue limit S_f can be increased by raising the strength of a material, either by the chemical composition of the alloy, or by a heat treatment which increases the hardness. Results for different C-steels and low-alloy steels were already presented in Section 2.5.2. Figure 2.11 illustrates a systematic increase of S_f for an increasing S_U, but a good deal of scatter is present. Similar graphs are presented here in Figures 6.8 for cast iron, Al-alloys and Ti-alloys. They show a similar proportionality between S_f and S_U again with considerable scatter. If the proportionality is written as

$$S_f = \alpha S_U \tag{6.3}$$

then Figures 2.12 and 6.8 indicate $\alpha \approx 0.5$ for steel, cast iron and Ti-alloys, but a lower value $\alpha \approx 0.35$ applies to the Al-alloys. In comparison to the tensile strength, Al-alloys are more fatigue sensitive. The α-value can be adopted to make a first estimate of S_f for unnotched material ($K_t = 1$) at $S_m = 0$. In view of the scatter shown in Figures 2.12 and 6.8, Equation (6.3) gives a first estimate only. Moreover, it should be kept in mind that the fatigue limit of unnotched specimens with $S_m = 0$ is not necessarily a good measure for the fatigue resistance of a material. It does not give an indication of the fatigue sensitivity if notches are present. This problem is discussed in Chapter 7.

It is noteworthy that the fatigue limit in Equation (6.3) is related to the tensile strength. It seems to be more logical to relate S_f to the yield stress $S_{0.2}$. The tensile strength is depending on the strain hardening of the material after substantial plastic deformation has occurred. The yield stress is more characteristic for the small plastic strain behavior. Although an increased S_U is usually attended by a higher $S_{0.2}$, the correlation between the yield stress and the tensile strength is not a proportional relation. The origin of the relation $S_f = \alpha S_U$ is of a historical nature, but the reader should be aware that its physical meaning is limited.

6.3.2 Mean stress effects

If S_m is increased and S_a remains the same, then S_{max} becomes larger. As a result, a larger stress is present to open microcracks or macrocracks. A shorter fatigue life and a lower fatigue limit should thus be expected as illustrated by the fatigue diagram in Figure 6.6. Two simple equations have been proposed for the constant N lines in a fatigue diagram which in the literature are labeled as the modified Goodman relation and the Gerber parabola, see Figure 6.9. As should be expected, the fatigue strength is reduced to zero if the mean stress is increased to the ultimate tensile stress S_U. Any small S_a cycle at $S_m \approx S_U$ should immediate lead to failure

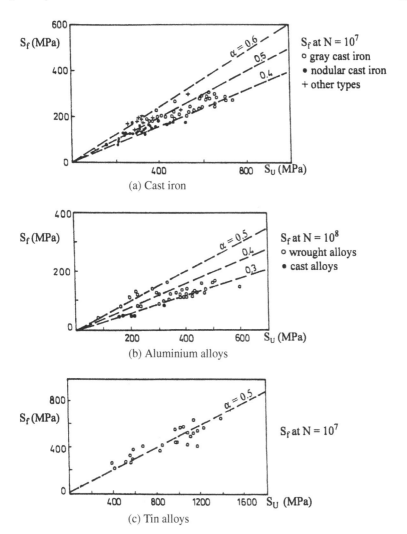

Fig. 6.8 Correlation between the fatigue limit and the tensile strength for different materials, see Figure 2.11 for steel [4].

because $S_{max} > S_U$. The modified Goodman relation assumes a linear decrease of the fatigue strength for an increasing S_m. In many cases this approximation is conservative which is true for the Ti-alloy in Figure 6.6. However, exceptions occur, especially for high-strength alloys with a low ductility, see Figure 6.10 for AISI-4340 steel heat treated to the very high S_U of 1830 MPa. The fatigue strength drops more rapidly then according to the modified Goodman relation. This may well be due to the presence of small inclusions as discussed in Section 2.5.2.

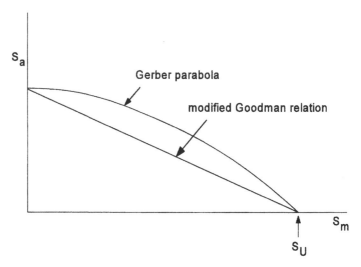

Fig. 6.9 Two approximations for constant N lines in a fatigue diagram.

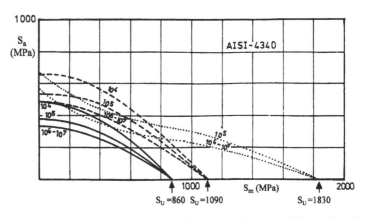

Fig. 6.10 Fatigue diagram for a low-alloy steel heat treated to three different S_U-values [3].

The Gerber parabola has its vertical axis along the S_a axis ($S_m = 0$). The parabola also passes through $S_m = S_U$ for $S_a = 0$. Denoting the fatigue strength for a certain fatigue life by S_N, the Gerber parabola is

$$\frac{S_N}{(S_N)_{S_m=0}} = 1 - \left(\frac{S_m}{S_U}\right)^2 \tag{6.4}$$

For the Ti-alloy in Figure 6.6 the Gerber parabola is unconservative. But for AISI4340 the parabola is a reasonable approximation in Figure 6.10 for the alloy heat treated to lower S_U-values. In general, the Gerber parabola reflects the effect of the mean stress better for ductile[11] materials and high N-values, thus also for the fatigue limit S_f. However, high-strength alloys with a low ductility are more sensitive for the mean stress. Schütz [5] analyzed this sensitivity for a variety of materials for which the fatigue limit was available at $S_m = 0$ and at $R = 0$ ($S_a = S_m$). With these two fatigue limits, he defined the mean stress sensitivity as the initial average slope of constant N lines, see Figure 6.11:

$$M = \tan \varphi = \frac{(S_N)_{S_m=0} - (S_N)_{R=0}}{(S_N)_{R=0}} \qquad (6.5)$$

A larger M implies a higher S_m sensitivity. Schütz collected data for unnotched and notched specimen and for different N-values including a high N-value associated with the fatigue limit. A systematic S_m sensitivity was observed for different groups of materials. The average curves of Schütz are presented in Figure 6.11. It clearly shows the increased mean stress sensitivity for materials with a higher tensile strength.

6.3.3 The size effect for unnotched specimens

A size effect implies that larger specimens may have a lower fatigue strength. A size effect on the fatigue limit of unnotched specimens has indeed been observed in experimental programs. As an example results of fatigue tests are shown in Figure 6.12a for three steel grades. A higher S_f is found for rotating bending specimens with a smaller diameter. Another example for a Cr-steel is given in Figure 6.12b with the same trend for rotating bending. It also shows a significantly lower S_f for cyclic tension/compression ($S_m = 0$). As pointed out in Chapter 2, the fatigue limit is primarily a question of some specific weak spots for crack nucleation at the material surface (or just below the surface). It is a matter of scatter of favorable sites for microcrack nucleation. The probability of having such weak spots is larger for a larger material surface area carrying the maximum stress cycle. In other words; it sounds logical that a size effect is present, and that larger specimen will exhibit a lower fatigue limit. Also, the critical material surface area of an unnotched tension specimen loaded under cyclic tension/compression is relatively large if compared to a cantilever rotating bending specimen. It thus

---

[11] A material is referred to as being ductile if it allows substantial plastic deformation without failure. The ductility is associated with a reasonable elongation until failure in a static tensile test. The opposite of ductile is "brittle".

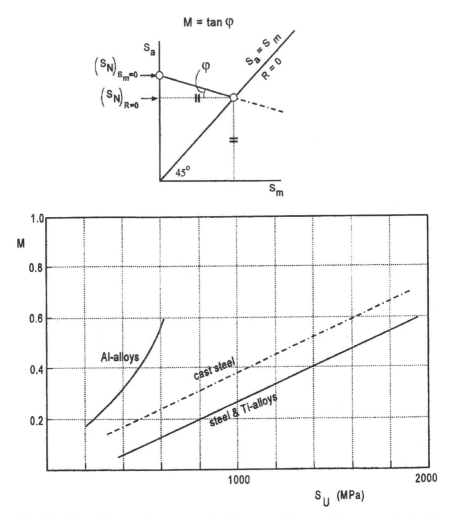

Fig. 6.11 Effect of the tensile strength on the influence of S_m on the fatigue strength S_N [5].

may be expected that the fatigue limit for cyclic tension in Figure 6.12b is lower than for cyclic bending.

Actually, the problem is more complex. Possible shapes for a flat and a cylindrical unnotched specimen are given in Figure 6.13. An unnotched specimen with $K_t = 1$ can theoretically be obtained in a prismatic specimen. However, specimens have to be clamped at both ends to transmit the load of the fatigue machine into the specimen. As a consequence, a stress concentration at the ends cannot be avoided. But the transition from the clamping area to the test section should be made as smooth as possible to

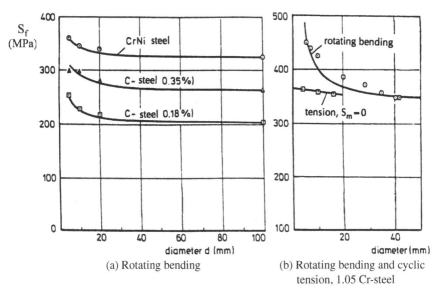

Fig. 6.12 Size effect on the fatigue limit of different types of steel [6, 7].

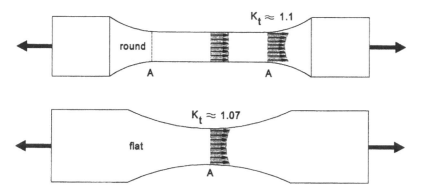

Fig. 6.13 Stress distributions in unnotched specimens.

keep K_t close to 1.0. The K_t-values for the two shapes shown in Figure 6.13 are 1.10 and 1.07 respectively. The round specimen in Figure 6.13 has a large cylindrical part, but the nominal stress in this part is about 10% lower than the peak stress in the two cross sections A. The peak stress in the flat specimen, $1.07S_{\text{nominal}}$, occurs in the minimum section A only. Fatigue failure will occur near this section A.

Another difference between the two specimens in Figure 6.13 should also be recognized. In the round specimen, the peak stress occurs all around the two cross sections A. However, in the flat specimen it occurs at the edges in cross section A. In the rectangular cross section, cracks often nucleate at the corners because the restraint on cyclic plasticity is minimal at those corners. It must be concluded that the fatigue strength, and in particular the fatigue limit is depending on *the size and the shape of the specimen*. It implies that the fatigue limit of unnotched specimens, which is considered to be a fundamental property of a material, is not really such a well defined property. The question may then be raised for which practical reasons should we be interested in fatigue properties of unnotched specimens? Several reasons for a limited significance of the fatigue limit can be mentioned.

(i)   *Material selection.* Because fatigue properties of unnotched specimens depend on the size and the shape of the specimen, material fatigue data of these specimens can only be used to obtain approximate ideas about the fatigue resistance of a material. But it should be kept in mind that basic data for unnotched specimens do not give indications about the fatigue notch sensitivity of a material.

(ii) *Comparative testing.* Because the fatigue properties of unnotched specimens depend heavily on the surface conditions of the specimens, comparative testing can certainly be done on unnotched specimen, e.g. to compare different surface treatments, e.g. nitriding of steel. Care should still be taken that the size and the shape of the specimen are sufficiently representative for the intended practical application of a surface treatment. In other words, a more relevant approach is to perform comparative tests on specimens with a notch geometry and a surface quality representative for the practical application.

(iii) *Prediction of fatigue properties of notched elements.* The old idea is that fatigue properties of notched elements can be predicted starting from the fatigue properties of unnotched specimens as basic material data. It assumes that the conditions for crack nucleation in notched elements and in unnotched specimens can be similar. In view of the preceding discussion on size and shape effects, this is no longer so obvious. Predictions of the fatigue properties of notched elements is a most relevant question for designing against fatigue. This topic is addressed in Chapter 7.

6.3.4 Type of loading, tension, bending, torsion

In the previous sections it was tacitly assumed that fatigue occurs under cyclic tension or cyclic bending. Fatigue under cyclic tension and under cyclic bending are not that much different. The critical stress of an unnotched

Table **6.1** Fatigue limit ratios.

Material	Mean value of τ_f/S_f
Steel	0.60
Al-alloys	0.55
Cu and Cu-alloys	0.56
Mg-alloy	0.54
Ti	0.48
Cast iron	0.90
Cast Al- and Mg-alloys	0.85

specimen in both cases is cyclic tension in the surface layer of the material. The stress gradient perpendicular to the material surface is different for tension and bending, but as discussed in Section 3.3 the more important stress gradients occur along the material surface.

The occurrence of fatigue under cyclic torsion was mentioned in Chapter 2, see Figures 2.31 and 2.32. Classical examples of fatigue under cyclic torsion are associated with axles and spiral springs. Nucleation of the first microcrack again occurs in slip bands carrying the maximum shear stress. This shear stress amplitude in an unnotched specimen loaded in tension is equal to half the tensile stress, or $\tau_a = S_a/2$. For an unnotched specimen, loaded under cyclic torsion, the maximum shear stress is equal to the shear stress on the specimen. As a first estimate, one might expect that the fatigue limit under cyclic torsion, τ_f, is half the fatigue limit S_f under cyclic tension in agreement with the Tresca yield criterion. However, as indicated in Figure 2.31, such slip bands under cyclic tension are also loaded by a tensile stress perpendicular to the slip bands, whereas this tensile stress is absent for cyclic torsion. As a result, the conversion of cyclic slip into a microcrack may be more difficult under cyclic torsion, and τ_f may be larger than $S_f/2$. Data from the book of Forrest [4] are given in Table 6.1.

The τ_f/S_f ratios for the wrought alloys, ignoring the value for Ti, are close to the ratio predicted by the Von Mises criterion ($0.577 = 1/\sqrt{3}$). Actually, physical reasons why that criterion should be applicable are questionable if crack nucleation is associated with cyclic slip. A noteworthy result is the high τ_f/S_f ratios for the cast alloys. Consider grey cast iron with the graphite flakes as the starter notches for microcracks. The flakes are depicted in Figure 6.14 as very flat defects with some random orientation. It implies that the maximum peak stress at the ends of the flakes will have an equivalent character for pure tension and for pure torsion. The fatigue limits S_f and τ_f might then be equal. The τ_f/S_f ratio of 0.9 is indeed close to 1.

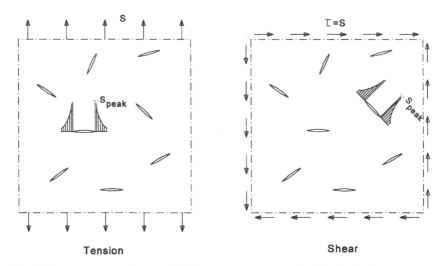

Tension   **Shear**

Fig. 6.14 Peak stress at flat material defects in cast iron (graphite flakes). Similarity between tension and shear, same peak stress.

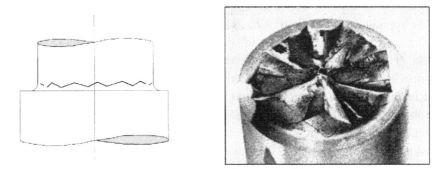

Fig. 6.15 Fatigue failure obtained under cyclic torsion ($\tau_a = 150$ MPa) in a round specimen with a fillet notch, material carbon steel (0.45%C). Factory roof type fracture surface [9].

Similarly, the high τ_f / S_f ratio of the cast light alloys may also be associated with defects in the material.

The mean stress under cyclic torsion, τ_m, should not be expected to have a significant influence on cyclic slip. If this is true, τ_m should not affect crack nucleation. As a consequence, the fatigue limit, τ_f, should hardly depend on the mean stress τ_m. Experiments, noteworthy by Smith [8], have confirmed this behavior. However, a problem arises as a finite life is involved. Fatigue cracks grow preferably in a direction perpendicular to the main principal stress. This leads to a spiral crack growth in a cylindrical bar as shown in Figure 2.32. The crack started from a surface defect.

A fatigue failure in a non-cylindrical specimen starts in the minimum cross section where the nominal shear stress has its maximum. Spiral crack growth then would imply that the crack front will move away from the minimum section to areas with a lower nominal shear stress. Moreover, crack growth in the depth direction is also difficult in view of the decreasing shear stress. The problematic crack growth stimulates crack nucleation at other places in the root area of the notch. As a result of more simultaneously growing fatigue cracks a so-called factory roof failure is observed. An example is shown in Figure 6.15. Due to the geometrical shape of the radial ridges, some interlocking between the two fracture surfaces occur under torsional load. The complexity of the phenomenon obstructs a reasonable approach to predictions of fatigue lives.

6.3.5 Combined loading

Fatigue under combined loading is a complex problem. A rational approach might be considered again for fatigue crack nucleation at the material surface. The state of stress at the surface is two-dimensional because the third principal stress perpendicular to the material surface is zero. The most simple combination of loads is biaxial tension which occurs in pressurized vessels. It consists of two perpendicular tensile stresses. Such a loading case cannot be easily simulated on simple specimens, especially for unnotched material.

Another relatively simple combination of different loads is offered by an axle loaded under combined bending and torsion. This combination can be simulated in experiments, and test programs were reported in the literature. Early tests were carried out by Gough et al. [10]. They considered the fatigue limit and found a systematic effect of the combination of a bending stress amplitude and a torsional stress amplitude for which they proposed the elliptical quadrant criterion:

$$\frac{S^2}{S_f^2} + \frac{\tau^2}{\tau_f^2} = 1 \tag{6.6}$$

In this equation, S_f and τ_f, are the fatigue limits for single load cases, i.e. for pure tension and pure torsion respectively. They are supposed to be material constants. The criterion apparently agreed with test results for different materials. Results for some types of steel are presented in Figure 6.16. Equation (6.6) was less successful for cast alloys.

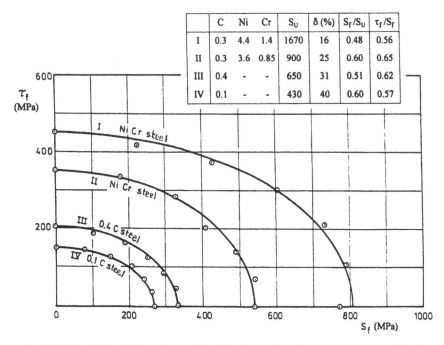

	C	Ni	Cr	S_u	δ (%)	S_f/S_u	τ_f/S_f
I	0.3	4.4	1.4	1670	16	0.48	0.56
II	0.3	3.6	0.85	900	25	0.60	0.65
III	0.4	-	-	650	31	0.51	0.62
IV	0.1	-	-	430	40	0.60	0.57

Fig. 6.16 The fatigue limit under combined tension and torsion (zero mean stress) [11].

The Von Mises criterion also predicts an elliptical quadrant equation for the 2D condition at the material surface. Actually, Equation (6.6) would become in full agreement with the Von Mises criterion if the ratio τ_f/S_f would agree with the Von Mises prediction (ratio 0.577). As said before, this is approximately true for several materials.

Combined loading in experimental programs has also been simulated by using tubular specimens which were loaded simultaneously in tension, in torsion and by internal pressure. Problems were encountered because the material used was not always isotropic. Tubular specimens are usually made of rod material which in many cases have a fibrous structure with elongated grains and impurities. Crack nucleation is sensitive to this kind of anisotropy. Secondly, crack growth observations were usually not made in such experiments. The test result was the number of cycles until failure or until a crack had penetrated through the full thickness of the wall of the tubular specimen. It should be realized that stress functions to account for combined stress conditions can only work for crack nucleation, i.e. for the fatigue life until the first microcrack has been created. As soon as crack growth occurs beyond the crack nucleation period, the stress condition becomes essentially different. Only if the crack growth period is very small compared to the crack nucleation period, functions like Equation (6.6) can be meaningful.

In the previous paragraphs it was tacitly assumed that the combined cyclic loads occur with the same frequency and phase angle. This obviously is true for the biaxially loading of a pressure vessel. However, a different situation can apply to dynamically loaded components subjected to two different types of cyclic loads. As an example consider an axle transmitting a torsional moment with slow variations of the magnitude of this moment. The same axle is simultaneously loaded by a high frequency bending moment. A case which may be relevant for axles of propellers. Other complex combinations can occur in several structures, e.g. motor cars.

Theories for combined fatigue loads with different frequencies and phase angles are discussed in the literature, see e.g. [12, 13]. If fatigue failures should not occur, a fatigue limit criterion must be adopted again. It may be recalled that the fatigue limit is associated with crack nucleation due to cyclic slip. Cyclic shear stresses at the material surface have to be considered. In theory, these stresses can be calculated for combined loading systems. However the orientation of the most critical slip plane for out-of-phase fatigue loads is not a priory known. Furthermore, a critical shear plane will not carry a cyclic shear stress only, but also a tensile stress which need not be in phase with the cyclic shear stress. This tensile stress can affect the nucleation of a microcrack in a slip band. Fatigue crack initiation under out-of-phase fatigue loads is still a topic of research which unfortunately is not easily validated by experimental data.

As soon as a fatigue crack has been initiated, the loading system for crack growth is significantly affected. The fatigue crack may be expected to grow perpendicular to the main principle stress, but unfortunately this stress can have a varying direction The prediction of crack growth becomes a complex problem.

6.4 Low-cycle fatigue

Low-cycle fatigue as a phenomenon has received much attention since the early work of Coffin and Manson in the fifties and the sixties. It became clear that low-cycle fatigue is a problem which is different from high-cycle fatigue. As pointed out before, the high-cycle fatigue mechanism on a macro scale occurs as an elastic phenomenon. However, in low-cycle fatigue, macroscopic plastic deformation occurs in every cycle.

Low-cycle fatigue can be relevant to structures that are subjected to small numbers of load cycles in their economic life. If it would be required to keep all stress levels below a fatigue limit, the structure may become very heavy without this being necessary. An example of a structure for which low-cycle fatigue can be important is a pressure vessel that is pressurized only a small number of times in many years. Other examples are power generator structures with an elevated operation temperature and significant thermal stresses. The number of on/off conditions can be low and low-cycle fatigue should be considered. Moreover, if thermal stresses occur due to differential thermal expansions, the nature of the loading is cyclic strain rather than cyclic stress.

Under low-cycle fatigue, failure can occur in a small number of cycles, say 1000 cycles or less. Small cracks are usually nucleated immediately. In view of the high stress level, final failure will occur when the cracks are still small. Periods of visible crack growth are hardly present. In the discussion on Figure 6.4, it was pointed out that low-cycle fatigue under constant-amplitude loading leads to a high plastic deformation in the first cycle followed by much smaller strain amplitudes in subsequent cycles. For that reason, it is instructive to study the low-cycle fatigue process in the laboratory by imposing constant strain cycles on a specimen. In general, this loading condition is also representative for the low-cycle conditions in structures. Such tests can be performed on closed-loop fatigue machines with a feedback signal obtained from the strain in the specimen.

The stress amplitudes under constant strain cycles can vary during successive cycles. This is illustrated in Figure 6.17. In the upper part of this figure, the strain cycles require an increasing stress amplitude. *Cyclic strain hardening* occurs which is more common for initially soft materials. In the lower part of Figure 6.17 the opposite occurs. The strain cycle can be maintained with a decreasing stress amplitude which is referred to as *cyclic strain softening*. It primarily occurs in materials which are already hardened to a significant level, either by a heat treatment or a deformation process. If cyclic deformations are applied to such a material, it can trigger structural changes which lead to some relaxation of the potential energy in the matrix of the material. The material becomes softer. In general, both cyclic strain hardening and cyclic strain softening stabilize to a constant level after a number of cycles, usually a low number compared to the fatigue life until failure. Some materials are practically stable almost immediately, which applies to several high-strength materials if the high strength is obtained by a heat treatment. Stabilizing is less predictable for soft materials and materials strain hardened by a deformation process.

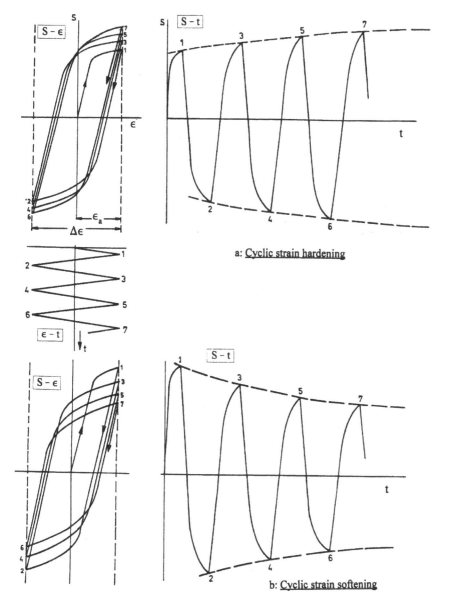

Fig. 6.17 Stress-strain loops during constant ε_a cycles of low-cycle fatigue, and the stress history for cyclic strain hardening and softening.

Coffin and Manson (independently) observed that the fatigue life under low-cycle fatigue conditions plotted as a function of the strain amplitude, ε_a, indicates a linear relation if plotted on a double logarithmic scale, see Figure 6.18. The relation can be written as:

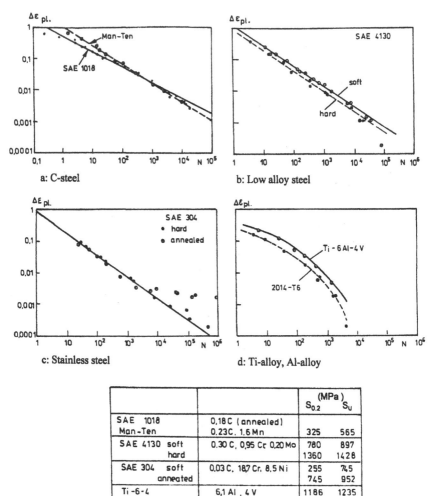

Fig. 6.18 Low-cycle fatigue curves, $N\text{-}\Delta epsilon_{pl}$ ($\Delta\varepsilon_{pl} = 2\varepsilon_{a,pl}$) [16–18].

$$\varepsilon_a N^\beta = \text{constant} = C \quad \text{or} \quad \varepsilon_a = CN^{-\beta} \tag{6.7}$$

This equation is known as the Coffin–Manson relation. As shown by Figure 6.18, the relation appears to be satisfactory for several materials with two exceptions in the lower right graph. The exponent β quite often is in the order of -0.5.

It is noteworthy that the upper horizontal asymptote of the S-N curve at $S_{\max} = S_U$ (see Figure 6.3) is no longer present in the ε-N diagrams. It thus seems logical to correlate low-cycle fatigue life to the strain amplitude. Note

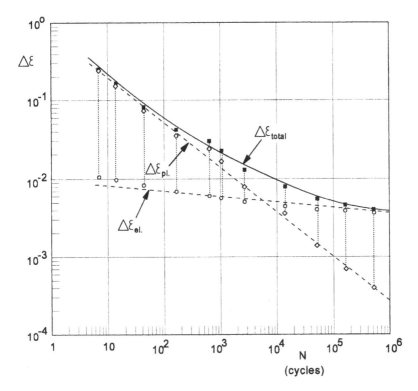

Fig. 6.19 Total strain range as the sum of the plastic and the elastic strain range. Material: AISI 4340 (annealed) [19]. (Range $\Delta\varepsilon = 2\varepsilon_a$)

the similarity between the Coffin–Manson relation and Basquin equation (Equation 6.2). Physical arguments underlying the Coffin–Manson relation appear to be questionable.

The Coffin–Manson relation can obviously not apply to high-cycle fatigue. The lower horizontal asymptote, i.e. the fatigue limit, is not covered by Equation (6.7).However, Manson and Hirschberg [15] considered also the elastic strain amplitude by a similar exponential relation to the fatigue life which lead to

$$\varepsilon_{a,\text{total}} = \varepsilon_{a,pl} + \varepsilon_{a,el} = C_1 N^{-\beta_1} + C_2 N^{-\beta_2} \tag{6.8}$$

Figure 6.19 shows the result for a stainless steel. The curve for $\varepsilon_{a,\text{total}}$ becomes non-linear with a tendency to bend towards a more horizontal direction at high endurances where the elastic strain amplitude predominates. Note that the description of the ε-N curve now asks for four material constants. Note also from Figure 6.19 that the elastic component ε_{el} for

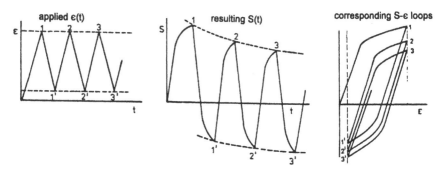

Fig. 6.20 Plastic shake-down, or stress relaxation.

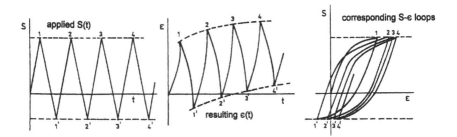

Fig. 6.21 Cyclic creep.

low lifetimes is negligible compared to the plastic one. Equation (6.8) might give a reasonable description of the ε-N curve of unnotched material if a fairly rapid stabilization of cyclic plasticity occurs. The question remains how useful it can be for prediction problems.

Plastic shake-down and cyclic creep

Two phenomena closely related to the occurrence of macroscopic cyclic plasticity should be mentioned here. Plastic shake-down can occur under cyclic plastic deformation if a positive mean strain is present, see Figure 6.20. During the cyclic deformations, the mean stress decreases, possibly to zero. This stress relaxation can occur because of dislocation rearrangements enabled by active cyclic slip. Residual stresses in a surface layer of a material can vanish if cyclic plastic deformation occurs in that layer. It generally requires relatively high cyclic stress levels.

A related phenomenon is cyclic creep occurring under a high cyclic stress with a positive mean stress, see Figure 6.21. The cyclic plastic deformations are partly used for permanent tensile deformations. As a dislocation mechanism, it is related to plastic shake-down because the cyclic slip tries to invoke plastic shake-down. However, under a cyclic tensile stress with a positive mean value, shake-down of the mean stress cannot be successful. The material becomes longer; creep occurs. If cyclic strain hardening is effective, as suggested in Figure 6.21, creep will stop after a number of cycles. If strain hardening does not occur, it may well lead to continued creep until failure.

6.5 Main topics of the present chapter

1. Fatigue limits and S-N curves for unnotched specimens are generally supposed to be basic material properties. However, it should be recognized that these properties depend on the size and the shape of the unnotched specimens used, and also on the surface finish of the specimens. This is especially relevant to the fatigue limit because this property is mainly controlled by nucleation of microcracks at the material surface.

2. For a group of similar alloys, the fatigue limit of unnotched specimens (S_{f1}) for $S_m = 0$ increases with the ultimate tensile strength of a material if the higher strength is obtained by changing the alloy composition or an other heat treatment. As a first estimate for the fatigue limit, the linear relation $S_{f1} = \alpha S_U$ can be used with a characteristic α-value for a group of similar materials.

3. A large part of the S-N curve can be approximated by the Basquin relation, $S_a^k N = \text{constant}$, a linear relation on a double log scale.

4. The fatigue properties on unnotched material can be described by fatigue diagrams. Lines for constant fatigue lives in such a diagram can be approximated by the Gerber parabola or by the modified Goodman relation. The Gerber parabola agrees more with the results of materials with a reasonable ductility whereas the modified Goodman relation is more applicable to the high-strength low-ductility materials.

5. The effect of the stress amplitude is more significant for the fatigue properties of a material than the effect of the mean stress, especially for high fatigue lives and the fatigue limit. The mean stress effect for different materials can be characterized by the slope factor M defined by Schütz (Figure 6.11). It indicates an increasing mean stress effect for an increasing strength of a material.

6. Fatigue is less significant for a compressive mean stress.
7. Fatigue under cyclic tension and cyclic bending is a similar phenomenon. However, under cyclic torsion it is different, both for crack nucleation and crack growth. The fatigue limit under cyclic torsion hardly depends on the mean shear stress.
8. For combined loading cases, e.g. cyclic bending and torsion, the fatigue limit can be described by an empirical relation. Complex problems are offered if combined loadings occur out of phase.
9. High-cycle fatigue and low-cycle fatigue refer to a significantly different behavior of the material. Under high-cycle fatigue, the material response is still macroscopically elastic. The fatigue life of specimens then is largely dominated by the crack initiation period while the crack growth period is relatively short. For low-cycle fatigue, macroscopic plasticity occurs in every cycle. Fatigue cycles should then be expressed in terms of strain amplitudes instead of stress amplitudes.

References

1. Schijve, J., *Significance of fatigue cracks in micro-range and macro-range.* Fatigue Crack Propagation, ASTM-STP 415 (1967), pp. 415–459.
2. Grover, H.J., Bishop, S.M. and Jackson, L.R., *Fatigue strength of aircraft materials. Axial load fatigue tests on unnotched sheet specimens of 24S-T3, 75S-T6 aluminum alloys and of SAE 4130 steel.* NACA TN 2324 (1951).
3. Grover, H.J., *Fatigue of Aircraft Structures.* U.S. Government Printing Office (1966).
4. Forrest, P.G., *Fatigue of Metals.* Pergamon Press, Oxford (1962).
5. Schütz, W., *View points of material selection for fatigue loaded structures.* Laboratorium für Betriebsfestigkeit LBF, Darmstadt, Bericht Nr. TB-80 (1968) [in German].
6. Buch, A., *Evaluation of size effects in fatigue tests on unnotched specimens and components.* Arch. Eisenhüttenwesen, Vol. 43 (1972), pp. 885–900 [in German].
7. Kloos, K.H., Buch, A. and Zankov, D., *Pure geometrical size effect in fatigue tests with constant stress amplitude and in programme tests.* Zeitschr. Werkstoftechn., Vol. 12 (1981), pp. 40–50.
8. Smith, J.O., *Effect of range of stress on fatigue strength.* University of Illinois, Engrg. Expt. Station Bulletin 334 (1942).
9. Leger, J., *Fatigue life testing of crane drive shafts under crane-typical torsional and rotary bending loads.* Schenck Hydropuls Mag., Issue 1/89 (1989), pp. 8–11.
10. Gough, H.J. and Pollard, H.V., *The strength of metals under combined alternating stresses.* Proc. Inst. Mech. Engrs, Vol. 131 (1935), pp. 3–103.
11. Gough, H.J. and Pollard, H.V., *Some experiments of the resistance of metals to fatigue under combined stresses.* Min. of Supply, Aero Res. Council, RSM 2522, Part I (1951).
12. Ballard, P., Dang Van, K. Deperrois, A. and Papadopoulos, Y.V., *High cycle fatigue and a finite element analysis.* Fatigue Fracture Engrg. Mater. Struct., Vol. 18 (1995), pp. 397–411.
13. Socie, D. and Marquis, G.B., *Multiaxial Fatigue,* John Wiley & Sons (1999).

14. Coffin Jr., L.F., *Low cycle fatigue – A review.* Appl. Mater. Res., Vol. 1, No. 3 (1962), p. 129.
15. Manson, S.S. and Hirschberg, M.H., *Fatigue behavior in strain cycling in the low- and intermediate-cycle range.* In Fatigue, An Interdisciplinary Approach, J.J. Burke, N.L. Reed and V. Weiss (Eds.). Syracuse University Press (1964), p. 133.
16. Coffin Jr., L.F. and Tavernelli, J.F., *The cyclic straining and fatigue of metals.* Trans. Metall. Soc. AIME, Vol. 215 (1959), pp. 794–807.
17. Brose, W.R., *Fatigue life predictions for a notched plate with analysis of mean stress and overstrain effects.* In Fatigue under Complex Loading, R.M. Wetzel (Ed.). SAE Advanced in Engineering, Vol. 6 (1977), pp. 117–135.
18. Smith, R.W., Hirschberg, M.H. and Manson, S.S., *Fatigue behavior of materials under stain cycling in low and intermediate life range.* NASA TN D1574 (1963).
19. Graham, J.F., *Fatigue Design Handbook.* Soc. of Automotive Engineers (1968).

General references

20. Sonsino, C.M., *Course of S-N-curves especially in the high-cycle fatigue regime with regard to component design and safety.* Int. J. Fatigue, Vol. 29 (2007), pp. 2246–2258.
21. Murakami, Y., *Metal fatigue: Effects of small defects and nonmetallic inclusions.* Elsevier (2002).
22. Stanzl-Tschegg, S.E. and Mayer, H.R., *Fatigue in the very high cycle regime.* Conference Proceedings. Institute of Meteorology and Physics, Vienna (2001).
23. Macha, E., Bedkowski, W. And Lagoda, T., *Multiaxial Fatigue and Fracture.* ESIS Publication 25, Elsevier (1999).
24. Suresh, S., *Fatigue of Materials*, 2nd edn. Cambridge University Press, Cambridge (1998).
25. Rice, C.R. (Ed.), *SAE Fatigue Design Handbook*, 3rd edn. AE-22, Society of Automotive Engineers, Warrendale (1997).
26. Hertzberg, R.W., *Deformation and Fracture Mechanics of Engineering Materials*, 4th edn. John Wiley & Sons, New York (1996).
27. Shiozawa, K. and Sakai, T. (Eds.), *Data Book on Fatigue Strength of Metallic Materials.* 3 Volumes. Elsevier, Amsterdam (1996).
28. *Fatigue and Fracture.* American Society for Materials, Handbook Vol. 19, ASM (1996).
29. *Fatigue Data Book: Light Structural Alloys.* ASM International (1995).
30. Dowling, N.E., *Mechanical Behavior of Materials. Engineering Methods for Deformation, Fracture, and Fatigue*, 3rd edn. Prentice-Hall (2006).
31. McDowell, D.L. and Rod, E. (Eds.), *Advances in Multiaxial Fatigue.* ASTM STP 1191 (1993).
32. Blom, A.F. and Beevers, C.J. (Eds.), *Theoretical Concepts and Numerical Analysis of Fatigue.* Proc. Conf. May 1992, Birmingham. EMAS (1992).
33. Klesnil, M. and Lukás, P., *Fatigue of Metallic Materials*, 2nd edn. Elsevier, Amsterdam (1992).
34. Brown, M.W. and Miller, K.J. (Eds), *Biaxial and Multiaxial Fatigue.* EGF Publication 3. Mechanical Engineering Publications (1989).
35. Boller, Chr. and Seeger, T., *Materials Data for Cyclic Loading.* Materials Science Monographs, 42, 5 Volumes. Elsevier, Amsterdam (1987).
36. Boyer, H.E., *Atlas of Fatigue Curves.* Amer. Soc. for Metals (1986).
37. Miller, K.J. and Brown, M.W. (Eds.), *Multiaxial fatigue.* ASTM STP 853 (1985).
38. Frost, N.E., Marsh, K.J. and Pook, L.P., *Metal Fatigue.* Clarendon, Oxford (1974).
39. Stephens, R.I., Fatemi, A., Stephens, R.R. and Fuchs, H.O., *Metal Fatigue in Engineering*, 2nd edn. John Willey & Sons (2000).

Chapter 7
The Fatigue Strength of Notched Specimens

Frequently used symbols

S nominal stress
σ local stress
S_{f1} fatigue limit of unnotched specimens
S_{fk} fatigue limit of notched specimens
K_t stress concentration factor
K_f fatigue strength reduction factor
ρ tip radius of notch
a^*, A material constants depending on S_U
S_U tensile strength

7.1 Introduction

Material fatigue properties obtained on unnotched specimens were discussed in the previous chapter. However, an engineering structure is not an unnotched specimen. On the contrary, various "notched" elements can always be indicated in a structure. Fatigue tests on notched specimens are necessary for two major purposes:

(1) To arrive at prediction methods for fatigue properties of structural elements.
(2) To carry out comparative fatigue tests to explore effects of different variables.

The first topic is covered in the present chapter, while the second topic is addressed in Chapter 13.

Engineering fatigue properties to be predicted are primarily associated with the fatigue limit and S-N curves. Predictions of the fatigue limit of a notched element is a more well defined problem than predictions of S-N curves. With respect to the fatigue limit, it is a matter of predicting whether a crack will be nucleated at the root of a notch, or whether that will not occur. For several engineering applications that is indeed a design criterion. It boils down to a prediction of a threshold stress level. However, for S-N curves, i.e. finite fatigue lives, the prediction problem is fundamentally different because the fatigue life to be predicted covers a crack initiation life and a crack growth life. Usually, the crack growth life for many structural elements, and also for fatigue specimens, is relatively short. It covers a small percentage of the total fatigue life. This aspect is generally disregarded in engineering prediction methods. As a consequence, it affects accuracies of prediction methods. It implies that predictions then give indications which should be handled with care, actually with engineering judgement.

In this chapter, fatigue properties of notched specimens are addressed first. The specimens are supposed to be representative for simple notches occurring in a real structure; notches with a well defined geometry for which the stress concentration factor (K_t) is available or can be calculated (e.g. holes, fillets, rounded corners, etc.). It starts with predictions of the fatigue limit based on the similarity of stress cycles in notched and unnotched specimens. Several variables are considered, the effect of a mean stress, different types of loading, and the effect of the quality of the material surface (surface finish). This is mainly restricted to effects on the fatigue limit. It is illustrated with some examples of predicting the fatigue limit of structural elements. Finally the prediction of S-N curve is addressed.

The present chapter is restricted to fatigue under constant-amplitude loading. Fatigue notch problems under variable-amplitude loading are discussed in Chapter 10. More complex notches occurring in various types of joints are covered in later chapters (Chapters 18 and 19). Fatigue of full-scale structures including additional variables and design problems are discussed in Chapter 20. The major points of the present chapter are summarized in the last section of this chapter.

7.2 The fatigue limit of notched specimens at $S_m = 0$

7.2.1 The similarity principle and the notch sensitivity

After fatigue was recognized in the previous century as a serious threat to the integrity of a structure, it was soon understood that stress concentrations aggravate the fatigue problem considerably. Many early fatigue experiments on the effect of notches on fatigue were carried out on small rotating beam specimens with circumferential notches with rather sharp root radii. This is remarkable because fatigue occurred in rather massive structures. Several early test programs were carried out on specimens with diameters of about 10 mm, a notch depth in the order of 1 to 2 mm, and notch root radii as small as 0.1 mm. Notches with such small root radii are now considered to be a poor fatigue design case, and not a realistic simulation of notches in a well designed structure. In the 1930s [1], it was already known from experiments that larger specimens with larger root radii, but the same K_t, could have a lower fatigue limit. In other words; a size effect on the fatigue limit of notched elements was discovered.

In order to understand and to predict the notch effect and the size effect on the fatigue limit, the definition of the fatigue limit must be recalled. It is the lowest stress amplitude which is still capable to nucleate a microcrack that can grow to failure, or, similarly the highest stress amplitude just not capable to create such a microcrack. Keeping this definition in mind, the similarity principle can be applied to compare fatigue in a notched specimen to fatigue in an unnotched specimen, see Figure 7.1. If a cycle with a stress amplitude S_a can create a microcrack in the unnotched specimen, the same cycle of the peak stress in the notched specimen, σ_{peak}, should also be capable to create a microcrack at the root of the notch. If S_a in the unnotched specimen is the fatigue limit (S_{f1}) of that specimen, then S_{peak} in the notched specimen should correspond to the fatigue limit of the notched specimen (S_{fk}). The similarity principle thus leads to:

$$S_{\text{peak}} = K_t S_{fk} = S_{f1} \quad \text{or} \quad S_{fk} = \frac{S_{f1}}{K_t} \tag{7.1}$$

It implies that the fatigue limit of an unnotched specimen should be divided by K_t to obtain the fatigue limit of the notched specimen. However, it has

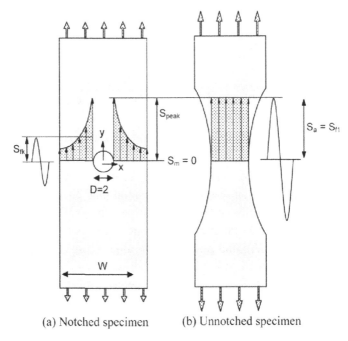

(a) Notched specimen   (b) Unnotched specimen

Fig. 7.1 Similarity principle: same S_{peak} at fatigue limit.

been shown in numerous fatigue tests that the reduction factor is smaller than K_t. The reduction factor obtained by experiments is denoted by the symbol K_f. It implies that

$$S_{fk} = \frac{S_{f1}}{K_f} \quad \text{or} K_f = \frac{S_{f1}}{S_{fk}} \tag{7.2}$$

K_f is also labeled as the fatigue notch factor. Experimental evidence thus indicated that

$$K_f \le K_t \tag{7.3}$$

Long ago it was often observed that $K_f < K_t$, especially for small specimens with high K_t-values, and the more so for low-strength materials, such as low-carbon steel. Although $K_f < K_t$ apparently limits the applicability of the similarity principle, it should be noted that the inequality is favorable. If K_f is smaller than K_t, it implies that a material is less notch sensitive than predicted by $K_f = K_t$. The notch sensitivity of a material as considered by Peterson [1] was defined by a factor q as

$$q = \frac{K_f - 1}{K_t - 1} \tag{7.4}$$

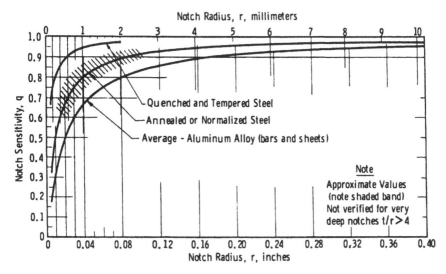

Fig. 7.2 The notch sensitivity of the fatigue limit ($S_m = 0$) of rotating beam specimens. Average results for different sizes (root notch) and material strength [1].

A high notch sensitivity is obtained if $K_f = K_t$ and thus $q = 1$. If there is no notch sensitivity, then $K_f = 1$ and thus $q = 0$. A variation of q from 0 to 1 corresponds to an increasing notch sensitivity. The factor q was used by Peterson [1] to illustrate both effects of the material strength and the size of specimens. A graph of Peterson is shown in Figure 7.2 with q-values between 0 and 1. It shows that q is larger for stronger materials and for larger specimens. In order to explain these trends the similarity principle as applied in Figure 7.1 must be reconsidered.

7.2.2 *The size effect on the fatigue limit of notched specimens*

Comments on size effects were already made in the discussions on Figures 3.7 to 3.9. It was pointed out that a size effect could depend on the amount of material with a high cyclic stress. For a notched specimen, it depends on stress gradients at the root of the notch. It was concluded in Section 3.3 that the variation of the stress along the material surface is very important because crack nucleation is a material surface phenomenon. This aspect was also considered in Chapter 6 (Section 6.3.3) in the discussion of a size effect on the fatigue limit of unnotched specimens. It was said that nominally unnotched specimens are always slightly notched. In spite of that, the area of highly stressed surface material of an unnotched specimen is generally larger than for a notched specimen. The straight forward similarity approach leading to $K_f = K_t$ does not give credit to the non-similarity of the highly stressed surface area of notched and unnotched specimens. As a consequence, a size effect can occur leading to a higher fatigue limit of a notched specimen than expected from $K_f = K_t$, and thus to $K_f < K_t$.

It is an intriguing but difficult question how the size effect on the fatigue limit of notched specimens can be accounted for in a quantitative way. The effect should be related to the probability of favorable locations for microcrack nucleation at the root of a notch. The question is whether the probability should be considered to be a volume effect (3D), or a surface effect (2D), or a corner edge effect (1D). The probability of having favorable sites for microcrack initiation can also depend on the type of material including inclusions and impurities. Although the presence of a size effect can be understood in a qualitative way, it must be admitted that K_f cannot be derived from K_t in a strictly rational way. Moreover, the fatigue limit of the unnotched specimen (S_{f1}) is also depending on the specimen type and size. Under such conditions, it is not surprising that semi-empirical relations were proposed to account for the size effect. Three different approaches were proposed in the literature which are discussed below. They are associated with the names of Peterson [1], Neuber [2] and Siebel [3].

Prediction method of Peterson

Peterson [1] considered the steep stress gradient in the direction perpendicular to the material surface, i.e. $d\sigma_y/dx$ at the root notch (see Section 3.3). He thought that the stress level σ_y at a small distance δ below the surface could be used as a criterion for the fatigue limit of a notched element. This stress level is approximately equal to:

$$\sigma_{\text{peak}} - \delta \left(\frac{d\sigma_y}{dx} \right)_{x=\rho} \tag{7.5}$$

Peterson proposed the following relation for the notch sensitivity defined in Equation (7.4):

$$q = \frac{K_f - 1}{K_t - 1} = \frac{1}{1 + a^*/\rho} \tag{7.6}$$

with ρ as the notch root radius and a^* as a material constant depending on the material. This equation cannot be derived by using Equation (7.5). But the equation agreed with average results of data on the fatigue limit

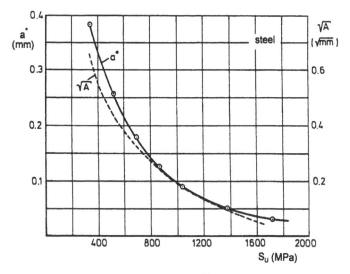

Fig. 7.3 Material constants a^* (Peterson [1]) and $\sqrt{A}$ (Kuhn and Hardrath [3]) to predict the fatigue limit ($S_m = 0$) of notched steel specimens.

for $S_m = 0$ which were available to Peterson in the 1950s. In Figure 7.2: $a^* = 0.063$ mm (0.0025 inch) for quenched and tempered steel (high strength), $a^* = 0.254$ mm (0.01 inch) for annealed or normalized steel (low strength), and $a^* = 0.51$ mm (0.02 inch) for Al-alloy sheets and bars. It is easily observed from Equation (7.6) that K_f will approach K_t for a small a^* and for a large root radius ρ. On the contrary, K_f becomes much smaller than K_t if ρ is very small (i.e. sharp notches with a small root radius). Peterson presented a^*-values for steels of different ultimate tensile strength levels. These empirically obtained values are plotted in Figure 7.3. The a^*-value is decreasing for an increasing S_U, and thus the notch sensitivity of the fatigue limit becomes more severe.

Prediction method of Neuber

Neuber is the author of a famous book on elastic stress distributions in notched elements (first edition in 1937, and translated in 1946 [2]). He noticed that the K_t-values obtained by elastic stress analysis highly overestimated the notch severity in experiments. Neuber then proposed an effective stress concentration factor, K_N, which is

$$K_N = 1 + \frac{K_t - 1}{1 + \dfrac{\pi}{\pi - \omega}\sqrt{A/\rho}} \tag{7.7}$$

Again ρ is the root radius, A is a material constant, and ω the opening angle of a V-notch. For $\omega = 0$ and considering that K_N is the fatigue reduction factor K_f, Equation (7.8) reduces to

$$q = \frac{K_f - 1}{K_t - 1} = \frac{1}{1 + \sqrt{A/\rho}} \tag{7.8}$$

Neuber discussed a relation between A and the average stress in a thin surface layer, but he did not give a derivation of Equation (7.8). Surprisingly enough, Neuber dropped the K_N equation in the second edition of his book (in 1958). However, Equation (7.8) was still evaluated by Kuhn and Hardrath [3] in the 1950s, and they found it to be a reasonably accurate engineering equation for different steels and Al-alloys. The value of A was considered to be a material constant which for steel varied in a way much similar to the variation of a^* of Peterson, see Figure 7.3. Kuhn and Hardrath give $A = 0.5$ mm (0.02 inch) for the Al-alloys 2024-T3 and 7075-T6.

The similarity and the difference between the equations of Peterson and Neuber are easily observed. The meaning of a^* and $\sqrt{A}$ is similar. However, the effect of ρ (size effect) appears in a different way as $1/\rho$ in the denominator of Equation (7.6), and $1/\sqrt{\rho}$ in Equation (7.8). Actually, the differences between the results obtained with the two equations are quantitatively fairly small. As an example, the ratio between the K_f-values are calculated for some different strength levels of steel with the values of a^* and $\sqrt{A}$ in Figure 7.3. Calculations for $K_t = 2.5$ and $\rho = 2.5$ mm then give the following ratios:

S_U (MPa)	400	800	1200	1600
(K_f) Neuber / (K_f) Peterson	0.91	0.95	0.97	0.99

For a low-C steel (K_f) Neuber is 9% lower than (K_f) Peterson, but for steels with a high-strength level, which are the more fatigue sensitive steels, the differences are practically negligible.

For engineering applications it is of some interest to know whether K_f is smaller than K_t, and how much smaller. This is not easily observed from Equation (7.8), but this equation can be rewritten as

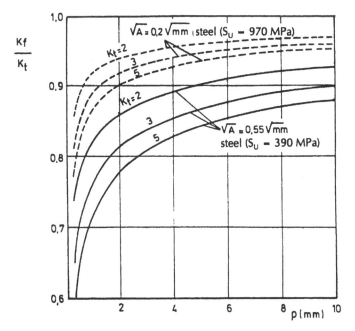

Fig. 7.4 The effects of the size (ρ), the K_t-value, and two different strength levels of steel on the ratio K_f/K_t according to the Neuber equation.

$$\frac{K_f}{K_t} = \frac{1 + \dfrac{1}{K_t}\sqrt{A/\rho}}{1 + \sqrt{A/\rho}} \qquad (7.9)$$

This ratio has been calculated for a low-strength steel ($S_U = 390$ MPa) and a high-strength steel ($S_U = 970$ MPa). The corresponding $\sqrt{A}$-values of Figure 7.3 were used. The K_f/K_t ratio was calculated for $K_t = 2, 3$ and 5 respectively. The results in Figure 7.4 indicate that K_f is significantly smaller than K_t if the root radius is small. For larger root radii and more fatigue sensitive high-strength steel, K_f/K_t is of the order of 0.95, i.e. the difference between K_f and K_t is about 5%. In other words; assuming $K_f = K_t$ can hardly be considered to add some hidden safety margin. Only for low-strength steel some appreciable advantage exists because then K_f is substantially smaller than K_t. In other words, the material is apparently less fatigue notch sensitive.

Prediction method of Siebel

Around 1950 Siebel and coworkers [4] proposed an other approach to predict notch and size effects on the fatigue limit at $S_m = 0$. The effects are assumed to be a function of the "relative" stress gradient at the root of the notch indicated by the symbol χ:

$$\chi = \frac{\left(\dfrac{dS_y}{dx}\right)_{\text{notch root}}}{S_{\text{peak}}} \tag{7.10}$$

The stress gradient was discussed in Chapter 3. Equation (3.10) is recalled here:[12]

$$\left(\frac{dS_y}{dx}\right)_{\text{notch root}} = \left(2 + \frac{1}{K_t}\right)\frac{S_{\text{peak}}}{\rho} = \alpha\frac{S_{\text{peak}}}{\rho} \tag{3.10}$$

with $\alpha = 2 + 1/K_t$. Combining the two equations gives

$$\chi = \frac{\alpha}{\rho} \tag{7.11}$$

As pointed out in Chapter 3, the variation of α is fairly small for technical notches with K_t in the range 2 to 5. Siebel et al. adopted $\alpha = 2$ for tension and bending. Instead of the ratio K_f/K_t, they considered the inverse ratio $n_\chi = K_t/K_f$. This ratio was proposed to be a function of the relative stress gradient χ. The function was empirically obtained for various materials and represented by

$$n_\chi = 1 + \sqrt{s_g \chi} \tag{7.12}$$

with the material constant s_g accounting for the type of material. The equation can be rewritten with $n_\chi = K_t/K_f$, $\chi = \alpha/\rho$ and $\alpha = 2$:

$$\frac{K_f}{K_t} = \frac{1}{1 + \sqrt{s_g \chi}} = \frac{1}{1 + \sqrt{A^*/\rho}} \tag{7.13}$$

with $A^* = 2s_g$ as the material constant. A comparison of Equation (7.13) with the Neuber equation (7.9) illustrates the similarity as well as the difference between the two prediction methods. In the Siebel equation (7.13), the ratio K_f/K_t depends on the size (ρ), but not on K_t whereas in the Neuber equation (7.11) the ratio depends on both. However, in the numerator of the Neuber equation $(1/K_t)/\sqrt{A}/\rho$ is usually much smaller than 1, and thus

---

[12] Siebel et al. considered the absolute value of the stress gradient. The minus sign of Equation (3.10) in Chapter 3 has been dropped for that reason.

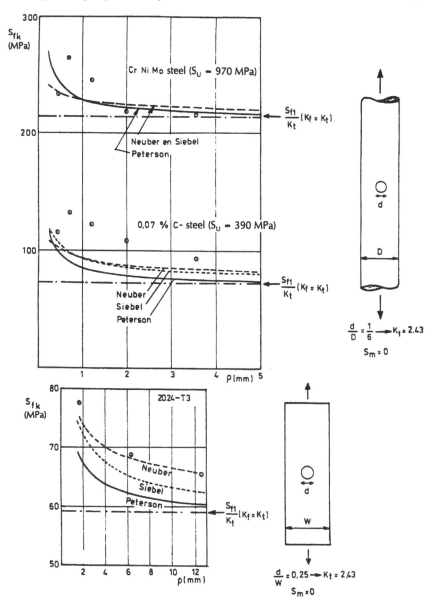

Fig. 7.5 Size effect on the fatigue limits ($S_m = 0$) of two steels [5] and an Al-alloy [6]. Comparison between three prediction methods.

both equations could still give approximately similar predictions This is illustrated in Figure 7.5 by empirical data for mild steel, a low-alloy steel and an Al-alloy. Only the size effect is checked by the results because K_t is not a variable in this figure. Two simple specimens are involved, a round bar with a transverse hole and a flat sheet specimen with a central hole. In all three cases an increasing fatigue limit occurs for smaller holes except for the smallest hole in both steels with $\rho = 0.4$ mm. This unsystematic result could be due to the drilling operation of such a small hole. Disregarding the results of the smallest hole in steel, Figure 7.5 illustrates a systematic size effect, but it also shows that highly accurate predictions were not obtained. Secondly, the predictions for the two steels are quite similar for the three prediction methods. The differences appear to be larger for the Al-alloy, but still within 10%. The agreement of the predictions with the test results is good for the low-alloy steel for the larger two holes ($\rho = 2.0$ and 3.6 mm). For the mild steel specimens, the fatigue limit is underestimated, slightly more so by the Peterson method. The underestimation is some 10 to 20% for the other two methods. The Neuber predictions for the Al-alloy are quite good whereas the other two methods give some underestimations.

7.3 The fatigue limit of notched specimens for $S_m > 0$

Local plastic deformation at the root of a sharp notch was not considered in the previous section on the fatigue limit of a notched specimen at $S_m = 0$. It is possible that some cyclic plastic deformation occurs in notched specimens of a soft material, but at the stress amplitude of the fatigue limit this will be stopped due to cyclic strain hardening.

The situation becomes different if a positive mean stress is present ($S_m > 0$). The stress cycle is no longer symmetric around a zero mean value. Figure 7.6 gives the stress distribution in a notched specimen for the maximum and minimum load of a cycle. The peak stress at the edge of the hole varies between $K_t S_{\max}$ and $K_t S_{\min}$ provided the material response is still elastic. The stress concentration factor K_t must then be applied to both $S_{\max}$ and $S_{\min}$. It also implies that the mean stress and the stress amplitude at the root of the notch are K_t times the net section values (see Figure 7.6):

$$S_{\max,\text{peak}} = K_t S_{\max}, \quad S_{\min,\text{peak}} = K_t S_{\min}$$

$$S_{a,\text{peak}} = K_t S_a, \quad S_{m,\text{peak}} = K_t S_m \tag{7.14}$$

If the same peak stress levels occur in an unnotched specimen, the same fatigue damage response should be expected according to the similarity principle. The same fatigue limit threshold stress levels should be applicable. The prediction of the fatigue limit of notched specimens employing the

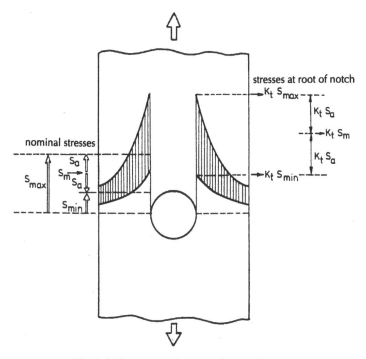

Fig. 7.6 Elastic behavior at notch root. $S_m > 0$.

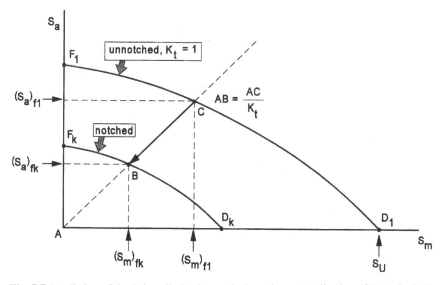

Fig. 7.7 Prediction of the fatigue limit of a notched specimen. Application of K_t to both S_a and S_m.

fatigue limit data of unnotched specimens is illustrated by Figure 7.7. The upper line in this figure gives the fatigue limit of the unnotched specimen, $S_{a,f1}$, as a function of the mean stress, $S_{m,f1}$. The subscript $f1$ refers to the fatigue limit of the unnotched specimen ($K_t = 1$), see point C in Figure 7.7. The subscript fk is used for the fatigue limit of the notched specimen. Point B on the line AC in Figure 7.7 is obtained by dividing $S_{a,f1}$ and $S_{m,f1}$ by K_t, which leads to $S_{a,fk} = S_{a,f1}/K_t$ and $S_{m,fk} = S_{m,f1}/K_t$. If C is a fatigue limit of the unnotched material, it implies that the corresponding fatigue limit of the notched specimen is obtained as point B. In this way, the entire line F_1-D_1 for the unnotched specimen is converted to the line F_k-D_k for the notched specimen.

The application of K_t to both S_m and S_a can lead to a reasonable approximation for the fatigue limit at low S_m-values. However, for a high S_m-value this cannot be true. The $S_{a,fk}$ line does not go to point D_k in Figure 7.7. In reality, the fatigue limit $S_{a,fk}$ line goes to zero (i.e. no cyclic load) at a stress level in the order of the static strength of the notched specimen. The notch effect on the static strength is rather small, at least if the material still has some ductility. In a static test on a notched specimens, significant plastic deformation occurs through the full cross section of the specimen if it is loaded to failure. The assumed elastic behavior is no longer valid. Empirical evidence has shown that the static strength of a notched specimen for moderate K_t-values is of the same order of magnitude as the material tensile strength. It implies that the fatigue limit line for the notched specimen for an increasing mean stress should converge to the point $(0, S_U)$.

For a relatively high mean stress on a notched specimen, the maximum peak stress at the root of a notch will exceed the yield stress. Some plastic deformation occurs and a plastic zone at the notch root is created, see Figure 7.8. The maximum stress would be $K_t S_{\max}$ if plasticity did not occur, but if the peak stress exceeds the yield stress the maximum peak stress is leveled off to a lower value, see S_{peak} in Figure 7.8. After reversion of the load ($S_{\max} \rightarrow S_{\min}$) elastic unloading will occur. Deformations during unloading are elastic deformations. This is the situation sketched in Figure 7.8 where the stress distribution at Smin is also indicated. Although the minimum stress on the specimen is still positive, a compressive stress can occur at the edge of the hole. Full unloading would further increase the compressive stress at the notch. The latter event was discussed in Chapter 4 on residual stresses. The elastic unloading in Figure 7.8 implies that the stress range at the root of the notch is not affected by the plastic deformation. As a result, $S_{a,\text{peak}}$ remains the same, but $S_{m,\text{peak}}$ is reduced by an amount $K_t S_{\max} - S_{\text{peak}}$. The

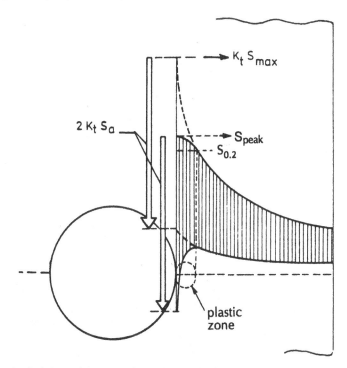

Fig. 7.8 Plastic deformation at S_{max} in the first cycle, followed by cyclic elastic deformation afterwards. $S_m > 0$.

reduction is advantageous for the fatigue limit S_{fk} although it complicates the prediction of S_{fk} with the similarity principle.

For the onset of root notch plasticity, the yield stress will be used as a criterion. In a notched flat specimen, plasticity at the root of the notch occurs if:

$$K_t S_{max} = K_t(S_m + S_a) > S_{0.2} \qquad (7.15)$$

The boundary line for elastic behavior of the notched specimen is thus given in the fatigue diagram by

$$K_t(S_m + S_a) = S_{0.2} \qquad (7.16)$$

This line with an angle of $45°$ to the axis of the S_a-S_m diagram is drawn in Figure 7.9. The location of the line depends on K_t. For low S_m-values, elastic conditions still exist up to point D in Figure 7.9, and the fatigue limit of the notched element (S_{fk}) is obtained with the method discussed before. However, for a larger mean stress, for instance in point B, the similarity conditions based on point C of the unnotched specimen curve is no longer

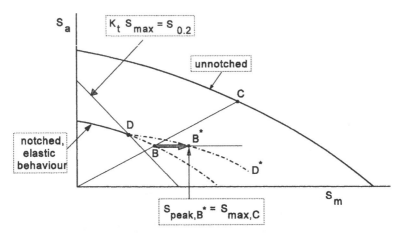

Fig. 7.9 Prediction of fatigue limit of a notched specimen if plastic deformation at the notch root has occurred.

valid due to some plastic deformation at the root of the notch. As long as unloading is still elastic, the amplitude $(S_a)_B$ is still equal to $(S_a)_C/K_t$, but point B shifts to the right because of notch root plasticity. Due to the increased mean stress it shifts horizontally to point B* where S_{peak} is again equal to S_{max} in point C of the unnotched specimen. The similarity of the stress cycle at the notch root with the stress cycle in the unnotched specimen is then restored. The prediction requires knowledge of the reduces peak stress due to notch plasticity. An estimate can be made by adopting the Neuber postulate discussed in Chapter 4 (Section 4.4). Anyway, after exceeding the yield limit in the root of the notch, the trend is that a further increase of the mean stress will give a smaller reduction of the fatigue limit then suggested with the dotted line DB. At still higher mean stresses on a notched element, plasticity will no longer be a local phenomenon at the root of the notch. Large-scale plasticity will occur, finally through the full net section when S_m is approaching the ultimate tensile strength S_U.

Examples of fatigue diagrams with fatigue limits of unnotched and notched specimens are given in Figures 7.10a and 7.10b for two different materials, a high-strength Al-alloy with a low ductility, and a relatively ductile low-alloy steel with a moderate strength respectively. The results for the unnotched fatigue limit, $S_{f1}(S_m)$, are noticeably different for the two materials. For the Al-alloy, the shape of the curve is approximately similar to the modified Goodman relation (Figure 6.9) whereas for SAE 4130 steel, it agrees much better with the Gerber parabola (Figure 6.9). Some further comments are relevant:

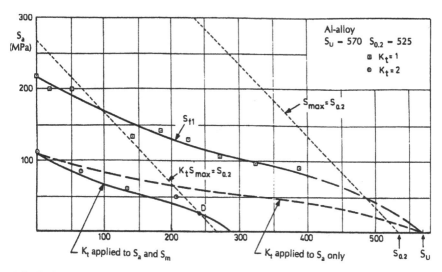

Fig. 7.10 (a) The fatigue limit for $K_t = 1$ and $K_t = 2$ for a high-strength Al-alloy [7, 8].

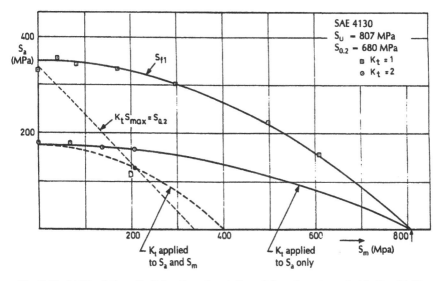

Fig. 7.10 (b) The fatigue limit for $K_t = 1$ and $K_t = 2$ for a low-strength Al-alloy [7, 8].

(i)   The mean stress obviously affects the fatigue limit, but the results are not fully systematic. Scatter of the test results around the trend lines can be observed, especially for the Al-alloy.

(ii) If K_t is applied to both S_a and S_m (Equation 7.14), the predicted fatigue limit for the notched Al-alloy specimens agrees reasonably well with the test results. This could be expected because plasticity at the root of the notch should not occur as long as $K_t S_{\max} < S_{0.2}$). However, the same trend is not observed for the SAE steel 4130. For this material it appears that application of K_t on the stress amplitude only is in good agreement with the experimental data. It may pointed out that this deviation from the simple similarity is a favorable behavior, because the experimental fatigue limits are larger than the theoretically expected values.

The similarity principle is theoretically sound and practically attractive. However, the above results show that the reality does not always follow this simple approach. As discussed in Chapter 6, the fatigue limit of the unnotched material (S_{f1}) can be sensitive to a size effect. In this chapter the size effect was recognized for the prediction of the fatigue limit of notched specimens tested with a zero-mean stress. However, it was not included in the above discussion on the mean stress effect for notched elements. Secondly, the effect of notch root plasticity as sketched in Figure 7.8 is a simplistic model. Notch root plasticity is not exclusively dictated by the yield stress $S_{0.2}$, especially not under cyclic loading. Other methods to estimate Speak at the root of a notch have been suggested in the literature including elasto-plastic finite element calculation. But even then, it is not realistic to expect accurate predictions of S_{fk} as a function of the mean stress. Predictions on the fatigue limit are addressed again in Section 7.7 where comments are made on applications.

7.4 Notch effect under cyclic torsion

The main fatigue loading cases are cyclic tension, cyclic bending and cyclic torsion. Differences between notch and size effects under cyclic tension and cyclic bending are not fundamentally different for predictions of the fatigue limit. However, the situation is different for cyclic torsion. This will be illustrated by two cases:

1. Axles with a stepped diameter or with a circumferential groove, see Figure 7.11. Such geometries are still axisymmetric.
2. Axles with a non-axisymmetric geometry. A simple case is a shaft with a transverse hole, see Figure 7.12.

In the first case, the analysis of the stress distribution is still relatively simple. As shown by Figure 7.11, the K_t-values for torsion are significantly lower than for tension and bending. Also stress gradients perpendicular to the material surface are lower. The semi-empirical methods on predicting notch

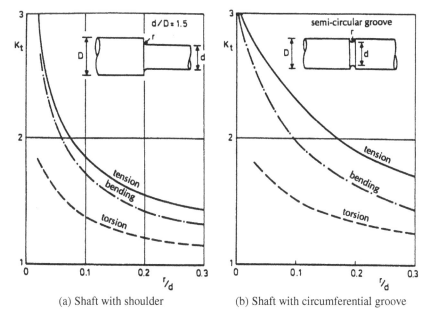

(a) Shaft with shoulder   (b) Shaft with circumferential groove

Fig. 7.11 Axisymmetric geometries [1].

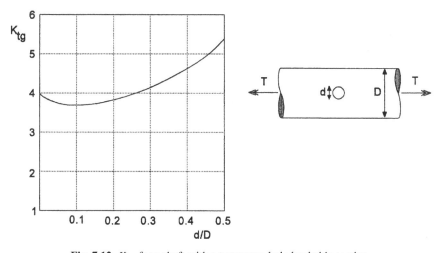

Fig. 7.12 K_{tg} for a shaft with a transverse hole loaded in torsion.

and size effects on the fatigue limit under cyclic tension (Section 7.2) cannot be adopted for this class of cyclic torsion problems because the empirical constants have not yet been obtained and verified. In view of the low stress gradients an obvious approach is to adopt simply $K_f = K_t$, also because K_t-values are relatively low. It should be recalled that the definition of K_t in this case is:

$$K_t = \frac{\tau_{\text{peak}}}{\tau_{\text{nominal}}} \quad \text{with} \quad \tau_{\text{nominal}} = \frac{T}{\frac{\pi}{16}d^3} \qquad (7.17)$$

where T is the torsional moment.

In the second case, the non-axisymmetric geometries, the problem is entirely different. In the simple case of a shaft with a transverse hole, the peak stress occurs at the edge of the hole where a uni-axial tensile stress is present. For that reason, it is more convenient to define the stress concentration factor in relation to the nominally applied shear stress on the gross section:

$$K_{tg} = \frac{S_{\text{peak}}}{\tau_{\text{nominal}}} \quad \text{with} \quad \tau_{\text{nominal}} = \frac{T}{\frac{\pi}{16}d^3} \qquad (7.18)$$

For $d \to 0$ the stress concentration factor in Figure 7.12 goes to the classical value: $K_{tg} = 4$ (see Section 3.5, Figure 3.20). In this second case, the notch and size effect as treated in Section 7.2 could again be valid.

Much more complex non-axisymmetric cases are offered by a shaft with a keyseat and a splined shaft, generally loaded under torsion. Peterson [1] offers some K_t data for these cases. But fatigue limit predictions are problematic.

Information on a mean stress effect (τ_m) on the fatigue limit under cyclic torsion of notched specimens is scarce. Because of the small effect of τ_m on the fatigue limit of unnotched specimens under cyclic torsion, it may be expected that the same is true for notched elements with an axisymmetric shape. However, for the second class of geometries (Figure 7.12), a similar mean stress as found for tension and bending should be applicable.

7.5 Notch effect on the fatigue limit for combined loading cases

Combined loading situations can vary from simple cases to rather complex ones. The most simple case occurs in a plate under plane biaxial loading with all cyclic loads occurring in-phase. The biaxial loading can include both tensile loads and shear loads, see Figure 7.13. For an open hole, it implies that the fatigue critical location is on the edge of the hole where a uniaxial stress is present. Stress analysis should indicate the most critical location and the magnitude of the local stress cycle, both amplitude and mean value. For a circular hole, this problem can be solved with equations presented in Chapter 3. Finite element calculation may be required for other types

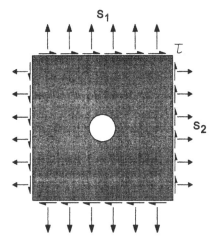

Fig. 7.13 Plate with a hole under biaxial loading.

of holes. The prediction of a fatigue limit then is a problem similar to the problems discussed in Sections 7.2 and 7.3.

The problem is more complex for combined bending and torsion on axisymmetric notch geometries such as shown in Figure 7.11. Assuming again that cyclic bending and torsion occur in-phase, the critical location at the root of the notch is cyclically stressed in a biaxial way. The stress concentration factors for bending and torsion are not the same. Peak stresses at the notch root for bending and torsion are $(K_t)_{\text{bending}} * S$ and $(K_t)_{\text{torsion}} * \tau$ respectively, with S and τ as the nominally applied stresses in the critical section. As discussed in Section 6.3.5, the elliptical quadrant criterion can then be used which should be written here as

$$\frac{(K_{t,\text{bending}}S)^2}{S_{f1}^2} + \frac{(K_{t,\text{torsion}}\tau)^2}{\tau_{f1}^2} = 1 \qquad (7.19)$$

It does not include a size effect, but ignoring this effect may be expected to be conservative.

Some other complications must be mentioned. Bending of an axisymmetric specimen induces a bending stress in the direction of the axis of the shaft, but at the same time it also leads to a tangential stress perpendicular to the bending stress, see Figure 7.14. The influence of this smaller stress is unknown from experiments. Usually it is ignored although it could be accounted for by some assumed yield criterion.

A complication is also offered by a shaft loaded under cyclic bending and a constant torsional moment, a situation that can occur in driving shafts.

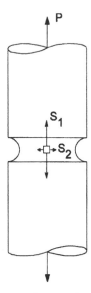

Fig. 7.14 Biaxial stress at the root of a circumferential notch

The question then is whether the constant τ has some influence on crack nucleation. It might be expected that the influence is small and can be disregarded.

Rather complex loading conditions occur when different cyclic loads are out-of-phase and occur with different cyclic frequencies. A crankshaft is an example of such a situation. It is obviously a notched element loaded in bending and torsion in a complex way. The variation of the stress distribution as a function of time can be calculated with finite-element techniques for a fatigue critical location of a structural element. The stress history at the material surface of a notch is still two-dimensional, and it will include both tensile stress and shear stress. In order to arrive at fatigue limit conditions, a sophisticated limit criterion is necessary. A prediction model has been proposed by DangVan [9].

7.6 Significance of the surface finish

As discussed in Section 2.5.5 the surface roughness has a significant effect on fatigue crack nucleation. As a result, the influence on the fatigue limit can be large. In the present chapter, the fatigue limit of notched specimens is discussed in relation to the fatigue limit of unnotched specimens. The similarity principle used for the predictions starts from the idea that the surface quality is good and similar for both unnotched and notched specimens. In general, unnotched specimens used in the laboratory experiments are manufactured with a high standard of surface quality. The unnotched specimen is provided with a smooth surface finish obtained by fine-turning or even fine-grinding. It should not change the material structure at the surface. A high-quality material surface finish is considered to be essential for acquiring elementary fatigue properties of a material. But it may then be questioned whether a similar surface quality is obtained in an industrial production process of a notched element. A high-quality surface is expensive. Surface qualities are depending on the production technology such as polishing, grinding, turning, contour milling, hole drilling, cutting, sawing, casting, forging, extruding, chemical milling, etc. The effect of different surface finish conditions on the S-N curve and the fatigue limit has been explored in many experimental investigations on specimens of different materials. In many cases this was done for the fatigue limit of unnotched specimens loaded under rotating bending, i.e. with $S_m = 0$. In order to characterize the effect of surface quality, a surface roughness reduction factor γ was introduced. The factor is defined as the ratio of the fatigue limit of unnotched specimens with a specific surface quality and the fatigue limit of the same specimens with a high quality of the surface:

$$\gamma = \frac{S_{1f,\text{specific surface quality}}}{S_{1f,\text{high-quality surface}}} \quad (7.20)$$

Results from the literature, see Figures 7.15 and 7.16, illustrate how γ can depend on the surface finish and on the static strength of the material. Values significantly below $\gamma = 1$ occur down to 0.5 in Figure 7.15, and to even lower values in Figure 7.16. Both figures show that an increasing strength of steel implies an increasing sensitivity to a poor surface finish. The surface reduction factor γ in Figure 7.15 is characterized by the average roughness depth, R_a in micrometers. Some data on the surface roughness effect for other materials than steel are presented in the literature, but the results are less abundant. Anyway, they show similar trends, i.e. an increasing sensitivity for surface roughness if the strength of a material is increased. This effect is most noteworthy for high-strength Al-alloys.

The question now is whether the same γ values should be applied to predict fatigue limits of notched specimens? Surface roughness implies a superposition of a roughness effect on a geometric notch effect. Regular notches like holes, fillets, etc. have macroscopic dimensions. However, the roughness notches are microscopic notches, but they can be important

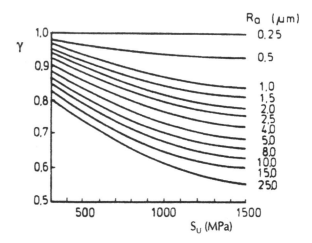

Fig. 7.15 Surface roughness reduction factor γ for steel as a function of R_a (average surface roughness) and the tensile strength (after data in German literature).

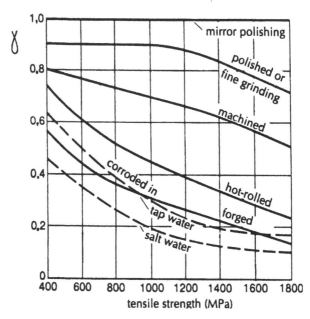

Fig. 7.16 Surface roughness reduction factor γ as a function of the tensile strength of steel [10].

for high-strength materials with a low-ductility. A superposition of notch effects was considered in the discussion on Figure 3.23 (Section 3.6). A multiplication of K_t-values was suggested:

$$K_t = (K_t)_{\text{notch}} * (K_t)_{\text{surface roughness}} \qquad (7.21)$$

This is not realistic because the effective root radii of the machining grooves are very small, and K_f should be rather small according to the notch size effect discussed in Section 7.2. On the other hand, it is known from experience in service that surface roughness can substantially reduce the fatigue quality of a structural element of a high-strength material. It is more realistic to apply corrections factors as shown in Figures 7.15 and 7.16 on the predicted fatigue limit to account for the surface quality:

$$S_{fk} = \gamma * (S_{fk})_{\text{predicted}} \qquad (7.22)$$

The γ factor as applied in this equation is a reduction factor. The values presented in Figures 7.15 and 7.16 are empirical factors based on results of comparative laboratory experiments. It remains a matter of engineering judgement to decide whether the selected γ-value can be representative for a specific problem. In case of a fatigue critical problem, more accurate information requires that fatigue experiments are carried out on notched specimens. The specimens must be fully relevant for the problem considered, i.e. representative with respect to the notch geometry (especially the root radius) and the surface finish obtained in production.

Two lessons to be learned from the above discussion are:

(i) The advantage of a higher strength material for a fatigue critical element in view of the higher fatigue properties of the material may be offset by the increased sensitivity for the quality of the material surface finish.

(ii) Information on the effect of the surface finishing quality of a product can be estimated (γ-values), but more reliable information requires carefully planned fatigue experiments.

The effect on fatigue of surface treatments, such as nitriding, anodizing, shot peening, etc., is not considered in the present chapter, but comments are presented in Chapter 14.

7.7 Discussion on predictions of the fatigue limit

The fatigue limit is an important material property for many structural elements subjected to large numbers of load cycles. Examples are rotating parts (machinery, engines, etc.) and elements for which vibrations can occur. Large numbers of cycles occur in the service load spectrum of such structural

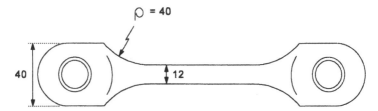

Fig. 7.17 Rod used in prediction example.

elements. In general, fatigue failures will not be acceptable which implies that the amplitudes of all cycles should remain below the fatigue limit. Although safety factors can be applied, as discussed later in Chapter 20, the question remains how accurate the fatigue limit can be predicted in the design stage of a structural element. In the previous sections important variables of the fatigue limit were discussed:

- the stress concentration factor K_t,
- the size of the notch (root radius ρ),
- the surface roughness (quality of surface finish), and
- the type of material.

The effects of these variables on the fatigue limit are understood in a qualitative way. At the same time, this qualitative understanding leads to the conclusion that highly accurate predictions cannot be expected. The question arises if reasonably accurate predictions or conservative predictions are possible. In view of this problem some prediction examples are discussed below.

Prediction examples

1. *Connecting rod, $S_m = 0$.*

A connecting rod of a mechanism as shown in Figure 7.17 should not be heavy in view of fast movements. For that reason, a high-strength low-alloy steel is selected with a tensile strength $S_U = 1000$ MPa. The end fittings are not considered here. They represent a pin-loaded joint (lug) discussed in Chapter 18. However, the cylindrical rod between the end fittings should not fail by fatigue. The maximum tensile load and the maximum compression load are supposed to be equal, and to occur many times. As a consequence, the fatigue limit for $S_m = 0$ is a relevant property to analyse the risk of fatigue failures. If the fatigue limit for unnotched material is not available, it can be estimated by Equation (6.3):

$$S_{f1} = \alpha S_U$$

Choosing of $\alpha = 0.4$ (to be safe, see Figure 2.12) leads to

$$S_{f1} = 0.4 \times 1000 = 400 \text{ MPa}.$$

A modest stress concentration occurs between the rod and end fittings. The shape of this configuration can be found in the book of Peterson [1], but the ratios of the dimensions in Figure 7.17 are outside the relevant graph in Peterson. However, starting from K_t-values for flat bars and knowing that K_t-values for round bars are lower, it may well be assumed that K_t will be in the order of 1.1 (good design).

Accounting for a notch size effect should not be done. K_f would only be slightly smaller than K_t by a few percent, see Figure 7.4. Moreover, the highly stressed material surface area of the connecting rod is relatively large. A first estimate of the fatigue limit of the critical section is

$$S_{fk}(S_m = 0) = 400/1.1 = 364 \text{ MPa}$$

The following step is to apply a reduction factor γ for the quality of the surface finish. Graphs as shown in Figsures 7.15 and 7.16 should be consulted. The production technique of the connecting rod should give a reasonably good surface finish. The surface roughness could be in the order of $R_a = 5$ μm. The γ-value in Figure 7.15 for $S_U = 1000$ MPa is about 0.75 while Figure 7.16 also suggests that this value could be applicable. Obviously, the choice is somewhat arbitrary. The fatigue limit is now further reduced to

$$S_{fk}(S_m = 0) = 0.75 \times 364 = 273 \text{ MPa}$$

This fatigue limit should now be compared to the maximum load amplitude of the load spectrum of the connecting rod. If the corresponding stress level is indicated by $S_{\text{max,spectrum}}$, it implies a safety margin:

$$\text{safety margin} = 273 \text{ MPa}/S_{\text{max,spectrum}}$$

Whether the safety margin is considered to be sufficient, or too small or too large, depends on a number of questions, such as: Is an occasional fatigue failure acceptable? Is the load spectrum really well known? These questions do not bear upon the accuracy of the fatigue limit prediction. The uncertainties of the prediction of the fatigue limit of the connecting rod

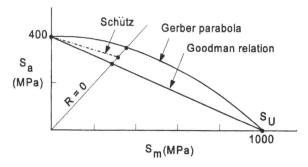

Fig. 7.18 Fatigue diagram for fatigue limit, second prediction example.

are partly a matter of scatter of S_{f1} and the assumptions made (values of α and γ), and for another part due to the variability of the connecting rod as an industrial product. These uncertainties can be accounted for by safety factors, but rational arguments to quantify these factors cannot be offered, other than experience and engineering judgement. If these uncertainties cannot be accepted for economical reasons, it must be advised to do relevant fatigue experiments. The most relevant fatigue tests are tests on the connecting rod itself.

2. Lifting rod element, $S_{min} = 0$, $R = 0$

The rod element has the same geometry as the connecting rod in Figure 7.17. Also the material is the same. The rod is loaded in tension only. Each applied load is followed by an unloading until zero. Because a failure is considered to be fully unacceptable, all applied loads should be below the fatigue limit. In the present case the fatigue limit must be estimated for $S_{min} = 0$ (or $R = S_{min}/S_{max} = 0$). The fatigue limit for unnotched material is again adopted as $S_{f1} = 400$ MPa for $S_m = 0$. But now $S_{min} = 0$, and a function $S_{f1}(S_m)$ must be adopted. In Figure 7.18, $S_{f1}(S_m = 0) = 400$ MPa and $S_U = 1000$ MPa are used to determine the Goodman relation and the Gerber parabola (Figure 6.9):

$$\text{Goodman relation:} \quad S_{f1} = 400 \left[1 - \frac{S_m}{1000} \right]$$

$$\text{Gerber parabola:} \quad S_{f1} = 400 \left[1 - \left(\frac{S_m}{1000} \right)^2 \right]$$

The line for $R = 0$, i.e. $S_f = S_m$, is indicated in Figure 7.18. The values of S_{f1} for $R = 0$ are found by substituting $S_{f1} = S_m$ in these equations, which

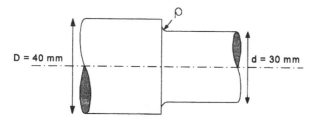

Fig. 7.19 Shaft of the third prediction example.

gives

$$S_{f1} = 286 \text{ MPa (Goodman diagram)}$$

$$S_{f1} = 351 \text{ MPa (Gerber parabola)}$$

These values are substantially different. The S_m-effect as indicated by the data analysis of Schütz, see Figure 6.11, will also be considered. For $S_U = 1000$ MPa, this figure gives $M = \text{tg}\,\alpha = 0.27$. It is easily derived from that figure that

$$S_{f1,R=0} = \frac{S_{f1,S_m=0}}{1 + M}$$

Substitution of M gives $S_{f1} = 400/1.27 = 315$ MPa, a value in between the values of the Goodman diagram and the Gerber parabola, but closer to the lower one. Because Figure 7.10b suggests that the Goodman diagram could well be conservative, the result obtained with the Schütz correction is suggested to be an acceptable estimate. The result should still be corrected for the small stress concentration ($K_t = 1.1$) and the same roughness reduction factor of the previous example ($\gamma = 0.75$). It leads to a fatigue limit:

$$S_{fk} = (315/1.1) * 0.75 = 215 \text{ MPa}$$

Because this is an amplitude the corresponding maximum stress for $R = 0$ is 430 MPa, a stress level well below the yield strength of the material.

3. Rotating bending of a shaft with a shoulder fillet

Transitions of the shaft diameter in rotating machinery occur for different reasons. An example was shown in Chapter 3 on stress concentrations, see Figure 3.16. The simple geometry to be considered here is given in Figure 7.19. Predictions are made for two steels, a low-strength C-steel ($S_U = 450$ MPa) and a high-strength low-alloy steel ($S_U = 1350$ MPa).

Table 7.1 Predicted fatigue limits.

Material	C-steel		Low-alloy steel	
S_U (MPa)	450		1350	
S_{f1} (MPa)	202.5		540	
$\sqrt{A}$ (mm) (Fig. 7.3)	0.50		0.10	
γ (Fig. 7.15)	0.88		0.72	
ρ (mm)	1	5	1	5
K_t	2.35	1.30	2.35	1.30
K_f	1.90	1.24	2.23	1.29
S_{fk} (MPa)	94	143	175	302

Two root radii (ρ) are considered, viz. a sharp radius ($\rho = 1$ mm) and a generous radius ($\rho = 5$ mm). The results of the analysis are compiled in Table 7.1.

S_{f1} was again estimated with $S_{f1} = \alpha S_U$ with $\alpha = 0.4$ for the high-strength low-alloy steel while a slightly higher value was used for the low-carbon steel, $\alpha = 0.45$. K_t-values are derived from a graph in Peterson's book [1]. The notch size effect was accounted for by the Neuber equation (Equation 7.9), and the surface roughness reduction factor γ was read from Figure 7.15 for a surface roughness characterized by an average groove depth $R_a = 4$ μm (reasonably good finish).

The fatigue limit results S_{fk} on the last line of the table show that increasing the root radius from 1 to 5 mm is beneficial for both materials. The gain is larger for the high-strength low-alloy steel, improvement factor 1.73 compared to 1.52 for the C-steel. This should be expected because of the higher notch fatigue sensitivity of stronger materials, compare K_f-values with K_t-values in the table.

The fatigue limits for the high-strength steel are about two times higher than the fatigue limits for the C-steel, whereas the tensile strength is three times higher. The gain on the fatigue limit is smaller than the gain in static strength. Whether the gain is worthwhile depends on various aspects to be considered by risk analysis and cost-effectivity arguments. In this respect, it is possible that the fatigue limit of the much cheaper C-steel shaft could be improved by some local material surface treatment at the root of the notch, e.g. by rolling (see Chapter 14).

Some general comments on predictions of the fatigue limit

The K_t-value cannot always be derived from data in the literature. In such cases, clever estimates can sometimes be made by considering similar geometries, see also Chapter 3. If accurate values are needed, FE calculations have to be made. The FE techniques are well developed, but care should be taken with mesh-refinements around the notch and with relevant boundary conditions. At the same time, it should be realized that great efforts to obtain accurate K_t-values become less meaningful if other uncertain effects are present and cannot be removed. The effect of the size of the notch (root radius ρ) is accounted for by empirical equations. It cannot be expected that a more rational method with a general validity will be available in the future. The theoretical problem is less serious for high-strength materials because $K_f = K_t$ appears to be realistic. The advantage of $K_f < K_t$ for low-strength materials could be estimated with empirical equations, but the reliability of such equations has certain limitations. The equations were developed primarily for various types of steel whereas the verification for other materials is less abundant, or even non-existent. Under such conditions, predictions on the fatigue limit with the simple methods used in the above examples should be earmarked as "estimates" rather than predictions.

A significant obstacle for predictions on the fatigue limit of notched elements is how to account for the quality of the material surface. It is questionable if this effect can be adequately accounted for by a single surface roughness parameter. True enough, the effect must be considered in predictions, especially because the effect can be large for high-strength materials. In this respect, it should also be recognized that the sensitivity for incidental mechanical surface damage as well as for corrosion damage can be significant (see the lower lines for corrosion damage in Figure 7.16). Designing against fatigue is not simply a question of materials with good fatigue properties. If a material is considered for a dynamically loaded structural element, it should always be tried to collect relevant fatigue data from the literature. With respect to notch and size effects and the significance of the material surface quality, it is also advisable to consult literature data banks with combinations of key words as: name of material, fatigue, notch effect, surface conditions, etc.

As pointed out before, fatigue tests can be desirable if insufficient realistic data are available and if the accuracy of predictions is unsatisfactory. The type of specimen should be as representative as possible for the fatigue problem concerned (notch size, notch shape, surface quality, material, see Chapter 13). With respect to predictions of the fatigue limit, it is noteworthy that the prediction methodology is historically based on the similarity between fatigue in notched specimens and unnotched specimens. It is an old dream that a fatigue limit for a notched element with e.g. $K_t = 3.0$ should be predictable from the fatigue limit data for the unnotched material, $K_t = 1.0$, which are supposed to be fundamental material data. However, it is more realistic to realize that it is a large extrapolation step to go from data for unnotched specimens with $K_t \approx 1$ to a real structure with geometrical notches with K_t-values in the order of 3.0. If data are available for $K_t = 2.0$ (a value mentioned here as an example), the extrapolation to $K_t = 3.0$ is significantly smaller. A better accuracy could thus be expected, but again due attention must be paid to size effects and surface roughness aspects.

7.8 The S-N curves of notched specimens

The prediction of the fatigue limit of notched elements discussed in previous sections is associated with problems of load spectra with large numbers of cycles while fatigue failures are not acceptable. The fatigue limit then is a most import property. In the present section, predictions of the S-N curve of notched elements are considered. These curves are relevant to problems where crack nucleation cannot be avoided. It then becomes of interest to know how many cycles it will take before failure occurs. A practical question is whether this fatigue life can be accepted as being sufficient for the structure concerned.

As pointed out in the Introduction of this chapter, the prediction of S-N curves is an essentially different problem compared to the prediction of the fatigue limit. The main reason is that S-N data are associated with two different periods of the fatigue life: (i) the crack initiation period including microcrack growth, and (ii) the crack growth period. As discussed in Chapter 2, the transition of the crack initiation period to the crack growth period cannot easily be defined numerically in crack size dimensions. The initiation period is supposed to be completed when microcrack growth is no longer depending on the material surface conditions. The cracks in this period are still microscopically small, and the growth of these cracks can be irregular depending on the type of material and the material structure. The stress intensity factor offers little hope for a reasonable prediction of the crack initiation period.

Another problem about predicting S-N curves is caused by the difference between low-cycle fatigue, say N up to 10^4 and high-cycle fatigue, $N >$

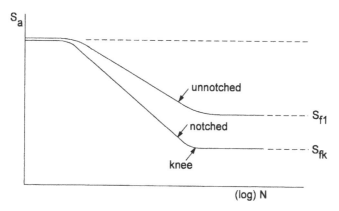

Fig. 7.20 The fatigue strength for notched and unnotched specimens.

10^5 cycles, see Figure 6.3. In the high-cycle fatigue range, the length of the crack initiation period is depending on surface conditions. However, in the low-cycle fatigue range, nucleation will occur immediately due to the high strain amplitudes.

The finite fatigue life in the high-cycle fatigue range of an S-N curve is mainly covered by the crack initiation period. This was discussed in Chapter 6 for unnotched specimens, but is also applies to notched specimens and structural elements with a limited cross sectional area. Visible cracks occur relatively late in the fatigue life. Nevertheless, it is problematic to arrive at some similarity concept for fatigue life predictions. An engineering approach to estimate S-N curves is based on empirical trends observed in relevant test programs. Some characteristic aspects of S-N curves should be recalled here, see Figure 7.20:

(i)   As discussed in Chapter 6, S-N curves have two horizontal asymptotes. The lower one corresponds to the fatigue limit ($S_a = S_f$). The upper horizontal asymptote is governed by the static strength: $S_a + S_m = S_U$, see Figure 6.3. The notch effect in a static test is rather small if not negligible. As a result of this relatively small effect, the upper horizontal asymptotes occur at approximately the same stress level for notched and unnotched specimens.

(ii) The location of the knee, i.e. the transition point of the S-N curve to the fatigue limit (Figure 7.20) depends on the type of material, but also on the presence of a notch, the size of the notch and the surface roughness. The knee for unnotched specimens of low-carbon steel is frequently assumed to occur at $N = 2 \times 10^6$ cycles. Al-alloys have a reputation that fatigue failures can still occur at fatigue lives up to 10^7 cycles. However,

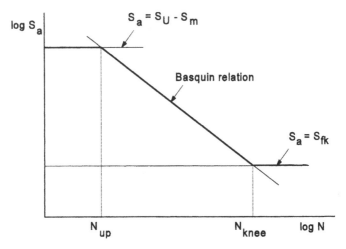

Fig. 7.21 Estimate of S-N curve.

a tendency has been observed that the knee of notched specimens does not shift to high N-values. The knee may occur at a fatigue life of 10^6 to 2×10^6 cycles, and sometimes it can be fairly sharp.

(iii) Between the upper part of the S-N curve and the fatigue limit the Basquin relation ($S_a^k \cdot N$ = constant) is a reasonable approximation, i.e. a linear relation in a $\log(S) - \log(N)$ graph.

With the above observations summarized in Figure 7.21, the following procedure can be adopted for estimating the S-N curve of a notched specimen. First determine the stress levels corresponding to the two horizontal asymptote: $S_a = S_U - S_m$ for the upper asymptote, and $S_a = S_f$ for the fatigue limit asymptote. Secondly, construct a Basquin relationship between the two asymptotes, i.e. a linear relation between S_a and N in a double log plot. The problem then is where this line must be located. The intersections with the two asymptotes are occurring at $N = N_{up}$ and $N = N_{knee}$, see Figure 7.21. Considering S-N data of various sources for notched specimens, two conservatively selected values are: $N_{up} = 10^2$ cycles and $N_{knee} = 10^6$ cycles.

As an example, S-N estimates will be made for two specimens of SAE 4130 for which S-N curves are available in [8]. The K_t-values of the two specimens shown in Figures 7.22a and 7.22b are 2.16 and 4.0 respectively. The S-N curves are estimated for $S_m = 0$. First the fatigue limit must be predicted which is done with the Neuber equation (Equation 7.9). The material constant $\sqrt{A} = 0.275 \sqrt{\text{mm}}$ for $S_U = 806$ MPa. With the root radii $\rho = 8.1$ mm and $\rho = 1.45$ mm respectively, and the fatigue limit of the

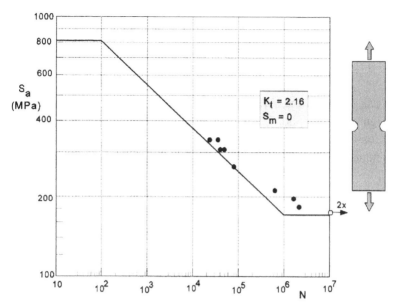

Fig. 7.22 (a) Predicted S-N curve for a mildly notched specimen, root radius $\rho = 8.1$ mm, material SAE 4130 steel.

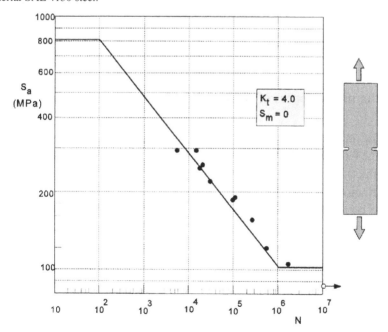

Fig. 7.22 (b) Predicted S-N curve for a sharply notched specimen, root radius $\rho = 1.45$ mm, material SAE 4130 steel.

unnotched material $S_{f1} = 350$ MPa (measured value in [8]), the results are:

$$K_t = 2.16 \quad \rightarrow \quad K_f = 2.06 \quad \rightarrow \quad S_{fk} = 350/2.06 = 170 \text{ MPa}$$
$$K_t = 4.0 \quad \rightarrow \quad K_f = 3.44 \quad \rightarrow \quad S_{fk} = 350/3.44 = 102 \text{ MPa}$$

With $N_{up} = 10^2$ and $N_{knee} = 10^6$ the S-N curve can then be drawn, see Figures 7.22a and 7.22b. The experimental results in these figures agree reasonably well with the predicted S-N curves. The calculated slope coefficients k of the Basquin relation ($S_a^k N = $ constant) are equal to 5.9 for the $K_t = 2.16$ specimen and 4.5 for the $K_t = 4.0$ specimen respectively. This order of magnitude is generally observed with a tendency to lower k-values for sharper notches.

It should be recognized that the above empirical approach is in fact an extrapolation from observed trends. Equally good estimates as shown by Figures 7.22a and 7.22b cannot generally be guaranteed. Conservative and non-conservative predictions are possible. The selection of a fairly low N_{up} (= 10^2) introduces a slight conservatism. In various cases $N_{up} = 10^3$ could give better approximations of the S-N curve. The prediction method illustrated by Figure 7.22 is an engineering method which leads to approximate estimates of an S-N curve. Improved predictions require fatigue experiments.

7.9 The major topics of the present chapter

1. *The fatigue limit of notched specimens for $S_m = 0$*
    The fatigue limit depends on K_t (notch effect) and the root radius ρ (size effect). The fatigue limit can be predicted by adopting the similarity principle in its most simple form, which is $K_f = K_t$. The prediction will be conservative in most cases, i.e. $K_f < K_t$. A reasonable prediction of the fatigue limit is possible with empirical equations to account for the notch effect, the size effect, and the strength of the material. The Neuber equation (Equation 7.8) gives reasonable estimates. However, for high-strength materials with a low ductility it is advised to adopt $K_f = K_t$.

2. *The fatigue limit of notched specimens, $S_m > 0$*
    Prediction of the fatigue limit is complicated because local plastic deformation at the root of the notch can level off the peak stress. The mean stress effect can be accounted for by adopting a Gerber parabola for ductile materials with a low or moderate strength level. For

high-strength low-ductility materials, the modified Goodman relation should be advised. In the latter case it is even more safe to apply K_t to both S_a and S_m. Useful indications on the mean stress effect are obtained by the methods of Schütz (Equation 6.5).

3. *The fatigue limit under cyclic torsion*
   For a shaft with a stepped diameter it is advisable to adopt $K_f = K_t$. The mean stress effect should be expected to be small.

4. *The surface finish effect on the fatigue limit of notched elements*
   Predictions on the fatigue limit should include the effect of the surface finish in addition to the notch and the size effect. Reduction factors (γ) of the literature are indicative.

5. *S-N curves of notched specimens*
   Predictions on the fatigue life until final failure are complicated because a finite fatigue life consists of a crack initiation period and a crack growth period. Estimations of S-N curves using the Basquin relation are possible.

6. Important variables for the prediction of the fatigue strength of a notched element are K_t, the size of the notch, surface finish and mean stress. Mechanistic aspects of these variables are reasonably well understood in a qualitative way. Because of this understanding, it is obvious that limitations on the accuracy of fatigue strength predictions should be present. Empirical trends are helpful in making engineering estimates. In fatigue critical cases, experiments are indispensable.

References

1. Peterson, R.E., *Stress Concentration Factors*. John Wiley & Sons, New York (1974).
2. Neuber, H., *Kerbspannungslehre*. Springer, Berlin (1937) (2nd edn., 1958). Translation: *Theory of Notch Stresses*. J.W. Edwards, Ann Arbor, Michigan (1946).
3. Kuhn, P. and Hardrath, H.F., *An engineering method for estimating notch-size effect in fatigue tests of steel*. Report NACA TN 2805 (1952).
4. Siebel, E. and Stieler, M., *Significance of dissimilar stress distributions for cyclic loading*. Zeitschr. VDI, Vol. 97 (1955), pp. 146–148 [in German].
5. Phillips, C.E. and Heywood, R.B., *Size effect in fatigue of plain and notched steel specimens*. Proc. Inst. Mech. Engrs., Vol. 165 (1951), pp. 113–124.
6. Landers, C.B. and Hardrath, H.F., *Results of axial-load fatigue tests on electropolished 2024-T3 and 7075-T6 aluminum-alloy-sheet specimens with central holes*. Report NACA TN 3631 (1956).
7. Grover, H.J., Bishop, S.M. and Jackson, L.R., *Fatigue strengths of aircraft materials. Axial load fatigue tests on notched sheet specimens of 24S-T3 and 75S-T6 aluminum alloys and of SAE 4130 steel*. Report NACA TN 2324 (1951).

8. Grover, H.J., Bishop, S.M. and Jackson, L.R., Fatigue strengths of aircraft materials. Axial load fatigue tests on notched sheet specimens of 24S-T3 and 75S-T6 alumimum alloys and of SAE 4130 steel with stress-concentration factors of 2.0 and 4.0. Report NACA TN 2389 (1951).

9. Dang-Van, K., *Macro-micro approach in high-cycle mutiaxial fatigue. Advances in multiaxial fatigue.* ASTM STP 1191, McDoell, D.L. and Ellis, R. (Eds.) (1993), pp. 120–130.

10. Juvinall, R.C., *Engineering Considerations of Stress, Strain and Strength.* McGraw-Hill, New York (1967).

Some general references

11. *SAE Fatigue Design Handbook*, 3rd edn. AE-22, Society of Automotive Engineers, Warrendale (1997).

12. *Fatigue Data Book: Light Structural Alloys.* ASM International (1995).

Chapter 8
Fatigue Crack Growth. Analysis and Predictions

8.1 Introduction

In Chapter 2 the fatigue life until failure has been divided into two periods:
(i) the crack initiation period, and (ii) crack growth period. Crack nucleation
and microcrack growth in the first period are primarily phenomena occurring
at the material surface. The second period starts when the fatigue crack
penetrates into the subsurface material away from the material surface.
The growth of the fatigue crack is then depending of the crack growth
resistance of the material as a bulk property. The two previous chapters,

Chapters 6 and 7, mainly deal with fatigue in the crack initiation period. The subject of the present chapter is fatigue crack growth in the second period. It could also be referred to as the growth of macro fatigue cracks.

Under which conditions is crack growth in the second period of practical interest? Obviously, the load spectrum should contain stress cycles above the fatigue limit in order to have a fatigue crack problem. Secondly, some macrocrack growth must be acceptable, but it should then be known how fast crack growth occurs. Two well-known examples are:

(i)   Crack growth in sheet material where the crack is growing through the full thickness of the material. An obvious example is fatigue crack growth in aircraft skin structures.

(ii) A second example is the growth of part through cracks, see Figure 5.3 where a corner crack or a surface crack starts at a hole. Part through cracks also occur as surface cracks in welded structures at the toe of a weld. In many practical cases, part through cracks are associated with massive components and thick plate structures.

Cracks can be acceptable for different reasons. It is possible that cracks do not have significant safety or economic consequences. The more serious cases arise if safety, or economy, or both are involved. That occurs in pressure vessels if a fatigue crack can induce an explosion. Another category of problems is related to fatigue crack growth monitored by periodic inspections. The purpose of the inspections is to discover fatigue cracks before they become dangerous. It then is necessary to know how fast cracks are growing in order to set timely inspection periods.

Fatigue crack growth experiments are described in Section 8.2, including the evaluation of crack growth data by using the stress intensity factor K and the similarity principle. Aspects of crack growth in different regions of ΔK are discussed in Section 8.3. The so-called crack closure phenomenon and plane strain/plane stress aspects are considered in Section 8.4. Crack growth fatigue properties of different materials are presented in Section 8.5. The problem of predicting fatigue crack growth is discussed in Section 8.6. It includes both through cracks and part through cracks. Fatigue crack growth under Variable-Amplitude loading is not covered in the present chapter. This problem is addressed in Chapter 11. The last section summarizes the major topics of the present chapter.

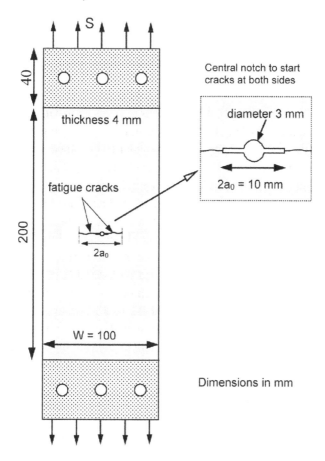

Fig. 8.1 Centre cracked tension (CCT) specimen for fatigue crack growth tests. Dimensions (mm) are given as an example.

8.2 Description of fatigue crack growth properties

8.2.1 Test results

A crack propagation test can be carried out on a simple sheet or plate specimen with a central crack as shown in Figure 8.1. The specimen is provided with a sharp central notch for rapid crack initiation. The crack starter notch in Figure 8.1 consists of a small hole with two saw cuts at both sides of the hole. Two fatigue cracks grow from either side of the starter notch. If the starter notches are made with a fret-saw which may be a convenient procedure, the tip of the notch has two sharp corners. As a result cracks can start simultaneously at both corners and the initial crack growth can be somewhat erratic. A starter notch with a single sharp notch should be preferred. Spark erosion is usually adopted for that purpose.

The lengths of a through crack (a) is measured from the centre line of the specimen to the tip of the crack. In general, the two cracks grow symmetrically, i.e. the length of both cracks remains approximately the same. The total tip to tip crack length can then be indicated by a single value ($2a$). It is assumed that the crack front is perpendicular to the plane of the specimen, which implies that the central crack has only one dimension, the crack length. The dimensions of the specimen in Figure 8.1 are given as an example, but other dimensions can be adopted, see the ASTM standard E647 [1].

The specimen must be clamped at the ends to apply the fatigue load in the testing machine. The clamping should ensure a homogeneous load distribution at the specimen ends. The specimen in Figure 8.1 is clamped at the ends between two steel plates fastened by a number of bolts. The steel plates are then mounted in the testing machine. Crack growth is recorded by periodic observations of the locations of the two crack tips. It can be done with simple means, e.g. a magnifying glass and some length scale attached to the specimen. More advanced crack length measurement techniques are also used, e.g. the potential drop technique (see Chapter 13) which allows automatic crack length measurements recorded by a computer.

The most simple representation of crack growth records is a graph with the crack length data plotted as a function of the number of cycles, see Figure 8.2a. In this figure, results are given for a low and a high stress level. Both curves start at the same initial crack length, a_0, with $2a_0$ as the tip to tip length of the crack starter notch (Figure 8.1). The slope of the crack growth curves is da/dN which is the crack growth rate, e.g. in mm/cycle. The slope as determined along the crack growth curves is plotted in Figure 8.2b to obtain the crack growth rate as a function of the crack length. The data in Figure 8.2 are fictitious. They are used here for illustration purposes.

8.2.2 The stress intensity factor and the similarity concept

As shown in Figure 8.2b, the crack growth rates for the high and low S_a are partly overlapping. It implies that similar crack growth rates occurred in the two tests, although at different values of the crack length. This phenomenon was recognized by Paris et al. [2] which led to the application of the similarity principle based on the stress intensity factor. As discussed

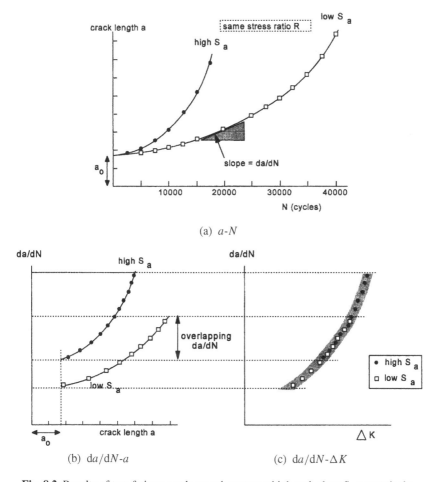

(a) *a-N*

(b) da/dN-a   (c) da/dN-ΔK

Fig. 8.2 Results of two fatigue crack growth tests at a high and a low S_a respectively.

in Chapter 5, the stress intensity factor K is a parameter indicating the severity of the stress distribution around the tip of a crack. If the cyclic stress varies between S_{min} and S_{max} then the corresponding stress intensity factor varies between K_{min} and K_{max}, see Figure 8.3. Because of $K = \beta S\sqrt{\pi a}$ (Equation 5.7 with β as the geometry factor) the stress ratio R is the same for the cyclic stress and the cyclic K-value:

$$R = \frac{S_{min}}{S_{max}} = \frac{K_{min}}{K_{max}} \qquad (8.1)$$

In order to explain the similarity principle for fatigue crack growth, two similar centre cracked tension (CCT) specimens are shown in Figure 8.4. In an ASTM Standard [1] this specimen is labeled as an M(T) specimen,

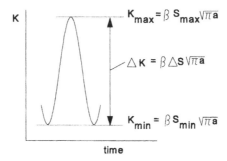

Fig. 8.3 The variation of the stress intensity factor in a load cycle.

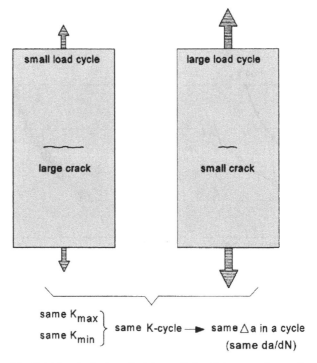

Fig. 8.4 The similarity principle applied to fatigue crack growth.

M from middle, T from Tension). One specimen with a large crack is loaded by a low cyclic stress. The other specimen has a small crack, but it is loaded by a high cyclic stress. The crack lengths and cyclic stresses are chosen in such a way that the corresponding K cycles are the same. In other words, the same $K_{\min}$ and $K_{\max}$ apply to both specimens. According to the similarity principle, the same fatigue process should then happen at the crack tips in both specimens. Thus the same crack extension, Δa, should occur because the same K cycle is applicable. The crack growth rate, i.e. the crack extension in a single cycle, should be the same. The conclusion is that the crack growth rate according to the similarity concept must be a function of K_{min} and K_{max} of the load cycle.

$$\frac{da}{dN} = f(K_{max}, K_{min}) \tag{8.2}$$

The K cycle can also be defined by ΔK ($= K_{max} - K_{min}$) and the stress ratio R ($= K_{min}/K_{max}$). The crack growth function in Equation (8.2) can thus be replaced by

$$\frac{da}{dN} = f(\Delta K, R) \tag{8.3a}$$

or

$$\frac{da}{dN} = f_R(\Delta K) \tag{8.3b}$$

It implies that da/dN is a function of ΔK, but this function depends on the stress ratio R. The da/dN results of Figure 8.2b for a high and low ΔS ($= 2S_a$), both obtained at the same R, are plotted as a function of ΔK in Figure 8.2c. The da/dN results of the two stress levels occur in the same scatter band, in accordance with Equations (8.2) and (8.3). The results of the two stress levels overlap in agreement with Figure 8.2b.

Results on fatigue crack growth rates in CCT specimens of an Al-alloy (2024-T3 Alclad) [3] are presented in Figure 8.5a. Tests were performed at two R-values (0.52 and -0.05), and at two S_a-values for each R-value. The results confirm that the crack rate data for different S_a-value, but the same R-value, are in the same scatter band in agreement with Equation (8.3b). However, for different R-values different scatter bands are obtained, again in agreement with Equations (8.2) and (8.3). Figure 8.5b is discussed in Section 8.4.1.

Figure 8.4 was used to illustrate the similarity of fatigue cracks in two similar specimens. However, the similarity of the shape of the specimen is not essential for the application of the similarity principle. One of the two specimens in Figure 8.4 could be a structural component. The shape is accounted for by the stress intensity factor K, or more in particular by the geometry factor β in $\Delta K = \beta \Delta S \sqrt{\pi a}$. Empirical results of a CCT specimen can thus be used to predict crack growth rates in a structure if ΔK as a function of the crack length in the structure can be calculated. This application of the similarity principle implies that results of a fatigue crack growth test on a simple CCT specimen can be used to predict fatigue crack growth in a structure. Instead of a CCT specimen, a compact tension

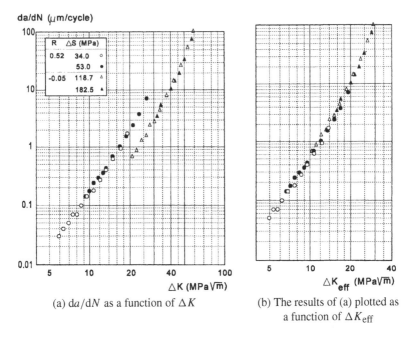

(a) da/dN as a function of ΔK (b) The results of (a) plotted as
                                                    a function of ΔK_{eff}

Fig. 8.5 The crack growth rate in sheet specimens of an Al-alloy (2024-T3) for two R-values and two S_a-values for each R-value [3].

(CT) specimen (see Figure 5.12) can also be used to determine the basic fatigue crack growth properties of a material. The larger and more simple CCT specimen should be preferred as discussed in Section 5.4.

Although the similarity principle leading to da/d$N = f_R(\Delta K)$ is a most versatile approach, it should be recalled that this principle does not say anything about the fatigue crack mechanism and about how much crack extension should occur in a ΔK cycle. The similarity principle only says that Δa in similar ΔK cycles should be the same. The amount of crack extension in a ΔK cycle must be determined in experiments, which can be carried out on simple specimens. The da/d$N = f_R(\Delta K)$ relation characterizes the *fatigue crack growth resistance of a material*.

8.2.3 Constant-ΔK tests

It can be desirable for fatigue research to keep experimental conditions as much constant as possible in order to study the effect of a single variable on the fatigue mechanism. This has led to constant-ΔK tests, i.e. fatigue crack growth tests in which ΔK remains constant during crack growth. Because ΔK increases in a CCT specimen tested under a constant stress cycle, the stress must be reduced during crack growth in order to keep $\Delta K = \beta \Delta S \sqrt{\pi a}$ at a constant level. The increase of $\beta \sqrt{\pi a}$ due to an increasing crack length must be compensated by a reduction of ΔS on the specimen. This can be done in computer controlled closedloop fatigue machines. It requires a continuous crack length measurement, which is usually done with a potential drop technique, see Chapter 13. The reduction of the cyclic load on the specimen is referred to as load shedding. It has been amply shown that the crack growth rate is approximately constant in constant-ΔK test as should be expected according to the similarity principle. However, it should be realized that a constant-ΔK test gives only one $\mathrm{d}a/\mathrm{d}N$-value for a single ΔK-value applied in the test. Such tests are not cost-effective for the purpose of obtaining an experimental $\mathrm{d}a/\mathrm{d}N$-ΔK correlation. But if the purpose of the test is to study the morphology of the fatigue fracture surface, it is advantageous to have a constant crack growth rate for a substantial part of the fracture surface. Also for research of other aspects constant-ΔK tests can be attractive. Some examples are presented in Chapter 11.

8.3 Fatigue crack growth regions

Fatigue crack growth results as shown in Figure 8.5 cover a range of ΔK-values and crack growth rates. It does not give indications about crack growth rates outside this range. More extensive experiments have shown that two vertical asymptotes occur in a $\mathrm{d}a/\mathrm{d}N$-ΔK graph, see Figure 8.6. The left asymptote at $\Delta K = \Delta K_{\mathrm{th}}$ indicates that ΔK-values below this threshold level are too low to cause crack growth. The other asymptote at the right-hand side occurs for a ΔK cycle with $K_{\max} = K_c$. It means that $K_{\max}$ reaches a critical value which leads to complete failure of the specimen. If $\mathrm{d}a/\mathrm{d}N$ is plotted as a function of ΔK on a double log scale, the function $\mathrm{d}a/\mathrm{d}N = f_R(\Delta K)$ is supposed to cover three different parts, indicated by I, II and III in Figure 8.6. The corresponding ΔK-regions are referred to as: (i) the threshold ΔK-region, (ii) the Paris-ΔK-region, and (iii) the stable tearing crack growth region.

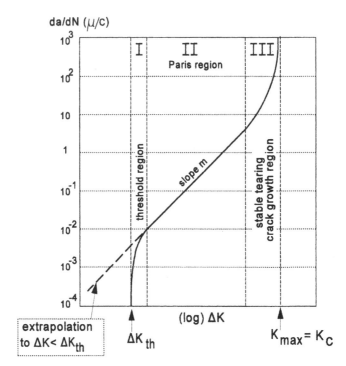

Fig. 8.6 Three regions of the crack growth rate as a function of ΔK.

The threshold region

The threshold region is not associated with the non-propagating microcracks discussed in Chapter 2 (Section 2.5.3). These microcracks were nucleated at the material surface, but due to microstructural barriers, they could not penetrate into the material. The cracks remained microcracks. The threshold value ΔK_{th} in this section is associated with macro fatigue crack growth. It implies that ΔK_{th} is concerned with fatigue cracks which have grown to a macroscopic size at a ΔK level above ΔK_{th}. If ΔK is then decreased, crack growth slows down and it is assumed that no further growth occurs if ΔK is below ΔK_{th}.

The question is whether a unique threshold level for crack growth, ΔK_{th}, does exist, and how this level could be determined in fatigue tests. Obviously, ΔK must be decreased in a test to obtain zero crack growth at the threshold level. However, this can be done in different ways, see Figure 8.7. The K reduction should be done in small steps, actually smaller steps than suggested in Figure 8.7. After each step, a large number of cycles (ΔN) must be applied to see whether the crack is still growing or has been arrested. If the crack

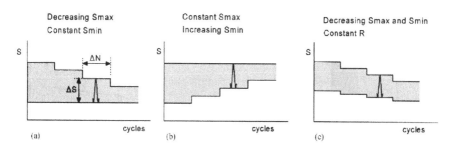

Fig. 8.7 Different ways to reduce ΔS in steps to determine ΔK_{th}.

does not growth any further, K of the last ΔS is just below ΔK_{th}. Load sequence (a) in Figure 8.7 is not advised because a reduction of S_{max} implies a reduction of K_{max} and thus a reduction of the crack tip plastic zone size. This can have a retardation effect on crack growth in the following ΔS period, an interaction effect to be discussed in Chapter 11. A retardation effect should not be expected in load sequence (b) of Figure 8.7, where S_{max} is kept constant and S_{min} is raised in small steps. If crack growth is arrested the corresponding R ratio (S_{min}/S_{max}) can be calculated. If ΔK_{th} should be determined for a fixed R ratio, both S_{max} and S_{min} have to be reduced simultaneously with the same percentage to keep the R-value at a constant value, see sequence (c) of Figure 8.7.

Experimental evidence has shown that ΔK_{th} is not a single material constant because it depends on the stress ratio R. Recommendations for ΔK_{th} tests are given in an ASTM Standard [1]. A problem of ΔK_{th} tests is to decide whether crack growth has come to a stop. According to the standard, this occurs if the crack growth rate is of the order of 10^{-10} m/cycle corresponding to a crack extension of just 1 mm (0.04") in 10^7 cycles, indeed a small extension. Such tests are time-consuming.

A different method to determine ΔK_{th} employs a special type of specimen with a decreasing K-value for an increasing crack length. The advantage is that a constant cyclic load can be used. The test should be continued until crack arrest occurs. As discussed in Chapter 5 (Section 5.4), crack edge loading gives a decreasing $K(a)$ function. Already in 1968, Figge and Newman [4] published results of such specimens in comparison to results of CCT specimen. Some of their crack growth curves are presented in Figure 8.8. The two crack edge loaded specimens tested at a different load level show indeed a decreasing crack growth rate for an increasing crack length which is in agreement with the decreasing K-value. Evaluation of the data of the three curves in terms of da/dN as a function of K indicated that

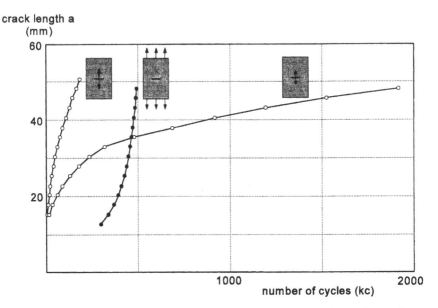

Fig. 8.8 Increasing crack growth rate in a CCT specimen, and a decreasing crack growth rate in two crack edge loaded specimens [4] (material Al-alloy 7075-T6, $R = 0.05$).

all results occurred in the same scatter band. This confirms the applicability of the similarity principle. Although it would be experimentally attractive to use specimens with a decreasing $K(a)$ function for research on the threshold behavior this is not reported in the literature.

It would be useful for crack growth predictions if the same correlation between da/dN and K could be applicable to the entire range from very small cracks to large cracks. The problem has attracted much attention for Al-alloys because experimental results have shown that the da/dN-ΔK correlation can apply to cracks as small as 100 μm [5]. Wanhill [6] obtained the results shown in Figure 8.9 with data of macrocracks and microcracks. The macrocrack test results are partly from decreasing ΔK tests. Apparently, a ΔK_{th}-value is found for the macrocracks, and the data for increasing ΔK and decreasing ΔK-values fit into the same scatter band. Wanhill also presented data on microcracks, see the data points in the same Figure 8.9. Apparently, the microcracks were growing with a relatively high growth rates at ΔK-values below ΔK_{th} of the macrocracks. Scatter was large, probably due to an initially erratic growth of microcracks discussed earlier in Chapter 2 (Section 2.5.3).

Two questions may now be raised: (i) Why do macrocracks stop growing at K-levels at which microcracks are growing, and (ii) what is the technical

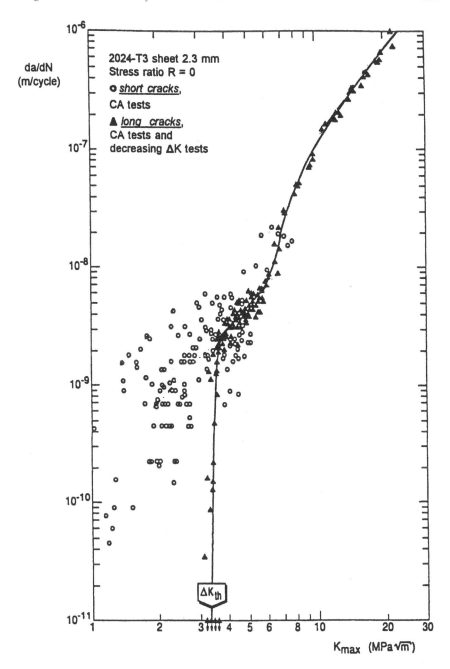

Fig. 8.9 Crack growth results of Wanhill for large and small cracks [6].

significance of ΔK_{th}? The first question appears to be a paradox, but as discussed in Section 2.5.3, cyclic slip can occur more easily at the material surface. It implies that the conditions are favorable for microcrack nucleation and initial growth. Furthermore, it should also be recognized that the stress intensity factor may not be a meaningful concept for a microcrack. This factor was defined to describe the stress distribution in the close proximity of the tip of a crack in a homogeneous material provided that the plastic zone is relatively small. These conditions are just not satisfied for microcracks with a size of 1 or 2 grain diameters. The literature on small cracks and crack growth at K-values below ΔK_{th} is rather extensive, see e.g. [29 and 34] in Chapter 2.

With respect to the second question, the crack driving force of a slowly growing macrocrack is low. If the effective ΔK is very low it becomes more difficult to activate the slip systems which should contribute to the fatigue crack extension mechanism at the crack tip. The mechanism becomes a more erratic process with deviating crack growth directions on a microlevel, and crack growth is impeded. Microstructural characteristics can become important again. Moreover, the crack driving force (K) for a tortuous crack front is lower than for a straight crack front. The very crack tip is not easily opened any more, and the crack stops growing. This process is not easily described in detail, while it can also be different for different materials. However, the engineering significance of ΔK_{th} may be limited as illustrated by the results in Figure 8.9. In view of the fatigue crack growth resistance of a material, it is more realistic to extrapolate the da/dN-ΔK relation in the Paris regime further down to low ΔK-values below the ΔK_{th}-value. The extrapolation is shown as a dotted line in Figure 8.6.

In the literature, it was occasionally proposed to adopt ΔK_{th} if a defect was found in a structure is service. A practical question is whether the crack or defect will grow any further under the anticipated service load spectrum. Safety or economic arguments may be involved. It might be assumed that further growth does not occur if ΔK_{th} is not exceeded, but this is not necessarily a safe assumption. The fatigue crack front or the edge of the defect need not be similar to the crack front occurring in the threshold experiment at the moment of crack arrest. Conditions for applying the similarity principle are not really satisfied. The threshold fatigue crack can be practically closed, whereas a defect may be an open interruption of the material, which can be propagated more easily than a fatigue crack. Take care.

The Paris region

According to Paris [7], the relation between da/dN and K can be described by a power function:

$$\frac{da}{dN} = C\Delta K^m \tag{8.4}$$

with C and the exponent m as material constants. The equation in a double log plot gives a linear relation: $\log(da/dN) = log(C) + m\log(K)$ with m as the slope of the linear function, see Figure 8.6. Equation (8.4) has some limitations. It does not account for the R-effect on crack growth, neither for the asymptotic behavior in regions I and III. Several alternative functions have been proposed in the literature to overcome this problem. Some are mentioned below. Forman [8] proposed the following:

$$\frac{da}{dN} = \frac{C\Delta K^m}{(1-R)(K_c - K_{max})} \tag{8.5}$$

Because of the term $(K_c - K_{max})$ in the denominator, da/dN will become very large if K_{max} is approaching K_c. The right asymptote in Figure 8.6 is thus taken care of by Equation (8.4). The effect of the stress ratio R is also accounted for by the term $(1 - R)$ in the denominator. But the ΔK_{th} asymptote is not yet included. This is obviously done in the numerator of an equation proposed by Priddle [9]:

$$\frac{da}{dN} = C\left[\frac{\Delta K - \Delta K_{th}}{K_c - K_{max}}\right]^m \tag{8.6}$$

More similar equations satisfying the two asymptotic conditions are possible, e.g. a similar function as Equation (8.6), but with different exponents m_1 and m_2 for the numerator and the denominator respectively. The value of ΔK_{th} in such equations is supposed to be a function of R, for which again functions were proposed in the literature, e.g. by Klesnil and Lukáš [10]:

$$\Delta K_{th} = A(1-R)^\gamma \tag{8.7}$$

Constants in the above equations depend on the type of the material. It should be understood that none of the formulas can claim a physical background. They are proposed to agree with trends observed in test results. It then requires some regression analysis to arrive at the values of the constants. It sometimes requires a K_c-value different from the measured fracture toughness. It is suggested in the literature that such equations are useful in view of prediction algorithms but this argument is not really appropriate, see also the discussion in Section 8.6. However, a useful indication can be drawn from the Paris relation (Equation 8.4) if the exponent m is known from experiments. Substitution of the K formula in the Paris relation gives

$$\frac{da}{dN} = C(\beta \Delta S \sqrt{\pi a})^m = C(\beta \sqrt{\pi a})^m \Delta S^m \tag{8.8}$$

If the design stress level on a structure is changed, the value $C(\beta \sqrt{\pi a})^m$ remains the same because it depends on the crack length only. A different design stress level affects ΔS but it leaves the R ratio unaffected. As a consequence, the crack growth rate is changed proportionally to ΔS^m. Because that is true for any crack length, the crack growth life is inversely proportional to ΔS^m. As an example, consider an increase in the stress level of 25%. For $m = 3$, which is the order of magnitude for several materials, the crack growth rate is increased with a factor $1.25^3 = 1.95 \approx 2$. It implies that the crack growth life in the Paris region is approximately halved.

Stable-tearing crack growth region

The crack growth rate in the stable-tearing crack growth region is high, in the order of 0.01 mm/cycle and above. Observations on the fatigue fracture surface in the electron microscope still reveal patches of fatigue striations caused by a number of successive cycles. However, the striation spacing is generally smaller than da/dN derived from the crack growth curve $a(N)$. Local areas of ductile tearing are observed between the patches of striations. Ductile tearing does not yet occur along the entire crack front, and for that reason crack growth is still a stable process. Further crack growth requires additional cyclic loading, but the stable ductile tearing areas indicate that unstable final failure is imminent. The crack growth life spent in this region is very short, which implies that its engineering significance is limited.

Predictions on the occurrence of the final unstable failure are apparently simple, because it should occur if $K_{max} = K_c$. By definition, K_c is the stress intensity factor causing final failure. However, usually K_c for structural materials is not a constant material property. Final failure for several materials with some ductility occurs when the remaining net section is plastically yielding over its entire width (net section yielding). In such cases, the value K_c as a stress intensity factor is meaningless. The stress intensity factor is based on elastic material behavior with small-scale yielding at the crack tip only. Moreover, in K_c experiments substantial stable crack extension can occur before the unstable situation is reached. It depends on the dimensions of the specimen.

8.4 Crack closure

8.4.1 Plasticity induced crack closure and ΔK_{eff}

A tension stress on a cracked specimen gives crack opening as discussed in Section 5.7 (see also Figure 5.18). Equations were presented for crack edge displacements. According to the equations based on elastic material behavior, crack opening displacements return to zero if the tension stress on a specimen is removed. A compression stress should induce negative crack edge displacements, which is physically impossible. The crack will be closed under compression, and the compressive stress can be transmitted through the cracked area.

In the late sixties, Elber [11] discovered that a fatigue crack under a tension load was already closed during unloading before the tension stress became zero. It implies that the crack tip is closed at a positive tension stress, which was an unexpected result at that time. This observation has significant consequences for crack growth predictions. How is it possible that crack closure occurs at a positive stress, and how can it be shown to occur in experiments. Furthermore, what are the consequences for fatigue crack growth predictions based on the similarity principle? A sheet specimen with a central crack is considered for a discussion on these questions. During cyclic loading, plastic deformation occurs at the crack tip. Although this is "small-scale yielding", it implies that a plastic zone is created when the stress goes from S_{min} to S_{max} ("uploading"). The size of the plastic zone at S_{max} is proportional to $(K_{max}/\sigma_{yield})^2$, see Equations (5.35) and (5.36). The plastic zone is plastically elongated in the loading direction. It becomes longer than it was before. As a consequence, the zone is loaded in compression during unloading, and reversed plasticity occurs. As pointed out by Rice [12], reversed plasticity requires a local stress increment in the reversed direction in the order of twice σ_{yield}. It implies that the reversed plastic zone size should be in the order of 1/4 of the plastic zone created during loading. Thus the reversed plastic zone is significantly smaller than the plastic zone obtained when S_{max} is reached as schematically indicated in Figure 8.10. Plastic deformation in the much larger area of the plastic zone outside the reversed plastic zone has occurred only during loading to S_{max}. This is called monotonic plasticity, and it is this monotonic plastic deformation which is causing a permanent elongation in the loading direction.

Crack tip plasticity occurs in every cycle and the crack is thus growing through plastic zones of previous cycles. As a result, plastic deformation

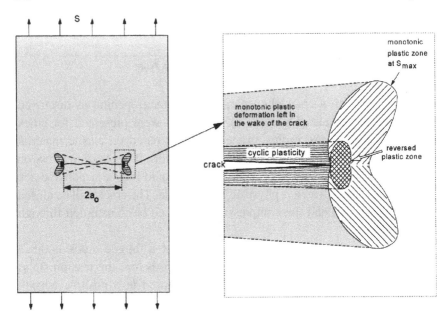

Fig. 8.10 Plastic zones of a growing crack leave plastic deformation in the wake of the crack.

is left in the wake of the crack. In this plastic wake field, the larger part of the material has been subjected to monotonic plasticity during uploading (Figure 8.10). A much smaller rim of material along the crack edges has been subjected to cyclic plasticity. The material in the plastic wake field is plastically extended in the loading direction. This explains why the crack can be closed during unloading while the specimen is still under a positive tensile stress, i.e. before the specimen is unloaded. The phenomenon is known in the literature as *plasticity induced crack closure*, sometimes labeled as the Elber mechanism.

Elber observed the occurrence of crack closure by measuring the crack opening displacement (COD) between points A and B at the centre of the specimen close to the crack edges, see Figure 8.11. He measured the COD as a function of the remote stress S. The COD-S record started with a non-linear part until $S = S_{op}$. Above this stress level, a linear relation was found as expected for an elastic behavior. Unloading from S_{max} to $S = 0$ occurred practically along the same COD-S record in the reversed direction as indicated in Figure 8.11. A specimen with a saw cut instead of a fatigue crack does not have a plastic wake field, and the COD-S record shows a fully linear behavior. If the fatigue crack and saw cut have the same length, the linear parts of the two COD-S records are parallel, see Figure 8.11. The

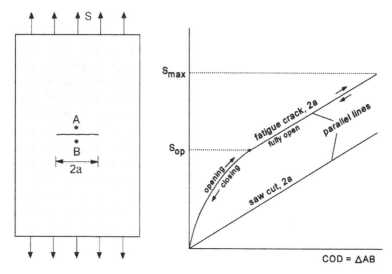

Fig. 8.11 Measurements of the crack opening displacement (COD) are confirming crack closure at a positive tensile stress.

same slopes indicate the same stiffness. Moreover, the slope agreed with the calculated stiffness for the specimen with an open crack. The fatigue crack thus must be fully open during the linear part of the COD-S record. For S lower than S_{op}, the slope of the non-linear part of the COD-S record becomes larger, the stiffness is higher, and the specimen behaves as if the crack is shorter. This is a consequence of the crack being partly closed due to the excess of plastically elongated material in the wake of the crack. Upon unloading, crack closure starts at the tip of the crack, which continues away from the crack tip at a decreasing stress level.

During cyclic loading, crack closure occurs if $S_{min} < S_{op}$ (Figure 8.12). The crack is partly (or fully) closed at S_{min}. The crack tip during loading is just fully open at $S = S_{op}$, and then remains open up to S_{max}. Upon unloading, the crack is open until crack closure starts at the crack tip. The corresponding closure stress level, S_{cl}, is usually assumed to be approximately equal to S_{op}. Although small differences between S_{op} and S_{cl} might occur, accurate measurements of these stress levels are difficult also because they are a transition point between a linear and a non-linear part of the COD-S record in Figure 8.11. The important aspect to be recognized is that the stress singularity at the tip of the crack, defined by the stress intensity factor K, is present as long as the crack tip is open. As soon as the tip is closed the stress singularity is no longer present at the tip of the crack. In other words, the large stress variations at the crack tip occurring if the crack

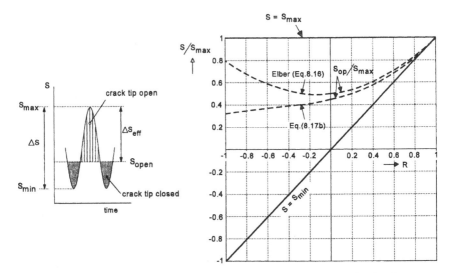

Fig. 8.12 Crack closure: the crack tip is open during a part of the load cycle.

is fully open do not occur any longer as soon as the crack tip is closed. Elber proposed that the stress variation will contribute to crack extension only if the stress singularity occurs at the crack tip. He defined an effective stress range ΔS_{eff} as shown in Figure 8.12. It implies that

$$\Delta S_{\text{eff}} = S_{\text{max}} - S_{\text{op}} \tag{8.9}$$

The corresponding effective stress intensity factor range is then

$$\Delta K_{\text{eff}} = \beta \Delta S_{\text{eff}} \sqrt{\pi a} \tag{8.10}$$

According to the concept of Elber, the fatigue crack growth rate depends on ΔK_{eff} only:

$$\frac{\mathrm{d}a}{\mathrm{d}N} = f(\Delta K_{\text{eff}}) \tag{8.11}$$

This relation includes the effect of the stress ratio R because the reversed crack tip plasticity depends on S_{min}, and as a consequence the plastic wake field of the crack depends on R. Elber carried out experiments on the Al-alloy 2024-T3 and found that S_{op} was approximately constant during a fatigue test. This empirical observation implies that S_{op} was independent of the crack length a and only dependent of the applied cyclic stress. Elber defined the ratio U as

$$U = \frac{\Delta K_{\text{eff}}}{\Delta K} \quad \left(= \frac{\Delta S_{\text{eff}}}{\Delta S} \right) \tag{8.12}$$

which is the percentage of the K-range (or ΔS-range) during which the crack tip is open. Elber's crack closure measurements on an Al-alloy (2024-T3) specimens indicated that the ratio U depends on the stress ratio R. Elber has used R-values in the range of 0.1 to 0.7 and he could describe the test results by a simple equation:

$$U = 0.5 + 0.4R \tag{8.13}$$

and thus

$$\Delta K_{\text{eff}} = (0.5 + 0.4R)\Delta K \tag{8.14}$$

The equation was used for Figure 8.5 to plot the results of Figure 8.5a as a function of ΔK_{eff} in Figure 8.5b. The graph confirms a very good correlation of da/dN with ΔK_{eff}. It still should be recognized that Equation (8.13) is a fit to empirical data obtained in tests carried out with R-values between 0.1 and 0.7. Extrapolation outside this range is not necessarily justified. Actually, Equation (8.13 cannot be valid for more negative R-values down to $R = -1.0$ ($S_m = 0$). This can easily be shown by considering the ratio $S_{\text{op}}/S_{\text{max}}$ as a function of R. With Equations (8.9) and (8.12):

$$\frac{S_{\text{op}}}{S_{\text{max}}} = \frac{S_{\text{max}} - \Delta S_{\text{eff}}}{S_{\text{max}}} = 1 - \frac{U\,\Delta S}{S_{\text{max}}} = 1 - \frac{U(S_{\text{max}} - S_{\text{min}})}{S_{\text{max}}} = 1 - U(1 - R) \tag{8.15}$$

Substitution of Equation (8.13) gives:

$$\frac{S_{\text{op}}}{S_{\text{max}}} = 0.5 + 0.1R + 0.4R^2 \tag{8.16}$$

This $S_{\text{op}}/S_{\text{max}}$ ratio is plotted in Figure 8.12 as a function of R. It shows that the ratio is increasing again for larger negative R-value. This is physically unrealistic. An improved function with a more realistic S_{op} behavior for negative R-values was proposed in [13]:

$$U = 0.55 + 0.33R + 0.12R^2 \tag{8.17a}$$

Substitution in Equation (8.15) gives

$$\frac{S_{\text{op}}}{S_{\text{max}}} = 0.45 + 0.22R + 0.21R^2 + 0.12R^3 \tag{8.17b}$$

The equation is also plotted in Figure 8.12 which shows a continuously decreasing S_{op} for a decreasing R-value. This trend should be expected because a decreasing R for the same S_{max} corresponds to a decreasing S_{min} (and also a decreasing S_m). The equation was checked in another test program on 2024-T3 sheet specimens including results for R-values between

−1 to 0.54 [14]. All results of da/dN as a function of ΔK_{eff} calculated with Equation (8.17a) were concentrated in a single scatter band. This is a useful result because it implies that crack growth data obtained at a certain R-value can be used to calculate crack growth data for other R-values.

As said before, the equation $da/dN = f(\Delta K_{\mathrm{eff}})$ implies that da/dN depends on ΔK_{eff} only. The similarity principle for fatigue crack growth should thus be rephrased as: a similar ΔK_{eff} in a cycle occurring in different specimens (or in a structure) should give the same da/dN in that cycle. Equations (8.14) and (8.17) suggest that some $U(R)$ function should be available. In the literature, such equations have been proposed for different materials [15]. In general, the equations were based on the analysis of crack growth data from simple specimens (CCT and CT), obtained for different R-values. It should be understood that such equations are empirical correlation functions to describe the R-effect on crack growth as obtained in fatigue crack growth tests. It may be recalled that the present discussion applies to fatigue crack growth under constantamplitude (CA) loading, while the description of the crack growth properties in terms of $da/dN = f(\Delta K_{\mathrm{eff}})$ presupposes that S_{op} remains at the same level during CA loading. Predictions of crack growth under variable-amplitude (VA) loading are also using the crack closure concept, but S_{op} under VA loading does not remain constant. The variation of S_{op} during VA loading must be predicted. The occurrence of plasticity induced crack closure is an important feature of modern prediction models on fatigue crack growth under VA loading. This topic is discussed in Chapter 11.

8.4.2 Plane strain/plane stress

Plasticity induced crack closure has also been the subject of elasto-plastic FEM calculations. Such calculations offer significant problems because of non-linear aspects including material plasticity as well as changing contacts between the fracture surfaces during crack closure and opening. Furthermore, the question whether plane strain or plane stress is applicable is another problematic issue which cannot be ignored because plastic zone sizes are essential for the plastic wake field of a crack (Figure 8.10). As discussed in Chapter 5, plane stress conditions with larger plastic zones are more significant for thin sheet material of ductile alloys; whereas cracks in thick plates of low-ductility material are better represented by plane strain conditions resulting in smaller plastic zones. It was confirmed by crack

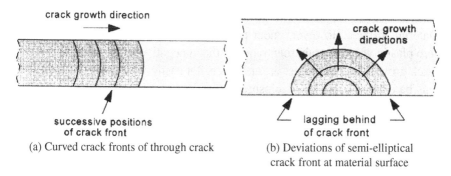

(a) Curved crack fronts of through crack

(b) Deviations of semi-elliptical crack front at material surface

Fig. 8.13 Effect of more crack closure at the material surface on crack front shapes.

closure measurements that S_{op} in thin sheet material could be significantly larger than in thick material for a similar stress cycle.

Interesting tests were done by Ewalds and Furnée [16] on centre cracked plate specimens of 2024-T3 Alclad, thickness $t = 10.2$ mm, width $W = 100$ mm, $S_{max} = 97$ MPa ($R = 0.1$). Cracks were grown to $a = 12$ mm and 18 mm respectively. Crack closure measurements indicated $S_{op} \approx 34$ MPa corresponding to $U = 0.72$ (crack fully open during 72% of the load cycle). The specimens were then made thinner by removing surface layers at both sides of the plate specimens. After a thickness reduction from 10.2 mm to 7.7 mm, the crack opening stress dropped from $S_{op} \approx 34$ MPa to 21 MPa ($U = 0.87$), and after a further thickness reduction to 3.75 mm, the result was $S_{op} \approx 19$ MPa ($U = 0.89$). Apparently, the first thickness reduction caused the major change of S_{op}. The conclusion was that crack closure occurs more at the material surface where plane stress conditions are relevant, and less at midthickness of the material where plane strain conditions prevail. Similar observations were made by Sunder [17] who developed an ingenious fractographic technique to measure S_{op} from striation patterns.

More crack closure at the material surface agrees qualitatively with the expected plastic zone sizes for plane stress and plane strain. It implies that fatigue cracks open first at midthickness and later at the material surface. Apparently, crack closure is a 3D mechanistic phenomenon. As a result of more crack closure at the material surface, the crack front is lagging behind where the crack front intersects the surface, which often leads to curved crack fronts for through cracks, see Figure 8.13a. It might be stated that crack closure is predominantly a surface phenomenon occurring under plane stress conditions only. The plastic wake field, see Figure 8.10, is practically elongated in the loading direction. Keeping in mind that the plastic volume strain should be zero, the elongation must be compensated by a negative plastic strain in the thickness direction (ε_z) which can occur indeed at the material surface. However, under pure plane strain conditions $\varepsilon_z = 0$. The zero plastic volume strain then requires that a negative plastic strain ε_x in the crack growth direction must occur. This is not easily visualized and it should thus be expected that crack closure is less important in thick plates than in thin sheets. However, crack closure measurements have shown that it is not absent under plane strain conditions.

A part through surface crack often shows a lagging behind of the crack front at the material surface, see Figure 8.13b. Also in this case it is due to more crack closure at the material surface. The shape of these cracks under cyclic tension usually is semielliptical, except at the ends where the crack front meets the free surface. In predictions on fatigue crack growth, these deviations of the crack front geometry are usually ignored.

Another complication by deviating crack front shapes is due to the shear lips at the material surface, see Figure 2.38. Shear lips do not occur in all materials, but they are observed for fatigue cracks in several steels and Al-alloys. However, a correlation with K for the same R-value, or with ΔK_{eff} for different R-values can still be satisfactory. The occurrence of shear lips does not necessarily upset the similarity condition if they occur in a similar way for the same K.

8.4.3 Thickness effect on fatigue crack growth

It was pointed out before that fatigue crack growth in thin sheet material can occur under predominantly plane stress conditions, while plane strain conditions prevail during crack growth in thick plates. A different crack closure behavior for the two states of stress can lead to a material thickness effect on fatigue crack growth. Plane strain conditions imply less cyclic crack tip plasticity. However, it also leads to a higher peak stress at the crack tip, which enhances the conversion of cyclic slip into crack extension. Empirical evidence must thus be considered to see whether there is a systematic thickness effect on fatigue crack growth. The literature on this question is not abundant. Moreover, if an effect is found, the results of different sources do not always show the same systematic trends. Fatigue crack growth in Al-alloy sheet material seems to be slower in thinner sheets, whereas the thickness effect for steel can be rather small. It should be kept in mind that the material structure of thick plates and thin sheets can be different depending on the production technique to obtain the various thicknesses. Anyway, it is advisable that crack growth data to be used for predictions should be obtained with specimens of a similar thickness as used in the structure for which predictions are required.

The thickness effect is considered again in Chapter 11 on fatigue crack growth under VA loading. It will turn out that a large thickness effect can occur under VA loading, much larger than observed under CA loading.

8.4.4 Other crack closure mechanisms

After the discovery of plasticity induced crack closure by Elber, other mechanistic possibilities were recognized for crack closure in the wake of the crack at a positive tensile stress on a specimen, noteworthy by the research of Ritchie and coworkers [18]. Oxide layers on the freshly created fatigue fracture surfaces have a larger volume than the originally uncorroded material. Another wake field phenomenon is related to the question whether the upper and lower fracture surfaces of a fatigue crack perfectly fit after closing of the crack during unloading. Fractographic observations have indicated that a mismatch between the topography of the upper and lower fracture surface of a fatigue crack can occur, especially if crack growth appears to be a complicated phenomenon. This can happen in more exotic materials with a complex material structure. But mismatching is also possible in more homogeneous materials . Moreover, small transverse displacements (out-of-plane, mixed mode I and III) may cause sliding contacts between both fracture surfaces of a fatigue crack. An example is contacts between shear lips of fatigue cracks in Al-alloys. The sliding contacts between the shear lips are causing black fretting products. Sliding contact should also be expected if the crack is growing with a tortuous crack front. It can be enhanced by crack branching and asperities between the two fracture surfaces. There are several phenomena which will affect the significance of K by some complex interaction between the two fracture surfaces. The associated effect on the stress intensity is sometimes labeled as *roughness induced crack closure.* Another term for various effects on crack tip opening and closing is *crack tip shielding* [18]. Anyway, it is important to realize that relevant da/dN-K data for prediction purposes should come from crack growth tests on identical material as the material of the structure, preferably with the same thickness.

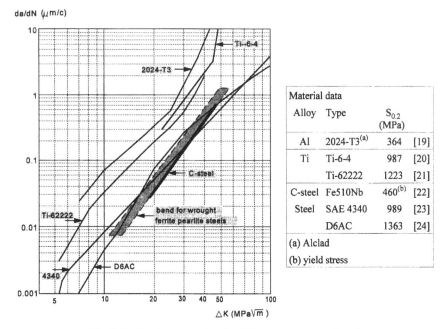

Fig. 8.14 Crack growth data for different materials, $R = 0$.

8.5 Crack growth data of different materials

As shown in Section 8.2, the fatigue crack growth properties of a material can be described by the correlation between da/dN and K for constant values of the stress ratio R (Equations 8.3a and 8.3b). The correlation can be presented in graphs or by empirical functions based on the test data (Section 8.3). Such a graph represents the correlation between the *crack driving force* (K or ΔK_{eff}) and the *crack growth resistance of the material* (da/dN). The crack growth resistance depends on the type of material, and for a certain type of alloy, on the strength level of the material as obtained by the material production including the heat treatment. It also can depend on the loading direction in relation to material directions (e.g. parallel or perpendicular to the rolling direction; anisotropy effect).

Illustrative examples of crack growth properties of materials are shown in Figures 8.14 to 8.16. Figure 8.14 shows crack growth results of several materials, one Al-alloy, two Ti-alloys, a low carbon steel, two high-strength steels, and a group of ferrite pearlite steels. The strength level of the materials is indicated in the table of Figure 8.14 by the yield stress, $S_{0.2}$. All crack growth data in this figure are for $R \approx 0$. The highest crack rates

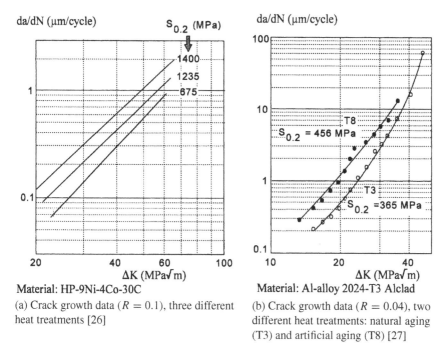

Material: HP-9Ni-4Co-30C

(a) Crack growth data ($R = 0.1$), three different heat treatments [26]

Material: Al-alloy 2024-T3 Alclad

(b) Crack growth data ($R = 0.04$), two different heat treatments: natural aging (T3) and artificial aging (T8) [27]

Fig. 8.15 Heat treatment effects on the crack growth properties of a high-strength steel and an Al-alloy.

occur in the aluminium alloy, although this 2024-T3 alloy has a relatively good fatigue crack growth resistance if compared to other high-strength Al-alloys. Crack growth rates in the two titanium-alloys are significantly higher than for the steels. Figure 8.14 also shows that the crack growth rates for steels with a highly different yield strength are not widely different. More data can be found in the literature which show the same trends. It implies that an increased yield stress obtained for low-alloy high-strength steel do not offer a substantial increase of the crack growth resistance. In other words, the fatigue crack growth resistance of steels with a very high $S_{0.2}$ can be relatively poor. This observation is of practical interest in view of consequences for selecting a high yield strength material for structural application. It was already pointed out in Chapter 7 that high-strength materials are also fatigue notch sensitive. It now appears that the fatigue crack growth resistance of these materials may also be critical. Designers should consider these aspects if they prefer to select a high-strength material for weight reduction or other reasons.

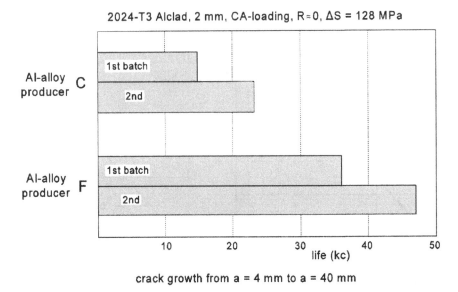

crack growth from a = 4 mm to a = 40 mm

Fig. 8.16 Comparison between crack growth lives of sheet specimen of different producers and different batches [27].

The high sensitivity for fatigue crack growth of strong materials is illustrated by two simple calculations using the data of Figure 8.14. The calculations will give an indication for comparing materials with a considerably different yield stress. The calculations are made for a small semi-circular surface crack with crack depth a, see Table 8.1. The stress intensity factor for the semi-circular surface crack is obtained with the Newman–Raju equations [25] assuming that the crack is small compared to all other dimensions of the component. The K solution for the deepest point of the semicircular crack then becomes

$$\Delta K = 0.723\Delta S\sqrt{\pi a} \tag{8.18}$$

In view of the comparison between different materials, the cyclic stress ΔS is chosen to be 1/3 of the yield stress $S_{0.2}$. The results can then be instructive for material selection. In the first calculation, the crack growth rate da/dN is calculated for the moment that the fatigue crack has reached a depth of 2.5 mm. The K-values for five different materials are obtained by substitution of 1/3 $S_{0.2}$ as ΔS in Equation (8.18). The crack growth rate can then be read in Figure 8.14. The results in the third column of Table 8.1 illustrate that a high $S_{0.2}$ leads to a high fatigue crack growth rate. The lowest crack growth rate is obtained for the low-carbon steel (mild steel) with a low yield stress.

Table 8.1 Semi-circular surface crack. Crack growth rate at a crack depth of 2.5 mm, and crack depth at a crack growth rate of 0.1 μm/cycle.

Alloy	$S_{0.2}$ (MPa)	da/dN (μm/kc) at $a = 2.5$ mm	a (mm) at $da/dN = 0.1$ μm/cycle
2024-T3	364	23	7.4
C-steel	460	2	23.3
SAE 4340	989	50	4.3
D6AC	1363	170	1.8
Ti-62222	1223	300	1.0

In the second type of calculations, the crack size is determined for which the crack rate becomes 0.1 μm/kilocycle. The corresponding K-values are read in Figure 8.14, and the related crack size then follows from Equation (8.18). As shown by the last column in Table 8.1, the crack rate of 0.1 μm/kilocycle in already obtained at a very small crack size for the alloys with a high $S_{0.2}$ whereas this occurs at a much larger crack size in the low-carbon steel (mild steel) and the 2024-T3 Al-alloy. To catch a crack during inspection of a structure when the crack length is smaller than a few millimeters is rather tricky.

The yield stress of several materials can be modified by changing the heat treatment. It should be expected that the crack growth properties are also affected. Two examples are presented in Figure 8.15. The yield stress of a high-strength steel was significantly increased from 675 MPa for the material in the annealed condition to values of 1235 and 1400 MPa by additional heat treatments. As shown by Figure 8.15a, the increased $S_{0.2}$ leads to higher crack growth rates. It means that the fatigue crack growth resistance is significantly reduced. A similar trend is observed in Figure 8.15b for the Al-alloy 2024. The alloy is generally used in the naturally aged condition (T3). A 25% increased $S_{0.2}$ is obtained by artificial aging (T8 condition), but at the same time the fatigue crack growth rates are about twice as large.

A most drastic effect of a heat treatment was reported in [28] for a Ti-alloy (Ti-8Al-1Mo-1V). One heat treatment (duplex anneal) caused a 10 to 40 times larger crack growth rate than an other heat treatment (beta annealed) (tests at $R = 0.1$). The structure of the Ti-alloys can be significantly changed by a heat treatments due to the occurrence of two phases with a different lattice structure (hexagonal α, bcc β). The plastic deformation behavior of the α and β phases are essentially different. If such large differences can occur, it is not surprising that the crack growth resistance can vary considerably by changing the material properties by a heat treatment.

A technically important aspect of the variability of fatigue crack growth properties is related to the question whether these properties are the same for nominally similar materials, i.e. similar according to the material specification. Differences may occur between materials obtained from different producers, or between different batches of material from the same producer. Indications on this issue are available from an investigation on 2024-T3 sheet material procured from seven different aluminium companies. Results in Figure 8.16 show crack growth lives (averages of three tests) for sheet material obtained from two producers and sheets of two different batches from each producer. Apparently, the aluminium company F has produced sheet material with approximately two times longer crack growth lives than company C. Batch to batch differences were observed for both industries. The results could not be correlated to differences between mechanical properties, grain size, or chemical composition. The same investigation revealed that fatigue crack growth for specimens loaded in the transverse direction (i.e. perpendicular to the rolling direction) was about 40% faster than in specimens loaded in the longitudinal direction. The $S_{0.2}$ yield stress in the transverse direction was some 10% lower than in the longitudinal direction. Similar directionality effects have been reported in the literature for other materials although empirical evidence is limited. Such effects could be expected if there is some obvious directionality in the structure of a material, such as a crystallographic texture, or a banded two-phase structure, e.g. pearlite bands in C-steel. The literature is not clear about this issue. A related effect could be due to a different grain size and grain shape. This effect is not easily studied on technical materials because different grain dimensions are obtained as a result of different production techniques and heat treatments. More differences are then involved.

Kage et al. [29] studied fatigue crack initiation and propagation in small specimens of low-carbon steel with two different grain sizes (15 and 50 µm). It turned out that the grain size effect was large for the first 200 µm of crack growth, i.e. in the crack initiation period. However, afterwards the additional life time to failure was not affected by the microstructure. A grain size effect may be possible for macrocrack growth as long as the cyclic plastic zone is not larger than the grain size. Cyclic crack tip plasticity could then be subjected to some constraint by grain boundaries. According to Wanhill [30], such structural effects are responsible for kinks in a da/dN-ΔK graph in the

Paris region. Results in Figure 8.14 already indicate that a single slope in the entire Paris region is not always obtained.

Ti-alloys are noteworthy sensitive for "structural sensitive crack growth", also in the Paris region. Stubbington and Gunn [31] obtained a flat and apparently structure-insensitive fatigue fracture surface during crack growth in Ti-6Al-4V specimens tested at a high K_{max} (19 MPa$\sqrt{m}$). Different slip systems are activated and contribute to the flat fracture surface. However, at a low K_{max} (11 MPa$\sqrt{m}$), crack growth occurred on preferred crystallographic planes, which caused a rough fracture surface. It is noteworthy that fracture toughness tests on specimens precracked at the low K_{max} indicated $K_{Ic} = 73$ MPa$\sqrt{m}$ whereas for precracking at the high K_{max} the result was $K_{max} = 49$ MPa$\sqrt{m}$, which is a 1.5 times lower value. Actually, a rough fracture surface implies a more tortuous crack front with a larger length of the crack front. The crack driving force (see dU/da in Section 5.9) per unit length of the crack front will be smaller than for a flat crack front. At the same time it should be expected that the crack growth resistance of the material for a tortuous crack front will be larger than for a flat crack front. It implies that the balance between the crack driving force and crack growth resistance is essentially different for structure-sensitive and non-structure-sensitive crack growth. In passing it may be said here that crack tip branching also leads to slower crack growth.

A message of this section is that fatigue crack growth resistance of a material with a certain chemical composition cannot be considered to be a unique property of the material. It depends on a variety of influencing factors with the heat treatment as a most prominent one. The crack growth properties can also be affected by the material production technique as a result of a different material structure. Furthermore, thickness effects and directionality effects are possible. A large amount of data can be found in the literature, but it should be checked if the experimental conditions of the literature data are relevant for the purpose of the application. Actually, more confidence about the crack growth resistance of a material can easily be obtained by a few crack growth experiments. A crack growth test on a simple CCT specimen is not expensive.

8.6 Prediction of fatigue crack growth

8.6.1 Some basic aspects

The prediction of fatigue crack growth in the present chapter applies to constant-amplitude (CA) fatigue loading. The problem on fatigue crack growth under variable-amplitude (VA) loading is discussed in Chapter 11.

The application of the similarity principle on the prediction of fatigue crack growth in a structure requires two different types of information:

1. Crack growth data: $da/dn = f_R(K)$, representing the *crack growth resistance of the material*.
2. The stress intensity factor as a function of crack length in the structure: $K(a)$, which accounts for the *crack driving force* in the structure.

The incremental number of cycles (ΔN_i) required for an incremental crack extension (Δa_i) at a crack length a_i is obtained as

$$\Delta N_i = \frac{\Delta a_i}{(da/dN)_{a=a_i}} \qquad (8.19)$$

The crack rate should follow from $da/dn = f_R(K)$. The number of cycles for crack growth from an initial crack length a_0 to a final crack length a_f is then obtained by integration of Equation (8.19):

$$N_{a_0 \to a_f} = \int_{a_0}^{a_f} \frac{da}{f_R(\Delta K)} \qquad (8.20)$$

If the Paris relation is applicable, then:

$$\frac{da}{dN} = C\Delta K^m = C(\beta \Delta S \sqrt{\pi a})^m \qquad (8.21)$$

Substitution in Equation (8.20) gives the crack growth life:

$$N = \frac{1}{C\Delta S^m} \int_{a_0}^{a_f} \frac{da}{(\beta \sqrt{\pi a})^m} \qquad (8.22)$$

In general, the integral in this equation must be solved numerically because β is also depending on the crack length. An interesting aspect was already mentioned in Section 8.3. The crack growth life N is inversely proportional to ΔS^m, independent of the value of the integral. This implies that the effect of changing the design stress level can simply be estimated. The example of Section 8.3 is repeated here. The value of the exponent m of the Paris relation is of the order of $m = 3$. An increase of the design stress level by a factor of 1.25 will thus lead to a reduction of the crack growth life by a factor $1.25^3 = 2$.

Comments on the initial crack length a_0

An illustrative and simple calculation can be made with Equation (8.22). Consider a small crack in a large structure, which can imply that $\beta \approx 1$. With $\beta = 1$, the integral of Equation (8.22) can be solved analytically:

$$
\begin{aligned}
N &= \frac{1}{C(\Delta S \sqrt{\pi})^m} \int_{a_0}^{a_f} \frac{da}{a^{m/2}} \\
&= \frac{1}{C(\Delta S \sqrt{\pi})^m} \cdot \frac{1}{m/2 - 1} \left[\frac{1}{a_0^{m/2-1}} - \frac{1}{a_f^{m/2-1}} \right]
\end{aligned}
\tag{8.23}
$$

The Paris relation for C-steel in Figure 8.14 is represented by

$$
\frac{da}{dN} = 1.294 * 10^{-12} * \Delta K^{3.40}
$$

with da/dN in m/cycle and ΔK in MPa$\sqrt{\text{m}}$. After substitution of $C = 1.294 * 10^{-12}$ and $m = 3.40$ in Equation (8.23), the crack growth life was calculated for three crack growth ranges from a_0 to a_f, see Table 8.2.

Table 8.2 Illustrative crack growth life predictions for a carbon steel.

a_0 (mm)	a_f (mm)	crack growth life (kc)	ratio
5	50	382	1
5	100	419	1.1
1	50	3781	10

With the crack growth life for the crack length range of 5 to 50 mm as a basis for comparison, it turns out that extending the range at the upper side from 50 mm to 100 mm increases the life by 10% only. The difference is small because the crack rate is relatively high between $a = 50$ mm and $a = 100$ mm. However, starting the crack growth range at 1 mm instead of 5 mm, increases the crack growth life about 10 times. This large increase is due to the very low growth rates between 1 and 5 mm. This observation is of great practical interest. It confirms that a large part of the crack growth life is consumed by crack growth of a very small and practically invisible crack. At the same time, the question must be considered whether the similarity principle can be applicable at a crack length as small as $a = 1$ mm because of grain size and elastic anisotropy. Also the validity of the Paris relation for such small cracks is not generally supported by empirical results.

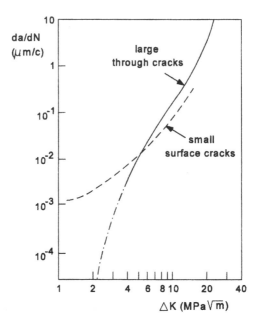

Fig. 8.17 Crack growth results of Pearson for microcracks and macrocracks in an Al-alloy (L65 ≈ 2014-T3) [32].

These questions are touching upon the problematic aspect of the prediction accuracy if small crack sizes are to be considered.

As discussed earlier, the physical meaning of the stress intensity factor for small cracks in the crack initiation period is rather limited, and application of the similarity principle may not be justified. A practical question then is: how large should cracks be to allow application of the similarity principle? Most research on this question was done on specimens of strong Al-alloys, probably because the application of fracture mechanics to fatigue crack growth in these alloys has been reasonably successful, also for small cracks. Pearson observed that small cracks in two Al-alloys mostly nucleated at inclusions and then showed a crack growth rate much higher than expected, see Figure 8.17 [32]. This graph agrees with the trend of Figure 8.9. In Pearson's graph, the curves for small cracks and large cracks intersect. At the intersection point the crack length of the small crack was about 0.13 mm (0.005"). For larger cracks, the similarity principle could give a more or less similar $da/dN \Delta K$ relation. Aluminium-alloys are a special group of materials because the elastic anisotropy of the aluminium crystal structure is low, and also because there are many slip systems with easy cross slip (see Chapter 2). It implies that the material can already behave as a homogenous material for small cracks. However, for other materials, and noteworthy for Ti-alloys, the similarity principle cannot be expected to be relevant for such small cracks as in Al-alloys. The application of the similarity principle for predictions of fatigue crack growth is questionable as long as smallcrack effects still occur, i.e. as long as the growth of small cracks is still a microstructurally sensitive phenomenon. The nominal stress intensity factor cannot be representative for the crack driving force, and also the crack growth resistance is not yet a well defined material characteristic. The minimum size of a crack for crack growth prediction using the $da/dN = f_R(\Delta K)$ relation for macrocracks depends on the type of material and its material structure. For certain materials it may be in the order of 1 to 3 mm.

8.6.1.1 Comments on the final crack length a_f

The prediction of the end of a crack growth life requires that the occurrence of the final failure is calculated. Usually it is assumed to occur if K_{max} reaches a critical value, K_{Ic} for plane strain conditions, and K_c for plane stress conditions. The problem is often associated with the determination of the residual strength of a structure or a specimen as a function of the crack length. It has received much attention in the literature because a static failure at the end of the fatigue life can be unacceptable in view of hazardous consequences. The assumption that a static failure will occur if $K_{max} = K_{Ic}$ seems reasonable for brittle materials. However, in many technical materials significant plastic deformation occurs during the final failure, sometimes even in the entire uncracked section. As said earlier, the stress intensity factor then becomes unsuitable to describe the stress severity at the crack tip. Predictions of the final failure are not discussed any further here, but the discussion on the results in Table 8.2 should be recalled. It was pointed out that the crack growth life until complete failure does not substantially change for different values of the size of the critical fatigue crack at the moment of failure (a_f) provided that it is large as compared to the size of the initial crack length (a_0). It implies that a conservative value of a_f can be used because of the small effect on the predicted crack growth life.

8.6.2 Crack growth predictions for through cracks

The prediction of crack growth of the through crack is discussed in this section by presenting two case histories. The first one is concerned with

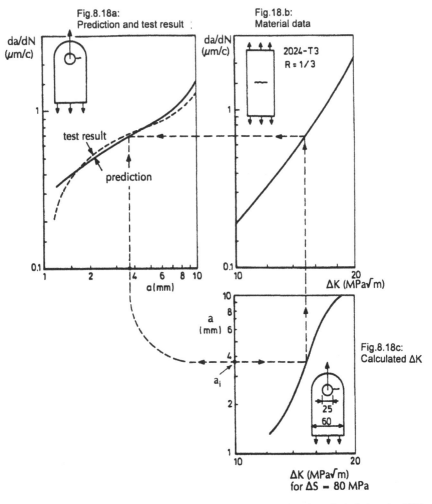

Fig. 8.18 Prediction of the growth rate of a fatigue crack at the edge of a hole in a lug [33].

fatigue of lugs. The second one is dealing with crack growth in a stiffened panel.

Prediction of crack growth in a lug

Lug type connections are frequently used in various structures. The load is transmitted to the top side of the hole in the lug by a pin or bolt. The lug in Figure 8.18 [33] was tested at a load ratio $R = 1/3$ and $\Delta S = 80$ MPa. A saw cut, depth 1 mm, was applied to the edge of the hole in order to

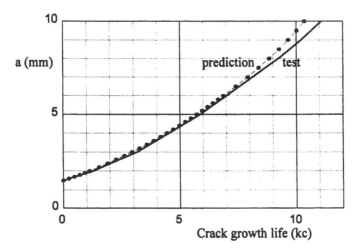

Fig. 8.19 Crack growth curves obtained by integration of da/dN results of Figure 8.17a for the lug specimen as compared to the test results.

start a crack through the full thickness of the lug (5 mm). Crack growth data for $R = 1/3$ were available for the lug material (Al-alloy 2024-T3). These data are presented in Figure 8.18b. Values of ΔK averaged from different sources are presented in Figure 8.18c. Figures 8.18b and 8.18c thus represent the crack growth resistance of the material and the crack driving force respectively. The prediction according to the similarity principle now occurs in a few simple steps:

1. Consider a certain crack length, e.g. $a = a_i$ (e.g. 3.7 mm in Figure 8.18c).
2. Read ΔK for the lug at $a = a_i$ in Figure 8.18c.
3. For this ΔK-value, read the corresponding da/dN in Figure 8.18b.
4. The da/dN-value is plotted in Figure 8.18a at $a = a_i$.

This procedure should be done for the range of a-values of interest for the prediction of crack growth in the lug. The predicted curve in Figure 8.18a shows a good agreement with the experimental results.

Some practical comments should be made:

(i) The prediction of da/dN in four steps for a range of a-values can be done by a computer program. It then should be easy if the da/dN-ΔK curve in Figure 8.18b and the $\Delta K(a)$-curve in Figure 8.18c are available as a mathematical function. Apparently, the Paris relation does not apply to Figure 8.18b. However, non-linear curves can always be represented by a number of linear pieces (usually on a log-log scale).

Computer calculations can then be made by using the linear pieces of a curve. It implies an interpolation along the linear pieces with a negligible loss of accuracy.

(ii) The predicted *a*-*N* crack growth curve must still be obtained by integration of the predicted $\mathrm{d}a/\mathrm{d}N$-*a* curve, which can be done by a numerical integration. The curve thus obtained is compared to the experimental result in Figure 8.19. The two curves practically coincide, which confirms that the similarity concept can give satisfactory predictions.

(iii) Tests were also carried out *without* the small saw cut as a crack starter. Natural cracks were nucleated along the bore of the hole. Fretting between the pin and the hole produced surface damage from which several crack nuclei were initiated. It took a substantial number of cycles before these small cracks joined to a single through crack. As far as crack growth could be observed, it occurred noticeably slower than for the through crack starting from the 1 mm saw cut. As a consequence, the crack growth from 1 mm to failure was about two times longer. The predicted crack growth result is thus conservative. The more complex crack initiation phenomenon in the hole of the lug under real fatigue conditions has caused a longer crack initiation period.

Prediction of crack growth in a panel with stiffener

Another illustrative example on the application of the similarity principle to fatigue crack growth was published by Poe [34]. He considered crack growth in sheet panels (Al-alloy 2024-T3, $R = 0.1$) with seven strip stiffeners attached to the sheet by a row of fasteners (dots in Figure 8.20). Crack growth was artifially started under the central stiffener. If the growing crack is approaching the adjacent stiffeners, the K-value is decreasing. The stiffeners are constraining the deformation of the sheet, which reduces the displacements in the sheet. It implies that crack opening is reduced, and K is lower than in an unstiffened panel.

Material crack growth data were also obtained on unstiffened panels, see the results in the upper scatter band of Figure 8.20. The stress intensity factors for cracks in the panel with stiffeners were calculated with a displacement compatibility technique. The similarity principle could then be applied to predict crack growth rates for the stiffened panel. Starting from the scatter band for the unstiffened panel, the lower scatter band in Figure 8.20 was predicted for the stiffened panel. It confirms that the stiffener

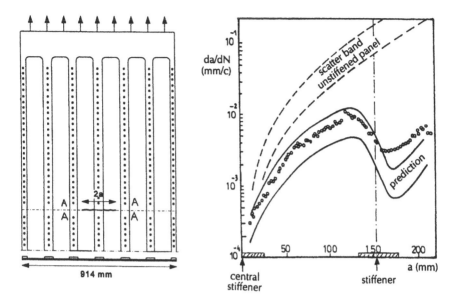

Fig. 8.20 Comparison between prediction and test results for fatigue crack growth in a panel with strip stiffeners [34] ($R = 0.1$, $S_{max} = 103$ MPa).

is reducing the stress intensity factor. The data points in Figure 8.20 are the test results. They follow the predicted scatter band until the crack tip is below the stiffener. After further crack growth (the stiffener remained unbroken) K increases again, but now the crack grows faster than predicted.

Also for this case some practical comments should be added:

(i) The prediction for the crack growth rate up to $a = 150$ mm is satisfactory. For $a > 150$ mm the crack growth rate is underestimated although the trend of the da/dN variation is still correct. Poe mentioned two possible sources for the underestimation [34]. If the crack tip is passing the stiffener, the load on the adjacent fasteners A (Figure 8.20) is high, and some plastic deformation occurs around the fastener holes. As a consequence, the restraining effect of the stiffeners on crack opening is less effective than predicted in the elastic K analysis, and thus da/dN will be larger. Secondly, the stiffeners are at one side of the sheet only, which implies that some bending occurs, and this has also been disregarded in the analysis.

(ii) In spite of some shortcomings in modeling the stiffened panel for calculation of K-values, the information of the effect of the stiffener on fatigue crack growth in the panel is instructive. It implies that such calculations can give useful indication for a design analysis. Poe

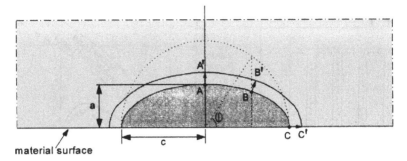

Fig. 8.21 Semi-elliptical surface crack.

already analyzed the effect of different stiffener materials, stiffener cross sections and spacing between stiffeners.

8.6.3 Crack growth prediction for part through cracks

The prediction of fatigue crack growth of a part through crack is a more complex problem because the stress intensity factor varies along the crack front. It implies that da/dN will also vary along the crack front. The semi-elliptical surface crack is a well-known example of a part through crack. It is a technically relevant type of crack, which can be initiated by surface damage, fretting corrosion or corrosion pits. The geometry of a semi-elliptical surface crack is defined in Figure 8.21 by the crack depth "a" and the semi-width "c". Two successive crack fronts are indicated. During crack growth, point A moves to A', and C to C'. Along the crack front the crack extension occurs perpendicular to the crack front, e.g. point B moves to B'. A small incremental crack extension is considered, which occurs in a small incremental number of cycles ΔN. According to the similarity principle, the result for points A and C is

$$\Delta a \ = \ \mathrm{AA}' = \Delta N \cdot (da/dN)_\mathrm{A} = \Delta N \cdot f_R(\Delta K_\mathrm{A})$$

$$\Delta c \ = \ \mathrm{CC}' = \Delta N \cdot (da/dN)_\mathrm{C} = \Delta N \cdot f_R(\Delta K_\mathrm{C}) \qquad (8.24)$$

For simplicity it may be assumed that the semi-elliptical character of the crack will be maintained during further crack growth. The crack growth rate should then be calculated for points A and C. Predictions are illustrated in Figure 8.22 by the number of crack fronts for fatigue crack growth in a plate of SAE 4340 steel. The K-values for points A and C of Figure 8.21 were

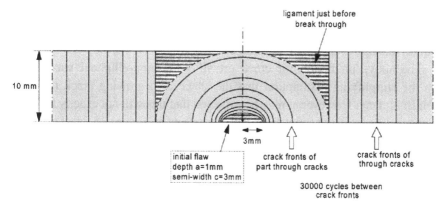

Fig. 8.22 Crack growth prediction of a surface crack in an SAE 4340 steel plate of 10 mm thickness ($\Delta S = 200$ MPa, $R = 0$, Paris constants $C = 1.82E\text{-}12$, $m = 2.67$, da/dN in m/c ΔK in MPa$\sqrt{\text{m}}$).

obtained by adopting equations of Newman and Raju [25] for $\varphi = 90°$ and $\varphi = 0°$ respectively. The Paris relation of the da/dN-ΔK results for SAE 4340 in Figure 8.15 was used. The numerical integration of Equation (8.24) was carried out with a small ΔN increment ($\Delta N = 100$ cycles). Crack growth started from an initially shallow flaw with an axes ratio $a/c = 1/3$, but during crack growth the shape of the crack front becomes less shallow. This should be expected because the value of K_A for a shallow crack is larger than K_C (see Figure 5.17). Crack growth in the depth direction (point A in Figure 8.21) will be faster than along the surface (point C).

The change of the semi-elliptical crack shape was confirmed in fatigue tests on plate specimens (7075-T6 material, thickness 9.6 mm) reported by Ichsan and Schijve [35]. The predictions on the crack growth rate were in good agreement with the test results as long as the crack depth was small. However, for deep cracks the agreement was poor. Crack closure measurements with the fractographic technique of Sunder [17] indicated that S_{op} was reduced due to plastic tensile deformation in the ligament between the crack and the back surface of the specimen. Accelerated crack growth through the entire ligament (see Figure 8.22) will rapidly occur which then leads to a substantial "break-through". This is an important phenomenon for pressure vessels because it leads to leakage. The possibility of an immediate unstable crack extension after a break-through must be considered [37]. If the residual strength is still sufficient to carry the load on the structure, the so-called leak-before-break criterion is satisfied and an explosion does not occur. Actually this applies more to liquid containers because the internal pressure rapidly decreases after some leakage. However, for gas containers the pressure reduction occurs much slower. For containers with an aggressive liquid, leakage is anyhow undesirable.

The above prediction for crack growth of a semi-elliptical was based on the simplifying assumption that the shape of a growing crack remains semi-elliptical. However, it may be expected that this will not exactly be true. Consider an arbitrary point B of the crack front in Figure 18.21. The local crack extension BB' can be calculated with the K-values of Raju–Newman. The predicted crack extension is then

$$\Delta b = \mathrm{BB}' = \Delta N \cdot (\mathrm{d}a/\mathrm{d}N)_\mathrm{B} = \Delta N \cdot f_R(\Delta K_\mathrm{B}) \qquad (8.25)$$

Point B' may be approximately on the ellipse of A' and C' but not exactly. An obviously more realistic approach is to calculate K-values along the entire crack front with FE techniques. It would also require that the FE analysis must be repeated after each crack size increment. This method was studied by Lin and Smith [36] for corner cracks at a hole. They used eight elements along the crack front. An automatic remeshing technique was developed. Crack extension occurred in at least 70 small steps. The Paris relation was used to calculate the local crack extension along the crack front. Calculations were made for three ratios between the hole radius and sheet thickness (r/t = 0.5, 1 and 3). The results indicated that the predicted crack fronts, starting from a small quarter elliptical crack, deviated from the quarter elliptical shape. However, the deviations defined in Figure 8.23 were still rather small and always negative (crack front less curved than the ellipse). The largest deviation, about 8%, occurred for the larger cracks which had almost reached thickness break-through. Apparently, the crack shape simulation with a quarter elliptical crack is still reasonable.

The previous problems about crack shapes are associated with structural elements loaded in tension. The through crack problem in plates has been mentioned as a simple problem if the crack front is perpendicular to the material surface. However, crack growth with an oblique crack front occurs in a plate loaded by combined tension and bending. This situation is present in joints with eccentricities. A tensile load then introduces secondary bending (see Chapter 18). Fawaz studied fatigue crack growth in open hole specimens loaded under combined tension and bending with approximately the same maximum and minimum stress for tension and bending ($S_{\max}$ = 90 MPa, $R \approx 0$) [38]. Observations on the crack front shape were made by testing several specimens until different percentages of the fatigue life. The specimens were then statically pulled to failure to open the fatigue cracks and to determine the crack front shapes. Illustrative results are shown in

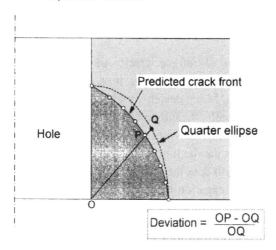

Fig. 8.23 Deviation of elliptical crack front shape as defined by Lin and Smith [36].

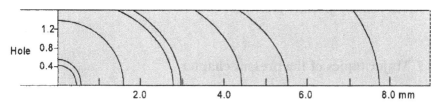

Fig. 8.24 Crack fronts observed by Fawaz on the fracture surface of a fatigue crack starting at a hole in sheet specimens loaded under combined tension and bending [38]. (Material 2024-T3 Alclad, $t = 1.6$ mm, $R \approx 0$).

Figure 8.24. After initial crack growth as a part through corner crack, the growth continued as a through crack with a curved crack front under oblique angles with the sheet surface. Data on K-values for oblique through cracks under combined tension and bending were not yet available. Fawaz has made FE calculations to obtain the required K-values. Crack growth predictions with these K-values gave a reasonable agreement with the fatigue test results. Later these calculations were expanded by Fawaz and Anderson [39] to cover tension and bending loads and configurations of oblique part through cracks and full through cracks starting form open holes and pinloaded holes. The results overlap the Raju–Newman data [40] but a much larger variety of crack sizes is covered. Moreover the present FE technique has led to improvements of the accuracy. The data bank of the results is available in AFGROW, a software package of the USAF [41].

8.6.4 A final comment

Nowadays the growth of fatigue cracks in a structure is no longer a mysterious phenomenon. It can occur in a structure subjected to large numbers of load cycles. Many design variables can affect the propagation of fatigue cracks . Fortunately, the influence of the variables is qualitatively well understood. Unfortunately, the present understanding also shows that the accuracy of predictions can be limited in many practical cases. With respect to the basic crack growth data the situation can be improved by carrying out some simple crack propagation tests on specimens of the same material as used in the structure. The specimen thickness should also be equal to the material thickness in the structure at the fatigue critical location. Secondly, if the stress intensity factor is not available in the literature, FE calculations should be considered. Basic understanding and engineering judgement are essential in many practical problems.

8.7 Major topics of the present chapter

Fatigue crack growth of macrocracks under constant-amplitude loading has been considered in the present chapter. The main aspects are summarized below:

1. The fatigue crack growth resistance of a material can be described as $da/dN = f(\Delta K, R)$. This function is obtained from fatigue crack growth tests on simple specimens. The results can be presented in graphs, equations and tables. They should be considered to be empirical data representing the fatigue crack growth resistance of a material.
2. Fatigue crack growth can occur in three regions of ΔK: (i) the threshold region, (ii) the Paris region, and (iii) the stable-tearing crack growth region. The technical relevance of the threshold region is problematic, due to a changing crack growth mechanism at low ΔK-values. The Paris relation, $da/dN = C \cdot \Delta K^m$, is an approximation of empirical results. Deviations of this linear relation (log-log scale) are observed.
3. The Paris relation is useful for estimating the effect of the design stress level on the crack growth life.
4. Plasticity induced crack closure (Elber) is an important phenomenon to understand the macrocrack growth behavior. Crack closure is more significant at the material surface in view of the plane stress situation,

and less important under plane strain conditions in thick sections of materials with a relatively high yield stress.

5. The crack closure concept has led to the ΔK_{eff} concept which is helpful in accounting for the stress ratio R. The $\Delta K_{\text{eff}}(R)$ equations are based on empirical evidence.

6. The fatigue crack growth resistance depends on the material, and more specifically on the yield strength of a material as obtained by the material production and heat treatment. Increasing the yield strength by a heat treatment usually leads to a reduction of the fatigue crack growth resistance. It is remarkable that the fatigue crack growth resistance of steels with a significantly different static strength is not so much different. Materials with a very high static strength usually have a relatively low fatigue crack growth resistance.

7. The predictions of fatigue crack growth in a structure is based on the similarity principle. The value of ΔK is used as the crack driving force to obtain the corresponding da/dN from the basic material fatigue crack growth data of the material. The most simple case is a prediction of crack growth of a through crack in a plate under cyclic tension. Curved crack fronts occur for corner cracks and for through cracks under combined tension and bending. Depending on the purpose of the predictions, some simplifications of the prediction procedure can be justified to obtain estimates of ΔK-values. However, more accurate predictions require that data on the K variation along the crack front. It may imply that FE calculations should be made, but information is also available in data banks.

8. The prediction of the crack growth life starting from a small crack length a_0 is strongly depending on the size of a_0. The predicted life increases substantially for smaller values of a_0. The prediction can be unconservative if crack growth at a_0 is still affected by the small crack phenomenon which can occur for $\Delta K < \Delta K_{\text{th}}$. Extrapolation of the Paris equation to low ΔK-values below the threshold value should then be considered.

References

1. *Standard test method for measurement of fatigue crack growth rates.* ASTM Standard E64791 (1991).
2. Paris, P.C., Gomez, M.P. and Anderson, W.E., *A rational analytical theory of fatigue.* The Trend of Engineering, Vol. 13 (1961), pp. 9–14.

3. Schijve, J., *Fatigue crack propagation and the stress intensity factor.* Faculty of Aerospace Engineering, Delft, Memorandum M-191 (1973).

4. Figge, I.E. and Newman, Jr, J.C., *Fatigue-crack-propagation behavior in panels with simulated rivet forces.* NASA TN D-4702 (1968).

5. Schijve, J., *Significance of fatigue cracks in micro-range and macro-range.* ASTM-STP 415, (1967) pp. 415–459.

6. Wanhill, R.J.H., *Durability analysis using short and long fatigue crak growth data.* Aircraft Damage Assessment and Repair. The Institution of Engineering, Australia (1991). Barton, Australia.

7. Paris, P.C. and Erdogan, F., *A critical analysis of crack propagation laws.* Trans. ASME, Series D, Vol. 85 (1963), pp. 528–535.

8. Forman, R.G., Kearney, V.E. and Engle, R.M., *Numerical analysis of crack propagation in cyclic-loaded structures.* J. Basic Engrg., Trans. ASME, Vol. D89 (1967), pp. 459–464.

9. Priddle, E.K., *High cycle fatigue crack propagation under random and constant amplitude loadings.* Int. J. Pressure Vessels & Piping, Vol. 4 (1976), p. 89.

10. Klesnil, M. and Lukáš, P., *Influence of strength and stress history on growth and stabilization of fatigue cracks.* Engrg. Fracture Mech., Vol. 4 (1972), pp. 77–92.

11. Elber, W., *The significance of fatigue crack closure. Damage tolerance in aircraft structures.* ASTM STP 486 (1971), pp. 230–242.

12. Rice, J.R., *The mechanics of crack tip deformation and extension by fatigue. Fatigue crack propagation.* ASTM STP 415 (1967), pp. 247–309.

13. Schijve, J., *Some formulas for the crack opening stress level.* Engrg. Fracture Mechanics, Vol. 14 (1981), pp. 461–465.

14. Van der Linden, H.H., *NLR test results as a database to be used in a check of crack propagation prediction models. A Garteur activity.* Nat. Aerospace Lab. NLR, TR 79121U, Amsterdam (1979).

15. Schijve, J., *Fatigue crack closure observations and technical significance.* Mechanics of Fatigue Crack Closure, Int. Symp., Charleston 1986. ASTM STP 982 (1988), pp. 5–34.

16. Ewalds, H.L. and Furnee, R.T., *Crack closure measurements along the crack front in center cracked specimens.* Int. J. Fracture, Vol. 14 (1978), pp. R53–R55.

17. Sunder, R. and Dash, P.K., *Measurement of fatigue crack closure through electron microscopy.* Int. J. Fatigue, Vol. 4 (1982), pp. 97–105.

18. Ritchie, R.O., *Mechanisms of fatigue crack propagation in metals, ceramics and composites: Role of crack tip shielding.* Mater. Sci. Engrg., Vol. A103 (1988), pp. 15–28.

19. Broek, D. and Schijve, J., *The influence of the mean stress on the propagation of fatigue cracks in aluminium alloy sheet.* Nat. Aerospace Lab. NLR, Report TR M.2111, Amsterdam (1963).

20. Crooker, T.W., *The role of fracture toughness in low-cycle fatigue crack propagation for high-strength alloys.* Engrg. Fracture Mech., Vol. 5 (1973), pp. 35–43.

21. Stephens, R.R., Stephens, R.I., Veit, A.L. and Albertson, T.P., *Fatigue crack growth of Ti-62222 alloy under constant amplitude and mini-TWIST flight spectra at 25°C and 175°C.* Int. J. Fatigue, Vol. 19 (1997), pp. 301–308.

22. Houdijk, P.A., *Effect of specimen thickness and specimen geometry on fatigue crack growth in Fe510Nb.* Faculty of Chemistry and Materials, Delft University of Technology (1993) [in Dutch].

23. Song-Hee Kim and Weon-Pil Tai, *Retardation and arrest of fatigue crack growth in AISI 4340 steel by introducing rest periods and overloads.* Fatigue Fracture Engrg. Mater. Structure, Vol. 15 (1992), pp. 519–530.

24. Liaw, P.K., Peck, M.G. and Rudd, G.E., *Fatigue crack growth behavior of D6AC space shuttle steel*. Engrg. Fracture Mech., Vol. 43 (1992), pp. 379–400.
25. Newman, J.C., Jr. and Raju, I.S., *Stress-intensity factor equation for crack in three-dimensional finite bodies subjected to tension and bending loads*. Fracture Mechanics, ASTM STP 791, Vol. 1 (1983), pp. 238–265.
26. Petrak, G.S., *Strength level effects on fatigue crack growth and retardation*. Engrg. Fracture Mech., Vol. 6 (1974), pp. 725–733.
27. Schijve, J. and De Rijk, P., *The fatigue crack propagation in 2024-T3 Alclad sheet materials from seven different manufacturers*. Nat. Aerospace Lab. NLR, Report TR M.2162, Amsterdam (1966).
28. Yoder, G.R., Cooley, L.A. and Crooker, T.W., *The effect of load ratio on fatigue crack growth in Ti-8Al-1Mo-1V*. Engrg. Fracture Mech., Vol. 17 (1983), pp. 185–188.
29. Kage, M., Miller, K.J. and Smith, R.A., *Fatigue crack initiation and propagation in a low-carbon steel of two different grain sizes*. Fatigue Fracture Engrg. Mater. Structure, Vol. 15 (1992), pp. 763–774.
30. Wanhill, R.J.H., *Low stress intensity fatigue crack growth in 2024-T3 and T351*. Engr. Fracture Mech., Vol. 30 (1988), pp. 233–260.
31. Stubbington, C.A. and Gunn, N.J.F., *Effects of fatigue crack front geometry and crystallography on the fracture toughness of an Ti-6Al-4V alloy*. Roy. Aero. Est., TR 77158, Farnborough (1977).
32. Pearson, S., *Initiation of fatigue cracks in commercial aluminium alloys and the subsequent propagation of very short cracks*. Engrg. Fracture Mech., Vol. 7 (1975), pp. 235–247.
33. Schijve, J. and Hoeymakers, A.H.W., *Fatigue crack growth in lugs and the stress intensity factor*. Fatigue Engrg. Mater. Structures, Vol. 1 (1979), pp. 185–201.
34. Poe, Jr., C.C., *Fatigue crack propagation in stiffened panels. Damage tolerance in aircraft structures*, ASTM STP 486 (1971), pp. 79–97.
35. Ichsan, S. Putra and Schijve, J., *Crack opening stress measurements of surface cracks in 7075-T6 Al alloy plate specimens through electron fractography*. Fatigue Fracture. Engrg. Mater. Structures, Vol. 15 (1992), pp. 323–338.
36. Lin, X.B. and Smith, R.A., *Fatigue shape analysis for corner cracks at fastener holes*. Engrg. Fracture Mech., Vol. 59 (1998), pp. 73–87.
37. Broek, D., *The Practical Use of Fracture Mechanics*. Kluwer Academic Publishers (1988).
38. Fawaz, S.A., *Fatigue Crack Growth in Riveted Joints*. Doctor Thesis, Delft University of Technology (1997).
39. Fawaz, S.A. and Andersson, B., *Accurate stress intensity factor solutions for corner cracks at a hole*. Engrg. Fracture Mech., Vol. 71 (2004), pp. 1235–1254.
40. Newman, J.C., Jr. and Raju, I.S., *Stress-intensity factor equation for crack in three-dimensional finite bodies subjected to tension and bending loads*. Fracture Mechanics, ASTM STP 791, Vol. 1 (1983), pp. 238–265.
41. Harter, J.A., *AFGROW Users Guide and Technical Manual,* AFGROW version 4.0012.15, AFRL-VA-WP-TR-2007 (2007).

Some general references

42. Pook, L.P., *Crack Paths*. Wit Press, Southampton (2002).
43. Socie, D.F. and Marquis, G.B., *Multiaxial Fatigue*. Society of Automotive Engineers (1999).

44. Wang, S.-H. and Müller, C., *A study on the change of fatigue fracture mode in two titanium alloys.* Fatigue Fracture Engrg. Mater. Structure, Vol. 21 (1998), pp. 1077–1087.

45. De Freitas, M. and Francois, D., *Analysis of fatigue crack growth in rotary bend specimens and railway axles. Fatigue Fracture Engrg. Mater. Structure, Vol. 18* (1995), pp. 171–178.

46. Carpinteri, A., *Handbook of Fatigue Crack Propagation in Metallic Structures.* Elsevier, Amsterdam (1994).

47. Blom, A.F. and Beevers, C.J. (Eds.), *Theoretical Concepts and Numerical Analysis of Fatigue.* Proc. Conf. May 1992, Birmingham. EMAS (1992).

48. Anderson, T.L., *Fracture Mechanics: Fundamentals and Applications.* CRC Press (1991).

49. Reuter, W., Underwood, J.H. and Newman, Jr., J.C. (Eds.), *Surface-crack growth: Models, experiments, and structures.* ASTM STP 1060 (1990).

50. Brown, M.W. and Miller, K.J. (Eds.), *Biaxial and Multiaxial Fatigue.* EGF Publication 3. Mechanical Engineering Publications (1989).

51. Newman, Jr., J.C. and Elber, W. (Eds.), *Mechanics of Fatigue Crack Closure.* ASTM STP 982 (1988).

52. Miller, K.J. and Brown, M.W. (Eds.), *Multiaxial Fatigue.* ASTM STP 853 (1985).

53. Pook, L.P., *The Role of Crack Growth in Metal Fatigue.* The Metals Society, London (1983).

54. ESDU Engineering Science Data. *Fatigue-Fracture Mechanics Data.* Vol. 2 (aluminium alloys) and Vol. 3 (Titanium alloys and steels). (1981–1999).

55. *Fatigue Crack Propagation*, ASTM STP 415 (1967).

56. Hudson, C.M. and Seward, S.K., *A compendium of sources of fracture toughness and fatigue crack growth data for metallic alloys. Parts I, II and III.* Int. J. Fracture, Vol. 14 (1978) pp. R151–R184, Vol. 20 (1982) pp. R59–R117, Vol. 39 (1989) pp. R43–R63.

57. McClung, R.C., *The influence of applied stress, crack length, and stress intensity factor on crack closure.* Metallurgical Trans., Vol. 22a (1991), pp. 1559–1571.

58. Wanhill, R.J.H., *Microstructural influences on fatigue and fracture resistance in high strength structural materials.* Engrg. Fracture Mech., Vol. 10 (1978), pp. 337–357.

59. *Short Crack Growth Behaviour in Various Aircraft Materials*, AGARD Report No. 767 (1990).

60. Schijve, J., *Difference between the growth of small and large fatigue cracks. The relation to threshold K-values.* Fatigue Thresholds, Fundamentals and Engineering Applications. Proc. Int. Conf. Stockholm 1981. EMAS Warley (1982), pp. 881–908.

Part II
Load Spectra and Fatigue under Variable-Amplitude Loading

Chapter 9
Load Spectra

9.1 Introduction

The fatigue loads on a structure in service are generally referred to as the load spectrum. The description of load spectra and methods to obtain load spectra are discussed in the present chapter. A survey of various aspects of fatigue of structures was presented as a flow diagram in introductory chapter (Chapter 1, Figure 1.2). A reduced diagram is presented here in Figure 9.1 to illustrate the significance of load spectra for fatigue design analysis of a structure. Without information on the anticipated load spectrum, the analysis of the fatigue performance of a structure is impossible. Furthermore, verification tests to support the analysis are often necessary for economic or safety reasons. The load spectrum must be consulted for planning such validation tests.

Sometimes the load spectrum is changed after a number of years by a modified use of the structure, which is different from the initial expectations. The load spectrum must then be considered again. Fatigue load spectra should also be reviewed if fatigue failures occur in service.

The load spectrum of a structure should give information about the load-time history, which is the variation of the load as a function of time,

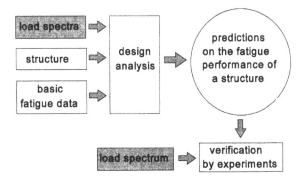

Fig. 9.1 Load spectra as input for the fatigue performance of a structure.

$P(t)$. The present knowledge of the fatigue phenomenon as it occurs in technical materials (see Chapter 2) clearly indicates that the significant points of a $P(t)$ load history are the maxima and minima, $P_{\max}$ and $P_{\min}$, see Figure 9.2. At these load levels, reversal of cyclic slip occurs in the material, either at the material surface or in the crack tip plastic zone. These reversals are decisive for the fatigue damage accumulation in a structure. Several practical questions arise:

1. Is it necessary to know the full sequence of all turning points of the load history?
2. Are all similar structures in service subjected to the same load history, or in other words, how unique is a certain load history for a structure?
3. Are small cycles of interest, or is the fatigue damage contribution of these cycles negligible?
4. Is it important whether loads are applied at a high or a low loading rate (wave shape)?
5. Long periods at zero load (rest periods, structure not in use) or long periods at a significant load level (average load in service if dynamic

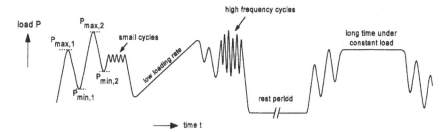

Fig. 9.2 Characteristic occurrences of a load-time history $P(t)$.

loads do not occur during that period), are these periods important for the fatigue damage accumulation?

The last two questions are pointing to problems of time dependent phenomena, e.g. corrosion, creep, or diffusion processes in the material which might affect the fatigue process. In the literature, these problems are frequently discussed as effects of the load frequency (cycles per minute) and the cyclic wave shape, see Section 2.5.7 (e.g. Figure 2.30).

Before the above problems can be discussed, an essential question is: Is the load history known which a structure will experience in service, or can it be estimated? Even more, how can the load history be described, and can it be measured? In the present chapter, load histories of different types of structure are discussed first (Section 9.2), which reveals essential differences between the statistical nature of load histories. Methods for the description of a load history and statistical compilations of load spectrum data are presented in Section 9.3. The determination of load spectra is discussed in Section 9.4. Service-simulation load histories are addressed in Section 9.5. The major aspects of the present chapter are summarized in Section 9.6.

9.2 Different types of loads on a structure in service

Which loads occur on a structure in service?
Answers to this question depend on the type of structure and how the structure is used. First some exemplary cases are discussed in a qualitative way to illustrate the variety of problem settings. Different types of loads can then be defined.

1. *Pressure vessel*

Many pressure vessels used in the industry and other production facilities are used in a simple way. The pressure is built up to a specific working level, maintained at that level, and then released to zero. If such a pressure cycle occurs about five times a day, the load spectrum contains approximately 40000 cycles in a life period of 20 years. Fatigue problems could arise. A number of questions can easily be formulated. Is the pressure always the same. Are the number of pressurization cycles user dependent? Is the duration of a pressure cycle important? Is the gas or liquid inside the pressure vessel aggressive? Anyhow, a number of questions to be considered if fatigue critical notches, usually inlets and joints, occur in the pressure vessel.

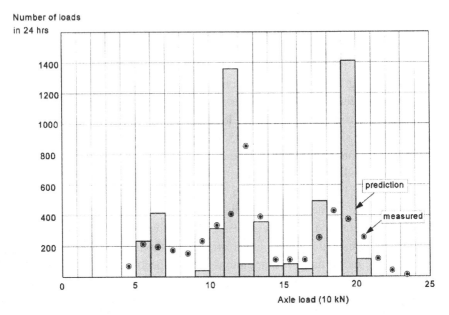

Fig. 9.3 (a) Axle loads on a railway bridge in 10 kN intervals.

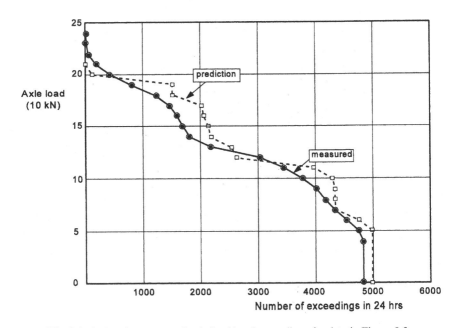

Fig. 9.3 (b) Load spectrum of axle load level exceedings for data in Figure 9.3a.

2. *Bridge*

The variety of bridges is large. A simple steel railway bridge is considered here. It is loaded in bending by each passage of a train. The load spectrum in a specific case of the Dutch Railways was depending on the number of train axles passing the bridge each day and the weight applied by the axles to the bridge. The load spectrum was predicted by considering the variety of trains that would use the bridge [1]. The prediction is shown as a bar chart (histogram) in Figure 9.3a, which gives the number of axle loads in intervals of 10 kN. The load spectrum was checked later, see the measured data in Figure 9.3a. It turns out that the scatter of the axle loads was larger than predicted. The same data are compared in Figure 9.3b by plotting the numbers of load level exceedings, i.e. the number of times that a specific load level is exceeded in 24 hours. Low load levels are exceeded many times, while high load levels occur less frequently. The results in Figure 9.3b show a reasonable agreement between the measured data and predictions. Some agreement should be present if the utilization of a structure is well defined and known in advance. This applies to the example of Figure 9.3 of trains used in accordance with a specified time table. However, for other moving vehicles such a prediction can be more difficult.

3. *Lamp post*

Modern aluminium street light posts are predominantly loaded by wind forces coming from different directions and varying intensities. For a lamp post as shown in Figure 9.4, it leads to bending and torsion load cycles with maxima stress levels near the base of the pillar. Usually, an opening is made in the pillar close to the base for making electrical connections. Although a cover is closing the opening, stress concentrations are present in that area and fatigue cracks have occurred. A correlation between the function of the street light and load spectrum does not exist. The load spectrum depends on the weather conditions, which should be described in statistical terms. Weather conditions depend on the geographical location. These conditions can be more severe along a sea cost where humidity and salt concentration can also adversely affect the fatigue behavior. Another obvious aspect involved is the dynamic response of the pillar on the wind fluctuations. It cannot be expected that the wind load spectrum on the street light and the stress spectrum at the fatigue critical location are linearly related. Dynamic response calculation techniques are well developed, but it may be advisable to measure the stress spectrum on a representative location of the structure.

Fig. 9.4 Lamp post in Pijnacker (the Netherlands).

4. *Motor-car* The load spectrum on a car can be very complex. It obviously depends on two major inputs: (i) the driver, and (ii) the condition of the roads to be used. A single load spectrum applying to all cars of the same type is impossible. Moreover, an average load spectrum applicable to most cars is meaningless. Fatigue failures are associated with severe driving and poor roads which applies to a small percentage of cars. However, a small percentage is still a large number of cars. It implies that a relatively severe load spectrum must be considered for the fatigue performance. The fatigue problem of motor-cars is also associated with the complexity of the structure with several components which can be fatigue critical. In addition, loads on a car act on the wheels in three different directions (x, y, z) with different frequencies and phase angles. Inertial forces on the flexible structure are also complex. All these conditions imply that a load spectrum cannot easily be defined. It is for these reasons that the motor-car industry is relying on experience, measurements and experiments.

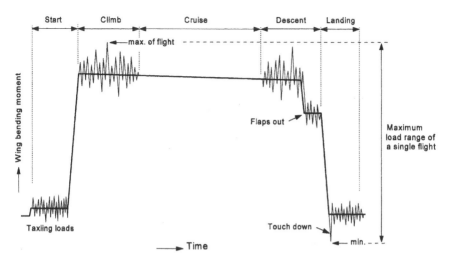

Fig. 9.5 Slow load variation of the wing bending moment during a single flight, with fast superimposed turbulence loads and ground loads.

5. *Wing of transport aircraft*

The aerodynamic lift on the wing of an aircraft is carrying the aircraft weight. The distributed lift on the wing exerts a bending moment with a maximum at the root of the wing. On the ground, the lift is zero and the aircraft is supported by the undercarriage. Each flight thus implies a cycle of the bending moment on the wing, see the heavy line in Figure 9.5. Bending of the wing introduces tension stresses in the lower wing skin structure and compression in the upper wing skin structure. The tension skin is well recognized as a fatigue critical part of the wing. The once per flight cycle on the tension skin is a very slow cycle with an almost quasi-static variation of the load. However, the wing is also subjected to much faster load cycles, see Figure 9.5. In flight, these cycles occur in turbulent air (gusty weather) predominantly during the climb and descent period at low altitudes. The turbulence at cruising altitude is usually very limited, and load variations are small (a small change due to fuel consumption). Also maneuver loads can be significant, depending on the type of aircraft. During take-off and landing, high-frequency cycles are introduced by runway roughness, touch-down on the ground, and spin up of the wheels. In addition to wing bending, torsional moments are also exerted on the wing. The loading picture is fairly complex, which is only schematically illustrated by Figure 9.5.

The above examples illustrate a variety of different loads. Two major types of characteristic groups of loads must be recognized:

1.   Deterministic loads.
2.   Stochastic loads.

A load is considered to be deterministic if it can be defined as a specific occurrence, from which it is known that it will occur with a magnitude that can be estimated. Deterministic loads should follow from the planned utilization of a structure. The load cycle of a pressure vessel is fully deterministic. Manoeuvers of ships and transport aircraft are predominantly deterministic. Many loads on a motor-car, a bridge, or a crane are predictable and have a deterministic character. However, depending on how such structures are used, loads cannot always be considered to be deterministic. Obviously, joyriding a car can lead to unpredictable loads.

Stochastic loads have an essentially statistical nature. They cannot be predicted to occur with a certain magnitude at a given moment. Good examples are wind forces on a street light pillar, forces exerted by waves of the sea on ships and drilling platforms, turbulence on an aircraft, and loads on motor-cars due to poor road conditions. A description of stochastic loads can only be done in a statistical way, i.e. in terms of the probability that something will happen. Stochastic loads are also referred to as random loads. In many cases, the statistical properties of stochastic loads are not very well known, although long-term measurements have provided useful data, e.g. for sea waves and wind forces.

Stochastic and deterministic loads can also occur simultaneously on the same structure. An example is shown in Figure 9.5 with random turbulence and runway roughness loads superimposed on the deterministic once-per-flight load cycle. The problem is how to combine these loads for fatigue evaluations. The superimposed loads increase the severity of the flight because the maximum load occurring during a flight becomes more severe, and the same is true for the minimum load. This aspect will be reconsidered in the following section on describing load histories.

Another aspect of random loads is that the intensity is not always the same. The statistical properties are not necessarily constant. This is easily understood by considering random loads depending on the weather conditions. Stormy weather can induce severe random loads on a lamp post, but more frequently occurring milder weather conditions can also contribute to fatigue. It has led to a second differentiation between load histories:

(i)   Stationary load histories.
(ii) Non-stationary load histories.

In the first case, the statistical properties do not vary as a function of time, whereas in the second case, these properties can vary during the service

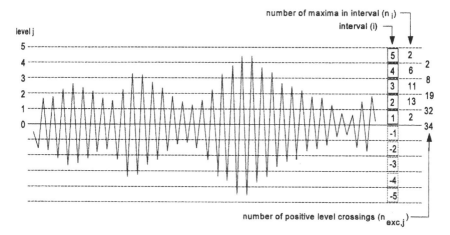

Fig. 9.6 Results of counting maxima of a symmetric load time history (varying amplitude).

usage of a structure. Although the terms stationary and non-stationary load histories are usually associated with stochastic loads, they can also apply to deterministic loads, e.g. by changing the use of a structure.

9.3 Description of load histories

Level crossing count methods

A load-time history is defined by a sequence of maxima and minima if time-dependent phenomena are not considered: $P_{\max,1}$, $P_{\min,1}$, $P_{\max,2}$, $P_{\min,2}$, etc. Such a sequence is usually reduced to a statistical representation in order to have a useful survey of the fatigue loads. In the past, several counting techniques were developed for this purpose based on counting level crossings for a number of load levels or counting peak values above a number of load levels. The historical development (see [2]) will not be followed here, but basic aspects of statistical count procedures are considered.

A simple load sequence is shown in Figure 9.6, a load signal with a varying amplitude. Approximately similar maxima and minima occur around a mean level, indicated as level 0. In view of the symmetry around this level, it is sufficient to consider the maxima only. Usually, load spectra are presented as numbers of peak values occurring above a load level j denoted as $n_{\mathrm{exc},j}$. In the present case, this number is equal to the number of positive level crossings (going from a minimum to a maximum) of load level j.

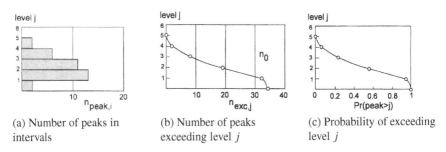

(a) Number of peaks in intervals

(b) Number of peaks exceeding level j

(c) Probability of exceeding level j

Fig. 9.7 Peak counting results of the load-time history sample in Figure 9.6.

Numbers are counted in Figure 9.6 for levels $j = 0$ to $j = 4$, see the numbers $n_{\text{exc},j}$ in the row to the far right of this figure. The number of peak values in an interval ($n_{\text{peak},i}$) is then obtained as the difference between the numbers of level crossings of the two enclosing levels of the interval:

$$n_{\text{peak},i} = n_{\text{exc},j=i-1} - n_{\text{exc},j=i} \qquad (9.1)$$

These numbers have been plotted in Figure 9.7a, which is a histogram of the number of peak loads in the intervals. The numbers of peak values above load level j are plotted in Figure 9.7b. A curve is drawn through these counting results. These exceeding numbers are normalized in Figure 9.7c by dividing $n_{\text{exc},j}$ by the total number of peaks (n_0) above the zero reference level ($j = 0$), see Figure 9.7c. The values obtained are related to the probability of a peak value occurring above level j, or:

$$\text{Pr}\,(\text{peak} > \text{level } j) = \frac{n_{\text{exc},j}}{n_0} \qquad (9.2)$$

In Figure 9.6, a short load-time history was used to illustrate the counting technique. A long load-time history with a stationary character will lead to an exceeding probability curve with a stationary character. In statistical terms, the curve becomes an estimate of the *probability function* of the occurrence of peak values. The bar chart of the number of loads in load level intervals is associated with the *probability density function*. Load spectra are usually presented as load exceeding curves as shown in Figure 9.7b. They must then be related to a certain time in service.

A second example of a load history is given in Figure 9.8. The load variation is no longer symmetric in this case, but a reference level can again be determined with alternating maxima and minima above and below this level respectively. Counting can occur in load intervals for the minima and maxima separately, which leads to two load spectra for the maxima and

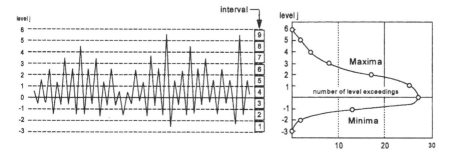

Fig. 9.8 A non-symmetric load-time history. Separate counts of maxima and minima.

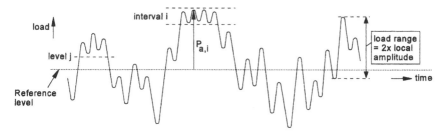

Fig. 9.9 An irregular load-time history.

minima as presented in Figure 9.8b. The number of maxima and minima must be equal (the number is 27 in Figure 9.8a). The two spectra for the maximum and the minimum peak values in Figure 9.8b give indications about the size of the positive and negative peak values, and about how often they occur. This information may be instructive for a first evaluation of the severity of a non-symmetric load spectrum, and also for comparing load spectra of different severities.

The two load-time histories in Figures 9.6 and 9.8 contain only maxima above the reference level and minima below this level. However, the situation is different for a more irregular load-time history as shown in Figure 9.9. In Figures 9.6 and 9.8, the number of positive level crossings of level j ($n_{exc,j}$) was equal to the number of peak values above level j. However, in Figure 9.9, level j in the first part of the load history is associated with one positive level crossings whereas the corresponding number of positive peaks larger than level j is equal to three. Actually, it is not difficult to see that the number of positive level crossings is equal to the number of maxima above that level reduced by the number of minima above that level. As a consequence, Equation (9.1) is no longer applicable and the numbers of peak values in an interval cannot be derived unambiguously from level crossing counts. Of course, the peak values can be counted in a number of intervals and the counting results can still be presented in statistical graphs. But it may be questioned whether this is meaningful. In Figure 9.9, four maxima are counted in interval i, but they are due to small load variations. These peak values cannot be associated with four loads with an amplitude $P_{a,i}$.

A load sequence as shown in Figure 9.9 is more irregular than the load sequences of Figures 9.6 and 9.8. The irregularity of a load-time history can be defined by an irregularity factor which is the ratio of the number of peak values and the number of level crossings of the reference level:

$$k = \frac{\text{number of peak values}}{\text{number of level crossings of the reference level}} \tag{9.3}$$

The irregularity factor is obviously equal to 1 for constant-amplitude loading, but also for a load-time history with an amplitude modulation in Figure 9.6. The factor remains equal to 1 in Figure 9.8 for the non-symmetric load-time history with alternately positive and negative load amplitudes. If the irregularity factor is equal to 1, the magnitude of load excursions with respect to the reference level can be indicated by a single load parameter. However, this is not possible for an irregular load-time history for which $k > 1$. The value of k in Figure 9.9 is 2.5 which implies a high irregularity. In such cases, an apparent need is present to consider load variations between successive peak values in terms of load ranges which is discussed later.

Flat and steep load spectra

If the irregularity of a load-time history is limited (i.e. with an irregularity factor not too much above 1), useful statistical data on peak loads can still be presented in the format of a one-parameter load spectrum. In such a case, the shape of the load spectrum is of interest. Two significantly different shapes are shown in Figure 9.10. As mentioned before in the discussion on Figure 9.7b, such spectra should be associated with certain periods in service, e.g. hours or years, or also the number of times of using the structure (missions). Load spectra should preferably cover long periods in order to be representative for the variability of the load-time history. The load level in the two fictitious spectra in Figure 9.10 (1000 hours in service) are expressed as a percentage of the maximum load occurring in that period. Figure 9.10 shows a steep load spectrum and a flat load spectrum. In a steep spectrum, the number of high loads is small and the number of low loads is large. As an illustration, high loads with a peak value exceeding 80% of the maximum

load level (% max. load)

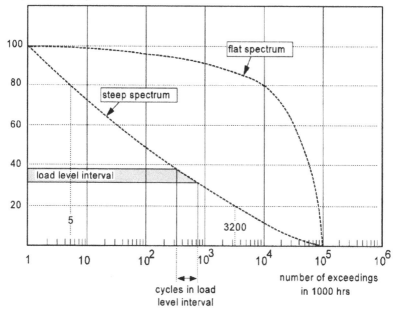

Fig. 9.10 Two different types of load spectra.

load of the steep spectrum in Figure 9.10 occur only five times, whereas the number of low loads with a peak value below 20% of the maximum load is $100000 - 3200 = 96800$ cycles which is 97% of all cycles. The opposite is true for the flat load spectrum. Again in Figure 9.10, the 80% load level is exceeded 10000 times, whereas the number of small cycles below 20% of the maximum load is relatively small: $100000 - 85000 = 15000$ cycles which is 15% of all cycles. Large and small cycles have special effects on fatigue as discussed in Chapters 10 and 11, see also Section 9.5.

Range counting methods

From a fatigue damage point of view, load amplitudes are more significant than mean loads. The amplitude is half the range between a minimum load and the subsequent maximum load. Load ranges represent important characteristic values of a load-time history exerted on a structure or applied in a fatigue tests. Load ranges of a load-time history can be counted, but since ranges are defined by a minimum and a maximum, a two-parameter counting methods must be adopted. Results can then be presented in matrix

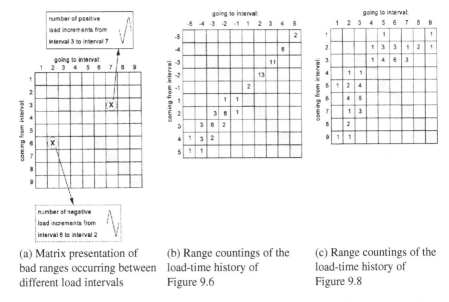

(a) Matrix presentation of bad ranges occurring between different load intervals

(b) Range countings of the load-time history of Figure 9.6

(c) Range countings of the load-time history of Figure 9.8

Fig. 9.11 Two-dimensional load range countings in matrix format.

format as is illustrated by Figure 9.11. A range is counted in the matrix at the corresponding interval in which the range was starting (listed at the left-hand side of the matrix) and the interval in which the range is completed (listed at the top side of the matrix). As indicated in Figure 9.11a, a positive load range coming from a minimum and going to a maximum, is counted in the upper right triangle of the matrix. Negative load ranges, coming from a maximum and going to a minimum, are counted in the lower left triangle of the matrix. The counting results of the load-history samples in Figures 9.6 and 9.8 are given in Figures 9.11b and 9.11c respectively. It should be noted that the counting results in Figure 9.11b are along a diagonal of the matrix. This should be expected because for this load-history (Figure 9.6) each peak load is followed by an opposite peak load of approximately the same magnitude. This is not true for the load history in Figure 9.8, which leads to more distributed counting results in the matrix of Figure 9.11c. The matrix is thus characteristic for the random nature of the load history. It is a two-parameter counting method, and for each range the mean value can easily be calculated because the minima and the maxima of the ranges are known. The accuracy is limited because counting of peak values does not indicate the exact location of a peak in an interval. However, smaller intervals can improve the accuracy.

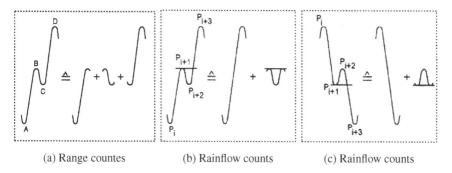

| (a) Range countes | (b) Rainflow counts | (c) Rainflow counts |

Fig. 9.12 Intermediate load reversal as part of a larger range.

The rainflow count method

In principle, range counting includes counting of all successive load ranges, also small load variations occurring between adjacent larger ranges. It might be thought that small load variations can be disregarded in view of a negligible contribution to fatigue damage. A fundamental counting problem arises if a small load variation occurs between larger peak values. This situation is illustrated in Figure 9.12. A two-parameter range counting procedure will count the ranges AB, BC and CD, and store this information in a matrix. Now, consider the situation that the intermediate range BC would not occur. Then, the large range AD would be counted only. Fatigue damage is related to load ranges. It should be expected that the fatigue damage of the large range AD alone is larger than for the three separate ranges AB, BC and CD. This has led to the so-called rainflow counting method of Endo [5].[12] The intermediate small load reversal BC is counted as a separate cycle and then removed from the major load range AD. This larger range can then be counted as a separate load range, see Figure 9.12b. If four successive peak values are indicated by P_i, P_{i+1}, P_{i+2} and P_{i+3}, the rainflow count requirement for counting and removing a small range from a larger range is

$$P_{i+1} < P_{i+3} \quad \text{and} \quad P_{i+2} > P_i \qquad (9.4a)$$

If the intermediate small load reversal occurs in a descending load range, see Figure 9.12c, the requirement is

$$P_{i+1} > P_{i+3} \quad \text{and} \quad P_{i+2} < P_i \qquad (9.4b)$$

---

[12] A similar eliminating concept for small intermediate ranges was described by Anne Burns in 1956 [6]. The Strain-Range-Counter developed by the Vickers aircraft industry was counting in accordance with this method.

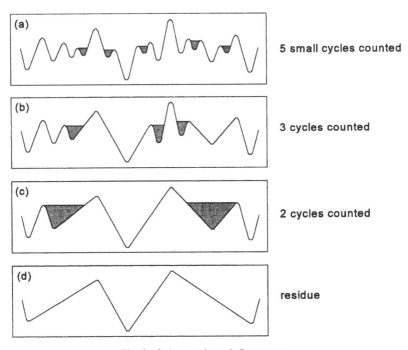

Fig. 9.13 Successive rainflow counts.

In words: the peak values of the intermediate small load reversal should be inside the range of the two peak values of the larger range. Successive rainflow counts are indicated in Figure 9.13. In Figure 9.13a five rainflow counts can be made. After counting and removing these small cycles, Figure 9.13b is obtained. In this figure again three rainflow counts can be made, but now of larger ranges. Removing these cycles lead to Figure 9.11c in which again two still larger load reversals can be counted and removed. In the final residue, Figure 9.11d, no further counts are possible. The ranges of the residue must be counted separately at the end of the counting procedure. The rainflow count results can be stored in a similar two-parameter matrix as discussed before (Figure 9.11).

The rainflow count procedure has found some support [7] by considering cyclic plasticity. A short load sequence is given in Figure 9.14a, which leads to counting two intermediate load reversals by the rainflow count method, as indicated in this figure. The corresponding plastic behavior is schematically indicated in Figure 9.14b, which could apply to local plasticity at the material surface during the initiation period, or to crack tip plasticity during crack growth. The intermediate load reversals $c1$ and $c2$ are causing hysteresis loops inside the major hysteresis of the major cycle between A and B. It is

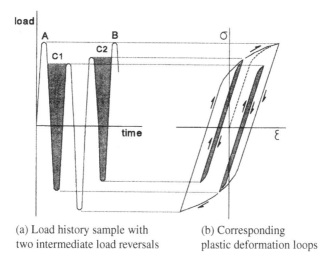

(a) Load history sample with
two intermediate load reversals

(b) Corresponding
plastic deformation loops

Fig. 9.14 Hysteresis loops associated with rainflow counts.

thus assumed that the intermediate plasticity loops do not affect the major loop. This reasoning gives somewhat speculative support to the rainflow counting method.

Some more comments on counting methods

As discussed in the previous text, statistical information of load-time history obtained by counting of level crossings, peak values, or ranges can be presented in a graph or a matrix. A graph represents a one-parameter distribution function while a matrix corresponds to a two-parameter distribution and thus gives more information. However, one significant aspect was not yet mentioned. Information about the sequence in which the counts were made is lost by these counting procedures. The matrix in Figure 9.11 collects numbers of ranges between successive peak values, but information about the sequence of the ranges is not obtained.

Some indirect information about sequences is retained in the rainflow count method. Each range counted by the rainflow procedure and stored in the matrix combines two peak values which may have been separated by intermediate load reversals in the original load-time history. However, these smaller ranges should have occurred between those two peak values in order to satisfy the rainflow count equation (9.4). Intermediate larger ranges did not occur because of the counting condition in the same equation. Anyway, the question must be considered whether the sequence is important

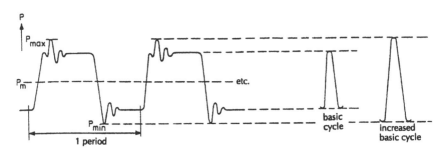

Fig. 9.15 Transient load fluctuations during on/off switching of machinery.

for fatigue predictions. The significance of sequences for fatigue damage accumulation is considered again in Chapters 10 and 11 on fatigue under variable-amplitude loading. Here, it should be emphasized that a most direct impression of a load-time history is obtained from records as presented here in Figures 9.6 and 9.8, and other ones discussed later.

The rainflow count method is generally suggested to give a better statistical reduction of a load time history defined by successive numbers of peaks and valleys if compared to the level crossing counts of Figures 9.6 and 9.8 and the range counts of Figure 9.11. Two main reasons for preferring the rainflow count method are: (i) an improved handling of small intermediate ranges, and (ii) an improved coupling of larger maxima and lower minima to range counts as will be illustrated below. In spite of this advantages, it cannot be said that the rainflow count method is fully based on rational arguments. It is easily recognized from Figure 9.13 that larger ranges counted by the rainflow method are separated by more intermediate smaller load ranges. The counting residue mentioned before consists of an increasing/decreasing series of positive peaks, and a decreasing/increasing series of negative peaks. These peaks may be separated by many intermediate peaks. It then becomes questionable whether counts of the residue are still meaningful for the evaluation of fatigue damage accumulation. It depends on the memory of the material during fatigue damage accumulation. The problem may be less serious for a flat spectrum with a large number of severe peak loads.

The significance of the basic idea of the rainflow counting method are illustrated by two simple examples. Figure 9.15 shows the load variation in a workshop machine due to switching on and off. The basic load cycle is a static block type load, but at the moment of switching some vibrational loads are introduced. The rainflow procedure will count several small load cycles, but instead of the deterministic static load cycle it also counts an increased basic load cycle between P_{max} and P_{min}, see Figure 9.15. Although the small

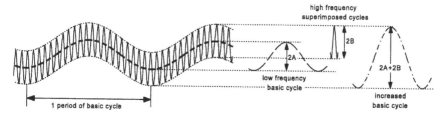

Fig. 9.16 High-frequency cyclic load superimposed on a low-frequency base line cycle.

vibrational load cycles could be practically harmless by themselves, they increase the severity of the basic load cycle. From a fatigue point of view, the material feels the increased basic cycle, and this increased load cycle is recognized by the rainflow counting method. A similar case was already discussed in relation to the superposition of random loads on a deterministic load cycle in Figure 9.5. The rainflow count method recognizes the largest cycle of the flight between the maximum peak in flight and the most severe downward load on the ground.

A second illustrative example is shown in Figure 9.16. A high-frequency load cycle (amplitude B) is superimposed on a low-frequency base line load cycle (amplitude A, frequency ω_1). According to the rainflow counting method, one cycle with an amplitude of $A + B$ will be counted in each period of the base line period, and that makes this cycle more damaging. This can be important depending on the damage done by the small superimposed high-frequency cycles. If ω_2 is much larger than ω_1, the number of the high-frequency cycles will contribute the major part to the fatigue damage, and the base line cycle is no more than a varying mean load, probably with a limited effect only.

Random Gauss process

Some types of random loads are caused by a stochastic random process. Turbulent air, in which an aircraft is flying, is supposed to be such a process. The same applies to random noise of a jet engine and water waves of the sea. It is often assumed that such processes are a random Gauss process which implies that the relevant variables have a normal distribution function (i.e. a Gaussian distribution). A random Gauss process is defined by a power spectral density function, $\phi(\omega)$, which fully describes its statistical properties. Examples are shown in Figure 9.17a. The power spectral density function shows how the energy of the signal is distributed as a function of

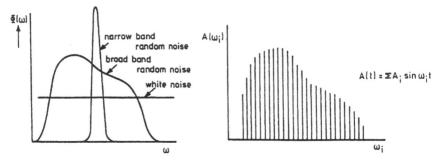

(a) Different types of energy density functions of a random signal

(b) Fourier series approximation of a random signal

Fig. 9.17 Energy density functions to define a random load Gauss process.

the frequency ω. This concept can be understood by considering a Fourier series with a very large number of terms and small differences ($\Delta\omega$) between the frequencies of successive terms, and coefficients A being a function of ω, see Figure 9.17b. The load-time history is then

$$A(t) = \sum A_i \sin(\omega_i t)$$

This sum gives a signal $A(t)$ which is approximately similar to random noise. It becomes a real random Gauss signal if $\Delta\omega \to 0$. The energy is proportional to the square of the amplitude:

$$\Phi(\omega) \propto [A(\omega)]^2$$

Two examples of a random Gauss signal and the corresponding power spectral density functions are shown in Figure 9.18. In Figure 9.18a, the energy is concentrated in a narrow frequency band and as a result the load-time history is somewhat similar to an amplitude modulated signal, in this case with a random modulation. This *narrow band random loading* is typical for resonance systems, which predominantly respond at one single resonance frequency if activated by some external random process covering a wider frequency band. The structure acts as a frequency filter to the excitation. The second example in Figure 9.18b shows a random signal covering a wider frequency band and the corresponding broad band random load signal shows a higher degree of irregularity. It was shown by Rice [9] that the distribution function of the peak values of a random Gaussian signal can mathematically be derived from the spectral density function $\Phi(\omega)$. This is also true for the irregularity factor k defined earlier as the

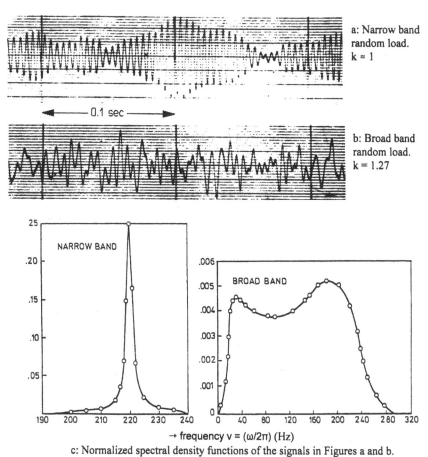

a: Narrow band random load. k ≈ 1

b: Broad band random load. k = 1.27

c: Normalized spectral density functions of the signals in Figures a and b.

Fig. 9.18 Two records of random load and the corresponding spectral density functions [8].

ratio of the number of peaks and the number of zero-crossings (crossing the level $A = 0$). As might be expected, this factor is almost equal to 1 for narrow band random loading, whereas it is larger for more irregularly varying signals, see Figure 9.18b. A mathematically closed form solution for distribution functions of ranges cannot be derived from the spectral density function.

The application of power spectral density techniques to the dynamic behavior of a structure is of interest. If a random Gaussian load is applied to a linearly elastic structure, then the stress in the structure is also a random Gaussian phenomenon. The spectral density function of the stress can be calculated from the spectral density function of the external load by using a transfer function, depending on the elastic properties of the structure. This problem of applied dynamics is outside the scope of the present discussion. However, it can be useful for discussions on random fatigue load spectra to know whether the loads are narrow band or broad band phenomena.

9.4 Determination of load spectra

Different types of fatigue loads and description techniques of load-time histories were presented in Sections 9.2 and 9.3 respectively. The question now is: How can a load spectrum be determined if the fatigue performance of a structure must be investigated? Two obvious approaches are: (i) by analyzing the use of the structure, and (ii) by measuring the loads on the structure if the structure, or a similar previous structure, already exists. In Section 9.2, five examples of structures under fatigue loads were discussed, which led already to several questions. The principal question is: which cyclic loads will occur? Sometimes, the categories of loads follow directly from the purpose of the structure, e.g. pressurization cycles of a vessel, and train passages on a railway bridge. The question is more difficult for moving vehicles. Different problems arise if vehicles can be used for several purposes under various conditions. As an example, this has been recognized for trucks to be used in countries with poor roads. The trucks for these countries are made stronger by the automotive industry, stronger than for countries with a modern road system. Stronger implies a more heavy truck and thus less cargo capacity.

Knowledge of relevant load spectra is also depending on experience of an industry with respect to the performance of their products in service. Much has been learned from case histories of fatigue failures occurring in service and accident investigations. In any case, the designer must use his imagination to see how the structure can be used. Perhaps he should consider abusive use as well. Depending on the consequences of fatigue failures in service, a worst case approach must be considered.

In general, it is difficult to define straight forward procedures how to obtain load spectra to be used for fatigue life predictions, and also for supporting experimental work if that appears to be necessary. Two steps in the load spectrum evaluation may be recognized: first, a qualitative approach, and second, a more quantitative approach.

9.4.1 The qualitative approach

Figure 1.1 shows a broken front-wheel of a heavy motorbike. The fracture surfaces of the spokes clearly indicated fatigue failures. The front wheel collapsed during braking before a railway level crossing which was closed. The police-officer driving the motorbike survived. It turned out that similar failures occurred in the same motorbikes of the police in other countries. The failure was not an incidental case, but a symptomatic one. The question then is, what is so special about the use of these motorbikes by the police? Analysis showed that the police were using the brakes very often, which led to a heavy moment on the front-wheel spokes even if the driving speed was low. Apparently, a predominant fatigue load occurs for a special group of users of the motorbike, but not for all users. It illustrates that different groups of users of a structure must be considered.

A subsequent question then is: which utilization is the most severe one on which a fatigue analysis should be based? The answer to this question can depend on economic and safety consequences of fatigue failures in service. The variety of economic consequences can be large. Related aspects are: repairs, replacements, inspections, structure not to be used before some remedial efforts are introduced, etc. Financial liability as well as the reputation of a product of the industry can also be involved. Sometimes consequences are self-evident. A pressure vessel should not explode as a result of fatigue cracks. Cables of a passenger lift in the mountains should never fail. In any case, designing against fatigue requires consideration of all possible *scenarios* of how a structure might be used and how it can fail by fatigue.

In general, the designer knows more or less the type of loads to which his structure is subjected in service, at least the regular type of loads. As an example, a list of loads on a transport aircraft is shown in Table 9.1 which also includes some numerical information. It should be noted that the data illustrate possible orders of magnitudes only. The numerical data can vary within fairly wide margins, depending on the type of aircraft and how it is going to be used. Moreover, the list of types of fatigue loads in the table is not necessarily complete.

Information on load spectra for aircraft is relatively abundant, but for various structures it may be difficult to set up similar lists. The type and character of known fatigue loads in service can usually be determined reasonably well. However, statistical information about the load spectra in

Table 9.1 Global information about different types of loads on an aircraft structure.

Type of load	Character	Number of cycle in the design life time	Period of 1 cycle
ground-air-ground transition cycle	deterministic	10^3 to 10^5	10 min to 10 hrs
pressure cycle of the cabin		10^3 to 10^5	10 min to 10 hrs
maneuvers		4×10^3 to 4×10^5	10 sec to 3 min
turbulence (gusts)	random	10^5 to 10^6	0.1 sec to 10 sec
taxiing loads		10^5 to 10^7	0.05 sec to 1 sec
acoustic loads		10^7 to 10^8	0.001 sec to 0.01 sec

many cases is non-existent. Measurements, quite often with strain gages, can then be most informative.

Buxbaum [10] presented an instructive example of measurements of the bending moment in an axle-spindle of a motor car, see Figure 9.19. The upper record suggests that two types of loads are superimposed. By a separation technique, it was shown to consist of a maneuver type of loading and a random vibrational load due to roughness of the road. Rainflow counts of the upper record can give a useful representation for fatigue damage analysis, but the separated signals give a much better impression of the two different types of loads acting simultaneously on the structure. Moreover, a design

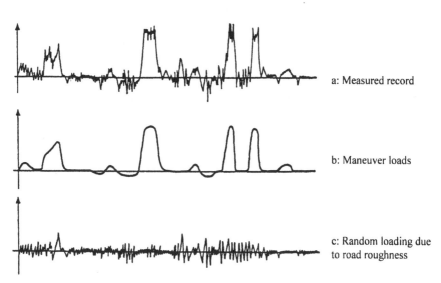

a: Measured record

b: Maneuver loads

c: Random loading due to road roughness

Fig. 9.19 Load record of an axle spindle of a motor-car, separated in two different types of loads [10].

modification for improving the fatigue performance of a structure may be effective for one type of loading, but not necessarily for other ones. For instance, a change of the resonance frequency of a structure and introducing more damping can significantly reduce the stress spectrum induced by random vibrations, whereas the stress spectrum of maneuver type loads is not substantially changed. The relation between the external load spectrum and the stress spectrum in the structure (the transfer function) can be different for different types of loads, in particular for random vibration loads compared to deterministic maneuver loads. The dynamic response of the structure should then be considered.

Unfortunately, fatigue failures occurring in service often have shown that not all significant fatigue loads on a structure were known before. An example is described by Griese et al. [11]. Fatigue problems occurred in the drive shaft of a heavy-plate rolling machine. The torsion moment on the shaft was measured. The results presented in Figure 9.20 show that significant load cycles occurred at the moment that the plate was entering the rollers, and also when the plate was leaving the rollers. Probably, such severe extra load cycles would not occur if rolling could be done slowly, almost quasi-statically. But slow rolling will be undesirable from an economic point of view. Anyway, it illustrates that it can be difficult to anticipate all loads which can occur in service.

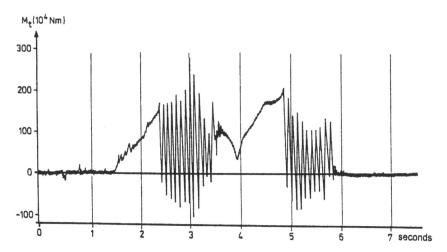

Fig. 9.20 Measured torsion moment on a driving shaft of a rolling machine for heavy steel plates [11].

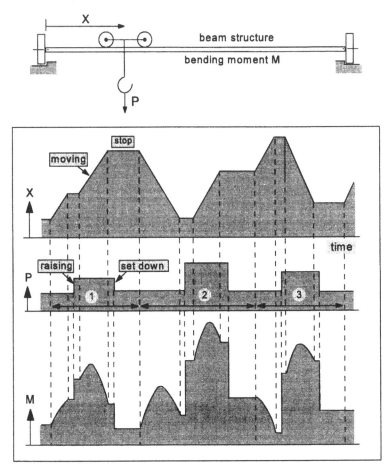

Fig. 9.21 Calculated load history for a traveling crane in a production hall [12].

9.4.2 The quantitative approach

Quantitative assessments of load spectra is a problematic issue for most structures. A simple case is a pressure vessel with known pressure variations and no other load fluctuations. Another illustrative example with still relatively simple conditions was discussed by Weiss [12]. He considered a traveling crane in an industrial production hall used for transportation of heavy items of different weights to various locations (x, y) in the hall, see Figure 9.21. A part with a certain weight is lifted by the crane and then transported to another location. The structure to be considered is the beam spanning the two rolling supports. The bending moment in this beam (M) depends on the weight (P) and the location of the lifting cart on the beam (x).

The top graph in Figure 9.21 shows that the crane is moving and makes stops for lifting up a weight or putting it down. The bending moment can be calculated from the weight P and the crane location x, see the lower graph of Figure 9.21. It should be noted that the three transport segments in this figure (1, 2 and 3) give rise to four load cycles. This example shows that the bending load cycles can be calculated, but it requires detailed information regarding the usage of the crane. Moreover, the load spectrum thus obtained presumes that no transient loads occur during the moment of lifting a weight or putting it down. Dynamic conditions are obviously more complex for a large outdoor crane with more kinematic possibilities, operating in a windy climate, and meeting inertia loads of moving parts. Measurements of load-time histories with strain gages are recommended.

With some imagination about how a structure will be used, a list can be made of the various types of cyclic loads acting on the structure. An example of such a list was already given in Table 9.1 for an aircraft structure for which predictions on load-time histories have received much attention. This certainly was stimulated by some dramatic aircraft crashes caused by fatigue cracks. The analysis starts with a so-called *mission analysis*, which implies that imaginary flights are made according to expectations for a transport aircraft, a military aircraft, or some other type of aircraft. The deterministic loads (maneuvers) often allow calculations about the magnitudes of the cyclic loads. It is much more difficult for random loads due to gusts and taxiing. Calculation techniques have been developed for the dynamic response of the structure, but the analysis is difficult, also because damping (aerodynamic damping for the wing) is a complication. If it is possible, calculations should be supplemented by measurements on an aircraft in service. Two samples of strain gage records for a wing in turbulent air are presented in Figure 9.22. The record of aircraft A with a slender wing and two jet engines attached to each wing shows vibrations with a frequency in the order of 2 Hz (period $\approx 1/2$ sec). This frequency corresponds to the first mode of wing bending vibrations. Such vibrations are hardly observed in the record for aircraft F with a relatively high bending stiffness and only one turboprop engine on each wing. The dynamic response of the wing structure is affecting the stress spectrum of the wing structure. Similar problems apply to the load records in Figure 9.19 for the motor-car axle. It is difficult to predict the random bending moment variation invoked by road roughness, partly because the statistical data on surface roughness may be questionable, but also because the dynamic response is a complex phenomenon. In addition, it is not really easy to predict the maneuver component of the car load history.

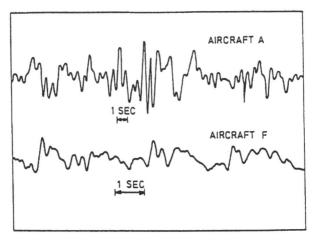

Fig. 9.22 Strain gage records of wing bending for two types of aircrafts flying in turbulent air [2].

Because quantitative evaluations of load spectra for a fatigue critical location in a structure can offer significant problems, a need for load-history measurements is easily recognized. Such measurements are not necessarily complicated. Some types of sensors can be used, but strain gages are most popular. Strain gages can be sealed very well and thus also be used over long periods in different environments. How to sample data depends on the problem to be investigated. If the purpose is to know more about the characteristic nature of certain load variations, then a continuous analogue record can be very instructive, see the samples in Figures 9.19, 9.20 and 9.22. However, if the nature of the loads in reasonably well understood, but the number and magnitude of the cycles are not very well known, then counting peaks and troughs is most useful. Small sized equipment and software for a counting analysis are commercially available for that purpose. It can provide the data in the two-parameter matrix format discussed before (Figure 9.11). It is also possible to store the full sequence of the peaks and troughs. Rainflow counting can be done, as well as other counting methods. Equipment was also developed for telemetric measurements on moving vehicles and rotating parts.

A typical problem is offered by the replacement of old bridges which may be older than 100 years, but which were originally designed with very high safety factors [13]. Measurements of load spectra are then essential to consider the question whether the bridge is still good enough for the present time (cheap solution), or should it be replaced by a new one (expensive solution).

Another illustrative example occurred when NLR (National Aerospace Laboratory, Amsterdam) wanted to test new load-history counting equipment. As a trial experiment, some strain gages were bonded on the wing struts of a training glider. Launching occurred by a winch. After landing, the glider was towed by a jeep to the original launching position on the airstrip. The measurement results, immediately available after the flight, indicated that just a single flight was sufficient to show that the larger cyclic loads did not occur in flight, but during towing of the glider after landing, due to transportation over a rough terrain.

Similar measurements on a new structure can rapidly and easily produce useful information about load spectra. It can show how a structure is really used. Furthermore, load spectrum measurements applied on an existing structure already in service for several years enables a comparison between the present load spectrum and the spectrum assumed in the design stage of the structure. Measurements are also carried out for load spectrum monitoring of military aircraft in order to analyze problems about life limitations and inspection periods. Indications about safety margins are then obtained.

9.5 Service-simulation fatigue tests and load spectra

Predictions on fatigue life and crack growth under Variable-Amplitude (VA) fatigue loading are discussed in Chapters 10 and 11 respectively. It will be explained that predictions made by the well-known Miner rule are not reliable, partly because of shortcomings of this rule, and also due to uncertainties about the required S-N curves. Service-simulation fatigue tests should then be considered. The load-time history in a service-simulation fatigue test must be a valid simulation of load histories which can occur in service. The problem is that the time scale in the test cannot be the same as in service because it would require an extremely long time for a single test. Some acceleration of the test is necessary in most cases. This may introduce a problem if some time-dependent phenomenon can effect the fatigue process.

The load-history sample in Figure 9.20 for a driving shaft of a rolling machine shows relatively fast cycles with a cyclic frequency of about 10 Hz, but most of the time the load is zero. If time-dependent phenomena do not affect fatigue, the load-history in Figure 9.20 can be compressed to the load

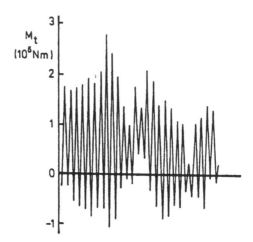

Fig. 9.23 Time compressed representation of the load history in Figure 9.20.

history shown in Figure 9.23, which can easily be applied in a computer controlled fatigue test.

A similar reduction of the time scale is adopted for fatigue tests of aircraft structures. The flight load profile for a single flight shown in Figure 9.5 in reality covers a period of 1 to 10 hrs, which is fully unacceptable for a service-load-simulation fatigue test. Time compression is obtained by applying all load variations in a short time, but maintaining the sequence of the successive maxima and minima, see Figure 9.24. Also the numerous taxiing loads may be omitted because they are supposed to have a negligible effect on fatigue in view of the low stress levels. However, the minimum load occurring during taxiing, the touch-down load in Figure 9.5, is maintained as the minimum load of the flight. This is necessary to obtain still the same fatigue damage of a flight according to a rainflow analysis of the load sequence.

Although a representative sequence of peaks is applied in a service-simulation fatigue test, the test in general will be an accelerated test. If fatigue in service occurs in a corrosive environment, results of service-simulation fatigue tests can be too optimistic. This problem is considered again in later chapters (Chapters 13, 16 and 20).

Service-simulation load histories in fatigue tests in most cases imply that a rather complex sequence of maxima and minima must be applied to a specimen or a structure. Fortunately, it is possible to apply complex load sequences in modern closed-loop electro-hydraulic fatigue machines which were introduced around the early 1970s. Such load sequences

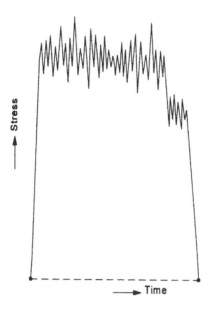

Fig. 9.24 Time compressed representation of the load history in Figure 9.5.

could not be realized in the older open-loop fatigue machines, which at best allowed block-program fatigue tests, discussed in Chapter 10. However, block-program tests are not recommended any more. Initially, service-simulation fatigue tests were more widely used for aircraft fatigue problems. These tests are also called *flight-simulation fatigue test* because a valid simulation requires flight-by-flight load sequences. As an example, a load history in such a test is shown in Figure 9.25. It applies to a civil aircraft wing structure. The full sequence includes 10 different types of flights. The stress-history was characterized by one specific stress level, for which the mean stress in flight (S_{mf} in Figure 9.25) was chosen. As discussed before, this load history is a combination of deterministic ground-air-ground cycles with superimposed gust loads (air turbulence) in flight. The ground-air-ground cycle can be calculated. The gust loads have to be derived from gust load statistics. Gust load spectra have been measured over long periods in service. The spectra are available as exceeding curves of the type shown in Figure 9.10 as a steep spectrum. Such a spectrum has to be broken down in spectra for different weather conditions occurring in different flights. It requires expertise about how this can be done in a statistically acceptable way [15].

Load-time histories for other types of structures require a similar analysis of the various missions of the structure in order to establish a

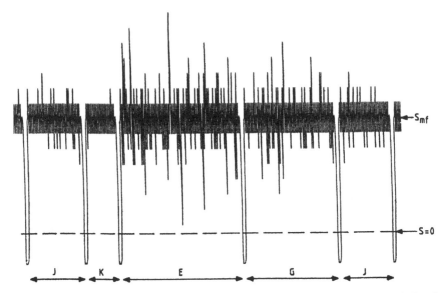

Fig. 9.25 Sample of a load history applied in flight-simulation fatigue tests [14]. Load spectrum of the Fokker F-28 wing structure. Five flights are shown with gust loads corresponding to different weather conditions.

simulation of load sequences as they would occur in reality. Obviously, load measurements in service can give most useful information for this purpose. Service-simulation fatigue tests are now carried out by various industries for several purposes discussed in Chapter 13.

If the load spectrum to be used in a service-simulation fatigue tests is available, problems can arise for a steep spectrum. The previous discussion on Figure 9.10 has revealed that a steep spectrum contains numerous low-amplitude cycles and a small number of high-amplitude cycles. If all low-amplitude cycles have to be included in a service-simulation fatigue test, the duration of the test would be very long. Very small cycles are therefore omitted from the load-time history in view of time efficiency. However, small cycles can contribute to fatigue damage, also when the amplitude is below the fatigue limit. Empirical evidence of representative specimens should give indications on this issue.

At the other end of a steep spectrum, the number of load cycles with a large amplitude is low. These cycles can considerably increase the fatigue life by introducing favorable residual stresses at notches. If the steep spectrum applies to wing bending of a transport aircraft, the high-amplitude gust loads occur in a flight in a most severe storm. Some aircraft of a fleet will meet this storm occasionally, but other aircraft will not. The fatigue life of the latter ones will be shorter. It is for this reason that the high-amplitude cycles in flight-simulation tests are truncated[13] to a lower amplitude level in order to avoid unconservative test results. The choice of the truncation level is a delicate question also for other types of structures subjected to a steep load spectrum. Fortunately, the problems of low-amplitude and high-amplitude cycles are much less important for a flat load spectrum. The topics are addressed again in Chapters 10 to 13.

9.6 Major aspects of the present chapter

The major aspects of load-time histories and load spectra discussed in this chapter are summarized below:

1. A load history applied to a structure in service is characterized by a sequence of successive maxima and minima (peaks) of the load on the structure. A load spectrum is a statistical representation of these maxima and minima, obtained by counting the numbers of peak values in load intervals, or counting the number of exceedings of load levels. The data can be presented in tables or graphs in which the magnitude of the load is indicated by a single load parameter. One-parameter load spectra can be instructive for general impressions of the spectrum severity and comparing load spectra of different severities.

2. Counting of load ranges between successive maxima and minima is also possible with results collected in a matrix. This is a two-parameter statistical representation of a load history which provides more information than a one-parameter counting method. The matrix gives information about load ranges between successive maxima and minima, but information about the sequence of these ranges is lost.

3. The rainflow count method is a range count method which counts intermediate smaller ranges separately. The rainflow count method should be preferred for a statistical analysis of load-time histories because it is more realistic in considering the fatigue damage of combined maximum and minimum loads.

4. Characteristic classifications of loads in service are: deterministic loads (especially maneuver type loads) and stochastic loads (in particular

---

[13] The terminology used here refers to truncation of high-amplitude cycles and omission of low-amplitude cycles. In the literature, these concepts are also called clipping of high-amplitude cycles and truncation of the low-amplitude tail of a load spectrum, respectively.

random loads). Another classification is stationary load spectra versus non-stationary load spectra.

5. Narrow band random loading looks like an amplitude modulated signal. Broad band random load has a more irregular character.

6. Continuous load-time records are most informative to show characteristic features of the load history which are not easily deduced from load counting results.

7. Load spectra for structures can vary from very simple (e.g. almost constant-amplitude loading of a pressure vessel) to rather complex (e.g. superposition of different types of loads from different sources with varying intensities and probabilities of occurrence).

8. Load spectra are essential for the analysis and predictions of fatigue critical structures. The spectrum of the stress in the structure is not linearly related to the load spectrum on the structure, depending on the dynamic response of the structure on external loads.

9. Assessments of load spectra for a structure should start with listing all types of loads occurring in service and their characteristic properties. Quantitative assessments of the spectra can be difficult due to lack of information. Load measurements should then be considered for which well developed techniques are available.

10. Quantitative information on load histories is also essential for planning service-simulation fatigue tests.

References

1. Private information Mr. van Maarschalkerwaart, Dutch Railways.

2. Schijve, J. *The analysis of random load-time histories with relation to fatigue tests and life calculations*. Fatigue of Aircraft Structures, 2nd ICAF Symposium, W. Barrois and E.L. Ripley (Eds.). Pergamon Press (1963) pp. 115–149.

3. Haibach, E., Fischer, R., Schütz, W. and Hück, M., *A standard random load sequence of Gaussian type recommended for general application in fatigue testing; its mathematical background and digital generation*. In: Fatigue Testing and Design. Vol. 2, Soc., Environmental Engineers, London (1976), pp. 29.1–29.21.

4. Schütz, W., *Standardized stress-time histories: An overview*. Development of Fatigue Load Spectra, STP 1006, ASTM (1989), pp. 3–16.

5. Matsuishi, M. and Endo, T., *Fatigue of metals subjected to varying stress – Fatigue lives under random loading*. Preliminary Proc. of the Kyushu District Meeting, The Japan Society of Mechanical Engineers (1968), pp. 37–40 [in Japanese].

6. Burns, A., *Fatigue loadings in flight: Loads in the tailplane and fin of a Varsity*. A.R.C. Technical Report C.P.256. London (1956).

7. Dowling, N.E., *Mechanical Behavior of Materials. Engineering Methods for Deformation, Fracture, and Fatigue*, 3rd edn. Prentice Hall (2006).

8. Hillbery, B.M., *Fatigue life of 2024-T3 aluminum alloy under narrow- and broad-band random loading*. ASTM STP 462 (1970), pp. 167–183.

9. Rice, S.O., *Mathematical analysis of random noise*. Bell System Tech. Journal, Vols. 23 and 24, (1944 and 1945).

10. Buxbaum, O., *Random load analysis as a link between operational stress measurement and fatigue life assessment*. ASTM STP 671 (1979), pp. 5–20.

11. Griese, F.W., Schöne, G., Schütz, W. and Hück, M., *Crack growth in torsion loaded axles of rolling machines during production*. Stahl und Eisen, Vol. 99 (1979), pp. 193–198 [in German].

12. Weiss, M.P., *Simulation and monitoring of loads in crane beams*. ASTM STP 671 (1979), pp. 208–221.

13. Yamada, K. and Miki, C., *Recent research on fatigue of bridge structures in Japan*. J. Construct. Steel Res., Vol. 13 (1989), pp. 211–222.

14. Schijve, J., Jacobs, F.A. and Tromp, P.J., *Fatigue crack growth in aluminium alloy sheet material under flight-simulation loading. Effect of design stress level and loading frequency*. Nat. Aerospace Lab. NLR, TR 72018, Amsterdam (1972).

15. de Jonge, J.B., Schütz, D., Lowak, H. and Schijve, J., *A standardized load sequence for flight simulation tests on transport aircraft wing structures*. Nat. Aerospace Lab. NLR, TR 72018, Amsterdam (1973). Also: Laboratorium für Betriebsfestigkeit LBF, Bericht FB-106 (1973).

Some general references

16. Murakami, Y., Mineki. K., Wakamatsu, T. and Morita, T., *Data acquisition by a small portable strain histogram recorder (Mini-Rainflow Corder) and application to fatigue design of car wheels*. Fatigue Design 1998, ESIS Publication 23, Elsevier (1999), pp. 373–383.

17. Heuler, P. and Schütz, W., *Standardized load-time histories – Status and trends*. In: Low Cycle Fatigue and Elasto-Plastic Behaviour of Materials. Elsevier (1998), pp. 729–734.

18. Rice, C.R. (Ed.), *SAE Fatigue Design Handbook*, 3rd edn. AE-22, Society of Automotive Engineers, Warrendale (1997).

19. Murakami, Y. (Ed.), *The Rainflow Method in Fatigue*. The Tatsuo Endo Memorial Volume. Butterworth Heinemann (1992).

20. Potter, J.M. and Watanabe, R.T. (Eds.), *Development of Fatigue Loading Spectra*. ASTM STP 1006 (1989).

21. Buxbaum, O., *Fatigue Strength in Service*. Verlag Stahleisen mbH, Düsseldorf (1986) [in German].

22. *Standard Practices for Cycle Counting in Fatigue Analysis*. ASTM Standard E 1049-85 (1985).

Chapter 10
Fatigue under Variable-Amplitude Loading

Symbols

CA   Constant-Amplitude
VA   Variable-Amplitude

> *Predictions are difficult, especially on the future.*
> *(Niels Bohr, Winston Churchill, Wim Kan)*

10.1 Introduction

Constant-amplitude (CA) fatigue loading is defined as fatigue under cyclic loading with a constant amplitude and a constant mean load. Sinusoidal loading is a classical example of CA fatigue loads applied in many fatigue tests. In the previous chapter on fatigue loads, it has been pointed out that various structures in service are subjected to variable-amplitude (VA) loading, which can be a rather complex load-time history, see several figures in Chapter 9. Predictions on fatigue life and crack growth should obviously be more complex than predictions for CA loading. The latter problem was discussed in Chapter 7 (Fatigue Lives of Notched Elements) and Chapter 8 (Crack Growth). In Chapter 7, the best defined problem was the prediction of the fatigue limit of a notched element. The fatigue limit is a threshold value of the stress amplitude. Stress amplitudes below this level do not lead to failure, while stress amplitudes above the fatigue limit lead to crack initiation and crack growth to failure. Rational arguments could be adopted for the predictions of the fatigue limit, by comparing fatigue limits of a structure to fatigue limits of simple unnotched specimens, but certain problems had to be recognized associated with the notch effect, size effect, surface effect and environmental influences.

For structures subjected to VA load cycles in service, it may be desirable that fatigue failures should never occur. It implies that all load cycles of the load spectra should not exceed the fatigue limit. The prediction problems is then restricted to the prediction of the fatigue limit as discussed in Chapter 7. However, this requirement can lead to a heavy structure and it can be unnecessarily conservative, especially if the number of more severe load cycles above the fatigue limit is relatively small. Moreover, a complete avoidance of fatigue is not always required. Failures after a sufficiently long life can be acceptable from an economical point of view, the more so if safety issues are not involved. Fatigue under VA load conditions is the subject of the present chapter. Possibilities for fatigue life predictions under VA loading are discussed, while predictions on crack growth under VA loading are covered in the following chapter (Chapter 11).

The discussion in Section 10.2 starts with considerations on the well-known Miner rule with its long lasting reputation. Reasons why and how this rule can be misleading are discussed. Results of various fatigue tests under VA loading are considered in Section 10.3. Alternative fatigue life prediction methods are reviewed in Section 10.4. A general discussion on the problems of fatigue life predictions for VA loading is presented in

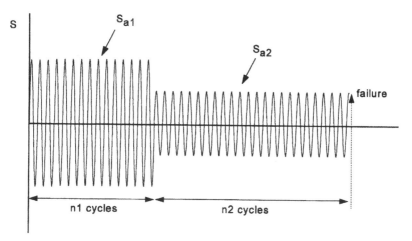

Fig. 10.1 A simple VA load sequence with two blocks of cycles.

Section 10.5. The major points of the present chapter are summarized in Section 10.6.

Several problems in this chapter are illustrated by test results of Al-alloys used in aircraft structures. The reason is that fatigue under VA loading has been extensively studied for these alloys in view of the significance of fatigue for the safety of aircraft. Load spectra measurements on aircraft in service have also widely been made for the same reason. It has stimulated the development of service-simulation fatigue tests. However, the experience of aircraft structures and materials is also relevant for other structures and materials, especially if these structures are made of relatively high-strength materials, which usually are fatigue sensitive.

10.2 The Miner rule

A specimen is fatigue tested under CA loading until a certain percentage of its fatigue life, say $x\%$. Fatigue damage must then be present in the specimen, because its original life (N) has been reduced to $(100 - x)\%$ of the fatigue life N. The damage may still be invisible, but it is present in the material of the specimen.

A most simple VA load history is presented in Figure 10.1. The amplitude is changed only once. Obviously, such a load history is not related to service load spectra, but it is considered here to discuss the basics of the Miner rule. Moreover, this simple VA load sequence was widely used in many older test programs to check the validity of this rule. In Figure 10.1, n_1 cycles at a stress amplitude S_{a1} are applied, followed by cycles with an amplitude reduced to S_{a2}. The test is continued until failure occurs after n_2 cycles at the lower amplitude. Two different blocks of load cycles are thus applied in this test. The problem is to predict n_2.

As early as 1924, Pålmgren [1] published the hypothesis which is now generally known as the Miner rule or the linear cumulative damage hypothesis. According to this rule, applying n_1 cycles with a stress amplitude Sa1 and a corresponding fatigue life endurance N_1, is equivalent to consuming n_1/N_1 of the fatigue resistance. The same assumption applies to any subsequent block of load cycles. Failure occurs if the fatigue resistance is fully consumed. For the two blocks in Figure 10.1, it implies that failure occurs at the moment that

$$\frac{n_1}{N_1} + \frac{n_2}{N_2} = 100\% \tag{10.1}$$

If more than two blocks are applied, this equation is generalized to read

$$\sum \frac{n_i}{N_i} = 1 \tag{10.2}$$

Pålmgren did not give any derivation, but he was in need of a rule in view of fatigue life calculations for ball-bearings under VA loading. He thus adopted the above simple assumption on fatigue damage accumulation. Langer in 1937 [2] postulated the same rule, but with a refinement, viz. the rule applies separately to the crack initiation period and to the crack growth period. Miner [3] in 1945 was the first one to propose a derivation of the linear cumulative damage rule. He assumed that the work that can be absorbed until failure has a constant value W, and in addition the work absorbed during n_i similar cycles is proportional to n_i. As a consequence, it implies that $w_i/W = n_i/N_i$. The criterion $\sum w_i = W$ then leads to Equation (10.2). Miner did VA experiments on unnotched specimens and a few riveted lap joints of 2024-T3 Alclad sheet material. He used two to four different blocks of load cycles in a test. He found $\sum n/N$-values varying from 0.61 to 1.45, but on the average reasonably close to 1.0. Since that time, the rule $\sum n/N = 1$ has frequently been quoted as the Miner rule, the linear cumulative fatigue damage rule, or sometimes as the Pålmgren–Miner rule, although it would be more correct to refer to the Pålmgren rule. After 1945, numerous VA fatigue test programs were carried out to verify the Miner rule. In many cases, significant discrepancies were found which are discussed in Section 10.3. Stimulated by such discrepancies, several new theories on VA fatigue were published. Unfortunately, similar to the validity of the Miner rule, the new theories did not have sufficient credibility from a physical point of view. Also, the experimental verification was quite often restricted to some specific test programs only. First, some essential comments will be made on a few principal inconsistencies of the Miner rule. Certain shortcomings of this rule must be understood in order to arrive at reasonable fatigue life considerations if VA load histories are applicable.

10.2.1 *Effect of load cycles with stress amplitudes below the fatigue limit*

If S_{a2} in the simple load sequence of Figure 10.1 is below the fatigue limit, then N_2 is infinite, and $n_2/N_2 = 0$. According to the Miner rule, the specimen will not fail because $\sum n/N = 1$ is never reached. In other words, according to the Miner rule, cycles with an amplitude below the fatigue limit are not damaging. This is physically inconsistent. Recall that cycles below the fatigue limit are unable to create a growing microcrack under CA loading, and thus can not cause failure, $N_2 = \infty$ and $n_2/N_2 = 0$. However, this argument is not relevant for VA loading. In the simple case of Figure 10.1, the cycles of the first block (n1 cycles with amplitude S_{a1}) can create a growing crack. The question then is whether cycles of the second block can propagate this crack. In view of the stress singularity at the tip of the existing crack, a contribution to further crack growth is possible, which can indeed lead to failure. This also applies for more complex load histories including load cycle amplitudes above and below the fatigue limit. The Miner rule ignores the fatigue damage contribution of cycles below the fatigue limit to crack growth, because $N = \infty$. However, these cycles can contribute to an increase of existing fatigue damage.

The significance of the damage contribution of such small cycles depends on the type of load spectrum. Two highly different load spectra, previously shown in Figure 9.10, are presented again in Figure 10.2, together with an S-N curve. According to the Miner rule, all cycles with an amplitude below the fatigue limit should be non-damaging. Figure 10.2 illustrates that this number is very large for the steep spectrum and relatively small for the flat spectrum. In the first case (steep spectrum) the damage contribution of the small cycles should be expected to be significant, whereas in the second case (flat spectrum) it could be small.

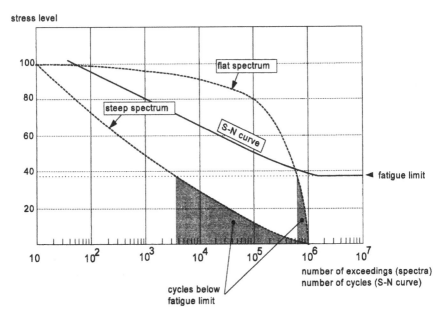

Fig. 10.2 The number of cycles below the fatigue limit depends on the shape of the load spectrum.

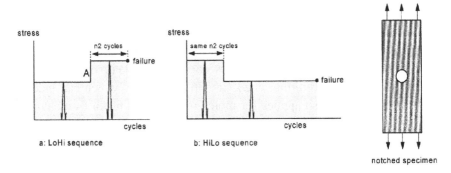

Fig. 10.3 Two different sequences of blocks in a simple VA load history ($R = 0$) applied to a notched specimen.

10.2.2 *Effect of notch root plasticity*

Two simple VA load sequences are shown in Figure 10.3. These sequences are applied to a notched specimen. The same two amplitudes are used in both sequences, but in the first sequence the test starts with the low amplitude (low-high sequence, or LoHi), and in the second one with the high amplitude (high-low sequence, or HiLo). The stress ratio is supposed to be zero ($S_{min} = 0$, or $R = 0$). The following case is considered. The peak stress at the root notch (σ_{peak}) exceeds the yield stress ($S_{0.2}$) only in the block with the high amplitude, whereas this does not occur in the block with the low amplitude. It implies that notch root plasticity did not occur in the LoHi sequence (Figure 10.3a) during the low-amplitude cycles of the first block of the LoHi sequence. However, in the other sequence (HiLo, Figure 10.3b), notch root plasticity occurs immediately in the first block with the high amplitude. In this case, compressive residual stresses at the root of the notch are present at the beginning of the second block with the low-amplitude cycles. This is favorable for fatigue in the second block. Although the number of high-amplitude cycles (n_2) as observed in the LoHi test is applied in HiLo test, the fatigue life will be larger in this test due to the favorable residual stress. In other words, the sequence of the two blocks is significant for the fatigue life. This sequence effect is not predicted by the Miner rule because the rule ignores any change of residual stresses induced by previous cycles.

Another instructive example of experimental results is shown in Figure 10.4, again for a two-block VA fatigue test. In this case, the mean stress is zero in both blocks. The first block with a high amplitude is causing cyclic plasticity at the notch root because $K_t \cdot S_{a1} > S_{0.2}$. A subtle difference exists between the two HiLo sequences of Figures 10.4b and 10.4c. In the

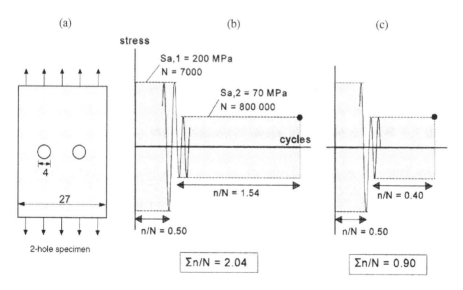

Fig. 10.4 Two-block VA tests on notched Al-alloy specimens [4]. Effect of compressive or tensile residual stress at the notch root in the second block.

first case the last maximum stress before reducing the amplitude is positive, and as a result it will leave a favorable negative residual stress at the notch. Consequently a longer life should be expected at S_{a2}. However, in the other case (Figure 10.4c), the opposite occurs. The last peak stress of the first block is negative, and it thus leaves an unfavorable residual tensile stress at the root notch, which shortens the fatigue life at S_{a2}. These trends are clearly reflected in the experimental results of $\sum n/N$-values of 2.04 and 0.90 respectively. As said before, this plasticity induced sequence effect is not recognized by the Miner rule.

10.2.3 Crack length at failure

The two load sequences of Figure 10.3 are considered again. Assume that a crack length $a = 2$ mm (as an example) is reached at the end of the first low-amplitude block (point A in Figure 10.3), while the crack at this low amplitude could grow until $a = 20$ mm until failure occurs. However, a change to the second high-amplitude block could lead to immediate failure, because a small crack is more critical at a higher stress level. It obviously would lead to $\sum n/N < 1$. In the reversed sequence of Figure 10.3b, consider a transition made shortly before failure in the first block of the high amplitude, i.e. at the corresponding n/N slightly smaller than 1. In view of the high amplitude, the crack will be small just before failure. After the transition, substantial crack growth at the low amplitude of the second block is still possible before failure at the lower amplitude occurs at a relatively large crack length. A value $\sum n/N > 1$ should be expected.

10.2.4 What is basically wrong with the Miner rule?

Any cumulative damage rule requires a definition of fatigue damage. Following the discussion on the fatigue process in Chapter 2, it appears that a fatigue damage concept should include the crack length, or the amount of fatigue cracking. It could still be microcracking, nonetheless decohesion in the material. Keeping this aspect in mind, the third objection to the Miner rule, based on crack length at failure, is easily understood. According to the Miner rule, the fatigue damage at failure is $\sum n/N = 1$, or 100%. However, the crack length at failure depends on S_{max} of the last load cycle. The Miner rule assumes that an S-N curve represents a curve of 100% fatigue damage.

That is not realistic because it is associated with a larger crack at a lower S_a-value, and a shorter crack at a higher S_a-value. An S-N curve is not a line of constant damage. Perhaps, this shortcoming of the Miner rule may not be so serious because the fatigue life spent in the macrocrack growth period is relatively short. However, the first objection (S_a below the fatigue limit being non-damaging) and the second one (residual stress effects at notches) can be very important. The fundamental shortcoming of the Miner rule is that fatigue damage is indicated by a single damage parameter only, viz. n/N, which accumulates from zero (pristine specimen) to 1 (failure).

As discussed in Chapter 2, fatigue cracking is an essential part of fatigue damage, starting with microcracking, followed by macrocracking. However, fatigue damage should be defined as embracing all changes in the material occurring as a result of the cyclic load. In addition to local decohesion (cracking), fatigue damage includes crack tip plasticity, local strain hardening in the crack tip zone, residual stresses around the crack tip, and for notched elements also macroplasticity at the root of the notch. It cannot be expected that all these conditions are uniquely interrelated irrespective of the magnitude of the cyclic load. As a simple example, the size of the crack tip plastic zone is not uniquely related to the size of the crack. This plastic zone at the same crack length will be larger for a high S_{max} than for a low S_{max}. Already in the crack nucleation period, microplasticity does not occur in the same way for any stress amplitude. The description of fatigue damage is complex. It certainly cannot be characterized by a single damage parameter. It implies that fatigue damage increments induced by load cycles will depend on the fatigue damaged condition of the material as caused by previous cycles. These effects are called *interaction effects*. The damage increment in a load cycle is depending on the size of the load cycle, but at the same time, it is affected by the damage caused by preceding cycles. The interaction effects are responsible for the *sequence effects* discussed before. As a result of the interaction effects it is difficult to arrive at a rational fatigue life prediction method for VA loading. Prediction models cannot be developed without simplifications. But several drastic simplifications have been proposed in the literature.

The most well-known simplified model of Miner was discussed before. Fatigue damage was supposed to linearly increase during CA cycles (see Figure 10.5a), also in blocks of cycles of VA loading, and without any interaction effects. In 1956 Shanley has proposed that fatigue damage should be defined as the size of a fatigue crack [5]. Crack initiation was supposed to start at the beginning of the fatigue life, and crack growth was supposed to be an exponential function, thus a non-linear function, see Figure 10.5b.

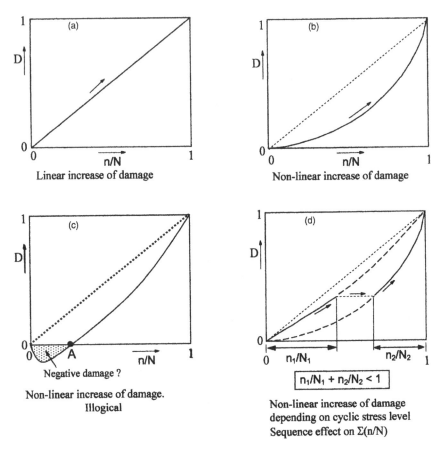

Fig. 10.5 Increasing fatigue damage characterized by a single damage parameter [6].

Shanley also thought that this function applied to all cyclic stress levels. It implies again that increasing damage during VA loading is moving along the damage curve from $D = 0$ to $D = 1$. Because Shanley also ignored interaction effects, his damage definition again leads to the Miner rule. Actually this applies to any non-linear damage function $D(n/N)$ if interaction effects are ignored.

A note about the $D(n/N)$ function should be made here. The function must be a monotonously increasing function. After it was recognized that a high load on a notched specimen can increase the fatigue life the term "negative damage" was used. The $D(n/N)$ function should show a curve with negative damage, see Figure 10.5c. After an initially decreasing part in the curve, it returns to $D = 0$ in point A. It should imply that the specimen is again undamaged condition, which appears to be illogical [6].

A different situation occurs if a damage function $D(n/N)$ is depending on the cyclic stress level, see Figure 10.5d. Now a sequence effects can occur. This is illustrated in this figure for a simple VA tests with two blocks of CA cycles, a stress history previously shown in Figure 10.1. During the first block, the damage parameter D increases along the upper curve. Changing over to the second block of small cycles implies a transition to the other damage curve to be followed until failure at $D = 1$. The sum of the two n/N contributions is obviously smaller than 1. The reversed sequence of the two blocks starting with the small cycles would lead to $\sum n/N > 1$.

Models with a non-linearly increasing single damage parameter have not led to an improved Miner rule giving reliable predictions with some general validity. Actually, the single damage parameter is in conflict with the present understanding of fatigue damage as it can accumulate under VA loading. The three shortcomings of the Miner rule discussed in the three previous sections are not removed by assuming a non-linear damage function.

In the previous sections it was shown that sequence effects could be explained as a consequence of plasticity effects, residual stress and the size of fatigue cracks. They all can contribute to the invalidity of the Miner rule. The problem of predicting fatigue lives under VA loading is a question of how fatigue damage is accumulating. The fundamental question is how fatigue damage should be defined. It is obvious that the microcracks or macrocracks must be part of the definition. At the same time, the discussion on the sequence effects indicate that plastic deformation and residual stresses should also be involved in the definition of fatigue damage. Although the significance of these aspects is evident in a qualitative way, it is also clear that a physical definition of fatigue damage in quantitative terms must be problematic. However for sure, fatigue damage cannot be defined by a single damage parameter.

In view of the complexity of defining the concept of fatigue damage, it is not strange that simplifications have been proposed for engineering purposes. And that was what Pålmgren did in 1923. Nevertheless, it is useful to realize that the Miner rule is a rather drastic simplification which sometimes can lead to substantially wrong predictions as will be discussed later. Some alternative prediction models are described in Section 10.4, but first trends observed in VA tests are discussed.

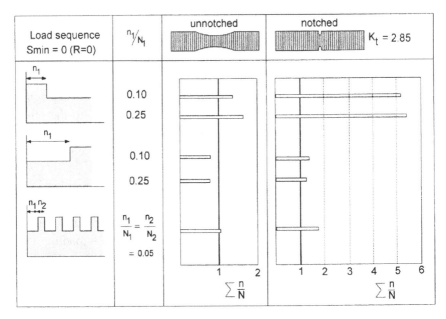

Fig. 10.6 Sequence effects in VA tests on unnotched and notched specimens of an Al-alloy (2024-T3) [7]. $S_{\min} = 0$.

10.3 Results of fatigue tests under VA loading

Large numbers of VA tests were carried out in the 1950s and 1960s to verify the Miner rule. Relatively simple load histories were used, partly because simple sequences could very well indicate possible deviations from $\sum n/N = 1$. Another argument was that the fatigue testing machines present in the laboratories in those days could apply only simple load sequences to specimens. More complex sequences as shown in Figures 9.23 and 9.25 were not yet possible. These load histories were applied later, starting around 1970, when closed-loop electro-hydraulic testing systems were introduced (see Chapter 13). In spite of the limitations of the older machines, much has been learned about fatigue damage accumulation under VA loading with simple load sequences.

Some illustrative test results of simple VA fatigue tests ($S_{\min} = 0$) are presented in Figure 10.6. Results of the notched specimens are considered first. Large values of $\sum n/N$ (>5) are found for the HiLo sequence. This should be associated with compressive residual stresses at the notch root introduced by the first block at a high $S_{\max}$ (= 10^3 MPa and $N = 185$ kc). In the LoHi sequence tests with the low $S_{\max}$ block applied first ($S_{\max} = $

64 MPa and $N = 1200$ kc), the damage contribution in this block is negligible. Fatigue occurred predominantly in the second block with the high amplitude, which leads to $\sum n/N$-values slightly above 1. An intermediate $\sum n/N$-value ($\sum n/N = 1.8$) was found for the third load sequence with alternating low and high S_{max} blocks.

Results of the unnotched specimens in the same figure show similar trends, but the deviations from $\sum n/N = 1$ are significantly smaller than for the notched specimens. Macroplasticity does not occur in the unnotched specimens, which excludes this mechanism for large sequence effects. In general, large $\sum n/N$-values have not been reported for unnotched specimens, but values significantly smaller than one are mentioned in the literature. In the older days, low $\sum n/N$-values were found for unnotched rotating beam specimens of steel.

Results of another test series on riveted lap joints are shown in Figure 10.7. The stress history consists of periods with blocks of four or five different amplitudes, applied in an increasing amplitude sequence, and superimposed on a positive mean stress (88 MPa). The number of cycles in one period is 433000 cycles. The result for series A is $\sum n/N = 1.3$. The stress history of series C was obtained from the history of series A by adding a small block (175 cycles) with a higher stress amplitude, which increased S_{max} in the test. In view of the positive mean stress, it implies that more favorable compressive residual stresses could be developed in test series C. As a result, the fatigue life was more than doubled; $\sum n/N$ increased from 1.3 to 2.9. If the Miner rule would be correct, then the life in series C should be slightly shorter than in series A because of some damage added by the more severe load cycles.

The blocks in each period of series A and B are the same, but a single severe peak load cycle is added at the end of each period is series B. This cycle starts with a large negative amplitude, followed by the same positive amplitude. The latter peak load introduced significant compressive residual stresses, and as a result the life is increased considerably (factor six times). However, the Miner rule predicts a very small life reduction caused by the small number of peak load cycles.

A large peak load cycle was also added in series D to the fatigue loading of series C. However, this cycle now starts with the positive amplitude followed by the negative one, which leads to a significant reduction of the fatigue life. Apparently, the negative peak load has introduced tensile residual stresses, and eliminated the compressive residual stresses of the preceding positive peak load. The residual stresses in the test series

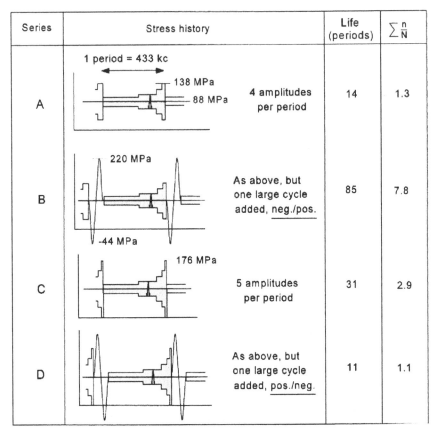

Series	Stress history		Life (periods)	$\sum \frac{n}{N}$
A	1 period = 433 kc 138 MPa 88 MPa	4 amplitudes per period	14	1.3
B	220 MPa -44 MPa	As above, but one large cycle added, neg./pos.	85	7.8
C	176 MPa	5 amplitudes per period	31	2.9
D		As above, but one large cycle added, pos./neg.	11	1.1

Fig. 10.7 Effect of stress history on fatigue life and $\sum n/N = 1$. Results of VA fatigue tests on riveted lap joints (material 2024-T3 sheet) [8]. $S_m = 88$ MPa.

of Figure 10.7 reveal a similar effect as discussed earlier in relation to Figure 10.4. The explanation of the highly different results in Figure 10.7 in terms of residual stress effects is probably not fully complete because a riveted lap joint is a complicated notched specimen. But the highly different fatigue lives clearly demonstrate large load history effects which are fully outside predictions by the Miner rule.

Gassner has introduced the so-called block-program fatigue test already in 1939 [9]. He varied the stress amplitude periodically around a constant mean stress, see Figure 10.8. Gassner recognized that fatigue tests should also be carried out with a variation of the amplitude in agreement with load spectra observed in service. The loading program in Figure 10.8 covers 500000 cycles in one program period distributed over eight amplitudes in agreement

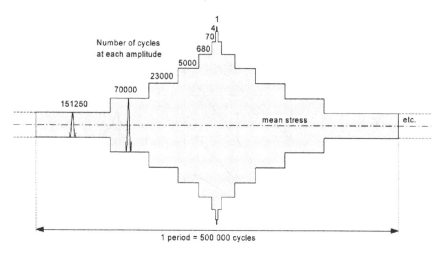

Fig. 10.8 Block program fatigue tests, introduced by Gassner [9], consisting of a low-high-low sequence of blocks of CA cycles.

with a load spectrum of aircraft wings. This low-high-low sequence (LoHiLo) of blocks of CA cycles is repeated until failure. The prime purpose was to adopt a load sequence that should be more representative for practical fatigue problems. However, the program fatigue test has also been used to check the Miner rule. Various test series are reported in the literature for block program tests with the same blocks in a period, but different sequences of the blocks. As an example, Figure 10.9 shows results of a NASA investigation on edge notched specimens ($K_t = 4.0$) [10]. It clearly illustrates that $\sum n/N$ depends on the sequences in which the blocks are applied. Secondly, it also shows that an increased mean stress leads to higher $\sum n/N$-values. In a symmetric spectrum of amplitudes around a tensile mean stress, the maximum stress in a period is a larger tensile stress than (the absolute value of) the minimum compression stress in the same period. This promotes the development of favorable residual compression stresses due to notch root plasticity, which then leads to higher $\sum n/N$-values as illustrated by the effect of the mean stress (S_m) in Figure 10.9.

Around 1970 closed-loop electro-hydraulic fatigue machines were introduced in fatigue testing laboratories. It then was possible to apply load time histories to specimens and structures, which could be defined as a computer command signal. Any sequence of load peaks, maxima and minima could be applied in a fatigue test, such as pure random loading histories (see Figure 9.18), or combinations of a deterministic load

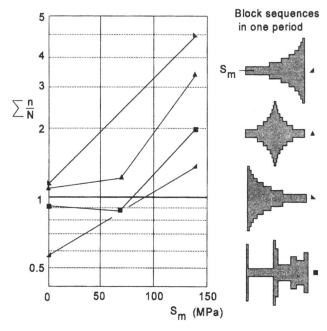

Fig. 10.9 Results of block-program fatigue tests of NASA [10]. Notched specimens, $K_t = 4$, material 7075-T6. Different block sequences and effect of mean stress.

(air-ground-air transitions) with a superimposed random loading as shown in Figure 9.25. The computer commands the fatigue load in the fatigue machine.

The sequence of load amplitudes during a random load history is significantly different from the sequence in block-program fatigue tests with the stress amplitude and the mean load remain constant during substantial numbers of cycles in each block. In a random load tests, and also in service, the amplitude varies from cycle to cycle. It must be expected that sequence effects and load cycle interaction effects are different in these two types of load histories. Experimental comparisons indeed showed different fatigue lives obtained under block-program loading and random load histories with the same load spectrum [11]. In general, fatigue lives in block program tests were longer than under equivalent random gust loading. Differences varied from relatively negligible to a 6 times longer life under block-program loading. It implies that a block-program fatigue test may give an unconservative indication of the fatigue life under more realistic random load sequences. The discrepancies between the results of the two types of fatigue tests have considerably affected the appreciation of realistic service-load-simulation fatigue tests. At the same time, the

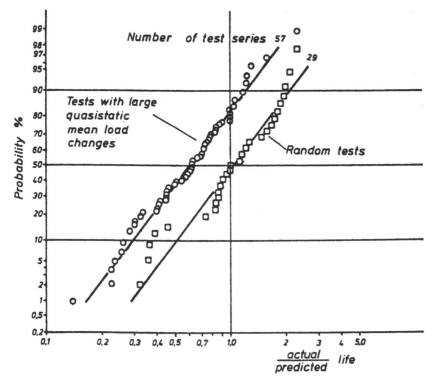

Fig. 10.10 Comparison between test results and predictions on fatigue lives in VA fatigue tests. Data collected by Schütz [12].

significance of the fatigue life indications obtained with the Miner rule were also questionable.

Schütz [12] made an analysis of $\sum n/N$-values obtained in various tests series reported in the literature. He considered two groups of load sequences, viz. sequences with large variations of the mean stress in addition to amplitude changes (57 test series, mainly non-randomized), and sequences with a constant mean stress and a random variation of the load amplitude (29 test series), see Figure 10.10. The (logarithmic) horizontal scale gives the experimental life divided by the Miner-predicted life. This ratio is the experimental $\sum n/N$-value. Along the vertical axis Schütz used the probability scale of the normal distribution function. It illustrates the scatter of $\sum n/N$-values obtained in different test programs.[14] In the large number of 57 test series, the average value is $\sum n/N = 0.6$, but $\sum n/N$ varies from

---

[14] It should be understood that this scatter is not illustrating scatter between similar tests, but rather scatter between average results obtained in tests with different VA load sequences.

0.15 to 2.0. In the random load tests, the average value is $\sum n/N = 1.05$, while individual values range from 0.3 to 3.0.

The question should be raised why such large deviations of $\sum n/N = 1$ can occur. As pointed out earlier, shortcomings of the Miner rule can be qualitatively understood. Low values of $\sum n/N$ are possible if many small cycles are present in the load history, and if a zero mean stress is applicable. High $\sum n/N$-values can be obtained if the load history has a positive mean stress, which promotes the possible occurrence of favorable residual stresses at notches (i.e. compressive residual stresses). The favorable residual stress effect is absent for unnotched specimens. Because of this qualitative understanding, it should not be a surprise that significant deviations from the Miner rule are observed.

10.4 Alternative fatigue life prediction methods for VA loading

10.4.1 Damage calculations and extrapolation of S-N curves below the fatigue limit

As pointed out earlier, an essential shortcoming of the Miner rule is due to ignoring damage contributions of load cycles with amplitudes below the fatigue limit. It then appears to be reasonable to extrapolate S-N curves to lower stress amplitudes with $S_a < S_f$ in order to assign some damage increments to cycles with amplitudes below the fatigue limit. Such an extrapolation is made in Figure 10.11, see line B. It implies that the Basquin relation, $S_a^k \cdot N =$ constant, is assumed to be applicable for all fatigue cycles with amplitudes below the fatigue limit. The effect of this extrapolation will be illustrated by life calculations with the Miner rule. The calculations are made for two load spectra, H_1 and H_2 shown in Figure 10.11. The calculation for spectrum H_1 is presented in Table 10.1. The first column of the table gives S_a-values for which the number of exceedings (spectrum for 5000 hrs) is read in Figure 10.11. The incremental number $n = \Delta H$ is the number of cycles for the corresponding S_a intervals. The fatigue life N for the median stress level of this interval is read from the S-N curves in the same figure. Damage increments n/N can then be calculated, which are presented in the last two columns of the table for S-N curves A and B respectively. The sum of these increments is $\sum n/N$ for 5000 hrs. The fatigue life until failure according to the Miner rule is then obtained as soon as $\sum n/N = 1$, which leads to

Stress level (MPa)

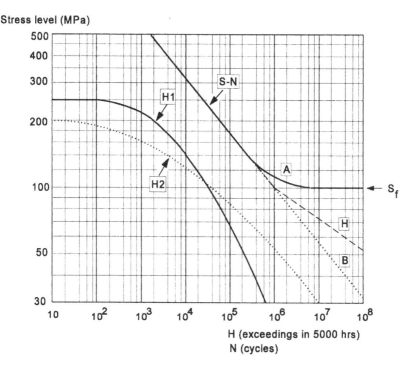

Fig. 10.11 Two load spectra and an S-N curve in a Miner calculation.

$$\text{Life} = \frac{1}{\sum \dfrac{n}{N}} * 5000 \text{ hrs} \qquad (10.3)$$

The results for the two S-N curves are shown at the bottom of Table 10.1. Smaller increments of S_a can be used, which makes the calculation more precise, but not necessarily more accurate because it is based on the validity of $\sum n/N = 1$ at failure. If the load spectrum and the S-N curve can be given in analytical format by $H(S_a)$ and $N(S_a)$, then $\sum n/N$ can be numerically calculated by integration of $dD = dH(S_a)/N(S_a)$ and again:

$$\text{Life} = \left(\frac{1}{D}\right) * 5000 \text{ hrs}$$

Calculated fatigue lives based on the S-N curves A and B for both spectra H_1 and H_2 are summarized in Table 10.2. The results confirm that the calculated fatigue life is significantly shorter if the Miner prediction is based on the extrapolated S-N curve B which then includes fatigue damage of cycles with $S_a < S_f$. The effect is much larger for spectrum H_2 with relatively

Table 10.1 Damage calculation for spectrum H in Figure 10.11.

Spectrum $H1$			S-N data			Damage calculation	
S_a	H	$n = \Delta H$	Average S_a of interval	Corresponding N (cycles)		n/N	n/N
(MPa)	(cycles)			Curve A	Curve B	Curve A	Curve B
250	100						
		450	240	30000	30000	0.015	0.015
230	550						
		700	220	43000	43000	0.0162	0.0162
210	1250						
		1150	200	62500	62500	0.0184	0.0184
190	2400						
		1900	180	95000	95000	0.02	0.02
170	4300						
		3200	160	150000	150000	0.0213	0.0213
150	7500						
		6000	140	260000	260000	0.0231	0.0231
130	13500						
		11500	120	600000	600000	0.0192	0.024
110	25000						
		21000	100	10^7	10^6	0.0021	0.021
90	46000						
		49000	80	-	2.4×10^6	-	0.0204
70	95000						
		125000	60	-	7.5×10^6	-	0.0167
50	220000						
		480000	40	-	40×10^6	-	0.012
30	700000						
		4300000	20	-	625×10^6	-	0.0069
10	5000000						
H = number of exceedings in 5000 hrs						$D_A = \Sigma n/N =$ 0.1353	$D_B = \Sigma n/N =$ 0.2150
			Calculated fatigue life: S-N curve A: Life = $(1/D_A) \times 5000$ = 37000 hrs S-N curve B: Life = $(1/D_B) \times 5000$ = 23000 hrs				

Table 10.2 Miner predictions.

S-N curve used	Fatigue life (hrs)		Life ratio (H_1/H_2)
	Spectrum H_1	Spectrum H_2	
A	37000	66000	0.56
B	23000	7000	3.3
	B/A = 0.62	B/A = 0.11	

many cycles below the fatigue limit. A comparison between the severities of spectra H_1 and H_2 is made in the last column of Table 10.2. It indicates that the predicted fatigue life for spectrum H_1 is almost 50% shorter than for spectrum H_2 if S-N curve A is used, whereas it is 3.3 times longer if S-N curve B is adopted. In view of the results of Table 10.2, it must be concluded that the Miner rule is suspect to indicate differences between the severity of different load spectra.

The damage contributions calculated in Table 10.1 are plotted in Figure 10.12. It illustrates the distribution of the damage increments of the various load intervals, which has a maximum at a certain stress level. This

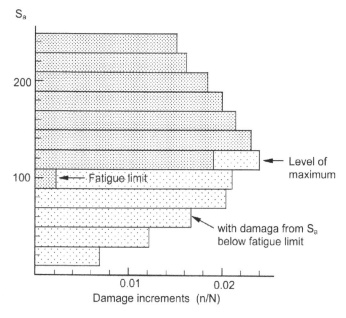

Fig. 10.12 Distribution of damage increments of a Miner calculation.

stress level has been labeled in the literature as the most damaging stress level of a load spectrum, but it should be recalled that this statement is entirely based on the Miner rule, and thus has a questionable meaning.

An other suggestions for extending the S-N curves below the fatigue limit was proposed by Haibach [13], which is line H in Figure 10.11. Haibach postulated that cycles with amplitudes above the fatigue limit will reduce the fatigue limit of the undamaged material. As a result, cycles with an amplitude $S_a < S_f$ are damaging. The damage of these cycles was accounted for by applying the Miner rule to a modification of the S-N relation by the extension with line H in Figure 10.11. According to Haibach, the slope of line H is related to the slope of the Basquin relation of the original S-N curve; $S_a^k \cdot N =$ constant. In this equation, k should be replaced by 2k-1. Of course, a prediction with this modified S-N curve is more conservative than the original Miner rule prediction, but less conservative than for the extended S-N curve with line B. Again, it must be recalled that the prediction with $\sum n/N$ remains a Miner rule prediction, which does not account for any interaction effect. Predictions must be considered with caution. The extrapolated S-N curves imply that an extra safety margin is introduced which intuitively seems to make sense, but it remains unknown how large this margin is, and also whether the prediction will be conservative, i.e. $\sum n/N \geq 1$.

10.4.2 The relative Miner rule

Schütz [12] considered the unconservative predictions of the Miner rule, and he introduced the idea that systematic unconservative predictions could be accounted for by replacing $\sum n/N = 1$ by a "relative Miner rule", $\sum n/N = q$, with $q < 1$. The value of q had to be selected by experience of VA tests with similar load-time histories relevant to the problem under consideration. The relative Miner rule can also be interpreted as using the Miner rule with a safety factor to account for possible unconservative life predictions. Although such a safety approach appears reasonable, Schütz also pointed out that realistic predictions require test results obtained under realistic load sequences to be applied to the structure or component itself.

10.4.3 Strain history prediction model

The previously discussed models try to predict the fatigue life under VA loading from fatigue lives obtained under CA loading, i.e. S-N curves. After it was realized that notch root plasticity could occur and would introduce a local residual stress distribution, it was tried to account for the real strain history at the root of the notch. It has led to fatigue predictions for VA loading based on the predicted strain history at the notch root [14, 15]. Such predictions came in focus when the Neuber postulate was developed for the prediction of residual strain and stress at the root of a notch. This postulate was discussed in Chapter 4 (Section 4.4). Furthermore, low-cycle fatigue experiments under constant-strain amplitudes have indicated an approximately linear relation between $\log \Delta \varepsilon$ and $\log N$ (the Coffin–Manson relation), discussed in Chapter 6 (Section 6.4). The procedures to be used for calculation of the strain history at the notch under VA loading, and for the subsequent life prediction are schematically shown in Figure 10.13. A detailed discussion of these procedures has been given by Dowling [15]. The main steps are mentioned here in order to see advantages and weaknesses of the approach.

In the first step (Figure 10.13a) the strain history $\varepsilon(t)$ is derived from the load history $P(t)$ by employing the Neuber postulate and the cyclic stress

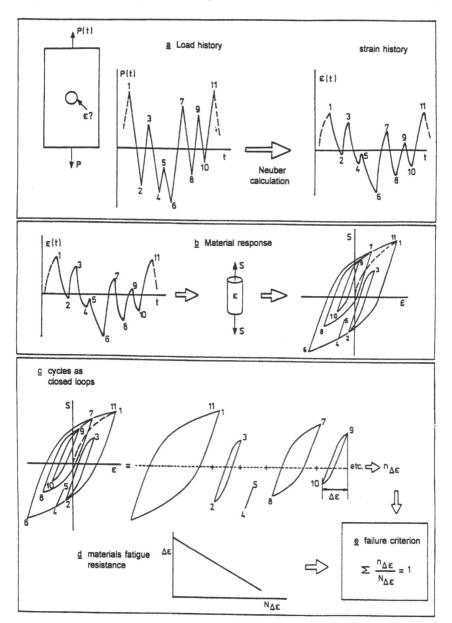

Fig. 10.13 Principles of the notch-root strain-based prediction model [14].

strain curve. In the second step, the σ-ε response of the material (at the root of the notch) is derived from $\varepsilon(t)$ (Figure 10.13b). It presumes a certain plastic hysteresis behavior based on the material memory for previous plastic deformation. In the third step (Figure 10.13c) the cyclic hysteresis history is decomposed into closed hysteresis loops. Each loop represents a full strain cycle. In the last step (Figures 10.13d and e) the $\Delta\varepsilon$-$N_{\Delta\varepsilon}$ curve is used as the material property characterizing the material resistance against low-cycle fatigue. The Miner rule is then adopted as the failure criterion.

The material properties required for the strain-history model are the cyclic stress strain curve and the Coffin–Manson relation. Both types of data are considered to be unique for a material. This is an advantage over the stress based S-N fatigue data, which depend on mean stress and surface quality. The surface quality is much less important for low-cycle fatigue, because the plastic strains are larger and thus mainly depending on the material bulk behavior. Low-cycle fatigue is no longer a surface phenomenon as it is for high-cycle fatigue. At the same time, limitations of the strain-history model are easily recognized. The failure criterion is again the Miner rule, for which physical arguments can hardly be mentioned. Secondly, crack initiation and crack growth are fully ignored. Moreover, the model is restricted to notched elements, for which a theoretical stress concentration factor has a realistic meaning. As a consequence, application to joints is generally impossible. It was emphasized by Dowling [15] that the merits of the model should be looked for in low-cycle VA problems. Actually, verification experiments are still rather limited.

A noteworthy comment should be made on the decomposition in Figure 10.13c. The individual cycles obtained are the same cycles as obtained with the rain-flow count method. It implies that this counting method finds some justification in the material memory for previous plastic deformation.

10.4.4 Predictions based on service-simulation fatigue tests

The application of the Miner rule to fatigue life predictions implies that results of CA tests are extrapolated to conditions of VA load-time histories. In view of the limited validity of the Miner rule, an obvious approach is to look for predictions based on results of "relevant" VA tests, i.e. results of service-simulation fatigue tests discussed in the previous chapter (Section 9.5). Such tests should imply a much smaller extrapolation. Due attention must then be paid to the notch effect, size effect, and effect of surface quality, aspects discussed in Chapter 8 (fatigue properties of notched elements). Specimens to be tested should be as much similar as possible to the real component, or the fatigue critical location of a structure. The most relevant service-simulation fatigue test is a test on the structure or a component itself. This is sometimes done, noteworthy in the automotive industry and the aircraft industry. However, if such a test meets with practical problems, a service-simulation fatigue test on a representative specimen can give useful indications on the fatigue performance of a structure.

An additional and important question is: which load-history should be used in service-simulation fatigue tests for prediction problems? The load spectrum, and also the load sequence in the test should be representative for conditions expected in service. A block-program fatigue test discussed before cannot be considered to be a good choice. Randomness of loads in service should be simulated in the test, which is experimentally very well possible with modern fatigue machines. Service-simulation fatigue tests for aircraft structures, referred to as flight-simulation fatigue tests, have been extensively used for various purposes. An example of the load history applied in such a test was shown in the previous chapter (Figure 9.25). The stress-history was characterized by one specific stress level, for which the mean stress in flight (S_{mf}) was chosen. If the design stress level of the structure is increased (or decreased), all cyclic stress levels are changed proportionally, and thus S_{mf} remains a characteristic stress level for the intensity of the stress history. Fatigue life results of flight-simulation fatigue tests, carried out at different values of the characteristic stress level S_{mf}, can then be plotted as a function of S_{mf}. Such fatigue life curves are presented in Figure 10.14. Similar plots can be made for other types of stress histories. Figure 10.15 shows a fatigue life curve for pure random loading with the root-mean-square of the stress (S_{rms}) as the characteristic stress level of the load history.

It is noteworthy that experimental results of service-simulation fatigue tests suggest a linear relationship between the stress level and the fatigue life plotted on a double log scale. It implies that the Basquin relation should again describe the effect of the cyclic stress intensity:

$$(S_{\text{characteristic}})^k \cdot \text{life} = \text{constant} \tag{10.4}$$

Moreover, it appears that the slope factor k is of the same order of magnitude as the slope factor for S-N curves. Intuitively, the latter observation about the effect of the stress level seems to be plausible, but a firm proof would require a valid model for fatigue damage accumulation, which in fact is not available.

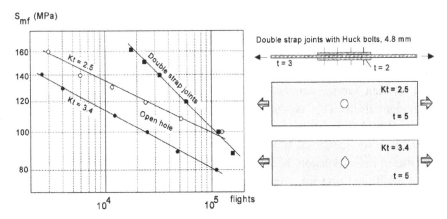

Fig. 10.14 Results of flight-simulation fatigue tests on notched specimens of an Al-alloy (2024-T3) [16]. The data points follow the Basquin relation (Eq. 10.4). Thickness t in mm.

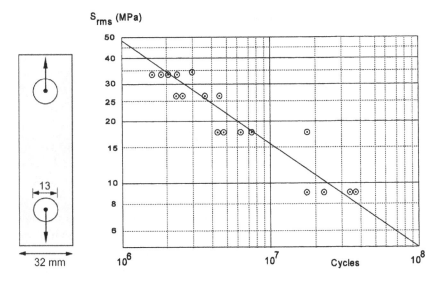

Fig. 10.15 Results of narrow band random load tests on lug specimens. Fatigue life as a function of the root-mean-square value of the stress amplitude; mean stress 108 MPa [17].

The analysis of a large number of flight-simulation tests on Al-alloys [18] indicated slope values around $k \approx 5$. The effect of the design stress level under service-simulating fatigue loading can be estimated with this slope factor. As an example, if $k = 5.0$, an increase of the design stress level with 20% would reduce the life with a factor equal to $1.2^5 = 2.5$. If k for fatigue life curves under VA loading is not available, a k-value can be adopted from S-N curves for a similar notch geometry and material.

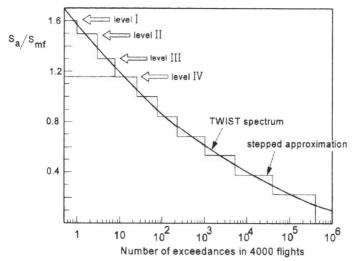

(a) The stress amplitude distribution of the TWIST load spectrum.
S_{mf} = mean stress in flight, see Figure 9.25

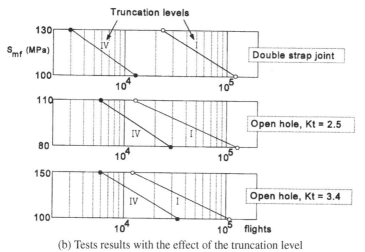

(b) Tests results with the effect of the truncation level

Fig. 10.16 The effect of truncating high-amplitude cycles of a steep spectrum (TWIST) in flight-simulation fatigue tests (similar specimens as in Figure 10.14) [15]. Large truncation effect.

In the previous chapter (Section 9.5), the significance of truncating high-amplitude cycles of a load spectrum was mentioned. As suggested, the influence could be large for a steep spectrum. This is illustrated by the results of Figure 10.16. The tests were carried out with the standardized TWIST spectrum, a spectrum for transport aircraft wing structures with mainly gust loads in flight. The spectrum given in Figure 10.16a is a steep spectrum. In the experiments it is approximated by the stepped function given in the figure. Truncation levels are numbered by Roman figures. Truncation at level I implies that the maximum level occurs once in 4000 flights. Truncation at the much lower level IV implies that the amplitude of cycles with a larger amplitude than level IV is reduced to level IV. As a consequence, the maximum amplitude becomes 72% of level I which now occurs 18 times in 4000 hours. The experimental results in Figure 10.16b show that the truncation has led to a significant reduction of the fatigue life until failure. This large truncation effect is not predicted by the Miner rule. On the contrary, the Miner rule predicts a slight increase of the fatigue life by the truncation instead of the significant reduction. Truncation of a steep spectrum is advised in order to obtain conservative test results. Under service conditions, the rarely occurring high amplitudes will not be encountered by all structures in service because of statistical variation of the load spectrum. The life-increasing effect of the rarely occurring high loads will thus not apply to all structures. The choice of a truncation level for a steep load spectrum remains a precarious question which should be answered by considering the variability of the load spectrum in service. Furthermore, it can be instructive to carry out service-simulation tests on representative specimens using some different truncation levels in order to explore the sensitivity for this variable.

Fortunately, the spectrum truncation problem is less important for flat load spectra with relatively many high-amplitude cycles. All structures will meet some high-amplitude cycles. Moreover, the relatively small number of low-amplitude cycles of a flat spectrum are less important for the accumulated fatigue damage. Experimental results in Figure 10.17 for a flat maneuver spectrum still show a systematic truncation effect, i.e. shorter fatigue lives for a truncated spectrum, but the effect is much smaller than in Figure 10.16 for a steep spectrum.

In spite of the truncation problem, results of service-simulation fatigue tests are much more significant than results obtained by Miner rule predictions based on S-N curves. The reliability of these predictions is not only depending on the doubtful validity of the Miner rule, but also on the estimated S-N curves. If tests for prediction problems are considered to be desirable, service-simulation tests are more instructive than constant-amplitude tests.

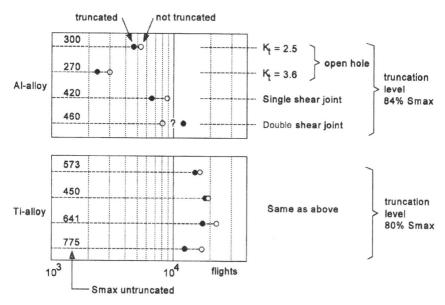

Fig. 10.17 Effect of truncating high-amplitude cycles of a flat maneuver load spectrum (FALSTAFF) in a flight-simulation fatigue test [19]. Specimens of an Al-alloy (2024-T3) and a Ti-alloy (Ti-6Al-4V). Smalll truncation effect.

10.5 Discussion of fatigue life predictions for VA loading

Results of VA fatigue tests clearly indicate that the Miner rule does not give accurate and reliable predictions on fatigue lives. The shortcomings can be understood. Two major objections are: (i) cycles with an amplitude below the fatigue limit do not contribute to fatigue damage, and (ii) the second one, notch root plasticity leads to interaction effects between cycles of different magnitudes which are not accounted for in the Miner rule. The first objection can be alleviated by extending S-N curves below the fatigue limit. The linear extrapolation curve B in Figure 10.11 is probably a conservative approach. But according to the present understanding of fatigue damage accumulation, it cannot be concluded that the problem is solved on rational grounds. The significance of the second shortcoming (notch root plasticity) can also be understood. Although it can be tried to account for this deficiency by sophisticated plasticity calculations, the Miner rule still ignores load cycle interactions. The only realistic approach for predicting fatigue lives of structural components is to rely on results of service-simulation fatigue tests. In view of this conclusion, the perspectives of the Miner rule ($\sum n/N = 1$) and service-simulation fatigue tests are reconsidered here in relation to different applications to fatigue prediction problems of VA loading. The following topics are addressed:

1. Life estimates for a specific component and the Miner rule.
2. Considerations on the effect of the design stress level.
3. Comparison between different options for design improvements (materials, notch geometries, joints, surface treatments).
4. Comparison of different spectra (different utilizations of the same structure).

10.5.1 Life estimates for a specific component and the Miner rule

The Miner rule $\sum n/N = 1$ can only give a rough estimate of the fatigue life. It is a kind of averaging the severity of the various load cycles of the load spectrum. Extrapolation of S-N curves to low amplitudes must be advised anyway. Predictions may still be considered to be reasonable if the deviation of the real life from the predicted life is no more than a factor 2, i.e. that it should be between 50% and 200% of the prediction. However, if a designer is using the Miner rule, he should realize that the accuracy of the prediction is not only depending on the Miner rule, but also on the relevance of the S-N curves used for the prediction. The designer may well keep in mind that the Miner rule may be more safe if a positive mean stress applies to the load spectrum and less safe for a zero mean stress. The latter case applies to rotating bending of axles where notch plasticity is generally not allowed. Furthermore, the applicability of the Miner rule will be more uncertain for steep load spectra and probably less uncertain for a flat load spectrum. Actually a flat load spectrum is more close to CA loading, and interaction effects will be less significant.

Whether an estimated life obtained with the Miner rule is sufficient to exclude the need for realistic fatigue tests is a matter of engineering judgement. Uncertainties about the S-N curve, specific service conditions (e.g. corrosion) and possible consequences of premature fatigue failures have to be considered. Safety factors can be adopted, which in view of scatter will be advisable anyhow (see Chapters 13 and 20). Beyond any doubt, more reliable predictions are obtained from relevant service-simulation fatigue tests.

10.5.2 *Considerations on the effect of the design stress level*

As pointed out earlier, the effect of the design stress level can be taken into account by using the Basquin relation, Eq. (10.4). A Miner calculation will not give a reliable indication. According to literature data, k-values in the Basquin relation will be in the order of 4 to 6. A fatigue life reduction factor due to an increased design stress level is conservatively estimated by using a relatively high k-value, while a life improvement factor due to a reduced design stress level is conservatively estimated by a relatively low k-value.

If the fatigue life is largely covered by macrocrack growth, a low k-value should be adopted in the range of 3 to 4. This conclusion is associated with the exponent of the Paris crack growth relation as will be discussed in Chapter 11.

10.5.3 *Comparison between different options for design improvements*

The fatigue resistance of a component can be improved by selecting an other material, or changing the geometry of the notch (a larger root radius to reduce the local stress concentration), or adopting a different surface treatment (e.g. shot peening). If the Miner rule would be used for such comparisons, it implies that first comparative CA tests must be carried out. These data should then be used for Miner calculations. This approach cannot be considered to be a clever solution for VA problem settings. Comparative service-simulation fatigue tests are by far the best method and the most efficient one. Of course, if fatigue failures are inadmissable, then the fatigue limit is the important property, and CA tests should be used.

10.5.4 *Comparison of different load spectra*

The problem of comparing different spectra is important if structures can be used in different ways. This is easily understood for aircraft, motor vehicles, cranes and various other structures as well. It then is desirable to know whether different usages of the same structure will lead to significantly different fatigue lives in service. Unfortunately, the Miner rule is unreliable to answer these questions. Some illustrations were presented in the previous sections. For instance, the effect of load cycles with a high S_{max} can have a large favorable effect, whereas the Miner rule predicts a small reduction. Furthermore, the discussion in Section 10.4.1 on the comparison between two different load spectra H_1 and H_2 in Figure 10.11 indicated that the Miner rule did not give an unambiguous result. It must be emphasized that experimental evidence is indispensable for the comparison of the severity of different load spectra. It should be obtained in realistic service-simulation fatigue tests on representative specimens.

10.6 Major topics of the present chapter

The major topics of this chapter are fatigue damage accumulation under Variable-Amplitude (VA) loading, the Miner rule and life predictions for VA loading. The present chapter is dealing with fatigue life until failure including the initiation period. Crack growth of macrocracks under VA loading is considered in Chapter 11. The more significant aspects of the present chapter are listed below:

1. The most simple method for fatigue life predictions is to use the Miner rule, $\sum n/N = 1$. Unfortunately, this rule is not reliable, because of some elementary shortcomings. Two important deficiencies are: (i) Cycles with a stress amplitude below the fatigue limit are supposed to be non-damaging. In reality, these cycles can extend fatigue damage created by cycles with amplitudes above the fatigue limit. (ii) Notch root plasticity leads to residual stresses which can affect the fatigue damage contribution of subsequent cycles. This interaction effect is also ignored by the Miner rule.

2. The Miner rule and several other prediction models assume that fatigue damage can be fully characterized by a single damage parameter ($\sum n/N$ in the Miner rule), which is physically incorrect. Fatigue damage also includes local plasticity and residual stress.

3. Results of $\sum n/N$-values at failure quoted in the literature vary from much smaller than 1 to significantly larger than 1. Small values are promoted by unnotched specimens and a zero mean stress. High values are prompted by notched specimens in combination with a positive mean stress, and also by steep load spectra (low numbers of severe load cycles). Residual stresses are important to explain high $\sum n/N$-values.

4. The sequence of different load cycles in a VA load history can have a large effect on $\sum n/N$ at failure. This is more true for block-program fatigue load sequences than for sequences with random

loads or mixtures of deterministic and random loads. Block-program tests should be discouraged for practical problems.

5. If fatigue life predictions are made with the Miner rule, it should be realized that the results are associated with uncertainties of the reliability of the rule, and also with the relevance of the S-N curves adopted. If the Miner rule is still adopted, it must be recommended to extrapolate S-N curves below the fatigue limit in order to assign fatigue damage contributions to small cycles below the fatigue limit. At best, the Miner rule gives a rough estimate of the fatigue life.

6. Improvements of life prediction methods have been proposed in the literature, but it is still questionable whether they offer a good solution for practical design problems.

7. The only reasonable alternative is to obtain test results of relevant service-simulation fatigue tests. Such tests should be relevant with respect to the material, notch configuration and material surface condition, as well as the load history expected in service.

8. Service-simulation fatigue tests are also recommended for comparative investigations of materials, surface treatments, joints and components in general, and also for comparing the severity of different load spectra. The Miner rule is not reliable for these purposes.

References

1. Pålmgren, A., *The fatigue life of ball-bearings*, Z. VDI, Vol. 68 (1924), pp. 339–341 [in German].
2. Langer, B.F., *Fatigue failure from stress cycles of varying amplitude*. J. Appl. Mech., Vol. 4 (1937), pp. A160–A162.
3. Miner, M.A., *Cumulative damage in fatigue*. J. Appl. Mech., Vol. 12 (1945), pp. A159–A164.
4. Wållgren, G., *Review of some Swedish investigations on fatigue during the period June 1959 to April 1961*. Report FFA-TN-HE 879, Stockholm (1961).
5. Shanley, F.R., *A proposed mechanism of fatigue failure*. Colloquium on Fatigue, Stockholm, 1956. W. Weibull and F.K.G. Odquist (Eds.), Springer, Berlin (1956), pp. 251–259.
6. Schijve, J., *Some remarks on the cumulative damage concept*. Minutes 4th ICAF Conference, Zürich (1956), paper 2.
7. Schijve, J. and Jacobs, F.A., *Fatigue tests on notched and unnotched clad 24 S-T sheet specimens to verify the cumulative damage hypothesis*. Nat. Aerospace Laboratory, NLR, Amsterdam, Report M.1982 (1955).
8. Schijve, J., *The endurance under program-fatigue testing*. Full-Scale Fatigue Testing of Aircraft Structures. Proc. ICAF Symposium, Amsterdam 1959, Pergamon Press (1961), pp. 41–59.

9. Gassner, E., *Strength experiments under cyclic loading in aircraft structures.* Luftwissen, Vol. 6 (1939), pp. 61–64 [in German].

10. Naumann, E.C., Hardrath, H.R. and Guthrie, E.C., *Axial load fatigue tests of 2024-T3 and 7075-T6 aluminum alloy sheet specimens under constant- and variable-amplitude loads.* NASA Report TN D-212 (1959).

11. Jacoby, G.H., *Comparison of fatigue lives under conventional program loading and digital random loading.* Effects of Environmental and Complex Load History on Fatigue Life. ASTM TP 462 (1970), pp. 184–202.

12. Schütz, W., *The prediction of fatigue life in the crack initiation and propagation stages. A state of the art survey.* Engrg. Fracture Mech., Vol. 11 (1979), pp. 405–421.

13. Haibach, E., *Modified linear damage accumulation hypothesis accounting for a decreasing fatigue strength during increasing fatigue damage.* Laboratorium für Betriebsfestigkeit, LBF, Darmstadt, TM Nr. 50 (1970) [in German].

14. Dowling, N.E., Brose, W.R. and Wilson, W.K., *Notched member fatigue life predictions by the local strain approach.* Fatigue under Complex Loading: Analysis and Experiments, R.M. Wetzel (Ed.), Advances in Engineering, Vol. 7, SAE (1977), pp. 55–84.

15. Dowling, N.E., *Mechanical Behavior of Materials. Engineering Methods for Deformation, Fracture, and Fatigue,* 3rd edn. Prentice-Hall (2006).

16. Schütz, D. and Lowak, H., *The application of the standardized test program for the fatigue life estimation of transport aircraft wing components.* Proc. ICAF Symp., Lausanne, Paper 3.64 (1975).

17. Kirkby, W.T., *Constant-amplitude of variable-amplitude test as a basis for design studies. Fatigue design procedures,* ICAF Symp., 1965, Munich 1965, E. Gassner and W. Schütz (Eds.), Pergamon Press (1969), pp. 253–290.

18. Schijve, J., *The significance or flight simulation fatigue tests.* Durability and Damage Tolerance in Aircraft Design, ICAF Symp. Pisa, 1985, A. Salvetti and G. Cavallini (Eds.), EMAS, Warley (1985), pp. 71–170.

19. Lowak, H., Schütz, D., Hück, M. and Schütz, W., *Standardized flight-simulation programme for fighters: FALSTAFF,* LBF-Report 3045, IABG Report TF, 568 (1976) [in German].

Some general references (see also [12] and [18])

20. Rice, C.R. (Ed.), *SAE Fatigue Design Handbook,* 3rd edn. AE-22, Society of Automotive Engineers, Warrendale (1997).

21. Schijve, J., *The accumulation of fatigue damage in aircraft materials and structures.* AGARDograph No. 157 (1972).

Chapter 11
Fatigue Crack Growth under Variable-Amplitude Loading

Symbols

CA	Constant Amplitude
VA	Variable Amplitude
OL	Overload
UL	Underload
a	crack length
β	geometry correction factor
C, m	constants in Paris relation
K	stress intensity factor

11.1 Introduction

Fatigue under Variable-Amplitude (VA) loading was discussed in the previous chapter. Key words of the discussion were: prediction of fatigue life until failures, the Miner rule and its shortcomings, fatigue damage of cycles with an amplitude below the fatigue limit, residual stress effects due to notch root plasticity, and service-simulation fatigue tests as an alternative to Miner rule predictions. The fatigue life was supposed to include the crack initiation period and the crack growth period until failure. It was tacitly assumed that the crack growth period was relatively short and could be disregarded. The present chapter is dealing with the growth of macrocracks under VA loading. The crack initiation period dealing with crack nucleation and microcrack growth is not addressed.

The propagation of macrocracks is a significant issue if fatigue cracks cannot be avoided, especially if safety or economy is involved. Dangerous situations can occur in pressure vessels, high-speed rotating masses (turbine disks, blades of wind turbines) and aircraft structures as some characteristic examples. Incidental cracks can be generated by a variety of conditions; such as surface damage, corrosion pits, material defects in welded joints, inferior production quality, etc. Furthermore, the fatigue life of a structure in service may cover many years. The occurrence of macrocracks can then be acceptable in order to avoid a low design stress level and a corresponding heavy structure.

Microcracks usually have a negligible effect on the ultimate strength of a structure, but macrocracks will substantially reduce the static strength. Obvious questions are: (i) How fast are these cracks growing? (ii) Is it possible to find the cracks by periodic inspections? Because of these arguments the prediction of fatigue crack growth under CA and VA loading has received much attention in the literature. Crack growth under CA loading was discussed in Chapter 8. Crack growth under VA loading is the subject of the present chapter. Significant interaction effects can occur during the growth of macro cracks, and as a consequence, sequence effects are also possible. However, there are essential differences with fatigue at notches. Notch root plasticity and fatigue damage increments of cycles below the fatigue limit were considered in the previous chapter. In the present chapter crack tip plasticity, crack closure and the significance of cycles with a low ΔK-value are discussed for VA-load histories.

Crack growth under simple VA-stress histories is discussed in Section 11.2 to illustrate and explain the occurrence of crack growth retardation and acceleration. Effects of the load history, material yield stress and material thickness are discussed. Plasticity induced crack closure is significant for explaining the interaction effects. Crack growth under complex VA-stress histories is discussed in Section 11.3. Crack growth prediction models for VA loading are considered in Section 11.4, followed by an evaluation in Section 11.5. The major topics of the present chapter are summarized in Section 11.6.

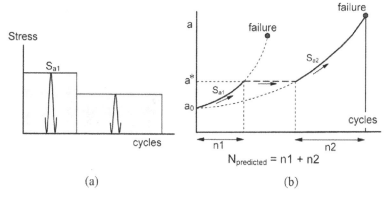

Fig. 11.1 Non-interaction fatigue crack growth in a two-block VA test.

11.2 Crack growth under simple VA-stress histories

Most investigations on fatigue crack growth under VA-stress histories were carried out on aircraft materials: strong Al-alloys, Ti-alloys and high-strength low-alloy steels. Significant work has also been done on low-carbon steels because fatigue crack growth is of interest to large welded structures. Many VA crack growth tests were carried out with simple load histories in order to see whether fatigue crack growth was delayed or accelerated by a change of the stress amplitude. The effect of single high load cycles was also abundantly studied. These high loads have a very large effect on crack growth. Although the simple load histories in these test series are different from service load histories, the experiments have significantly contributed to the present understanding of interaction and sequence effects during fatigue crack growth. The understanding is essential for developing crack growth prediction models for VA loading discussed in Section 11.4.

Non-interaction behavior

Fatigue crack growth under VA loading would be simple if crack growth in every cycle was dependent on the severity of the current cycle only, and not on the load history in the preceding cycles. A most simple case to be considered here is crack growth under a load history with two blocks of CA cycles as shown in Figure 11.1a. Fatigue crack growth under the high-stress amplitude S_{a1} starts from an initial crack length a_0 until a crack length a^* is obtained. The stress amplitude is then reduced to S_{a2} and the test is continued

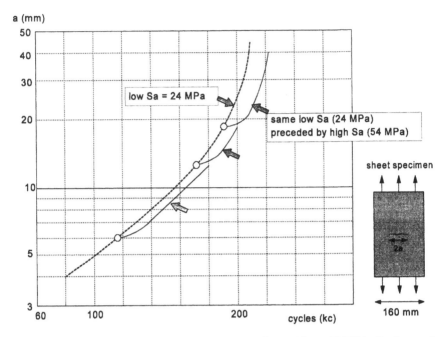

Fig. 11.2 Grack growth test on 2024-T3 Al-alloy specimens ($S_m = 80$ MPa). Crack growth retardation after transitions from a high amplitude to a lower amplitude [1].

until failure. The CA crack growth curves for both amplitudes are shown in Figure 11.1b. If crack growth at the second amplitude would not depend on how the crack has grown until a^*, the initial growth along curve 1 is continued along curve 2, and the total crack growth life is $n_1 + n_2$. This prediction implies that n_1 and n_2 are predicted in the same way as it should be done for CA loading as discussed in Chapter 8. Unfortunately, experimental evidence has shown that crack growth in the second block is affected by crack growth in the first block. An illustration of this interaction effect is shown in Figure 11.2. After the transition of a high amplitude (54 MPa) to a low amplitude (24 MPa), crack growth was retarded during a crack growth increment of 1 to 3 mm. The crack growth curves then were again parallel to the original crack growth curve of the low amplitude. The crack growth rate was no longer reduced and the retardation was over.

Effects of OL cycles

Much larger retardations are observed in tests with a high peak load added to CA loading. Such high loads are frequently called overloads

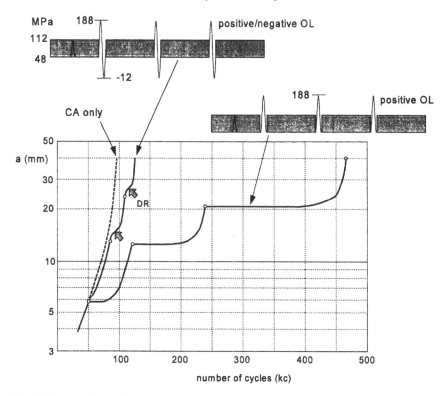

Fig. 11.3 The effect of overloads (OL) on fatigue crack growth in sheet specimens of the aluminium alloy 2024-T3 [1] ($S_m = 80$ MPa). (DR = delayed retardation)

(OL). Figure 11.3 shows the crack growth curve for a test with three high OLs. After the application of each peak load, a most significant delay of fatigue crack growth occurred. The original crack growth life with the CA loading only was just below 100 kc. However, with the three peak loads the crack growth life was almost 500 kc. These crack growth retardations are generally attributed to plasticity induced crack closure (the Elber mechanism) discussed in Chapter 8 (Section 8.4.1). The crack closure phenomenon is a consequence of plastically elongated material left in the wake of the crack by previously created crack tip plastic zones, see Figure 8.10. This phenomenon implicates crack closure at a positive gross stress. A load cycle with a high maximum stress causes a relatively large plastic zone and thus will leave more plastic deformation in the wake of the crack after further crack growth. It increases the crack tip opening stress level, S_{op}, and thus reduces the effective stress intensity range (ΔK_{eff}). The lower ΔK_{eff} explains the crack growth retardation in Figure 11.2. However,

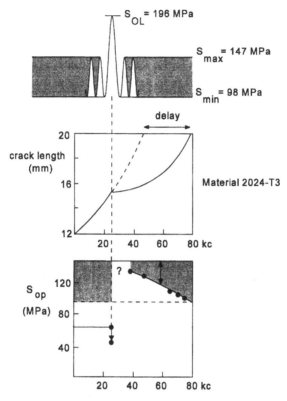

Fig. 11.4 Crack growth retardation after an OL and the relation with S_{op} [2].

the peak loads in Figure 11.3 have caused a relatively large plastic zone, and as a result much more crack closure when the crack tip was penetrating into these zones. This concept was confirmed by some elementary tests carried out by Arkema [2]. Crack closure measurements were made during a CA test ($R = 0.67$) with a single OL as shown in Figure 11.4. The delay caused by the OL can easily be observed from the crack growth curve. Crack closure measurements carried out before the application of the OL indicated $S_{op} \sim 62$ MPa. Immediately after the OL, the S_{op}-level was reduced to about 45 MPa because the OL is opening the crack by crack tip plasticity. Crack closure measurements after some further crack growth indicate a significantly increased S_{op}-values above S_{min} of the CA cycles. As a result, reduced ΔS_{eff}-values occur, and crack growth retardation is observed. After S_{op} decreased to $S_{op} = S_{min}$, the crack growth delay was over because crack closure no longer occurs during the following CA cycles. Although the crack closure measurements may have a limited accuracy, the trend of Figure 11.4 is considered to be correct.

Figure 11.3 also shows a crack growth curve of a specimen subjected to three OLs, but with the OL immediately followed by a large negative amplitude load, also called underloads (UL). There is still some crack growth retardation, but the large effect of the OLs was drastically reduced. Again the results can be explained by considering plasticity induced crack closure. The OL creates a relatively large crack tip plastic zone which also leads to crack tip blunting. Because of the open crack tip after the OL, reversed plastic deformation occurs at the crack tip during the application of the UL until the crack tip is closed again. In this reversed plastic zone, residual tensile stresses are present and crack growth retardation should not be expected. On the contrary, some crack growth acceleration is possible in this small reversed plastic zone. However, as discussed in Section 8.4.1, the reversed plastic zone is smaller than the monotonic plastic zone of the preceding OL, see Figure 8.10. After the crack tip has passed the reversed plastic zone and enters the monotonic plastic zone of the OL cycle, some crack growth retardation is possible again. This phenomenon is called delayed retardation because it required some crack extension before it becomes effective. Delayed crack growth retardation is addressed again later.

Difference between cracks and notches

An essential difference between cracks and notches should be noted here. During crack growth, compressive loads can reduce the favorable effect of tensile peak loads, but it does not necessarily lead to a reduced crack growth life. Macro cracks can already be closed during a decreasing load when the load is still a tension load. A closed crack is no longer a stress raiser, and significant negative plastic strains in the crack tip plastic zone do not occur. However, if a notch is loaded under compression, the notch is not closed. It remains a stress raiser, and notch root plasticity can introduce unfavorable residual tensile stresses with an adverse effect on fatigue life as discussed in the previous chapter (Section 10.2.2).

Effect of material thickness

Crack growth retardation after an OL should depend on the size of the plastic zone because crack closure is induced by crack tip plasticity of the OL.

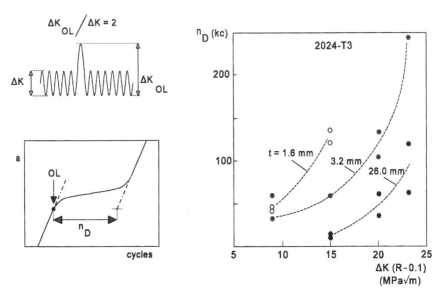

Fig. 11.5 Effect of material thickness on the crack growth delay period after a single OL. Constant-ΔK tests on 2024-T3 specimens. Results of Mills and Hertzberg [3].

Recall, the size of the plastic zone is different for plane strain and for plane stress, see Section 5.8. In a thin sheet, the state of stress at the crack tip is predominantly plane stress; whereas, in a thick plate it is predominantly plane strain. Plastic zones are larger in thin sheets. It then should be expected that the retardation effects are different for fatigue cracks in thin sheets and thick plates. This was confirmed by results of Mills and Hertzberg [3], see Figure 11.5. They carried out crack growth tests with a constant ΔK instead of a constant ΔS. It requires that S_{max} and S_{min} are continuously reduced during crack growth to maintain a constant R-value. This can be done automatically during a fatigue test, see Chapter 13. As a result of the constant ΔK, a constant crack growth rate da/dN is observed. An OL cycle inserted in a constant-ΔK test systematically reduces the crack growth during a period of crack growth retardation. After some crack growth, the growth rate returned to its original constant value. The delay period (n_D cycles) can then be defined in a simple way, see the inset figure in Figure 11.5. Two trends are obvious from the test results: (i) The delay period is larger for thinner material (larger plastic zone) for a given ΔK, see, for instance, the results in Figure 11.5 at $\Delta K = 15$ MPa$\sqrt{\text{m}}$. (ii) The delay period increases at higher stress intensities of the OL (larger plastic zones). These trends agree with

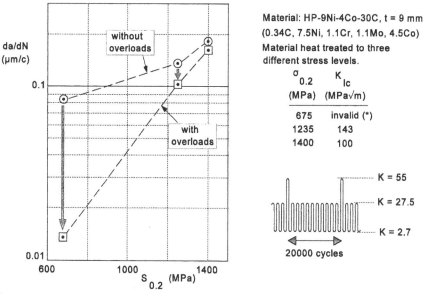

*A valid K_{Ic} cannot be measured if the plastic zone is too large compared to the specimen thickness, see ASTM Standard E399.

Fig. 11.6 Effect of material yield stress on crack growth retardation by periodic OL cycles. Results of Petrak [4].

the effect of the plastic zone size on crack closure and thus on crack growth retardation.

Effect of material yield stress

An instructive example is shown in Figure 11.6 by the results of Petrak [4] for a high-strength steel. The material was heat treated to three different yield stress levels. Petrak also carried out constant-ΔK tests, but he introduced periodic OL cycles after each 20000 cycles. In tests without peak loads, the crack growth rate was larger if the steel was heat treated to a higher yield stress. Apparently, the alloy was more sensitive for fatigue crack growth at a higher strength of the material (recall the discussion in Chapter 8). The periodic OL cycles reduced the crack growth rate. The reduction was large for a low yield stress material (larger plastic zone) and much smaller for the high yield stress material (small plastic zone). Also, this observation confirms the significance of the plastic zone size of OLs for subsequent crack growth retardation.

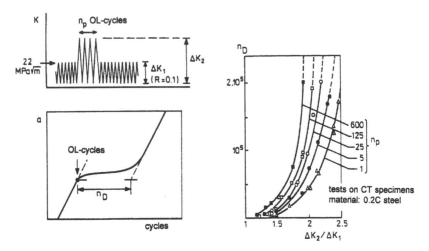

Fig. 11.7 The effect of the number of OL cycles (n_p) on the crack growth delay period (n_D). Results of Dahl and Roth [5].

Block of OL cycles

As discussed above, one OL cycle can give a considerable delay of crack growth. However, it has also been observed that more OL cycles give a larger delay. Illustrative results for a low-carbon steel are presented in Figure 11.7. Dahl and Roth [5] also carried out constant-ΔK tests and adopted the same delay period definition as Mills and Hertzberg (Figure 11.5). The test results show that the delay period is larger for higher OLs. However, it is noteworthy that larger numbers of OL cycles systematically increased the delay period. Crack extension occurs during the OL cycles. More OL cycles thus will leave more plastic deformation in the wake of the crack behind the crack tip. This explanation is based again on the Elber crack closure mechanism. The phenomenon of a larger delay by more OL cycles is called a multiple OL effect.

Delayed retardation

Retardation after an OL may require some crack growth before a maximum reduction of the crack growth rate is obtained, see Figure 11.8. Delayed retardation was observed in several investigations, but the observation is not easily made because this delay occurs during a small crack length increment only. As discussed before in relation to Figure 11.3, an OL can lead to a blunted and open crack tip. This will reduce S_{op} and facilitates the

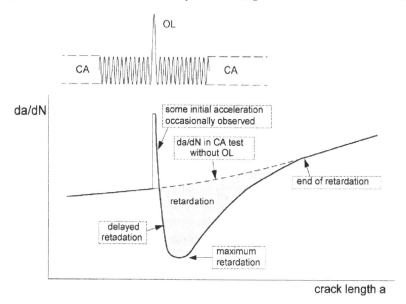

Fig. 11.8 Delayed retardation after an overload (OL).

very beginning of growing into the new crack tip plastic zone of the OL. Moreover, some crack growth acceleration can then occur due to residual tensile stress in the reversed plastic zone (Figure 8.10). The residual tensile stress promotes crack tip opening. The acceleration directly after the OL is observed during a small crack length increment. Further crack growth into the OL plastic zone meets with increasing crack closure as a result of residual compressive stresses introduced by the OL in the monotonic plastic zone (Figure 8.10). A decreasing crack growth rate is observed. Because the minimum crack rate does not occur immediately after the OL, the term "delayed retardation" was introduced. After passing the minimum crack growth rate and some substantial further crack extension, the growth rate returns to the normal CA crack growth rate of the CA base line cycles. The size of the OL plastic zone is important for the crack extension during which retardation is observed. But the end of retardation does not necessarily occur at the edge of this plastic zone as sometimes suggested for crack growth prediction models. Moreover, it should be recalled that the boundary of the plastic zone is not an accurately defined concept. The plastic zone estimates of Equations (5.35) and (5.36) are actually primitive estimates, based on one single dimension and disregarding the shape of the plastic zone. Furthermore, the plastic zone size estimates adopt $S_{0.2}$ as a criterion for the boundary between elastic and plastic material around the crack tip. However,

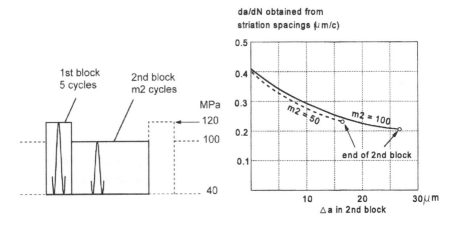

Fig. 11.9 Delayed retardation in second block of low-amplitude cycles. Sheet specimen of the Al-alloy 2024-T3.

a yield stress $S_{0.1}$ (0.1% plastic deformation) as sometimes used in the UK, would lead to larger calculated plastic zone sizes.

The more reliable indications on delayed retardation should come from striation observations. In this respect Al-alloys are instructive because they show striations better than most other alloys. Illustrative results have been obtained in tests with periodic blocks of larger and smaller cycles, see Figure 11.9 [6]. Five larger cycles were followed by 50 or 100 smaller cycles. The striation spacings indicated a decreasing crack growth rate. Delayed retardation apparently occurred, but a minimum crack growth rate was not yet reached before the next block of high-amplitude cycles was applied.

Crack growth retardation by crack closure or residual stress in the crack tip plastic zone?

Dahl and Roth [5] raised the question whether crack growth delay after an OL is due to crack closure, or whether there is also an effect of the residual compressive stress in the plastic zone ahead of the crack tip. An interesting CA experiment was carried out by Błazewicz [7]. He made ball impressions on 2024-T3 sheet specimens before the crack growth test was started, see Figure 11.10. As a result of the plastic deformation, residual compressive stresses were present in a zone between the ball impressions. This caused crack growth retardation. The delay was small during the growth through the zone between the impressions, but it was significant at a later stage.

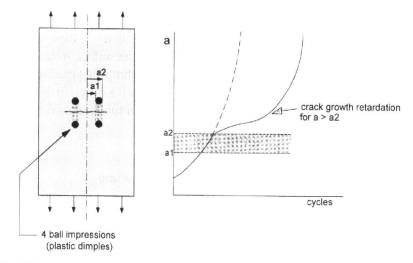

Fig. 11.10 Crack growth retardation in CA test by residual stress created in the wake of the crack by ball impressions. Measurements of Błazewicz [7].

It suggests that crack growth retardation should primarily be explained by the crack closure and opening phenomenon. In terms of the crack growth mechanism, it is logical that the crack must be opened before crack extension can start. The retardation for $a > a_2$ in Figure 11.10 thus indicates that crack tip opening was less than normal. This is a consequence of the excess of material in the wake of the crack. The retardation cannot be related to residual stresses ahead of the crack tip.

More crack closure at the material surface

At the surface of a material, the crack tip is loaded under plane stress conditions. Depending on the material thickness, the state of stress at mid thickness is approaching plane strain conditions. The plastic zone size under plane stress is significantly larger than under plane strain. As discussed in Chapter 5 (Section 5.8) the Irwin plastic zone size estimates are $r_p = (1/\alpha\pi) \cdot (K/S_{0.2})^2$, with $\alpha = 1$ for plane stress and $\alpha = 3$ for plane strain. It thus should be expected that crack closure is more significant near the material surface, and will occur to a lesser degree at mid thickness. This is confirmed by FEM calculations [8], but there is also experimental confirmation [9, 10]. McEvily [11] studied crack growth after an OL in Al-alloy specimens (6061), which gave a significant crack growth delay. In another test, he reduced the thickness of the specimen from 12.7 to 6.3 mm immediately after the OL by removing plate surface material on both sides of the specimen. A much smaller crack growth retardation was then found as compared to the crack growth delay without removing surface material. As already discussed in Chapter 8 (Section 8.4.2), it confirms that crack closure occurs predominantly at the material surface, and much less at mid thickness of a material. This implies a significant complication for prediction models for crack growth under VA loading.

Incompatible crack front orientation under VA loading

Interaction effects during fatigue crack growth under simple VA-load histories as discussed before, were qualitatively explained by considering the occurrence of crack closure. It was recognized that plane strain/plane stress problems were present, which implies that crack closure is predominantly occurring at the material surface. Plane stress is associated with a free surface. The same is true for the occurrence of shear lips discussed in Chapter 2 (Section 2.6 and Figure 2.38). If shear lips are developed under CA loading, which occurs for several technical materials, it can lead to a transition from a tensile-mode crack to a shear-mode crack. The development of shear lips depends on the crack driving force, ΔK. It implies that shear lips occur earlier at high stress amplitudes. The maximum width of the shear lips is half the material thickness. As a consequence, the full transition from the tensile mode to the shear mode is more evident in thin sheet material.

Under VA loading, the development of shear lips and the transition to the shear mode can imply incompatible crack front orientations, a topic rarely covered in the literature. A simple example is shown on the fracture surface in Figure 11.11. The fatigue cracks at both sides of a central notch (width 3 mm) were growing during cycles with $S_a = 50$ MPa. An increasing shear lip width can be observed. The crack were already fully in the shear mode when a block of low-amplitude cycles ($S_a = 25$ MPa) was introduced. It caused a narrow bright band on the fracture surface (arrows in Figure 11.11). The normal fracture mode of the low-amplitude cycles under CA loading at that particular crack length is the tensile mode (with minute shear lips). This is not compatible with the existing shear mode introduced by the high-amplitude cycles. A tendency to return to the tensile mode could be observed which gave the bands a faceted appearance. The growth rate in the bands was eight times lower than recorded in normal CA-tests at the same crack length, which is a significant retardation.

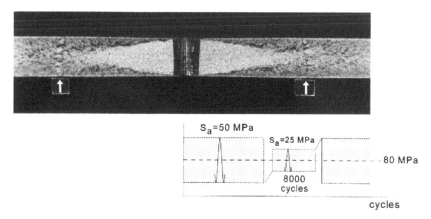

Fig. 11.11 Fatigue fracture surface of an Al-alloy (2024-T3) specimen with a central crack, thickness 4 mm. Crack growth in the first block of cycles produced shear lips until the entire crack front was in the shear mode. Crack growth in the second block with low-amplitude cycles then occurred with an incompatible crack front orientation [12].

The reverse case is perhaps more relevant, i.e. when high-amplitude cycles occur between many low-amplitude cycles. The fracture surface then can be largely in the tensile mode, whereas the failure mode corresponding to the nominal ΔK cycle of the high-amplitude cycle in a CA test may be the shear mode, or at least a mixed mode with significant shear lips. In elementary tests [12], such large cycles produced dark bands on the fracture surface, and a growth rate far in excess of the corresponding CA results. In this case the incompatible crack front orientation caused an increased crack growth rate during the high-amplitude cycles.

Another incompatibility can occur if different crack growth mechanisms apply to low and high stress amplitudes. Some materials, for instance Ti-alloys, can exhibit structurally sensitive crack growth at low amplitudes, and a more regularly flat crack growth at a high amplitude. If low amplitudes and high amplitudes occur, a mismatch (incompatibility) can occur, and thus lead to interaction effects. An interesting proof was given in fracture toughness tests of Stubbington and Gun [13]. Fatigue crack growth at a low K_{max}-value (10.8 MPa$\sqrt{m}$) produced a rough fatigue fracture surface (structurally sensitive crack growth), whereas crack growth at a high K_{max} (19.1 MPa$\sqrt{m}$) gave a flat fracture surface (structurally insensitive crack growth). In the former case, the rough crack in a static test gave a fracture toughness $K_{Ic} = 72.9$ MPa$\sqrt{m}$, whereas in the latter case, the flat crack gave a significantly lower $K_{Ic} = 49.3$ MPa$\sqrt{m}$. A similar phenomenon can occur in other materials during fatigue crack growth after an OL. The initial crack growth after the OL occurs extremely slowly in a plastic zone with severe plastic deformations and possibly void formation. Fractographic observations in the electron microscope have shown a rather distorted crack growth path after the OL. On a micro scale, the crack front is rather irregular and crack growth is more difficult, not only due to plasticity induced crack closure, but also because the crack driving energy must propagate a more complex and longer crack front.

11.3 Crack growth under complex VA-stress histories

The discussion in the previous section on crack growth under simple VA-stress histories has shown that plasticity induced crack closure (Elber mechanism) is the major mechanism responsible for interaction effects during the growth of macro cracks. Complications are associated with plane stress/plane strain aspects and with possible incompatibilities between crack growth mechanisms for low and high stress amplitudes. Although the interaction effects usually lead to crack growth retardation, crack growth acceleration is sometimes possible. All these observations can also occur under complex VA-stress histories. The following topics are discussed in this section:

- Sequence effects.
- Thickness effects.
- Initial fast crack growth at a notch.
- Truncation effects.

Sequence effects

Ryan [14] studied fatigue crack growth in a high-strength D6AC steel under Lo-Hi-Lo program loading at $R = 0$, see Figure 11.12. As a result of the stepwise amplitude changes, fatigue bands could be observed on the fracture surface for each block of the symmetric program. The crack growth rates could then be calculated from the measured width of the bands. It turned out that the crack growth rate was smaller in the Hi-Lo part of a period than in the Lo-Hi part, see the results in Figure 11.12. This figure also contains the da/dN curve for CA loading which confirms that crack growth retardation occurred in the Hi-Lo part of the period, and crack growth acceleration in the Lo-Hi part, including the maximum amplitude of the program. In view

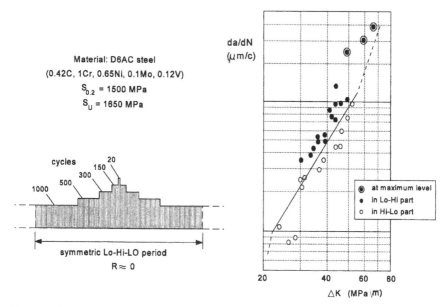

Fig. 11.12 Sequence effects during crack growth in a high-strength steel under program fatigue loading. Results of Ryan [14].

of the $R = 0$ condition, the block of the 20 largest cycles should be expected to have a favorable effect on crack growth in the following cycles with a lower $S_{\max}$. In terms of crack closure, S_{op} during the decreasing S_a cycles will be higher than in CA tests. During the increasing part of the sequence, S_{op} will be lower and the crack growth rate will be higher.

Another illustrative test program was carried out on fatigue crack growth in 2024-T3 sheet material [15]. The load spectrum used was based on a spectrum for air turbulence on a wing of a transport aircraft wing. The spectrum contains 40000 cycles with seven different amplitudes. The spectrum with the same load cycles was used with two random sequences and some programmed sequences, see Figure 11.13. The random sequences were applied with full cycles, i.e. each cycle consisted of two half cycles with the same amplitude. However, in one random load history, each cycle started with the positive half cycle followed by the negative one. In the other random load history this sequence was reversed. As shown by the results in the upper part of Figure 11.13, the difference between the crack growth lives of the two random load test series was small.

In the program fatigue tests with the full spectrum in one period (40000 cycles), the results in the lower part of Figure 11.13 indicate two remarkable trends. (i) There is a systematic sequence effect with the longer crack growth

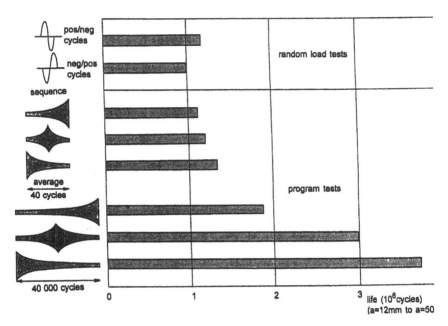

Fig. 11.13 A comparison between fatigue crack growth lives under random loading and different types of program loading [15]. Sheet material 2024-T3, $S_m = 69$ MPa.

life for the Hi-Lo sequence, a shorter life for the Lo-Hi sequence, and an intermediate life for the Lo-Hi-Lo sequence. The explanation given for the fractographic results of Ryan (Figure 11.12) also applies to these results. (ii) Even more remarkable, and actually disturbing, the crack growth lives were considerably longer than for the random sequence, i.e. about three times longer for the Lo-Hi-Lo sequence. It implies that the program fatigue tests give an unconservative result for a random sequence! Fractographic observations indicated that the fracture surfaces of these program fatigue tested specimens were more irregular than the fracture surfaces obtained under the random sequences. It then should be concluded that fatigue crack growth is just not the same phenomenon for these programmed and random sequences of the same cycles.

In the third group of tests, programmed sequences were used again, but now with a much shorter period (average of 40 cycles). In order to accommodate the full 40000 cycles spectrum in these tests, not all periods could be similar. Some periods contained also the rarely occurring high-amplitude cycles, while other periods did not. The sequence of the periods of different severity was random again. The crack growth life results were of the same order of magnitude as for the fully random

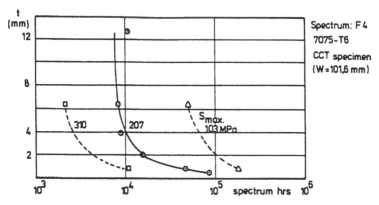

Fig. 11.14 Material thickness effect on crack growth life under flight-simulation loading. Results of Saff and Holloway [16].

sequence. The fracture surface appearance was also similar. The important lesson is that a load spectrum with a random load sequence in service should not be simulated by a programmed sequence with a long period. A programmed sequence with a long period is an artificial simulation, which as a simplification cannot be accepted. A similar conclusion was already drawn in the previous chapter on fatigue life problems including the crack initiation period.

Thickness effect

Saff and Holloway [16] carried out flight-simulation fatigue tests with a load spectrum based on a maneuver loads (F-4 aircraft). Crack growth was observed in center cracked tension specimens of different thicknesses varying from 0.5 to 12.7 mm. For a maneuvre spectrum the maximum stress occurring in the flight-simulation test is used as the characteristic stress level. Three stress levels were used. The results in Figure 11.14 clearly show a systematic thickness effect, i.e. lower endurances for thicker material. As discussed in the previous section, an increased thickness leads to more plane strain at the crack front, and thus to smaller plastic zones, less crack closure and less crack growth retardation. As a consequence, the crack growth lives are smaller. The life for the thin sheet material was about 10 times(!) longer than for the thick plate material. A similar thickness effect was also found in flight-simulation tests with a gust spectrum [17].

A thickness effect on fatigue crack growth is sometimes observed in CA tests, but in general the effect is larger for VA-load histories. It must be

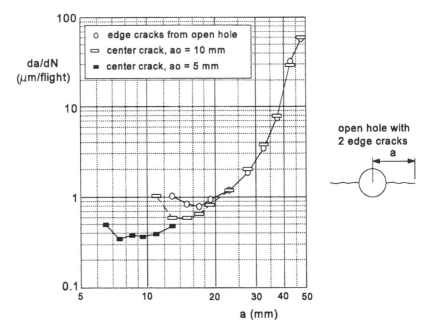

Fig. 11.15 Initial fast crack growth at the edge of a hole. Flight-simulation loading with gust load spectrum (F-28) [18]. Sheet material 2024-T3, $S_{mf} = 69$ MPa.

concluded that representative crack growth experiments should not only be based on a realistic load history, but a relevant material thickness should also be used.

Initial fast crack growth at a notch

Figure 11.15 shows the crack growth rate of a crack starting at the edge of an open hole (radius 5 mm) in a specimen loaded by a flight-simulation load history. It is quite remarkable that the crack growth rate initially decreased to a minimum at $a = 17$ mm. Decreasing of the crack rate occurred in spite of a nominally increasing stress intensity associated with an increasing crack length. After some further crack growth the expected increase of the crack growth rate occurred. Crack growth under the same flight-simulation loading was also recorded for cracks which were not initiated at the edge of a hole. They started as an extension of a fatigue crack obtained under CA loading at a low load level to have a low crack opening stress level (S_{op}). Crack growth during subsequent flight-simulation loading started again with a decreasing crack growth rate, see the results of two tests in Figure 11.15. After passing a minimum growth rate, the increasing parts of all tests nicely line up along the same growth rate curve. It should be concluded that the initial decrease of the growth rate should be associated with the development of a plastic wake field behind the crack tip. A substantial plastic wake field did not yet exist in the first period of the crack growth life. It requires some crack growth during which the maximum peak loads of the load spectrum can build up significant plasticity in the wake of the crack.

The behavior of an initially decreasing crack growth rate has been observed in various flight-simulation fatigue tests (e.g. [19]), but an initially decreasing growth rate of hole edge cracks under CA loading was also observed by Broek [20]. However, for VA loading the phenomenon can be more important, especially for steep load spectra with low numbers of high-amplitude cycles (see Figure 9.10), and less important for materials with a relatively high yield stress because of smaller crack tip plastic zones. Anyway, an important lesson to be learned from the above experience is that artificial fatigue crack starters may not give realistic information of the early crack growth rates.

Load spectrum truncation effect

In the previous chapter (Section 10.4.5) it was discussed that truncation of rarely occurring high load amplitudes can have a large effect on the fatigue life, especially for steep load spectra with rarely occurring high loads. This is also true for the propagation of macro cracks, probably still more than for notch fatigue problems. A steep load spectrum is applicable to wing structures of transport aircraft. Figure 11.16 shows results of crack growth tests on Al-alloy 2024-T3 sheet specimens (width 100 mm, thickness 2 mm) loaded at a mean stress in flight $S_{mf} = 70$ MPa. The gust spectrum is a rather steep spectrum, which implies that high loads are relatively rare, see Figure 11.16a. The highest gust level is reached once in 4000 flights. The lower levels II, III, IV and V are reached 3, 8, 26 and 78 times per 4000 flights respectively, while the total number of cycles in 4000 flights is about 400000. As shown by the crack growth curves in Figure 11.16b, truncation of the high amplitudes to lower levels significantly reduced the crack growth life. The crack growth life for truncation level V is about 7 times shorter than the life for truncation level II. Tests at truncation level I, the highest one in Figure 11.16a, were stopped because the growth rate was very low.

Similar truncation effects on fatigue crack growth have been found for different Al-alloys, a Ti-6Al-4V alloy and a high-strength steel (survey

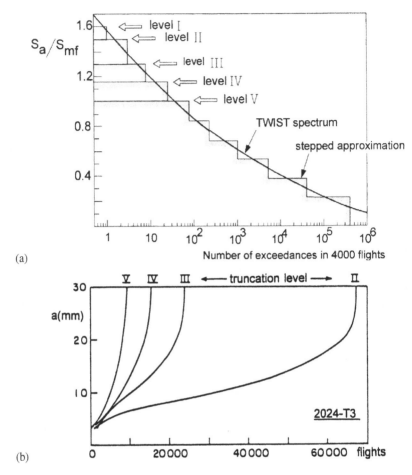

(a)

(b)

Fig. 11.16 Effect of spectrum truncation on crack growth in 2024-T3 Al-alloy sheet specimens under flight-simulation loading [21].

in [22]). However, it appears that the truncation effect is smaller for the 7075-T6 alloy, which could be associated with the higher yield stress of the material ($S_{0.2} \approx 475$ MPa for 7075-T6 and $S_{0.2} \approx 350$ MPa for 2024-T3). Plastic crack tip zones are smaller and as a result the retardation effects can be less significant. The important point is that truncation effects can occur. These effects were also noted for less steep spectra. The choice of a suitable truncation level must be given due attention if service simulation fatigue tests are carried out. A high truncation level, i.e. including load cycles with a very high amplitude, will give longer fatigue lives and lower crack growth rates. These results can be unconservative because not all structures in service will meet these high-amplitude loads.

11.4 Crack growth prediction models for VA loading

The literature on prediction models for fatigue crack growth under VA loading is extensive. Observations on crack growth retardation after OLs, and the occurrence of crack closure have stimulated the development of several prediction models. In general, these models predict fatigue crack propagation as a cycle-by-cycle process. In view of striation observations, this seems to be logical. Crack growth is assumed to be a summation of Δa-values, or

$$a_n = a_0 + \sum_{i=1}^{i=n} \Delta a_i \qquad (11.1)$$

The value of a_n is the crack length after n cycles, a_0 is the initial crack length, and Δa_i is the crack extension Δa in cycle number i. Some comments related to Equation (11.1) should be made:

(i)   The crack extension in a cycle is an equilibrium between a "crack driving force" and a "crack growth resistance", see the discussion in Section 8.6.1. The crack driving force in prediction models is expressed in terms of the stress intensity factor. This factor can only be defined if there is a crack of some length. It implies that the prediction with Equation (11.1) should always start with a finite crack length; a_0 cannot be zero. As a consequence, the crack nucleation life cannot be predicted by fracture mechanics methods based on K-values.

(ii)   Further to the previous remark, an additional comment must be made on the minimum value of a_0 which can be used in crack growth predictions. As discussed in Chapter 2, the transition from the crack initiation period to the crack growth period is dependent on the type of material and the structure of the material. As long as the growth of a microcrack is still a structurally sensitive phenomenon at the material surface, predictions based on K-values are not justified. This problem was discussed earlier in Section 8.6.1 on fatigue crack growth under CA loading.

(iii) The crack extension in a cycle is per definition the crack growth rate in that cycle. As a consequence, the prediction should be a cycle-by-cycle prediction of the crack growth rate. This occurs in the prediction models still to be discussed. But, models differ in the way of predicting the crack growth rate in every cycle.

(iv)   Equation (11.1) assumes that the size of the crack is fully defined by a single size parameter, the crack length a. As discussed in Chapter 8 for a part-through crack with a curved crack front, the shape of the crack must be accounted for by a variation of K along the crack front.

11.4.1 Non-interaction model

The most simple prediction model is the non-interaction model. Crack growth in each cycle is assumed to be independent of the preceding load history which has created the crack. This option was already discussed in Section 11.2 and illustrated with Figure 11.1. It was pointed out that this approach is physically unrealistic because considerable interactions have been reported. Crack growth retardations and accelerations can occur during VA loading. Retardations are the more likely phenomena. Thus, if interaction effects are ignored, it may be expected that non-interaction predictions are conservative. Prediction can even be highly conservative for a steep load spectrum with occasional high loads, and probably less conservative for a flat spectrum with many high amplitude cycles.

The non-interaction approach leads to a simple numerical summation with Equation (11.1) and values $\Delta a = \mathrm{d}a/\mathrm{d}N$ as obtained in CA tests. Such crack growth data should be available as a function of ΔK, including the stress ratio effect:

$$\Delta a_i = f(\Delta K_i, R) \qquad (11.2)$$

The equation was previously discussed as Equation (8.3) in Chapter 8. The crack growth data can be available as an analytical function (see Section 8.3), but CA crack growth data in tabular format can also be used with interpolation between these data. Computer algorithms can be written to obtain crack growth results with Equations (11.1) and (11.2). This should not be difficult if interaction effects in the VA-stress history are supposed to be absent.

One restriction should be made on using CA crack growth data. Contributions of cycles with a very low ΔK are supposed to be zero if the ΔK-value is below the nominal threshold value ΔK_{th} of macro-cracks, see Figure 8.6. The growth of cracks in the threshold region was discussed in Section 8.3, and it was pointed out that the determination of a ΔK_{th}-value has a problematic character. This value is obtained in experiments with a decreasing crack growth rate until the crack driving force can no longer overcome the crack growth resistance. Under these threshold conditions it can lead to an erratic crack growth mechanism, which may not be representative for crack growth under VA loading. It may be expected that cycles with $\Delta K < \Delta K_{th}$ can still contribute to crack growth because these cycles were not preceded by some erratic crack growth mechanism occurring in threshold experiments. It is recommended to extrapolate the $\mathrm{d}a/\mathrm{d}N$-ΔK data in the Paris region to lower ΔK-values in the threshold region as shown in Figure 8.6. Although the extrapolation has no solid physical background it still appears to be more reasonable and safer than ignoring this aspect. The problem has a certain similarity to the problem of extrapolating S-N curves below the fatigue limit for fatigue life predictions with the Miner rule for VA-load histories.

11.4.2 Interaction models for prediction of fatigue crack growth under VA loading

Different prediction models for fatigue crack growth under VA loading are associated with different concepts about crack tip plastic zones and crack closure. Three groups of models are listed in Table 11.1. The early yield zone models, do not consider crack closure. They account for interaction effects by assumptions about plastic zone sizes. Later the crack closure models include predictions on the occurrence of crack closure, but still with assumptions about crack closure stress levels during VA loading. The strip yield models are the most advanced models. Crack closure stress levels are obtained by calculations rather than assumptions.

Originally, the prediction models were mainly verified for through cracks in sheet and plate specimens of aluminium alloys, but later experiments were also done on other materials. The models are considered to be applicable for high-strength alloys with a limited ductility. Actually, these materials are the most fatigue critical materials. Mild steel is not in this category. Due to its special yielding behavior and its high ductility, mild steel is a class of materials of its own. Fatigue crack growth in low-C steel under VA loading is becoming an increasingly relevant problem because of welded structures. Essential features of the three types of models listed in Table 11.1 will be briefly summarized.

Table 11.1 Three categories of crack growth prediction models.

Type of model	Crack closure used?	Crack closure relation
Yield zone models	no	–
Crack closure models	yes	empirical
Strip yield models	yes	calculated

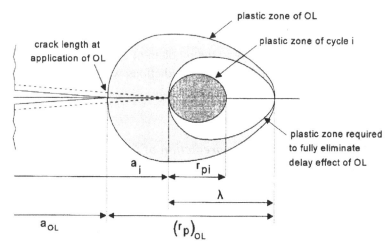

Fig. 11.17 Plastic zone size definition used in the models of Willenborg [23] and Wheeler [24].

Yield zone models

The models of Willenborg [23] and Wheeler [24] were proposed to explain crack growth delays caused by high loads. The models consider the plastic zone sizes indicated in Figure 11.17, but the concepts are different. In both models, it was recognized that new plastic zones are created inside the large plastic zone of an OL. Moreover, the possibility was considered that these new plastic zones could be large enough to extend beyond the OL plastic zone. The Willenborg model starts from a strange assumption, which implies that crack growth delay after an OL is due to a reduction of $K_{\max}$ instead of a reduction of ΔK_{eff}. This seems to be physically incorrect. Crack closure in the model is supposed to occur only if $K_{\min} < 0$. From a mechanistic point of view, the Willenborg model does not agree with the present understanding of crack closure.

Wheeler introduced a retardation factor γ defined by

$$\left(\frac{\mathrm{d}a}{\mathrm{d}N}\right)_{\text{VA}} = \gamma \cdot \left(\frac{\mathrm{d}a}{\mathrm{d}N}\right)_{\text{CA, same K-cycle}} \tag{11.3}$$

The amplification factor γ is assumed to be a power function of the ratio r_{pi}/λ_i, with $r_{p,i}$ as the current plastic zone size created by the cycle considered, and γ as the distance between the crack tip and the edge of the OL plastic zone, see Figure 11.17.

$$\gamma = \left(\frac{r_{p,i}}{\lambda}\right)^m \tag{11.4}$$

If $r_{p,i}$ is large enough to be equal to λ, then $\gamma = 1$ and according to Equation (11.3) the delay effect of the OL is gone. The exponent m is an empirical constant dependent on the type of the VA-load history. It must be determined by VA-load experiments for each load history of interest.

The Willenborg model and the Wheeler model can predict crack growth retardation only ($\gamma < 1$), not acceleration. After an OL, the maximum retardation occurs immediately. Delayed retardation is not predicted. Modifications of the two models have been proposed in the literature, which has led to more empirical constants. Plasticity induced crack closure is not considered. It appears that both models have fundamental limitations in predicting the crack growth behavior under VA loading discussed in Sections 11.2 and 11.3.

Crack closure models

The crack closure models account for the occurrence of plasticity induced crack closure (Elber mechanism). The values of S_{op} in each cycle of the stress history is predicted. The current value depends on the preceding fatigue crack growth and corresponding plastic wake field of the fatigue crack. A cycle-by-cycle variation of S_{op} as shown in Figure 11.18 must be predicted. The effective stress range in a cycle is $\Delta S_{eff} = S_{max} - S_{op}$. The corresponding ΔK_{eff}-range in cycle i becomes

$$\Delta K_{eff,i} = \beta_i \Delta S_{eff,i} \sqrt{\pi a_i} \qquad (11.5)$$

with β_i as the geometry correction factor, depending on the momentary crack length a_i. The predicted crack extension in the cycle is:

$$\Delta a_i = \left(\frac{da}{dN}\right)_i = f(\Delta K_{eff,i}) \qquad (11.6)$$

If the Paris relation should be valid, the equation becomes

$$\Delta a_i = C(\Delta K_{eff,i})^m \qquad (11.7)$$

Crack growth then follows from the summation of Δa_i according to Equation (11.1).

The main question about the crack closure models is how the variation of S_{op} is predicted for a VA-laod history. Four crack closure models proposed in the literature are briefly commented upon:

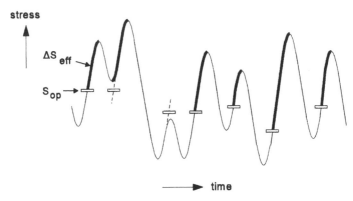

Fig. 11.18 VA loading with cycle-by-cycle variations of S_{op}.

- the ONERA model [25],
- the CORPUS model [26],
- the modified CORPUS model [27], and
- the PREFFAS model [28].

The models were developed primarily for aircraft fatigue problems with applications to flight-simulation load histories, such as illustrated in Figure 9.25. The variation of S_{op} during the flight-simulation load history depends on the preceding load history. It implies that information, characteristic for the previous load history, must be stored in a memory file. The characteristic information is associated with the larger positive and negative peak loads. These loads have introduced significant plastic zones and reversed plasticity which can increase or decrease S_{op} of later cycles. There are significant differences between the models, which will not be discussed here in detail. The reader is referred to the original publications and a survey presented by Padmadinata [27, 29]. However, some comments are made to illustrate essential features of these models.

The PREFFAS model is the simplest model. The CORPUS model of De Koning [26] is the most detailed one. De Koning also presents the most explicit picture about crack closure between the crack flanks. He assumes that it does occur at the larger plastic zones in the wake of the crack left by plastic deformation of the more severe loads. The somewhat protruding zones are called "humps". A hump can be "flattened" by later downward loads which implies a reduction of S_{op}. Various differences between the models are associated with assumptions made for the plane strain/plane stress transition during crack growth, calculation of the plastic zone sizes, empirical equations for calculating S_{op} (Elber type relations), decay of S_{op}

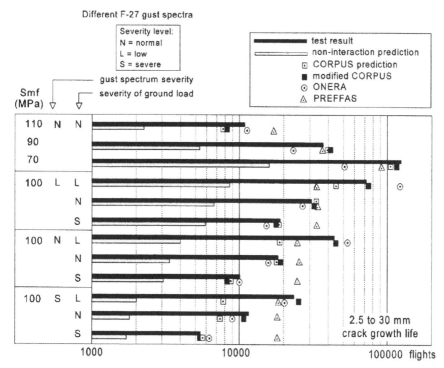

Fig. 11.19 Comparison between test results and prediction on fatigue crack growth life under flight-simulation loading (compiled with data of [27]). Sheet specimens of 2024-T3 Alclad, thickness 2 mm.

during crack growth, multiple overload effect, and method of deriving S_{op} from the previous load history. An analysis and comparison of the models was made by Padmadinata [27, 29] with extensive verifications primarily for realistic flight-simulation load histories and test results of the two Al-alloys, 2024-T3 and 7075-T6. Simplified flight-simulation tests were also included. As an example, predicted fatigue lives for crack growth from 2.5 to 30 mm under realistic flight- simulation loading are compared to test results in Figure 11.19. The test variables include the stress level, characterized by the mean stress in flight (S_{mf}), gust spectrum severity, and severity of the ground load during landing. The test results show that the effects of these variables agrees with expectations; i.e. more severe conditions lead to shorter crack growth lives.

Non-interaction predictions are also shown in Figure 11.19, which unfortunately is not always done in model verifications. A comparison between non-interaction predictions and the test results is made in order see whether significant interactions occurred. The results in Figure 11.19 clearly indicate the occurrence of significant interaction effects. The test results are much larger than the non-interaction predictions; on average about five times larger crack growth lives. This could be expected because the load spectrum in the flight-simulation fatigue tests was a gust spectrum, a steep spectrum with rarely occurring high loads. Figure 11.19 also shows that the prediction of all crack closure models are reasonably close to the test results. Some more comments on the predicted results can be made:

• The PREFFAS model does not predict any effect of the severity of the ground load, because all negative loads are clipped to zero. However, the test results show a systematic effect of the ground stress level.

• The predictions of the CORPUS and ONERA models are fairly close to the test results. The test results indicate a significant reduction of the crack growth life for a more severe ground stress level. This trend in not always correctly predicted by the CORPUS model if the gust spectrum is severe. In the latter case, the maximum downward gust load, occurring only once in 2500 flights, is a larger compressive load then the ground load occurring in every flight. This rarely occurring gust load overrules the negative effect of the ground load. This was the reason to modify the CORPUS, which led to the modified CORPUS model [27]. The model is still largely the same as the original CORPUS model, but it introduces a modified memory for downward loads to give a better prediction for the test conditions in the flight-simulation tests [27, 29].

The CORPUS model was verified by Ichsan [30] for semi-elliptical surface cracks in plate specimens of Al-alloys. The analysis included a variation of S_{op} along the crack front. A satisfactory agreement was found between predictions and test results including the development of the crack front shape.

The agreement between predictions and test results, as illustrated by Figure 11.19 appears to be promising, but it must be admitted that the agreement is largely limited to crack growth under specific aircraft load spectra and typical aircraft materials. Furthermore, although the models are based on the physically relevant crack closure phenomenon, they still include several plausible assumptions and adjustment in order to come to a better correspondence between prediction and empirical proof. An analysis of the models indicates some typical issues not yet resolved, such as:

• Crack growth retardation after OLs is predicted, but delayed retardation is not predicted. According to the models, the maximum retardation starts immediately after the overload, which is in conflict with empirical evidence.

- Plane strain/plane stress transitions are included in the CORPUS and the ONERA model, although not in the same way. It leads to a thickness effect, but variations along the crack front are averaged. The transition is not included in the PREFFAS model, but the model requires empirical data for the OL effect representative for the thickness considered.
- Multiple OL effects are accounted for in the CORPUS and the ONERA model, although not in the same way. The CORPUS model predicts an increasing S_{op} during the beginning of stationary flight-simulation loading, which is necessary to predict the initially decreasing crack growth rate (Figure 11.15).
- Incompatible crack front orientations and related phenomena are not covered.

Strip yield models

The previous crack closure models are based on the occurrence of crack closure in the wake of the crack. Assumptions had to be made to account for crack closure under VA loading, but plastic deformation in the wake of the crack is not calculated. This was done in some FE studies [31, 32] which confirmed the occurrence of crack closure and simple interaction effects to be expected in qualitative agreement with empirical observations. Because such calculations cannot be made for many cycles, the Dugdale strip yield model [33] was adopted to calculate the plastic zone size and the plastic extension of the material in this zone. This type of work was started by Führing and Seeger [34, 35]. Quantitative strip yield models were proposed by Dill and Saff [36, 37], Newman [38], De Koning et al. [39], and Wang and Blom [40]. Later modifications were proposed by Bos [41] and Skorupa and Machniewicz [42].

In the Dugdale plastic zone model, plastic deformation occurs in a thin strip with a rigid perfectly plastic material behavior. Plastic extensions of strip elements in the plastic zone are calculated by considering the opening of a fictitious crack with a tip at the right edge of the plastic zone in Figure 11.20. The crack opening depends on the remotely applied load and the yield stress applied in the plastic zone on the fictitious crack tip flanks. Because the crack grows into the plastic zone, a plastic wake field is created, which can induce crack closure at positive stress levels. The models are rather complex, which is a consequence of the non-linear material behavior and the occurrence of closing and opening of the crack. Reversed plastic deformation in the wake field can occur when the crack is closed and locally

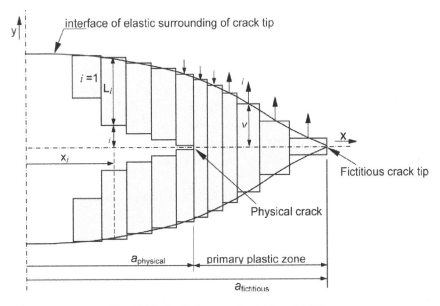

Fig. 11.20 Crack tip in a strip yield model [43].

under compression. Stresses and deformations for the strip elements are solved iteratively by considering compatiblity conditions along the fictitious crack surface. Plane strain/plane stress transitions are included by changing the yield stress used in the Dugdale model. It has led to a so-called plastic constraint factor α, which is used to tune the predictions to be in agreement with experimental data. Several predictions are reported for both simple tests with overload/underload cycles and flight-simulation tests. In general the agreement is considered to be good.

Strip yield models are not discussed here in detail, but some remarks are made, partly in comparison to the crack closure models:

- Empirical equations on crack closure levels are replaced by calculation of S_{op} as a function of the history of previous plastic deformations. Elber's assumption that $U(R)$ is independent of the crack length is no longer necessary.
- Delayed retardation is predicted [43].
- In the strip yield model of De Koning, his concept of primary and secondary plastic zones is introduced [44], which accounts for large Δa-values of peak load cycles, see the discussion in Section 11.5.
- Multiple OL effects should occur in a strip yield model if the modeling is sufficiently refined.

- The plane strain/plane stress transition is still covered by assumptions.
- Incompatible crack front interactions are not covered.

Strip yield models are superior to the crack closure models because the physical concept has been improved. The calculation of the crack driving force, i.e. ΔK_{eff}, is based on calculations of the history of the plastic deformations in the crack tip zone and in the wake of the crack. Several problems of these models still require further analysis. The models have not yet widely been verified.

11.5 Evaluation of prediction methods for fatigue crack growth under VA-load histories

Engineering aspects

Some predominant characteristics of fatigue crack growth under VA loading discussed in the previous sections are: (i) Significant interactions can occur. Cycles with a relatively high S_{max} can lead to a substantial reduction of the crack growth rate in subsequent cycles. (ii) Crack growth accelerations are possible, but crack growth retardations are generally overruling the acceleration. As a consequence, a non-interaction prediction may be expected to be a safe and conservative prediction. As mentioned in the introduction of this chapter, crack growth prediction as an engineering problem are relevant if fatigue cracks can affect the safety or economic use of a structure. A non-interaction crack growth prediction will then give a first indication about the possible duration of the crack growth period before a complete failure can be anticipated. It should be recalled that the result of such a prediction is depending on the crack growth rate data of the material, the stress-intensity factors used, and the load spectrum. The specimens used to obtain the basic crack growth data under CA loading should be representative for the conditions of the structure. It implies that the type of material should be the same, and preferably also the thickness of the material. The stress intensity factors, if not available in the literature, should be estimated or calculated. The load spectrum in service must also be available, either by analysis or measurements.

An important and practical question is whether a non-interaction prediction can be considered to be satisfactory. Two issues should be mentioned here. First, crack growth in service may be faster than predicted, due to a corrosive environment or an other time dependent phenomenon. This problem was already addressed in Chapter 8 because it also applies to crack growth predictions under CA loading. Safety factors can be used to account for environmental effects, while exploratory tests on the environmental effect can be carried out for further guidance. The problem is considered again in Chapter 16. Second, another obvious problem is associated with the load spectrum. It is very well possible that a non-interaction prediction is highly conservative, the more so if infrequent high loads are part of the load spectrum. Engineers do not like to be conservative if it is not necessary. The following problem can arise: a non-interaction prediction gives a result with an unsatisfactory margin of safety, whereas a prediction accounting for crack closure gives a much longer crack growth life which could be sufficient. Whether this result is acceptable depends on the reliability of the prediction model. In such a case, it must be recommended to verify the prediction by selected service-simulation fatigue tests.

Aspects of the validity of prediction models

Prediction models for fatigue crack growth under VA loading have been discussed in Section 11.4. It was said that the empirical verification is often limited. Furthermore, the verification is sometimes presented as a comparison between predicted and empirical crack growth lives, a

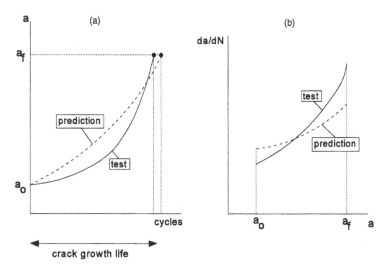

Fig. 11.21 Comparison between test result and prediction. Good agreement between crack growth lives, poor agreement between crack growth rates.

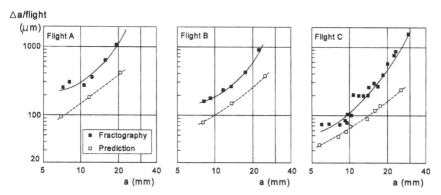

Fig. 11.22 Values of Δa in the most severe flights (A, B and C) of a flight-simulation fatigue test (miniTWIST) on an Al-alloy sheet specimen (2024-T3). Fractographic observations of the electron microscope (SEM) compared to predictions by the modified CORPUS model [45].

comparison used in Figure 11.19. However, as illustrated by Figure 11.21, a good agreement between crack growth lives (Figure 11.21a) can also be a result of an initially too high prediction of the crack growth rate da/dN, which is compensated later by a too low prediction (Figure 11.21b). Verifications of prediction models should include a comparison of predicted and empirical crack growth rates. A still more detailed comparison can be made by fractographic striation measurements in the electron microscope. Such a comparison was made for fatigue crack grown under flight-simulation loading [45]. The crack extension in each of the more severe flights could be determined in the SEM by fractographic analysis, which required some skill and experience. As shown by the results in Figure 11.22, the crack extension in the more severe flights was considerably larger than predicted by the modified CORPUS model, although the agreement between the macroscopic crack growth rates, predicted and measured, was good. This is not inconsistent because the number of most severe flights in a flight-simulation test is small. An incorrect prediction for these severe flights has a small effect on the overall prediction. It implies that the crack growth prediction was satisfactory from an engineering point of view, but not from a physical point of view. A physical verification of a prediction model also requires a comparison on a microscopic level. Recall that such observations were essential to reveal the occurrence of delayed retardation in Figure 11.9.

Two arguments can be considered to explain the discrepancy for the severe flights in Figure 11.22; first, the larger Δa for primary plastic deformation discussed below, and second, an incompatible crack front orientation.

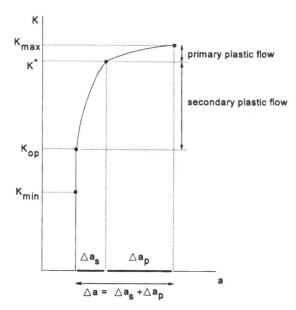

Fig. 11.23 Different crack increments during secondary and primary plastic deformation of crack tip plastic zone. Concept of De Koning [44].

As part of the CORPUS model, De Koning [26] introduced primary plastic deformation and secondary plastic deformation. Primary plastic deformation occurs at the crack tip if plastic deformation penetrates into elastic material which has not yet seen any plastic deformation by previous load cycles. Secondary plastic deformation refers to crack tip plasticity that remains inside a primary plastic zone. More recently De Koning and Dougherty [44] have proposed that crack extension during primary plastic deformation is much more effective than during secondary plastic deformation. In a load cycle, plastic deformation and crack extension will always start with secondary plastic deformation, see Figure 11.23.[15] If S_{max} is high enough, primary plastic deformation will occur after $K > K^*$. Crack extension, as long as $K < K^*$, will occur in agreement with a Paris type equation and a related crack growth mechanism. Above K^* crack extension will occur as a kind of stable crack growth under a quasi-statically increasing load. This could be a sound idea, which is supported by some empirical evidence, including fractographic observations [46, 47].

It may be expected that crack growth prediction models for VA loading will see further developments in the future. Also, more empirical verification

---

[15] In [44] a threshold level, slightly above K_{op}, is introduced. It is omitted here because it is not essential for the present discussion.

programs should validate wider applications to more load spectra and materials. This should also show which empirical constants are essential for tuning prediction models for specific applications. In view of the present qualitative understanding about interaction effects and the fatigue mechanism, the development of a generally valid and quantitative prediction model is still a problem for the future.

11.6 Major topics of the present chapter

The present chapter is dealing with fatigue crack growth of macrocracks under VA loading.

1. Significant interaction effects can occur during fatigue crack growth under VA loading. It implies that the crack growth rate (da/dN) in a cycle is dependent on the load history of the preceding cycles, and it is not necessarily the same as in a CA test.

2. A load cycle with a high S_{max} (an overload, OL) can significantly reduce the crack growth rate in subsequent cycles (positive interaction effect). A load cycle with a low S_{min} (an underload, UL) can slightly increase the crack growth rate, while it can also reduce the retardation effect of previous OLs (negative interaction effects). In general, the positive interaction effects will overrule the negative ones during crack growth under service load spectra. As a result, non-interaction predictions for fatigue crack growth under VA loading will usually give conservative results.

3. Plasticity induced crack closure is a significant phenomenon to explain interaction effects. Experiments with simple VA-load histories have essentially contributed to understanding these effects. Similar interaction effects occur during more complex service simulating load histories.

4. Larger interaction effects occur in materials with a relatively low yield stress and in thinner materials, due to larger crack tip plastic zones.

5. Three types of prediction models for fatigue crack growth under VA loading have been proposed in the literature; yield zone models, crack closure models and strip yield models. The yield zone models do not agree with the present knowledge of interaction effects. The crack closure models account for the occurrence of plasticity induced crack closure and the transition of plane strain to plane stress. The strip yield models are the most sophisticated models, which include calculations

of plasticity induced crack closure. These models predict delayed retardation after an OL. The calculation algorithm of the strip yield models is complicated.

6. Empirical verification of prediction models is unfortunately rather limited. Verifications should not be restricted to predictions on crack growth lives, but should include predictions on crack growth rate as a function of crack length. Non-interaction predictions should also be made to indicate whether significant interaction effects occurred. Fractographic observations are recommended for investigations on prediction models.

7. Crack growth predictions for practical engineering problems should be validated by service-simulation fatigue tests.

References

1. Schijve, J., *Fatigue crack propagation in light alloy sheet material and structures.* Advances in Aeronautical Sciences, Vol. 3, Pergamon Press (1961), pp. 387–408.

2. Schijve, J., *Observations on the prediction of fatigue crack propagation under variable-amplitude loading.* Fatigue Crack Growth under Spectrum Loads, ASTM STP 595 (1976), pp. 3–23.

3. Mills, W.J. and Hertzberg, R.W., *The effect of sheet thickness on fatigue crack retardation in 2024-T3 aluminum alloy.* Engrg. Fracture Mech., Vol. 7 (1975), pp. 705–711.

4. Petrak, G.S., *Strength level effects on fatigue crack growth and retardation.* Engrg. Fracture Mech., Vol. 6 (1974), pp. 725-733.

5. Dahl, W. and Roth, G., *On the influence of overloads on fatigue crack propagation in structural steels.* Paper Technical University Aachen (1979).

6. Ling, M.R. and Schijve, J., *Fractographic analysis of crack growth and shear Lip development under simple variable-amplitude loading.* Fatigue Fract. Engng Mater. Struct., Vol. 13 (1990), pp. 443–456.

7. Schijve, J., *Four lectures on fatigue crack growth.* Engrg. Fracture Mech., Vol. 11 (1979), pp. 176–221.

8. Chermahini, R.G., Shivakumar, K.N. and Newman, Jr., J.C., *Three dimensional finite-element simulation of fatigue-crack growth and closure.* Mechanics of Fatigue Crack Closure. J.C. Newman, Jr. and W. Elber (Eds.), ASTM STP 982 (1988), pp. 398–413.

9. Grandt, A.F., *Three-dimensional measurements of fatigue crack closure.* NASA-CR-175366, Washington (1984).

10. Sunder, R. and Dash, P.K., *Measurement of fatigue crack closure through electron microscopy.* Int. J. Fatigue, Vol. 4 (1982), pp. 97–105.

11. McEvily, A.J., *Current Aspects of Fatigue.* Appendix: Overload Experiments. Fatigue 1977 Conference, University of Cambridge (1977).

12. Schijve, J., *Fatigue damage accumulation and incompatible crack front orientation.* Engrg. Fracture Mech., Vol. 6 (1974), pp. 245–252.

13. Stubbington, C.A. and Gunn, N.J.F., *Effects of fatigue crack front geometry and crystallography on the fracture toughness of an Ti-6Al-4V alloy.* Roy. Aero. Est., TR 77158, Farnborough (1977).

14. Ryan, N.E., *The influence of stress intensity history on fatigue-crack growth.* Aero. Research Lab., Melbourne. Report ARL/Met. 92 (1973).

15. Schijve, J., *Effect of load sequences on crack propagation under random and program loading.* Engrg. Fracture Mech., Vol. 5 (1973), pp. 269–280.

16. Saff, C.R. and Holloway, D.R., *Evaluation of crack growth gages for service life tracking.* Fracture Mech., R. Roberts (Ed.), ASTM STP 743 (1981), pp. 623-640.

17. Schijve, J., *Fundamental and practical aspects of crack growth under corrosion fatigue conditions.* Proc. Inst. Mech. Engrs., Vol. 191 (1977), pp. 107–114.

18. Unpublished results, National Aerospace Laboratory NLR, Amsterdam.

19. Wanhill, R.J.H., *The influence of starter notches on flight simulation fatigue crack growth.* Nat. Aerospace Lab. NLR, Amsterdam, Report MP 95127 (1995).

20. Broek, D., *Elementary Engineering Fracture Mechanics*, 4th edn. Martinus Nijhoff Publishers, the Hague (1985).

21. Schijve, J., Vlutters, A.M., Ichsan, S.P. and ProvoKluit, J.C., *Crack growth in aluminium alloy sheet material under flight-simulation loading.* Int. J. Fatigue, Vol. 7 (1985), pp. 127–136.

22. Schijve, J., *The significance of flight simulation fatigue tests.* Durability and Damage Tolerance in Aircraft Design, A. Salvetti and G. Cavallini (Eds.), EMAS, Warley (1985), pp. 71–170.

23. Willenborg, J. Engle, R.M. and Wood, H.A., *A crack growth retardation model using an effective stress concept.* AFFDL-TR71-1, Air Force Flight Dynamic Laboratory, Wright-Patterson Air Force Base (1971).

24. Wheeler, O.E., *Spectrum loading and crack growth.* J. Basic Engrg., Vol. 94 (1972), pp. 181–186.

25. Baudin, G. and Robert, M., *Crack growth life time prediction under aeronautical type loading.* Proc. 5th European Conf. on Fracture, Lisbon (1984), pp. 779–792.

26. de Koning, A.U., *A simple crack closure model for prediction of fatigue crack growth rates under variable-amplitude loading.* Fracture Mechanics, R. Roberts (Ed.), ASTM STP 743 (1981), pp. 63–85.

27. Padmadinata, U.H., *Investigation of crack-closure prediction models for fatigue in aluminum sheet under flight-simulation loading.* Doctor Thesis, Delft University of Technology, Delft (1990).

28. Aliaga, D. Davy, A. and Schaff, H., *A simple crack closure model for predicting fatigue crack growth under flight simulation loading.* Durability and Damage Tolerance in Aircraft Design, A. Salvetti and G. Cavallini (Eds.). EMAS, Warley (1985), pp. 605–630.

29. Padmadinata, U.H. and Schijve, J., *Prediction of fatigue crack growth under flight-simulation loading with the modified CORPUS model.* Advanced Structural Integrity Methods for Airframe Durability and Damage Tolerance, C.E. Harris (Ed.). NASA Conf. Publ. 3274 (1994), pp. 547–562.

30. Ichsan S. Putra, *Fatigue crack growth predictions of surface cracks under constant-amplitude and variable-amplitude loading.* Doctor thesis, Delft University of Technology (1994).

31. Newman Jr., J.C. and Armen, H., *Elastic-plastic analysis of a propagating crack under cyclic loading.* AIAA J., Vol. 13 (1975), pp. 1017–1023.

32. Ohji, K., Ogura, K. and Ohkubo, Y., *Cyclic analysis of a propagating crack and its correlation with fatigue crack growth.* Engrg. Fracture Mech., Vol. 7 (1975), pp. 457–463.

33. Dugdale, D.S., *Yielding of steel sheets containing slits.* J. Mech. Phys. Solids, Vol. 8 (1960), pp. 100–104.

34. Führing, H. and Seeger, T., *Structural memory of cracked components under irregular loading.* Fracture Mechanics, C.W. Smith (Ed.). ASTM STP 677 (1979), pp. 1144–1167.

35. Führing, H. and Seeger, T., *Dugdale crack closure analysis of fatigue cracks under constant amplitude loading.* Engrg. Fracture Mech., Vol. 11 (1979), pp. 99–122.

36. Dill, H.D. and Saff, C.R., *Spectrum crack growth prediction method based on crack surface displacement and contact analysis.* Fatigue Crack Growth under Spectrum Loads, ASTM STP 595 (1976) pp. 306–319.

37. Dill, H.D., Saff, C.R. and Potter, J.M., *Effects of fighter attack spectrum and crack growth.* Effects of Load Spectrum Variables on Fatigue Crack Initiation and Propagation, D.F. Bryan and J.M. Potter (Eds.). ASTM STP 714 (1980), pp. 205–217.

38. Newman, Jr., J.C., *A crack-closure model for predicting fatigue crack growth under aircraft spectrum loading.* Methods and Models for Predicting Fatigue Crack Growth under Random Loading, J.B. Chang and C.M. Hudson (Eds.). ASTM STP 748 (1981), pp. 53–84.

39. Dougherty, D.J., de Koning, A.U. and Hillberry, B.M., *Modelling high crack growth rates under variable amplitude loading.* Advances in Fatigue Lifetime Predictive Techniques, ASTM STP 1122 (1992), pp. 214–233.

40. Wang, G.S. and Blom, F. *A strip model for fatigue crack growth predictions under general load conditions.* Engrg. Fract. Mech., Vol. 40, (1991), pp. 507–533.

41. Bos, M.J., *Development of an improved model for the prediction of fatigue crack growth in helicopter airframe structure.* Proc. 24th ICAF Symposium, Naples, 16–18 May 2007, Vol. 1, L. Lazzeri and A. Salvetti (Eds.) (2007).

42. Skorupa, M. and Machniewicz, T., *Some results on the strip yield model performance.* Paper presented at the seminar held at TU Delft, 12 June 2007. To be published.

43. de Koning, A.U. and Liefting, G., *Analysis of crack opening behavior by application of a discretized strip yield model.* Mechanics of Fatigue Crack Closure, J.C. Newman, Jr. and W. Elber (Eds.). ASTM STP 982 (1988), pp. 437–458.

44. de Koning, A.U. and Dougherty, D.J., *Prediction of low and high crack growth rates under constant and variable amplitude loading.* Fatigue Crack Growth under Variable Amplitude Loading, J. Petit et al. (Eds.). Elsevier (1989), pp. 208–217.

45. Siegl, J., Schijve, J. and Padmadinata, U.H., *Fractographic observations and predictions on fatigue crack growth in an aluminium alloy under miniTWIST flight-simulation loading.* Int. J. Fatigue, Vol. 13 (1991), pp. 139–147.

46. Schijve, J., *Fundamental aspects of predictions on fatigue crack growth under variable-amplitude loading.* Theoretical Concepts and Numerical Analysis of Fatigue, A.F. Blom and C.J. Beevers (Eds.). EMAS (1992), pp. 111–130.

47. Schijve, J., *Fatigue crack growth under variable-amplitude loading.* Fatigue and Fracture, American Society for Materials, Handbook Vol. 19, ASM (1996), pp. 110–133.

Some general references (see also [46, 47])

48. Skorupa, M., Machniewicz, T. and Skorupa, A., *Applicability of the ASTM compliance offset method to determine crack closure levels for structural steel.* Int. J. Fatigue, Vol. 29 (2007), pp. 1434–1451.

49. Skorupa, M. and Skorupa A., *Experimental results and predictions on fatigue crack growth in structural steel.* Int. J. Fatigue, Vol. 27 (2005), pp. 1016–1028.

50. Skorupa, M., *Load interaction effects during fatigue crack growth under variable-amplitude loading. A literature review. Part I: Empirical trends. Part II:*

Qualitative interpretations. Fatigue Fract. Engrg. Mater. Struct., Vol. 21 (1999) pp. 987–1006, Vol. 22 (1999), pp. 905–926.

51. Newman, Jr., J.C., Phillips, E.P. and Everett, Jr., R.A., *Fatigue analysis under constant- and variable-amplitude loading using small-crack theory.* NASA/TM-1999-209329 (1999).

52. Harter, J.A., *Comparison of contemporary fatigue crack growth (FGG) prediction tools.* Int. J. Fatigue, Vol. 21 (1999), pp. S181–S185.

53. Newman, Jr., J.C., Wu, X.R., Swain, M.H., Zhao, W., Phillips, E.P. and Ding, C.F., *Small-crack growth and fatigue life predictions for high-strength aluminium alloys. Part II: Crack closure and fatigue analyses.* Fatigue and Fracture of Engineering Materials and Structures, Vol. 23 (1999), pp. 59–72.

54. Wu, X.R., Newman, Jr., J.C., Zhao, W., Swain, M.H., Ding, C.F., and Phillips, E.P., *Small-crack growth and fatigue life predictions for high-strength aluminium alloys. Part I: Experimental and fracture mechanics analis.* Fatigue Fract. Engrg. Mater. Struct., Vol. 21 (1998), pp. 1289–1306.

55. Mitchell, M.R. and Landgraf. R.W. (Eds.), *Advances in Life Time Predictive Techniques*, ASTM STP 1122 (1992).

56. Newman, Jr., J.C. and Elber, W. (Eds.), *Mechanics of Fatigue Crack Closure*, ASTM STP 982 (1988).

57. *Fatigue Crack Growth under Spectrum Loads*, ASTM STP 595 (1976).

Part III
Fatigue Tests and Scatter

Chapter 12
Fatigue and Scatter

12.1 Introduction

Scatter is an inherent characteristic of mechanical properties of structures and materials. This also applies to fatigue properties. The fatigue lives of similar specimens or structures under the same fatigue load can be significantly different. Also, the fatigue limit of one and the same material is subjected to scatter. In the literature, statical aspects of fatigue of structures and materials are well recognized, but the implications for engineering problems are not always clear. In this chapter, various sources for scatter of fatigue are discussed first (Section 12.2). These sources can be essentially different for the crack initiation period and crack growth period. The description of the statistical variability is addressed in Section 12.3, including how to obtain experimental information about scatter. A special issue is scatter of the fatigue limit. Various engineering aspects of scatter are discussed in Section 12.4. The major topics of this chapter are recalled in Section 12.5.

12.2 Sources of scatter

The fatigue life covers a crack initiation period and a crack growth period. Differences between the fatigue mechanisms in the two periods are discussed

Table 12.1 Various sources of scatter.

Aspects considered	Possible sources of scatter	
	Laboratory test series	Structures in service
Material	Material structure	Material from different batches and manufacturers
Production	Specimen production	Production of structures over years
	Specimen surface quality	Surface quality of fatigue critical notches in structure
Fatigue load	Type of fatigue load (CA, VA) Test frequency Accuracy of test equipment	Load spectra in service Different users of structure Fatigue life covers years
Environment	Laboratory environment, controlled temperature and humidity	Service environment, possibly aggressive
Personal aspects	Skill of laboratory technicians	Different users

in Chapter 2. It was shown that crack initiation, including the first microcrack growth, is primarily a material surface phenomenon. As a result, the crack initiation period is strongly dependent on the material surface conditions. In the second period, the crack growth period, crack extension is hardly depending on the material surface condition, but predominantly on the crack growth resistance of the material as a bulk property of the material. It should thus be expected that sources of scatter are different for the crack initiation period and the crack growth period. It is generally recognized that the fatigue life of the initiation period is much more sensitive to various influences. Scatter of this life period can be large. Fatigue crack growth in the second period shows a limited variability. In the present chapter, scatter of fatigue life is considered, but some attention is also paid to scatter of crack growth.

Statistical information about fatigue properties is mainly coming from laboratory investigations, and not from service experience. It thus is useful to consider various potential sources for scatter for both types of circumstances. A survey is given in Table 12.1.

More aspects could have been listed in the table, but the purpose here is to illustrate that essentially different circumstances can occur in service and in the laboratory. As a result, the variability of fatigue properties in service cannot be simply related to scatter observed in the laboratory. Actually, in laboratory investigations, it is generally tried to eliminate various sources of scatter because the purpose is to study a fatigue problem without obscuring the findings by scatter of the data. It implies that specimen production is done most carefully from a single batch of material, aiming at a uniform and fine surface quality. Furthermore, fatigue tests are carried out under closely controlled conditions.

Effects of different circumstances of industrial production and structures in service have been studied in laboratory investigations, e.g. the effect of the surface finish quality. Such investigations have contributed to understand various practical sources of scatter. The understanding of scatter goes back to the fatigue mechanisms discussed in Chapter 2.

12.3 Description of scatter

Some statistical concepts used in this chapter are: distribution function, probability density function, mean value, standard deviation and probability level. These concepts are discussed in text books on statistics. Definitions are also presented in ASTM STP 91-A [1] and the ESDU data sheets [2].

Fatigue life results of an old investigation [3] with a large number of tests on unnotched rotating beam specimens ($S_m = 0$) are presented in Figure 12.1. Scatter is very large in these test series, and it may well be

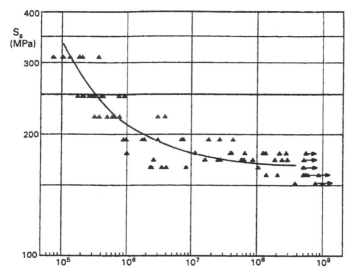

Fig. 12.1 Much scatter in an older test program of rotating beam fatigue tests on unnotched 7075-T6 specimens [3].

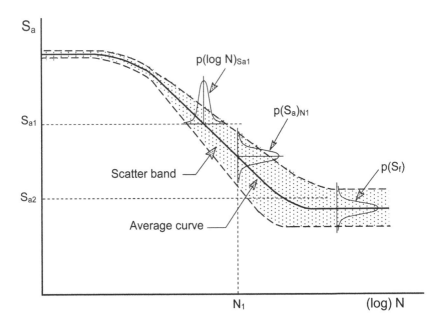

Fig. 12.2 Narrow scatter band at a high S_a, wide scatter band at a low S_a.

expected that several unfavorable conditions have contributed. Results of a much smaller number of experiments, previously shown in Figure 6.3, do not show a similarly large scatter. However, both figures show that scatter is less at high stress amplitudes, and larger at low stress amplitudes. In other words, the scatter band is narrower at a high S_a and wider at a low S_a. The wider band near the fatigue limit is also the result of specimens that do not fail after a very high number of load cycles (so-called run-outs). This scatter behavior is depicted in Figure 12.2. With reference to Chapter 2, the variation of scatter can be understood. At a high S_a-value, surface condition are less important for crack nucleation because microcracks are initiated early in the fatigue life. It is followed immediately by further crack growth. As a result, scatter will be relatively low. However, at a low S_a-value, crack nucleation and the first microcrack growth is meeting with material structural barriers. Nucleation can be dependent on local surface inhomogeneities and small surface irregularities, or even slight surface damage. These surface conditions can vary from specimen to specimen, and have a significant effect on the duration of the initiation period. As a result, more scatter is found at high endurances.

Statistical Distribution of Fatigue Life

Information on the distribution function of the fatigue life, N (or $\log N$) should be obtained by carrying out a large number of similar experiments at the same stress level. The probability density function, $p(\log N)_{S_{a1}}$ in Figure 12.2, represents the distribution of $\log(N)$ at the selected stress amplitude S_{a1}. Similarly, $p(S_a)_{N_1}$, also depicted in Figure 12.2, is the probability density function of the fatigue strength at the selected fatigue life N_1. A special function of the latter type is the distribution function of the fatigue limit, S_f. Scatter of the fatigue limit is of engineering interest if all cycles of a load spectrum have to be kept below the fatigue limit.

An intriguing question is: what is the distribution function of the fatigue strength? Especially, the distribution function of fatigue life has received considerable attention in the literature. Unfortunately, the function cannot be derived on the basis of physical arguments. In general, the function is simply assumed, or adjusted to experimental data of large test series. Constants in the function are derived from experimental data. Two popular distributions are the normal or Gaussian distribution and the Weibull distribution. The well-known normal distribution function is

$$P(x) = \frac{1}{\sigma\sqrt{2\pi}} \int_{-\infty}^{x} e^{-\frac{1}{2}[\frac{v-\mu}{\sigma}]^2} dv \tag{12.1}$$

with μ as the mean value and σ as the standard deviation of the variable x, while v is the integral variable. Consideration of scatter of fatigue lives usually start with assuming that the variable of interest is the logarithm of the fatigue life, $x = \log(N)$. The values of μ and σ are estimated from the results of a series of m similar experiments (a statistical sample) by calculating

$$\mu = \frac{1}{m}\sum_{1}^{m} x_i \quad \text{and} \quad \sigma = \sqrt{\frac{\sum_{i}^{m}(x_i - \mu)^2}{m-1}} \tag{12.2}$$

The normal distribution covers an interval from $-\infty$ to $+\infty$. The lower limit implies $x = \log(N) = -\infty$, or $N = 0$. A zero fatigue life is physically impossible. This discrepancy does not occur in the three-parameter Weibull distribution function:

$$P(x) = 1 - e^{-[\frac{x-x_0}{a}]^b} \tag{12.3}$$

In this equation, x_0 is the location parameter (values of x lower than x_0 are impossible, x_0 is the lower limit), a is the scale parameter, and b is the shape parameter. These three constants are determined from the test results by optimizing the correlation between the test results and Eq. (12.3). It requires an iterative calculation procedure [4].

The normal distribution function and the Weibull distribution function both have an upper limit at $x = \infty$, which is physically a strange result. However, the upper limit is of less practical interest. The lower limit, where probabilities of failure are low, is more significant for engineering problems associated with safety factors.

The validation of the two above distribution functions requires large test series. It is difficult to discriminate between the two distribution functions if the number of test data of nominally similar tests is small. An illustration is presented in Figure 12.3 with results of a series of 18 similar experiments on an unnotched specimen. A probability of failure $P(x)$ must be allotted to each result. For that purpose, the results are ranked in an increasing order of magnitude with rank numbers from $i = 1$ to $i = m$, see the inset table in Figure 12.3. The statistical estimate for $P(x)$ is:[16]

$$P(x_i) = \frac{i - \frac{1}{2}}{m} \tag{12.4}$$

The results are plotted in Figure 12.3 on so-called normal probability paper. The vertical scale of this paper is transformed to obtain the normal distribution as a linear relation with a slope dependent on the standard deviation σ. The test results in Figure 12.3 agree reasonably well with the normal distribution function. However, the results agree equally well with the Weibull distribution function also shown in the same figure by a dotted line. Differences between the two functions occur for $P(\log N) < 0.01$ and $P(\log N) > 0.99$, i.e. in the tails of the distribution function. In Figure 12.3, considering only 18 tests, the number of results is too low to discriminate between the two functions. The distribution function is of special interest at very low probabilities of failure, where the Weibull distribution shows a lower limit, whereas the normal distribution does not. If larger numbers of tests are carried out, it might be expected that a tendency to a lower limit of the distribution function becomes more evident. Even then, the question remains how to generalize this observation to practical conditions.

Some more warning comments must be made. First, experiments can never prove that a certain distribution function is applicable. At best, it can be shown that such a function agrees with test data. Second, in different test series, different distribution functions may give the best fit to

---

[16] A slightly better estimate, proposed by Rossow [6], is $P(x_i) = (3i - 1)/(3m + 1)$.

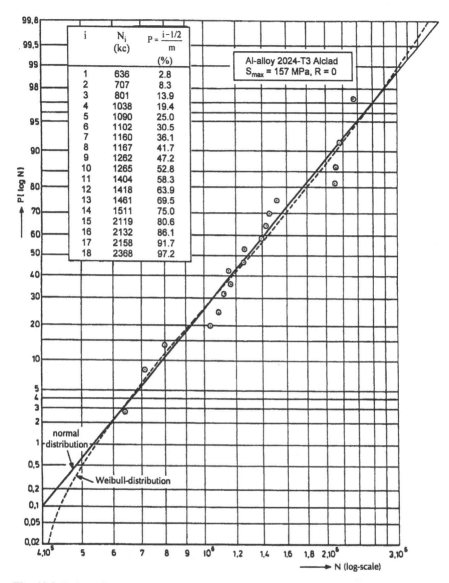

Fig. 12.3 Fatigue lives obtained in a series of 18 similar tests on unnotched specimens, plotted on normal probability paper [5].

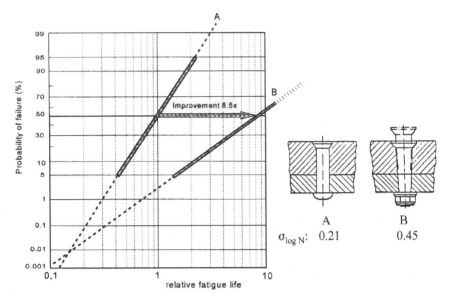

Fig. 12.4 Statistical fatigue life results in comparative tests on two types of fasteners [7].

the test data. Third, the Weibull distribution function is more flexible for data fitting because it has three constants instead of two for the normal distribution function. The third constant, x_0, accounts for the lower boundary of statistically possible fatigue lives, which seems to be more realistic. A lower boundary, N_0, could also be added as a third variable to the normal distribution by defining $x = \log(N - N_0)$ as the statistical variable instead of $x = \log(N)$. However, this is rarely done. Finally, it is physically unrealistic that a fatigue life distribution is continuous until $x \rightarrow \infty$, although this is not so much of practical interest. Low fatigue lives, rather than infinite fatigue lives, are the leading issue in statistical considerations on the fatigue performance of structures in view of the probability of failure of the structure.

Two Case Histories

If a substantial number of similar tests is carried out, the log mean value of the fatigue life is assumed to be a characteristic average value, and the standard deviation of $\log(N)$ gives an indication on scatter in the tests. In comparative test series, the mean values can show different effects for the variables investigated. However, statistical problems can arise if scatter is large. In a comparison between two types of fasteners, see Figure 12.4,

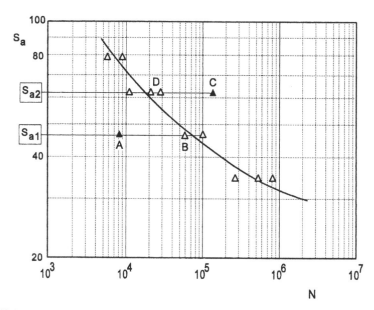

Fig. 12.5 An exceptionally low and high test result (A and C, respectively) in a series of experiments.

scatter was large indeed, see the standard deviations ($S_{\log N}$) in this figure. The taper-lok fastener (B) was developed to be an improvement of the conventional fastener (A). And indeed, the mean life in a test series of 10 specimens was significantly improved, 8.5 times. The results plotted on normal probability paper suggested a reasonable agreement with the normal distribution function, but in view of the number of 10 specimens in each series, this covers a range of probabilities of failure from 5 to 95% only. Extrapolations to lower probability of fatigue are made by the dashed lines in Figure 12.4. They intersect at $P = 0.01\%$, which implies that the taper-lok fastener then should become inferior. Obviously, this conclusion entirely depends on the validity of the assumed normal distribution function, which is questionable. Moreover, was the scatter observed in these test series really representative for scatter to be expected in mass-production? The major lesson to be learned from this test series is that the observed scatter is unacceptable. As long as this scatter cannot be reduced, a conclusion on improvements cannot and should not be drawn.

An other problem associated with scatter is illustrated by the results in Figure 12.5. Test results are plotted with the purpose to arrive at an average S-N curve. The curve in this figure was drawn by hand, while disregarding data points A and C, which are far outside the scatter band of the other results. These unusual data are a matter of concern. The low endurance of data point A could be due to incidental surface damage of the specimen, which would invalidate the test result. This should be checked by examining the specimen fracture surface, e.g. under a binocular optical microscope at magnifications from 10 to 50 times. The high endurance of data point C is not easily explained. If the specimen is a notched specimen, it could be due to an unintentionally applied high load at the beginning of the fatigue test. In statistical terms, it is hard to believe that results A and C are part of the population of the other fatigue lives. Such results can be omitted. An alternative approach is to consider median results of similar tests. In Figure 12.5, the median at stress amplitude S_{a1} is result B, and at S_{a2}, it is about D. Note that approximately the same S-N curve would be drawn when using these median values. Again, statistics is not a tool to solve problems, but to describe scatter. Problems as discussed above must be judiciously handled with understanding of possible influences.

Statistical Distribution of the Fatigue Strength and the Fatigue Limit

If a sufficiently large number of specimens is tested at the same cyclic stress level, a $P(\log N)$ curve can be plotted, a curve as shown in Figure 12.3. Fatigue lives, corresponding to specific $P(\log N)$-values, can be read from such graphs. Characteristic values of $P(\log N)$, used in discussions on failure probabilities, are $P(\log N) = 0.01, 0.05, 0.10$ and 0.50, corresponding to probabilities of failure of 1, 5, 10 and 50% (mean value) respectively; or to probabilities of survival of 99, 95, 90 and 50% respectively. If this is done at several stress levels, S-N curves for certain probabilities of failure can be drawn, see Figure 12.6. Such curves are referred to as P-S-N curves. These curves can only be drawn if large numbers of fatigue tests are carried out at several stress levels, which is expensive. Usually, an average S-N curve is drawn through a limited number of data, see Figure 6.3 as an example. The average curve is associated with $P = 50\%$, i.e. the average probability of failure.

A vertical cross plot of P-S-N curves gives a distribution function of the fatigue strength, $P(S_a)_N$ for a certain fatigue life N. The probability density function of such a function was already depicted in Figure 12.2 for N_1. It should be noted that a direct experimental determination of $P(S_a)_N$ is impossible. The result of a fatigue test is a fatigue life for an applied stress level, but it cannot be a stress amplitude for a selected life.

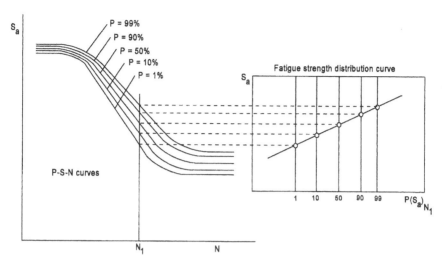

Fig. 12.6 P-S-N curves with a cross plot to obtain a fatigue strength distribution for a constant fatigue life.

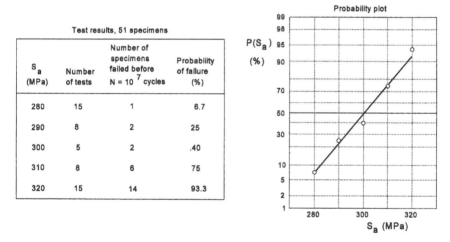

Test results, 51 specimens

S_a (MPa)	Number of tests	Number of specimens failed before $N = 10^7$ cycles	Probability of failure (%)
280	15	1	6.7
290	8	2	25
300	5	2	.40
310	8	6	75
320	15	14	93.3

Fig. 12.7 The Probit method to determine the distribution function of the fatigue limit [1].

The P-S-N curves for low S_a-values, close to the fatigue limit, are difficult to obtain by such a cross plot. Several specimens with a very high endurance are not tested until failure, but stopped at a selected high number of cycles, e.g. 10^7. As a result, the distribution function of the fatigue limit cannot be derived in this way. An alternative procedure is the Probit method [1]. Specimens are tested at a number of stress levels around the anticipated fatigue limit. Tests are stopped at a high N-value to be associated with the fatigue limit, say $N = 10^7$ cycles. Results in the table of Figure 12.7 are

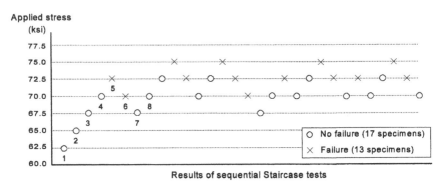

Fig. 12.8 The Staircase method to determine the fatigue limit [1].

shown as an example. At five stress levels the percentage of failed specimens is recorded, which is the estimated probability of failure at that stress level. These probabilities are then plotted on probability paper to obtain the distribution function of the fatigue limit associated with $N = 10^7$ cycles. The approximately linear relation indicates a mean value at $P(S_f) = 50\%$, equal to 300 MPa. The standard deviation derived from the slope is 14 MPa, which is 4.7% of the mean value.

The fatigue limit can also be determined by the so-called Staircase method which requires fewer specimens than the Probit method. Specimens are tested until a high number of cycles to be associated with the fatigue limit, e.g. $N_{sf} = 10^7$ cycles. The first specimen is tested at the estimated stress level of the fatigue limit. If failure does not occur, the test is stopped after N_{sf} cycles, and a second specimen is tested at a higher stress level. However, if failure occurs before N_{sf} cycles are applied, then the second test is carried out at a lower stress level. This procedure is sequentially followed with a number of specimens, see Figure 12.8. The increment between the stress levels is constant (2.5 ksi = 17 MPa in Figure 12.8). A non-failure test is followed by a test at a higher stress level, but if failure does occur, it is followed by a test at a lower stress level. As a result, the stress levels used are going up an down around the fatigue limit for N_{sf} cycles. The mean value and the standard deviation of the fatigue limit can ban be calculated from the test results, see [1]. The accuracy of the mean value is reasonable if some 20 to 30 specimens are used, but this is not true for the standard deviation. The number of specimens is still fairly large although smaller that for the Probit method. A rough estimate of the fatigue limit can be obtained with just a few specimens with a method described in Chapter 13 on testing techniques (see Figure 13.2).

It should be mentioned here that a determination of P-S-N curves requires many specimens. The same is true for the determination of the distribution function of the fatigue limit. It implies that such test programs are very expensive.

12.4 Some practical aspects of scatter

As discussed in Section 12.2, scatter in laboratory experiments and in service can occur for essentially different reasons. Furthermore, data on scatter in laboratory experiments are almost exclusively associated with CA loading whereas load spectra in service frequently apply to VA loading. Some practical consequences are illustrated in this section:

1. The fatigue limit and safety factors.
2. Scatter and VA loading.
3. Scatter of fatigue crack growth.
4. Scatter in different structures of the same type.
5. Symptomatic or incidental fatigue failures in service.
6. Scatter depending on how a structure is used.

Fully rational answers to practical questions about scatter cannot be given, but understanding the fatigue behavior of a structure offers some indications about the significance of scatter.

The Fatigue Limit and Safety Factors

Scatter of the fatigue life is mainly depending on the crack initiation period as explained before. If crack initiation can easily occur, scatter should be expected to be small. This applies to structures with sharp notches (high K_t, poor design). However, if crack initiation is difficult, scatter may offer problems. This can apply to structural elements designed to be free from fatigue failures for a long service life. All cycles of the load spectrum should then remain below the fatigue limit with a certain margin of safety. The situation is relevant to parts of engines, machinery, helicopter components, etc. Such parts are usually made of high-strength alloys, which in general are sensitive to notches and the quality of surface finish. The prediction of fatigue limits was discussed in Chapter 7 with emphasis on accounting for surface and size effects. If these effects are included in the predictions, and perhaps supported by supplemental test verifications, it must be recommended to apply a safety factor on the fatigue limit in order to account for possible scatter. In such a way an allowable design stress level can be obtained. The problem is how to arrive at a reasonable safety factor. This is a delicate problem, and statistical knowledge is not really helpful to select a safety factor. It could be assumed that the fatigue limit has a normal distribution with an estimated standard deviation. A low probability of failure must then be adopted, e.g. $P = 0.001$ (0.1%). This value is often mentioned in safety analyses, but if a designer is asked whether it is acceptable that a failure of an element in a production of 1000 elements could occur, the answer will be negative.

For a normal distribution of the fatigue limit, $P = 0.001$ implies that the estimated fatigue limit must be reduced by 3.09 times the standard deviation σ_{S_f}. For $P = 0.0001$ (0.01%), it would be 3.72 times σ_{S_f}.[17] In the previous example of determining the distribution function of S_f (Figure 12.7), the standard deviation was about 5% of the fatigue limit. The fatigue limit should thus be reduced by $3.09 \times 5 = 15.5\%$ and $3.72 \times 5 = 18.6\%$ respectively. The latter value corresponds to a safety factor of $1/(1-0.186) = 1.23$ on the fatigue limit. This safety factor is not impressive. Engineering judgement is indispensable in arriving at a more realistic value. It must then be recognized which sources of scatter can occur, e.g. variations with respect to the quality of the material, production, surface treatments, and possibly other sources of variability. Furthermore, how sensitive is the component for incidental surface damage. Also, are failures absolutely unacceptable, or could a few failures be allowed economically? As a suggestion, a safety factor of 1.5 could be considered to be a conservative factor for high-quality material and production, provided that quality control is assured.

Scatter and VA Loading

An informative series of tests was carried out in the late fifties [8] with results on scatter in CA and VA tests of riveted lap joints. At that time, service-simulation fatigue tests could not yet be carried out. Several types of program fatigue tests, including tests with different types of OL cycles, were carried out. The results reported in Figure 10.6 were part of this investigation. Each type of test was repeated ten times for CA loading and seven times for VA loading. Scatter was characterized by the standard deviation of $\log(N)$. Values of $\sigma_{\log N}$ are shown as a function of the fatigue life in Figure 12.9. The open data points are from the CA tests. The value

---

[17] These factors can be found in a table on the normal distribution function.

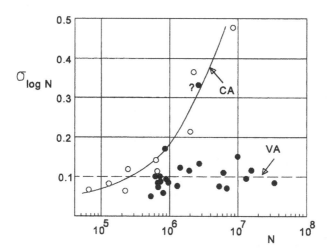

Fig. 12.9 Scatter in fatigue tests on riveted lap joints under CA loading and VA loading [8].

of $\sigma_{\log N}$ obviously increases for an increasing fatigue life, corresponding to a lower S_a-value. This observation agrees with the previous discussion on Figure 12.2. The results for the VA tests are remarkable. All values of $\sigma_{\log N}$ are centered around 0.1, without a noticeable dependence on the fatigue life. It is noteworthy that the average value of $\sigma_{\log N}$ for the VA tests is of the same order of magnitude as $\sigma_{\log N}$ in the CA tests for the largest amplitude of the VA loading. It suggests that the highest amplitude of the load spectrum is responsible for the amount of scatter. This observation is logical because scatter is mainly dependent on the variability of the crack initiation period. Cycles with the larger amplitudes of a VA load spectrum have a predominant effect on the occurrence of the initial crack nucleation. As a consequence, these more severe cycles regulate scatter in the VA tests.

Similar observations on scatter in tests with random or programmed sequences were reported in the literature. Mann [9] suggested that the maximum load of a spectrum will induce local plasticity, which will "smooth out" the influence of small inhomogeneities in the material, and cause load redistributions in the fatigue-critical regions of a built-up structure resulting in a more uniform fatigue response. As pointed out by Jacoby and Nowack [10], life predictions with the Miner rule overestimate the scatter under periodic VA-load histories and apparently gives too much weight to larger scatter of low stress amplitudes. Actually, scatter under VA loading cannot be calculated from scatter observed under CA loading.

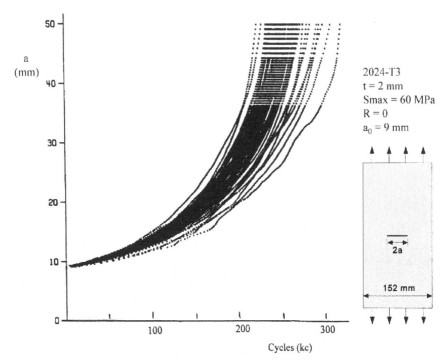

Fig. 12.10 Crack growth curves from 68 similar tests [11]. Scatter is small.

Scatter of Fatigue Crack Growth

Various publications on fatigue crack growth confirm that scatter of the fatigue life, including the crack nucleation period, can be large. By contrast, scatter of fatigue crack growth of visible cracks is generally found to be low. A frequently cited investigation was carried out by Virkler et al. [11], who carried out 68 similar crack growth curves on Al-alloy sheet specimens. The crack growth curves are shown in Figure 12.10. Cracks were started by a small spark eroded central notch. Scatter of the crack initiation was eliminated by normalizing the growth curves on an initial crack length of 9 mm. Figure 12.10 shows that most crack growth curves are concentrated in a narrow band with only a few curves for somewhat slower growth. Considering the shortest and longest crack growth lives in 68 tests, the standard deviation of log(crack growth life) is about 0.03. Standard deviations cited in the literature for the fatigue life, including the nucleation period, $\sigma_{\log(N)}$, are frequently in the range of 0.10 to 0.20 and even larger values are reported (see e.g. the values in Figure 12.9 for CA loading). The standard deviation for the 18 data in Figure 12.3 is $\sigma_{\log(N)} = 0.163$. Thus, the value $\sigma_{\log(N)} = 0.03$ for the crack growth results in Figure 12.10 must be considered to be small.

As explained earlier, scatter is low for fatigue crack growth because it is not dependent on surface conditions. Crack growth resistance then becomes a material bulk property, depending on the material structure. The crack growth resistance can be fairly uniform in a single plate or sheet, and even in plates or sheets from the same batch of material. Unfortunately, statistically significant differences have been found between crack growth lives of nominally similar material from different material manufacturers, and also between different batches of the same manufacturer. This problem was discussed in Chapter 8 and illustrated by the results in Figure 8.16.

The low scatter of fatigue crack growth is advantageous for investigations on crack growth. As little as two specimens for each test condition of a test program may be sufficient, if they show quantitatively the same crack growth curve. In case of doubt, testing of a third specimen is advisable.

Scatter in Different Structures of the Same Type

A service-simulation fatigue test on a full-scale aircraft structure of a transport aircraft with two turboprop engines indicated that some modifications of the wing structure had to be introduced. However, 13 wings were already manufactured and became available for a fatigue test program [12]. The center section of the tension skin (length 8.31 meters) was tested under various load sequences including VA load histories and CA loading. In general, two tension skins were tested under the same fatigue load history. A fatigue critical location occurred in the skin at the top of fingertip reinforcements. A total number of 40 similar fingertips loaded to the same stress level were present in the tension skin located near two skin joints. Inspections on fatigue crack nucleation were made by X-ray picture which allowed a determination of the crack initiation life until a crack with a length of 3 mm was present at the fingertips. These fatigue lives could be determined for a number of fingertips. Results of two similar tests are shown in Figure 12.11 plotted on log-normal probability paper. The tests could not be continued until cracks were present at all fingertips because cracks at other fingertips had grown too far and induced panel failure. However, the results in Figure 12.11 are sufficient to estimate $\sigma_{\log N}$-values of the two similarly tested tension skins. Both values are fairly low, 0.06 and 0.086. More important, the results also indicate that the fatigue lives of the fingertips in the two nominally similar skin structures are not

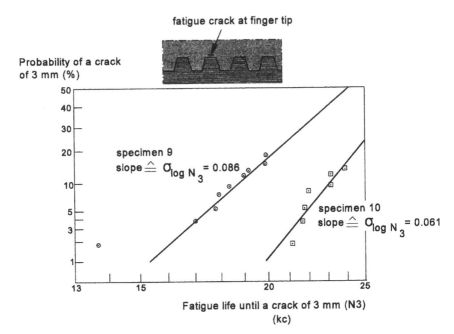

Fig. 12.11 Statistical results from two similar full-scale fatigue tests on an aircraft tension skin. The probability of failure for the crack initiation fatigue life of a crack (3 mm) at 40 similar finger tip reinformcementsin the tension skin. Scatter in each tension skin, and differences between two similar tension skins [12].

statistically identical. The fatigue lives in one tension skin (specimen 10) are systematically larger than for the other tension skin (specimen 9). In other words, the fatigue life distribution function of similar structural details in one structure was different from the distribution function of the other structure. The question then is why this could occur. The answer in this case is not known, but the difference may be due to the material (different batches or even different material producers) and also differences in the production of the two structures. These sources of scatter are avoided in simple specimens prepared for laboratory test programs, but they can apply to industrial products.

Symptomatic or Incidental Fatigue Failures in Service

Williams reported on a catastrophic fatigue failure at a bolt hole in the spar of a Freighter 170 aircraft after 13000 flights in 1957 [13]. Other aircraft of the same type were then inspected to see whether a similar fatigue crack

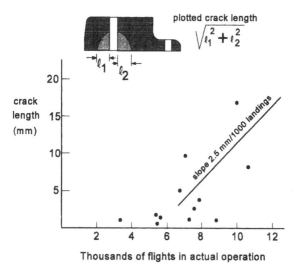

Fig. 12.12 The length of cracks found at a fatigue critical bolt hole in a lower wing spar cap of 12 aircraft [13].

occurred at the same bolt hole. Such an inspection is necessary, also because it should be made clear whether the fatigue crack is a symptomatic problem or an incidental case. Similar cracks were found in several other aircraft with a length shown in Figure 12.12. Apparently, the size of the cracks was larger for older aircraft, i.e. older in numbers of flights. Fatigue cracking thus was a symptomatic phenomenon, but scatter around the average trend can be seen in the graph. It is noteworthy that the average crack growth rate is about 2.5 mm per 1000 flights. In this case, scatter of fatigue crack nucleation and growth between similar aircraft can also be due to a different load spectrum depending on the utilization of the aircraft.

Scatter Depending on How a Structure Is Used

An example of different load spectra for the same aircraft structure is shown in Figure 12.13. The Fokker F28 aircraft of several operators were equipped with a counting accelerometer in the center of gravity of the aircraft. This apparatus counts the number of exceedings of certain acceleration levels which gives an indication of the severity of the load spectrum on the aircraft. The main loads on the aircraft are gust loads in turbulent air. The counting results in Figure 12.13 for four acceleration increments (Δn) are normalized in numbers per flight. Two design load spectra for different take-off weights

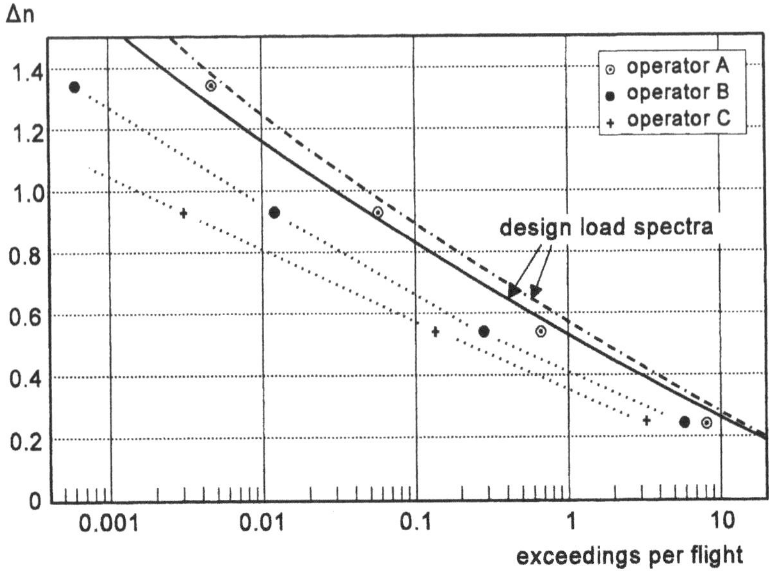

Fig. 12.13 Load spectra for the Fokker F28 aircraft measured in service by three operators with counting accelerometers. Comparison to design spectra.

of the aircraft are also shown in this graph. Counting results are presented for three aircraft operators. It turns out that operator A is encountering the most severe load spectrum which is close to the design load spectra. This operator was flying frequently over high mountains causing a severe load spectrum. Operator B is apparently flying in milder climatic conditions. Operator C meets the most benign load spectrum. Actually, operator C had one single aircraft which was used as an executive aircraft, and executives do not fly in poor weather conditions. Anyway, scatter depending on how a structure is used can be significant for many structures as pointed out in Chapter 9 on load spectra.

12.5 Major topics of the present chapter

1. Scatter of fatigue lives is mainly scatter of the crack initiation period. This period is easily affected by different conditions at the material surface. Scatter of macrocrack growth is significantly smaller.
2. Scatter of fatigue lives is less at high stress amplitudes and more significant at low amplitudes near the fatigue limit.

3. Several important sources of scatter in laboratory investigations and in service are essentially different (Table 12.1). Scatter in service is hardly predictable from scatter observed in laboratory investigations.

4. As a consequence of the previous conclusion, the application of safety factors on the fatigue limit of a structural component is a difficult question.

5. In a statistical analysis of a fatigue problem, a distribution function is usually assumed, but it is difficult to prove that application of the function is reliable.

6. Scatter under VA loading is predominantly controlled by the larger load cycles of the load spectrum. Scatter under VA loading cannot be deduced from scatter data obtained in CA tests.

7. Low scatter is promoted by sharp notches (poor design). Significant scatter is possible for long fatigue lives of carefully designed structural elements (low K_t-values) of high-strength materials. Accounting for scatter should then occur by adopting a suitable safety factor on the design stress level. Selection of this factor requires engineering judgement of all possible sources which can contribute to scatter of the structure in service.

References

1. *A Guide for Fatigue Testing and the Statistical Analysis of Fatigue Data*, 2nd edn. ASTM STP No. 91-A (1963).
2. *The Statistical Analysis of Data from Normal Distributions, with Particular Reference to Small Samples*. ESDU Fatigue Series. No. 91041 (1991) and *An Introduction to the Statistical Analysis of Engineering Data*. No. 92040 (1992).
3. Hardrath, H.F., Utley, E.C. and Guthrie, D.E., *Rotating-beam fatigue tests of notched and unnotched 7075-T6 aluminum-alloy specimens under stresses of constant and varying amplitudes*. NACA Technical Note D-210 (1959).
4. Schijve, J., *A normal distribution or a Weibull distribution for fatigue lives*. Fatigue Fract. Engrg. Mater. Struct., Vol. 16 (1993), pp. 851–859.
5. Schijve, J. and Jacobs, F.A., *Fatigue tests on notched and unnotched clad 24 S-T sheet specimens to verify the cumulative damage hypothesis*. Nat. Aerospace Laboratory, Amsterdam, Report M.1982 (1955).
6. Rossow, E., *A simple slide-rule approximation of normal probability percentages*. Qualitätskontrolle, Vol. 9, No. 12 (1964), pp. 146–147 [in German].
7. Schütz, D., *Planning and analysing a fatigue test programme*. Fatigue Test Methodology, AGARD Lectures Series No. 118: paper 2 (1981).
8. Schijve, J., *The endurance under program-fatigue testing*. Full-Scale Fatigue Testing of Aircraft Structures, F.J. Plantema and J. Schijve (Eds.), Pergamon Press (1961), pp. 41–59.

9. Mann, J.Y., *Scatter in fatigue life – A materials testing and design problem.* Materials, Experimentation and Design in Fatigue, E. Sherratt and J.B. Sturgeon (Eds.), West Bury House (1981), pp. 390–423.

10. Jacoby, G.H. and Nowack, H., *Comparison of scatter under program and random loading and influencing factors.* STP 511, ASTM (1972), pp. 61–72.

11. Virkler, D.A., Hillberry, B.M. and Goel, P.K., *The statistical nature of fatigue crack propagation.* Trans. ASME, J. Engrg. Mat. Technol., Vol. 101 (1979), pp. 148–153.

12. Schijve, J., Broek, D., de Rijk, P., Nederveen, A. and Sevenhuysen, P.J., *Fatigue tests with random and programmed load sequences with and without ground-to-air cycles. A comparative study on full-scale wing center sections.* Nat. Aerospace Lab. NLR, Amsterdam, Report TR S.613 (1965).

13. Williams, J.K., *The airworthiness approach to structural fatigue.* Fatigue Design Procedures, E. Gassner and W. Schütz (Eds.), Pergamon Press (1969), pp. 91–138.

Some general references

14. Marquis, G. and Solin, J. (Eds.), *Fatigue Design and Reliability.* ESIS Publication 23, Elsevier (1999).

15. Veers, P.S., *Statistical considerations in fatigue.* Fatigue and Fracture, American Society for Materials, Handbook Vol. 19, ASM (1996), pp. 295–302.

16. Maennig, W.-W., *Planning and evaluation of fatigue tests.* Fatigue and Fracture, American Society for Materials, Handbook Vol. 19, ASM (1996), pp. 303–313.

17. Schijve, J., *Fatigue predictions and scatter.* Fatigue Fract. Engrg. Mater. Struct., Vol. 17 (1994), pp. 381–396.

18. Tanaka, T., Nishijima, S. and Ichikawa, M. (Eds.), *Statistical Research on Fatigue and Fracture.* Elsevier Applied Science (1987).

19. *Statistical analysis of linear or linearized stress-life (S-N) and strain-life (ε-N) fatigue data.* ASTM Standard E739-80 (reapproved 1986).

20. Little, R.E. and Jebe, E.H., *Manual on statistical planning and analysis.* ASTM STP 588 (1975).

21. Heller, R.A. (Ed.), *Probabilistic Aspects of fatigue.* STP 511 (1972).

Chapter 13
Fatigue Tests

Experiments never lie,
but you should ask the right question!

13.1 Introduction

Fatigue tests are carried out for different purposes. The engineering objectives are the determination of fatigue properties of materials, joints, structural elements, etc., including comparisons of different design options. Research objectives of fatigue tests are concerned with understanding of the fatigue phenomenon and its variables. Research objectives and engineering objectives may be complementary.

The variety of fatigue test programs reported in the literature is large, and the number of publication is steadily growing. Different types of fatigue loads, specimens, environments, and test equipment are used. Fatigue tests generally require significant experimental effort and time, which implies that these tests are more expensive than simple tests of several other mechanical properties. Experiments on fatigue problems are supposed to answer questions, while empirical answers are assumed to be more convincing than a theoretical analysis. The saying is: "Experiments never lie". But, if an experiment is not correctly planned to answer the question under consideration, the result can be a right answer to a wrong question.

In view of large investments in experimental fatigue programs, a careful planning of a test program, experimental procedures, and evaluation of the results is required. Planning should always start with an explicit definition of the problem to be investigated. Another saying is: "A fully detailed definition of the problem is already half the solution".

The purpose of the present chapter is to summarize various aspects of planning fatigue test programs. Obviously, a program will be different for ad hoc problems of the industry and for general research on specific subjects. Different purposes of fatigue tests are indicated in Section 13.2 followed by separate sections on specimen selection, fatigue test procedures, and evaluation of test results (Sections 13.3 to 13.5 respectively). Crack growth experiments are considered in a separate section (Section 13.6). The major points of the chapter are again listed in a final section (Section 13.7).

13.2 Purposes of fatigue test programs

The extensive literature on fatigue problems illustrates the large variety of purposes of fatigue investigations. Some categories are:

- Collecting data on material fatigue properties for material selection by the designer.
- Investigations on effects of different surface finishes and production techniques.
- Investigations on joints and other structural elements.
- Investigations on environmental effects.
- Investigations on crack nucleation and crack propagation.
- Verification of fatigue prediction models.

Although other lists can be compiled, it is obvious that the choice of experimental variables will depend on the type of investigation to be carried out. Major variables to be selected are: (i) type of specimen, (ii) fatigue loads, and (iii) testing procedures.

The main purpose of an investigation can be to compare fatigue properties for different conditions, e.g. different surface conditions. It implies comparative fatigue tests. In other test series, the main objective is a determination of specific fatigue properties for a single condition, e.g. the determination of crack growth properties of a material. In this case, it is not a comparative investigation. Last, but not least, tests may have an ad-hoc nature because of questions of industrial applications.

Other investigations are carried out in view of a common interest to know more about the fatigue behavior of materials and structures under certain conditions. This category comprises many fatigue research programs published in the literature. Obviously, various circumstances can affect the choice of test specimen, fatigue load and testing procedures.

13.3 Specimens

Some elementary types of fatigue specimens are presented in Figure 13.1. The specimen with a central hole can be characteristic for radii occurring in a structure. Figure 13.1 also shows three simple types of joints, each having some special characteristic features. The lug joint specimen is representative for load transmission by a bolt or pin. Fretting corrosion can occur inside the hole. In the riveted lap joint, a tension load introduces bending, while fretting between the two sheets can also be important. The welded butt joint is the most simple type of a welded joint. Joints are discussed in more detail in a separate chapter (Chapters 18 and 19) because they are frequently the most fatigue critical elements of a structure.

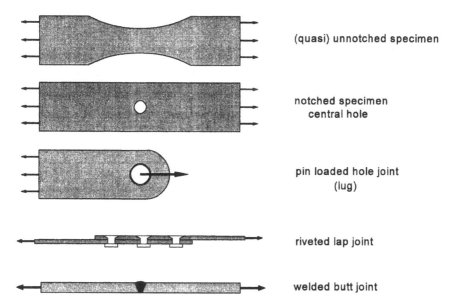

Fig. 13.1 Different types of simple fatigue specimens.

Unnotched specimens

Two aspects of unnotched specimens should be considered. First, fully unnotched specimens with $K_t = 1$ do not exist. Unnotched specimens still have an area where the material is carrying a slightly higher stress than the nominal stress, see the discussion on Figure 6.13. This area is important for size effects. If results of CA tests on unnotched specimens should be used for fatigue predictions or comparative experiments on certain different surface conditions, small specimens should be avoided. In view of possible size effects, these specimens can give misleading results, i.e. a higher fatigue strength or a longer fatigue life.

A second aspect of unnotched specimens is associated with the specimen cross section. The specimens shown in Figure 13.1 are usually manufactured from plate or sheet material. These specimens have a rectangular cross section with corners, but unnotched specimens can also have a cylindrical shape with a circular cross section without corners. Unnotched specimens produced from extrusions or rod material are easily manufactured with a circular cross section, while specimens of plate and sheet material are quite often made by contour milling without reducing the thickness. But even if the thickness is reduced, an unnotched specimen has machined edges with corners at the two specimen surfaces. These corners are a preferential site for fatigue crack nucleation. Theoretically, this should be expected because cyclic slip occurs more easily in grains at the corner. In addition, corners are machined by cutting, which can produce a different machining quality at the corners. Finally, 90° corners are easily damaged, and even minor scratches at the corner can promote local crack initiation. This is particularly true for high-strength alloys which are usually sensitive to the surface finish quality. The initiation of cracks at corners of cross sections can be prevented, or at least discouraged, by smoothing the corners with emery paper.

It is obvious that fatigue tests on unnotched specimens cannot give an indication of the material notch sensitivity, a property of relevant engineering significance. Three other specimens in Figure 13.1 are more informative for this purpose. However, unnotched specimens can be advantageous for problems related to the quality of the surface finish. The material surface is very important for fatigue crack nucleation as discussed in Chapter 2. It implies that a fatigue limit is particularly sensitive for the quality of the material surface as obtained by production techniques used in the industry. The same is true for special surface treatments such as nitriding of steel, see the list in Figure 2.21 and also Chapter 14. It was explained in Chapter 2 that surface effects are relatively small at high stress amplitudes, and much

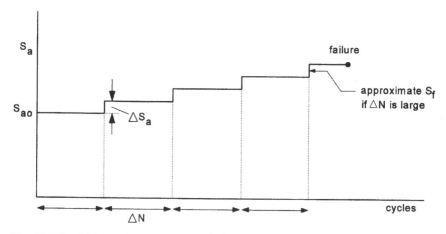

Fig. 13.2 Load history in a step test to obtain an approximate fatigue limit with a single specimen. A small ΔS_a and a large ΔN-value should be adopted.

more significant at low stress amplitudes, see Figure 2.23. Adverse effects can substantially reduce the fatigue limit, while other surface treatments can improve the fatigue limit. Such effects can be indicated by CA tests on unnotched specimens. If an important effect is expected, it is desirable to perform more comparative tests at stress levels close to the fatigue limit. In the previous chapter, it was discussed that many tests may then be necessary in view of scatter. Especially the determination of the fatigue limit requires a large number of specimens. An approximate determination of the fatigue limit can be done with less specimens in so-called step tests in which the stress amplitude is increased with small steps, ΔS_a in Figure 13.2 [1]. The initial S_{a0} should be below an expected fatigue limit. In each step of S_a, a large number of cycles should be applied, e.g. $\Delta N = 2 \times 10^7$ cycles. If failure does not occur at a certain S_a, the amplitude may be expected to be below the fatigue limit. If a specimen fails, the last S_a should be just above the fatigue limit and the previous S_a just below the fatigue limit. A comparison between different surface conditions is then based on the failure stress and number of cycles spent at the failure stress. The estimate of S_f is indicated in Figure 13.2.

Comparative tests to determine surface effects give qualitative indications on these effects. As an example, it can show the favorable effect of shot peening with different peening intensities. If such results give promising improvements for a certain peening intensity, the application to a specific structural element still needs a more realistic verification by experiments. This should preferably be done in service-simulation fatigue tests on notched specimens with a geometry representative for the structural element. It is possible that improvements in notched specimens are quantitatively smaller than in the experiments on unnotched specimens.

Notched specimens

Fatigue properties of candidate materials are considered by a designer for the purpose of selecting a suitable material for a dynamically loaded structure. The question is which fatigue properties must be considered? Fatigue properties provided by the material manufacturer are quite often limited to the fatigue limit of unnotched material for $S_m = 0$ without mentioning the type and size of the specimen used to determine this fatigue limit. Sometimes a broader evaluation of fatigue properties is made by determining S-N curves for $S_m = 0$ or $S_{\min} = 0$ ($R = 0$), generally on unnotched specimens. However, this information does not give indications on the notch sensitivity of a material. Fatigue tests on notched specimens should provide more relevant data. Comparative experiments on candidate materials should preferably be performed on notched specimens. Obviously, a service-simulation fatigue test on the real structural element is the best solution, but that is not easily done in a preliminary stage of a structural design. However, service-simulation fatigue tests on simple notched specimen shown in Figure 13.1 can provide useful information for material evaluations.

If the fatigue limit is a crucial property of a structural element in service, tests on notched specimens should be made with low stress amplitudes. Arguments presented earlier for unnotched specimens are applicable again.

If newly developed materials become commercially available for structural application, all kinds of material properties must be available. Various aspects of durability properties should be known. This includes fatigue properties of specimens with technically relevant notches loaded under various types of fatigue loads characteristic for service load spectra.

Structures

Tests on full size structures or structural elements are realistic tests with respect to the test item. Tests on small structural elements with relatively simple load transmissions to these elements can still be carried out in standard fatigue testing machines. Larger structures, e.g. an automobile, a

Fig. 13.3 Full-scale simulation test on the chassis and coach work of a motorcar with 12 electro-hydraulic actuators providing vertical, lateral and longitudinal inputs at four corners of the test vehicle. (Courtesy MTS)

truck, or an aircraft structure, require more specialized test equipment. In general, the structure is then loaded with a number of hydraulic cylinders with electro-hydraulic closed loop systems monitored by a computer. Depending on the size and the complexity of the loads on the structure, the number of hydraulic cylinders can vary from a small number, e.g. four cylinders, to large numbers in the order of 100 for full-scale fatigue tests on an aircraft structure. A test set-up for a motorcar is shown in Figure 13.3.

Obviously, it would be inconsistent to apply a simple load sequence on a realistic full-scale structure. The load history to be applied should also be realistic. Service-simulation load histories can be applied by computer controlled equipment. Equipment for such purposes can be built up, but it is also commercially available.

In the automotive industry, aircraft industry, and some other industries as well, complex tests are carried out on new designs. In general, the automotive industry is producing large numbers of vehicles, and in principle, fatigue failures in service are unacceptable. A full-scale test with a conservatively selected service-load history should reveal any weakness of the structure in order to modify the structure before it goes into mass production. In fact, such a test is primarily done to see whether all parts of the structure are properly functioning without any deterioration of any part of the structure after a long testing time. In the aircraft industry, full-scale tests are also carried out to prove the safety of the aircraft and to satisfy airworthiness regulations. The occurrence of fatigue cracks in aircraft structures can be accepted, provided that fail-safety is demonstrated. However, also in aerospace, a full-scale fatigue test is carried out to reveal deficiencies of the structure, which then require design modifications.

A noteworthy type of full-scale tests on parts of a car is done in the automotive industry. Many parts of a car are obtained from different sources, and also in different years. Checking the fatigue quality of delivered components is desirable. Variations of the properties of a product are possible, both with respect to material and production technique. These sources of variability and their effects on the fatigue endurance is checked by fatigue experiments in small tests systems. Such tests should run as fast as possible, while usually a simple but conservative load history is used. It is quality control by fatigue experiments.

13.4 Fatigue test procedures

Specimen production

In general, it should be tried to avoid scatter in fatigue test series as much as possible in order to accurately reveal experimental trends of the results. It starts with considerations about specimen production. All specimens for a test program must be made in exactly the same way. Schütz [2] once wanted to repeat fatigue tests on a riveted lap joint. New specimens were ordered from the same industry according to the same drawing. He obtained fatigue lives about three times longer than obtained in the previous test series. Schütz observed from the dimensions of the driven rivet heads that riveting of the second series of specimens was done with a significantly larger rivet squeeze force. It has a highly favorable effect on the fatigue life. This is an extreme example of large differences between nominally similar specimens. It should be recommended that investigators visit the workshop to ascertain that all specimens are made in the same way, rather than just sending a drawing to the shop. It is also advisable to prepare a substantial number of spare specimens, which later may turn out to be needed for additional tests in view of unexpected test results.

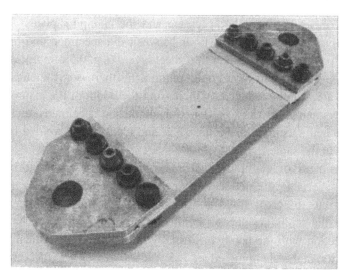

Fig. 13.4 Clamping of a sheet specimen with bolts to steel plates assembled outside the testing machine. Mounting in the machine occurs with a single pin at each end into a clevis.

Clamping of specimens

Modern fatigue testing machines are provided with standard grips which allow an easy installation of specimens in the machine if the specimens have flat ends. Due attention should be given to a correct alignment of the specimen in the testing machine to assure that the central axis of the specimen coincides with the loading axis of the machine [3]. It is recommended to check the alignment on each new type of specimen by strain gage measurements.

Several types of specimens cannot be loaded by the standard clamping method of the fatigue machine. This obviously applies to the lug type specimen in Figure 13.1, which requires a clevis to hold the pin to introduce the load on the hole of the lug. Sheet specimens are often clamped between steel plates with a row of bolts, see Figure 13.4. The bolts are sufficiently torqued to obtain a friction grip connection. The specimen with these plates can then easily be mounted in the fatigue testing machine with a pin-connection. A homogeneous load distribution on the specimen can thus be obtained. Because of the load transmission from the plates to the specimen, a significant stress concentration occurs at the edges op the clamping plates. Moreover, fretting corrosion between the plates and a metal specimen can occur near these edges. If no precautions are made, fatigue failures may be initiated at these edges. Fretting is avoided by preventing metallic contact between the specimen and the clamping plates. This can be done by placing a very thin layer between the specimen and the plates (e.g. wax paper). Also, bonding thin sheet layers (e.g. 0.5 mm thickness) on the clamping areas can be a good solution. Furthermore, the stress in the clamping area can be reduced by increasing the width of the clamping area of the specimen. This must be done for unnotched specimens anyhow, see the unnotched specimen in Figure 13.1, and also for the welded butt joint specimen in the same figure.

The problem of clamping failures is much less serious for notched specimens because the net section at the notch is smaller than the gross area, and thus the gross stress at the clamping is lower than the net stress. Moreover, the notch introduces a significant stress concentration. As a result, the net section with the notch will usually be more fatigue critical than the cross section at the clamping edges.

Sequence of tests of a test program

Fatigue test programs are usually defined in tables, indicating the number of specimens to be tested for each condition to be explored. The test program should start with a single test for each condition. It is possible that these first test results will indicate that the test program must be reconsidered.

Fatigue tests to determine an S-N curve should start with a test at a high stress amplitude. If the tests are started with a low amplitude, the amplitude may be below the fatigue limit. The specimen will not fail (run-out) after a long testing time. The only information gained is that the stress amplitude was below the fatigue limit without knowing how much.

If scatter at a high stress level appears to be low, which is usual, more specimens can be saved for later tests at low amplitudes. Such decisions require an immediate evaluation of the results of each test after it has been completed. Postponing the analysis until all tests have been completed is not a clever approach.

Service-simulation fatigue tests

Service-simulation fatigue tests and results of these test were discussed in Chapters 9 to 11. It was pointed out that load histories for such tests can be generated by computers, which are also regulating the load history in the test. A major problem is how to obtain a load spectrum for the problem to be investigated; and subsequently, how to compose a sequence of minima and maxima of the load spectrum. The approach is different for comparative tests and test associated with a specific design topic of a structure.

Comparative service-simulation fatigue tests can be used to investigate the effect of several variables (materials, surface conditions, production variables). As discussed before, some effects can be explored in CA tests on unnotched specimens. However, if fatigue problems are concerned with specific structural fatigue problems, quantitative indications on the effects are desirable. Service-simulation fatigue tests should then be used. The load history and type of specimen in such comparative tests should be representative for the structure under consideration. As an example, Schütz [4] mentions that the effect of coining (a Douglas technique to introduce favorable compressive residual stress around holes) increased the fatigue life ten times in CA tests. However, under a realistic flight-by-flight load history of the F-104 aircraft, the life increased by a factor of two only.

For several types of structures, service-simulation load histories have been designed which are characteristic for these structures. Standardized load histories are listed in Table 13.1.[18] The load histories are useful for general investigations related to these structures. However, verification of the fatigue performance of a specific structure should not be done with one of the standardized service-simulation load histories. Instead the spectrum and the load sequence must then be related to the real structure as used in service. As discussed in Chapter 9, it can be difficult to develop such a load history. Measurements under service conditions would be most useful, but that is possible only if the structure, or another similar structure is available. If full real-time load histories are recorded, the signal can be used in the fatigue machine. This is done in the automotive industry. Loads on a car are measured at several points with accelerometers or strain gages under realistic service conditions representative of intensive and practical use of the car. In a complex test set-up, the signals are simulated, but it can require a complex computer program to achieve exactly the same load history as measured. Such simulations should include the dynamic behavior of the car including the chassis. The purpose is not only to obtain fatigue lives, but also to observe the response of the structure and to find deficiencies if any.

---

[18] The aerospace and wind turbine blade load histories are available on a CD from the National Aerospace Laboratory NLR, Amsterdam.

Table 13.1 Survey of the standardized service-simulation load histories [5, 6].

Year	Name	Load history for:
1973	TWIST	Transport aircraft lower wing skin
1976	FALSTAFF	Fighter lower wing skin
1977	GAUSSIAN	Random loading
1979	miniTWIST	Shortened TWIST
1983	HELIX/FELIX	Helicopter main rotor blades
1987	ENSTAFF	Tactical aircraft composite wing skin
1987	Cold TURBISTAN	Fighter aircraft engine, cold engine disks
1990	Hot TURBISTAN	Ditto, hot engine disks
1990	WASH	Offshore structures
1990	CARLOS	Car components
1991	WISPER/WISPERX	Horizontal axis wind turbine blades

13.5 Reporting about fatigue test results

The evaluation of results of a test program should include a description of the material, specimens, experiments and results. The material is characterized by its composition, heat treatment, material structure, and mechanical properties. Specimens are described by dimension, workshop practice, and surface finish. Experimental details include the testing machine, clamping of specimens, stress levels, numbers of specimens, test frequency and environment (temperature and humidity); and in addition, a description of special techniques used in the experiments. The environment is often labeled as lab air of room temperature (RT). However, the humidity of lab air can vary from very dry to rather humid. Systematic effects on fatigue crack growth in a aluminium alloys and high-strength steel have been reported, even with different results obtained in summer and winter time, which was attributed to a humidly effect [7]. Incomplete descriptions of test conditions can imply that significant information cannot be retrieved any more after some years when a re-evaluation of the test results appears to be desirable.

Fractographic analysis

The evaluation of the test results should reflect how much has been learned form the experiments. The minimum information of a fatigue test is to present the fatigue life until failure. However, a fatigue test until failure has produced a fatigue fracture surface. Fractographic analysis of a fracture can reveal valuable information contributing to understanding fatigue test results

Fig. 13.5 Semi-elliptical surface cracks in a fatigue failure of a riveted lap joint of sheet material. Crack nucleation occurred away from the rivet hole, see arrows.

and their significance. It is strongly recommended to examine visually the fracture surface of each specimen with the unaided eye and a magnifying glass (six to eight times). Observations will reveal if crack nucleation occurred at the material surface or subsurface, and how many cracks were nucleated. It will also show if crack nucleation started at the most critical section of a notched specimen or away from this section. An example of the latter case is given in Figure 13.5. It shows the initial part of a fatigue failure in a riveted lap joint. Obviously, crack nucleation did not start at the rivet hole, but away from the hole. The question then is why this could happen in spite of a good fatigue performance. Such observations are essential for understanding the fatigue behavior of this type of joints. Another example, associated with an unexpectedly low fatigue test life, was discussed in the previous chapter (Figure 12.5). Fractographic observations can sometimes explain why a poor fatigue result occurred because unintentional surface damage was present.

Other useful information is related to the size of the fatigue crack at the moment of failure in comparison to the size of the quasi-static final fracture (see Figure 2.33). This gives an indication about the fatigue crack sensitivity of a material. The usefulness of fractographic observations was illustrated by several examples in Chapter 2 and the discussion in Section 2.6. Contributions of fractographic analysis, including electron microscopy, in order to analyze fatigue damage accumulation under VA loading, was discussed in Chapters 10 and 11.

An evaluation of the results of a fatigue test program without fractographic observations should be considered to be incomplete. Unfortunately, fractography is ignored too many times in publications in the open literature.

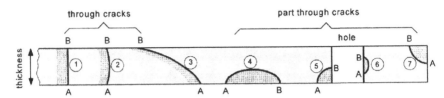

Fig. 13.6 Different shapes of crack fronts of AB through cracks and part through cracks. ① Straight crack front, ② slightly curved crack front, ③ oblique crack front, ④ surface crack, ⑤ corner crack at notch, ⑥ surface crack in notch, ⑦ edge corner crack.

13.6 Aspects of crack growth measurements

Crack growth records are data obtained by measuring the increasing size of a crack during a fatigue tests. Unfortunately, the crack front inside the material cannot be observed. Various crack front shapes are shown in Figure 13.6. Only the ends of the crack front at the free surface, points A and B, can be observed visually. Several measuring techniques have been developed for crack growth test programs. Some topics are discussed below:

1. Automation of crack length measurements to facilitate and speed up crack growth measurements. It also enables crack growth tests with a constant ΔK during a test.
2. Fracture surface analysis. It can provide information on crack front shapes, but also on local crack growth rates.
3. Crack closure measurements.

These subjects are briefly discussed, primarily for indicating experimental possibilities. More detailed information about the large variety of measurement techniques should be drawn from the literature.

Crack length measurements

In the early days, the crack length was only measured by visual observations, but this method is still frequently used. A kind of a ruler is attached to the specimen just below the crack path, see Figure 13.7. The location of the crack tip is then read from this ruler. The advantage of the visual observation method is that it is simple and does not require much preparation time. The accuracy is generally satisfactory.

Observations are improved by using a binocular microscope with a small magnification, e.g. 15 times, and a hair line system to locate the crack tip. Accuracies in the order of 0.1 to 0.2 mm can be achieved. Automatic

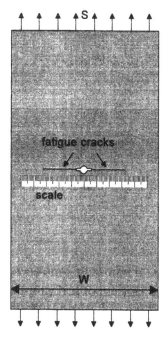

Fig. 13.7 Specimen with a simple scale for making visual crack growth records.

measurements can be performed by taking pictures at certain time intervals. A full crack growth record is obtained by video recording. Techniques have been developed to project such images on a monitor during an experiment. It is an interesting experience to see real-time opening and closing of the crack in each cycle.

Automated crack length measurements have been promoted by the introduction of the potential drop (PD) technique, see Figure 13.8. An electric current is passing through the specimen. The potential difference ($V_{1,2}$) is measured between two points at both sides of the fatigue crack (P_1 and P_2), and as a reference, also between two points (P_3 and P_4) in an undisturbed area of the specimen remote from the fatigue crack ($V_{3,4}$). The ratio $V_{1,2}/V_{3,4}$ is a measure for the crack length from which the length can be calculated. A theoretical function is available for this purpose [9], but it is advisable to determine the correlation also empirically by some calibration tests. Both DC and AC electrical currents are used in commercially available PD apparatus. With the DC option, the electric current is more uniform through the thickness of the material. It implies that the crack length of a slightly curved crack front, number 2 in Figure 13.6, is averaged. The AC method implies that the electric current occurs more along the material

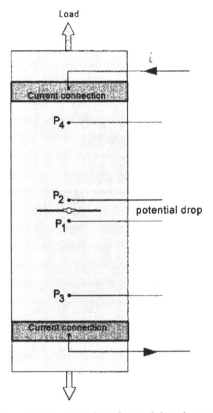

Fig. 13.8 Potential drop technique for crack length measurements.

surface, depending on the AC frequency. Both methods have advantages and disadvantages, see [10]. The potential drop measurement should be made at the maximum of the load cycle because the crack must be fully open. At lower stress levels, metallic contact between the upper and lower flank of a fatigue crack can occur which will conduct electrical current through the crack instead of around the crack. A false crack length indication is then obtained.

Results of crack length measurements consists of data pairs (a_i, N_i), i.e. the crack length a_i and the corresponding number of cycles N_i, with i as a rank number. A plot of a_i as a function of N_i gives the crack growth curve. The most simple procedure to calculate the crack growth rate as a function of the crack length is defined by

$$a = \frac{a_i + a_{i+1}}{2} \quad \text{and} \quad \frac{\mathrm{d}a}{\mathrm{d}N} = \frac{\Delta a}{\Delta N} = \frac{a_{i+1} - a_i}{N_{i+1} - N_i} \tag{13.1}$$

This equation is a simple averaging method between successive data pairs. More sophisticated data fitting evaluations have been proposed [8], but the improvement of the data representation may be small.

The electrical signal of the PD crack length measurements can be directly processed in the computer to calculate the crack length and crack growth rate. The advantages is that crack growth tests can run without personal attendance. The computer can also automatically stop a test at a preselected value of the crack length.

In computer controlled crack growth tests, it is usual to adopt a constant ΔN-value between successive PD crack length measurements. The interval ΔN may not be too large in order to have enough data points when the crack is growing fast, i.e. in the last part of the test. However, this will produce a large number of data points in the beginning of the test when the crack is still growing slowly. The crack length increment Δa may then be of the same order of magnitude as the inaccuracy of the measurements, and artificial scatter will be introduced due to too many data.

An other important advantage of the PD method is the possibility to perform crack growth tests with a constant ΔK-value. If a crack is growing under CA loading, the ΔK-value increases. In order to keep ΔK constant, the load on the specimen must be reduced during crack growth. This is called load-shedding. The constant ΔK equation for a center cracked specimen is

$$\Delta K = \beta \cdot \Delta S \sqrt{\pi a} = \Delta S \sqrt{\pi a / \cos(\pi a / W)} = \text{constant} \qquad (13.2)$$

After measurements of small crack length increments, the increase of the square root function is calculated. The increase is then balanced by a small reduction of ΔS to keep ΔK constant. This can be automatically done by the load-control program of the computer by readjusting S_{max} and S_{min} applied to the specimen in order to maintain both ΔK and the stress ratio R ($= S_{min} / S_{max}$) at a constant value.

Techniques for crack length measurements, and also for crack closure measurements, are different for $M(T)$ specimens (specimens with a central crack) and $C(T)$ specimens. Advantages and disadvantages were discussed in Section 5.4. It appears that the results for $M(T)$ specimens are more reliable and relevant. A disadvantage of the $C(T)$ specimen was that cracks in this type of specimen are also opened by a significant bending moment on the specimen. In general, this does not occur with fatigue cracks in a structure.

Crack growth data published in the literature are frequently presented as a function of da/dN on the vertical axis and ΔK on the horizontal axis. It is hiding the size of the observed cracks. It must be advised that crack growth curves $a(n)$ should also be given, i.e. graphs with the crack length as a function of the applied load cycles.

Fracture surface analysis

Crack front shapes of oblique through cracks and part through cracks can be observed on the fatigue fracture surface with some special test procedures. A destructive method is to interrupt a fatigue test before final failure occurs. If the specimen is then pulled to failure by static loading, the fracture surface will reveal the size and shape of the crack. If the development of the crack shape is the purpose of the investigation, a number of similar notches in a single specimen can be profitable. Due to scatter of fatigue crack nucleation, cracks of different sizes will be present in the specimen and can thus be observed after opening of the cracks.

A different approach is to perform fatigue tests with VA loading. Depending on the VA load history, macroscopic growth bands will appear on the fracture surface. An example was shown in Chapter 2 (Figure 2.37). Cracks started from two side notches of a specimen loaded alternately by two different blocks of cycles. Crack front shapes are easily observed from the growth bands, and average crack growth rates in the bands can be deduced from band width measurements. The dark bands correspond to the high-amplitude cycles, and the light bands to the low-amplitude cycles.

So-called marker load cycles have also been used in CA tests. Marker load cycles are applied at a cyclic stress level which should leave some markings on the fracture surface, but the marker load cycles should give a negligible contribution to crack growth. Furthermore, the marker load cycles should not affect subsequent crack growth during the base-line cycles of the CA loading. Ichsan [11] used very small marker load cycles as shown in Figure 13.9. The faint bands of these cycles could be used for a reconstruction of the crack front shapes during crack growth of a semi-elliptical surface crack in a thick plate, see the results in the figure.

Fractographic analysis in SEM introduces another possibility based on striation observations. A marker load history, applied by Piascik and Willard [12] and later used by Fawaz [13] and De Rijck [14], is shown in Figure 13.10. The marker loads have the same Smin as the base-line cycles, but a lower S_{max}. During the period of marker load cycles, small blocks of

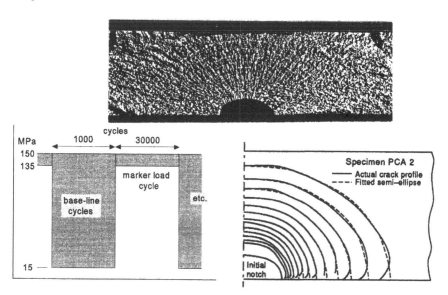

Fig. 13.9 Fractographic analysis of crack front shapes of a surface crack by using marker loads. Plate thickness 9.6 mm, Al-alloy 7075-T6, CA loading [11].

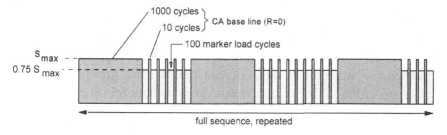

Fig. 13.10 Marker load history adopted by Piascik and Willard [12].

base-line cycles were inserted. These blocks could be seen as single lines in the SEM. The small blocks of 10 base-line cycles are applied in alternating numbers of 5, 9 and 3, see Figure 13.10. Counting these numbers in the SEM gives extra information for the correlation between the location on the fracture surface and the fatigue life in the test. A digital marking procedure was previously used by Sunder [15]. Careful fractography in the electron microscope can give valuable information, but it requires experience and a patient investigator.

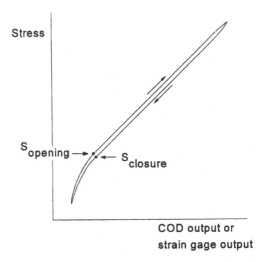

Fig. 13.11 Hysteresis of crack closure measurements. Differences between S_{op} and S_{cl}?

Crack closure measurements

Crack closure during fatigue crack growth was discussed in Section 8.4. The concept was used to describe the effect of the stress ratio R on crack growth under CA loading. In Chapter 11, the occurrence of plasticity induced crack closure was used again to explain several crack growth interaction effects under VA loading. Crack closure plays a major role in crack growth prediction models.

Since Elber detected crack closure in 1968, various methods have been developed to measure the stress level at which a crack is fully opened at the crack tip during loading, S_{op}, and the stress level at which closure starts at the crack tip, S_{cl}. A survey of methods was given in [16]. Some advanced techniques are based on optical observation of the crack tip with a microscope. In the more well known methods, a compliance technique is adopted, either by using a small displacement meter to measure crack opening displacement (COD), or strain gages bonded on a specimen close to the crack tip to measure the strain response as affected by crack closure. In both cases, a partly non-linear record is obtained as schematically shown in Figure 13.11. Some problems can be encountered. The record may show hysteresis. It may also suggest that crack opening and crack closure do not start at exactly the same stress level. These stress level are the transition points between the non-linear part and linear part of the record. It is not always fully clear at which stress level this occurs. Sometimes, the transition

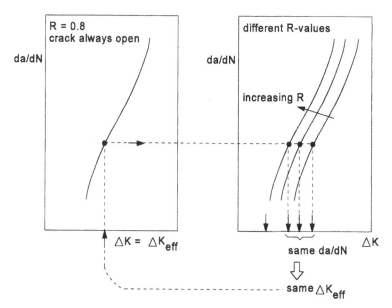

Fig. 13.12 Deviation of ΔK_{eff} in CA tests from da/dN results. Method adopted by Zhang et al. [18].

is more easily observed during unloading (S_{cl}) than during loading (S_{op}), while the latter one is theoretically of greater interest. Experience has shown that measurements close to the crack tip produce more well defined records than measurements taken at a distance of the crack tip. Anyhow, measurements confirm that crack closure occurs indeed, but it must be admitted that crack closure measurements offer problems with respect to reproducibility, accuracy and interpretation. It should be recalled from the discussion in Section 8.4.2 that crack closure stress levels under plane stress should be expected to be higher than under plane strain because of larger plastic zone sizes. As a consequence, the crack closure stress level will vary along a crack front with more crack closure at the material surface, and less crack closure away from the surface.

Measurements on crack closure along the crack front are difficult. Sunder and Dash developed a fractographic method [17], which was used by Ichsan [11] for semi-elliptical surface cracks as shown in Figure 13.9. The method, based on striation spacing measurements, confirmed that more crack closure occurred at the material surface, while crack closure along the major part of the crack front was more limited. Part-through cracks growing in the thickness direction will meet an approximate plane strain condition because of significant restraint on displacements along the crack front.

A method to obtain crack closure stress levels without crack closure measurements was adopted by Zhang et al. [18]. The method is illustrated in Figure 13.12. It starts from the assumption that crack closure is absent at high R ratios, i.e. stress cycles with S_{min} close to S_{max}. The crack is then fully open during the entire load cycle, and thus $\Delta K_{eff} = \Delta K$. Crack closure occurs at lower stress ratios. The value of ΔK_{eff} is now obtained by a cross plot as shown in Figure 13.12. The basic idea is that equal da/dN-values obtained at different R-values correspond to the same ΔK_{eff}-value. It thus can be studied how $U = \Delta K_{eff}/\Delta K$ depends on the stress ratio R. Zhang et al. [18] found an excellent agreement for crack growth data of an Al-alloy (7475-T7351). The method appears to be logical, but it should be recognized that the agreement is promoted by the ΔK_{eff} calibration procedure.

13.7 Main topics of this chapter

Several comments on planning and carrying out programs of fatigue tests are discussed in the present chapter. These comments will not be summarized here, but a few specific recommendations are recalled below:

1. The selection of specimens and fatigue loads should be carefully considered in relation to the purpose of the fatigue test program. Different options will apply to engineering problems and research investigations.
2. After completing a fatigue test, the results of the test should be immediately evaluated in order to allow a reassessment of subsequent tests.
3. An evaluation of the results of a fatigue test should always include a fractographic analysis of the fatigue fracture surfaces.

References

1. Bellows, R.S., Muju, S. and Nicholas, T., *Validation of the step test method for generating Haigh diagrams for Ti-6Al-4V.* Int. J. Fatigue, Vol. 21 (1999), pp. 687–697.
2. Schütz, W., *Fatigue strength of single shear riveted joints of aluminum alloys.* Laboratorium für Betriebsfestigkeit (LBF), Darmstadt, Bericht Nr.F-47, 1963 [in German].
3. *Standard Practice for Verification of Specimen Alignment under Tensile Loading.* ASTM Standard E 1012-89 (1989).

4. Schütz, W., *Fatigue Life Prediction – A Somewhat Optimistic View of the Problem*. AGARD Lectures Series No. 62 (1973).

5. Potter, J.M. and Watanabe, R.T. (Eds.), *Development of Fatigue Loading Spectra*. ASTM STP 1006 (1989).

6. ten Have, A.A., *Wisper and Wisperx. A summary paper describing their background, derivation and statistics*. Nat. Aerospace Lab. NLR, Amsterdam, Report TP 92410 (1992).

7. Le May, I., *Symposium summary and an assessment of research progress in fatigue mechanisms*. Fatigue Mechanisms, J.T. Fong (Ed.), ASTM STP 675 (1979), pp. 873–888.

8. *Test Method for Measurement of Fatigue Crack Growth Rates*. ASTM Standard E 647-99 (1999).

9. Johnson, H.H., *Calibrating the electric potential method for studying slow crack growth*. Mater. Res. Standards, Vol. 5 (1965), pp. 442–445.

10. Saxena, A. and Muhlstein, C.L., *Fatigue Crack Growth Testing*. ASM Handbook, Vol. 19 (1996) pp. 168–184.

11. Ichsan S. Putra and Schijve, J., *Crack opening stress measurements of surface cracks in 7075-T6 Al alloy plate specimens through electron fractography*. Fatigue Fract. Engrg. Mater. Struct., Vol. 15 (1992), pp. 323–338.

12. Piascik. R.S. and Willard, S.A., *The characteristics of fatigue damage in the fuselage riveted lap splice joint*. NASA/TP-97-206257 (Nov. 1997).

13. Fawaz, S.A., *Fatigue crack growth in riveted joints*. Doctor thesis, Delft University of Technology (1997).

14. de Rijck, J.J.M., *Stress analysis of fatigue cracks in mechanically fastened joints. An analylitical and experimental investigation*. Doctor thesis, Delft University of Technology (2005).

15. Sunder, R., *Binary coded event registration on fatigue fracture surfaces*. Nat. Aero. Lab., Bangalore. Report TM-MT-8-82, 1982.

16. Schijve, J., *Fatigue crack closure observations and technical significance*. Mechanics of Fatigue Crack Closure, ASTM STP 982 (1988), pp. 5–34.

17. Sunder, R. and Dash, P.K., *Measurement of fatigue crack closure through electron microscopy*. Int. J. Fatigue, Vol. 4 (1982), pp. 97–105.

18. Zhang, S., Marissen, R., Schulte, K., Trautmann, K.K., Nowack, H. and Schijve, J., *Crack propagation studies on Al 7475 on the basis of constant amplitude and selective variable amplitude loading histories*. Fatigue Engrg. Mater. Struct., Vol. 10 (1987), pp. 315–332.

Some general references

19. Heuler, P. and Schütz, W., *Standardized load-time histories – Status and trends*. Low Cycle Fatigue and Elasto-Plastic Behaviour of Materials, K.-T. Rie and P.D. Portella (Eds.), Elsevier (1998), pp. 729–734.

20. *1997 Annual Book of ASTM Standards*. Section 3. Metal test methods and analytical procedures. Vol.03.01, Metals.

21. Maennig, W.-W., *Planning and evaluation of fatigue tests*. Fatigue and Fracture, American Society for Materials, Handbook Vol. 19, ASM (1996), pp. 303–313.

22. Amzallig, C. (Ed.), *Automation in Fatigue and Fracture: Testing and Analysis*. ASTM STP 1231 (1994).

23. Ruschau, J.J. and Donald, J.K. (Eds.), *Special Applications and Advanced Techniques for Crack Size Determination*. ASTM STP 1251 (1993).

24. Marsh, K.J., Smith, R.A. and Ritchie, R.O. (Eds.), *Fatigue Crack Measurement: Techniques and Applications.* Engineering Materials Advisory Services (EMAS) (1991).

25. Marsh, K.J. (Ed.), *Full-Scale Fatigue Testing of Components and Structures.* Butterworth 1988.

26. Cullen, W.H., Landgraf, R.W., Kaisand, L.R. and Underwood, J.H. (Eds.), *Automated Test Methods for Fracture and Fatigue Crack Growth.* ASTM STP 877 (1985).

27. Beevers, C.J. (Ed.), *Advances in Crack Length Measurement.* Chameleon Press, Warley (1982).

28. Hudak, S.J. and Bucci, R.J. (Eds.), *Fatigue Crack Growth Measurement and Data Analysis.* ASTM STP 738, (1981).

29. *Fatigue Test Methodology.* AGARD Lecture Series No. 118, 1981.

30. Beevers, D.J., *The Measurement of Crack Length and Shape during Fracture and Fatigue.* Chamelon Press, London (1980).

31. Swanson, S.R., *Handbook of Fatigue Testing.* ASTM STP 566 (1974).

Part IV
Special Fatigue Conditions

Chapter 14
Surface Treatments

14.1 Introduction

Fatigue cracks generally start at the free surface of a material. As a consequence, the conditions of the surface are most significant for the fatigue behavior of a structure. The importance of these conditions was recognized long ago, if not in the laboratory, it was by practical experience. Corrosion pits, fretting corrosion, nicks and dents became well-known sources of fatigue problems. Surface treatments to improve the fatigue resistance were also developed a long time ago.

As discussed in previous chapters, material surface conditions are important for fatigue crack nucleation, and thus affect the crack initiation period of the fatigue life. Major influences on the fatigue limit and fatigue strength under high-cycle fatigue are expected, see the discussion in Section 2.5.5. Aspects of surface conditions affecting the fatigue performance are briefly described in Section 14.2. It includes surface treatments, surface roughness and residual stress in surface layers. Some practical consequences are considered in Section 14.3. Topics of the present chapter are summarized in Section 14.4.

It is not the purpose of this chapter to describe surface treatment techniques and the structural changes introduced in the surface layers of the material. This information should be found in handbooks [1].

14.2 Aspects of surface treatments

Each production process creates its own characteristic material surface. This is obvious for machining. But it equally applies to forging and casting if no additional machining is done. Furthermore, plates, sheets, rolled profiles (I-beams, U-profiles, etc.) and extrusions are frequently used in a structure without any machining of the surface. It implies that the original rolled, forged or extruded surface is still present in the final product.

In addition to the surface conditions as obtained in a production process, surface treatments are applied in the industry for several purposes, such as (i) protection against corrosion, (ii) improvement of fatigue properties, (iii) rectification of a poor surface quality obtained by normal production methods, (iv) improved wear resistance, and (v) surface appearance.

The variety of surface treatments used in the industry for different materials is large. Some treatments are applied to various materials (e.g. shotpeening), other ones are typical for steel or Al-alloys. The significance of surface treatments with respect to fatigue is usually associated with three properties of material surface layers:

1.   Fatigue resistance of the surface layer material.
2.   Surface roughness.
3.   Residual stress in the surface layer.

All three can be modified by surface treatments, both in a positive and negative sense. Moreover, surface treatments affect more than a single property. Nitriding of steel improves the fatigue resistance of the surface layer, but is also introduces residual compressive stress. Aspects of surface treatments are discussed below.

Fatigue resistance of the surface layer material

A surface layer with a decreased fatigue resistance can occur in C-steel if decarburization occurs at the material surface during a heat treatment. The lower C-content implies a soft surface layer in which cyclic slip is relatively easy. This promotes fatigue crack initiation and the fatigue limit will be reduced.

A similar phenomenon occurs in sheet material of the stronger Al-alloys (e.g. 2024-T3 and 7075-T6), which are provided with a soft pure aluminum layer[19] at both sides of the sheet, thickness about 5% of the total thickness

---

[19] Pure Al for 2024-T3 and Al + 1% Zn for 7075-T6.

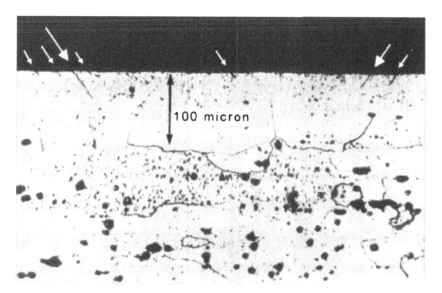

Fig. 14.1 Slipband microcracks in soft cladding layer (thickness 100 μm) of 2024-T3 sheet material.

each. The cladding layers are applied for corrosion protection of the strong core material. The soft cladding layer has a low resistance against crack nucleation along slip bands, see the microcracks in Figure 14.1. These cracks easily grow through the cladding layer. Penetration into the elastic core material is more difficult, but it does occur. As a consequence of the cladding layers, the fatigue limit for $R = 0$ can be reduced by more than 50%. This sounds rather dramatic. However, a corrosion pit in bare sheet material without a cladding layer can be equally disastrous. The same is true for surface damage by fretting corrosion. The effect of the cladding layer is relatively small at low endurances because microcracks are nucleated early in the fatigue life, also in the bare (unclad) material.

Anodizing is another surface treatment applied to light alloys. It can also cause a decrease of the fatigue strength in the high-cycle fatigue range. Anodizing produces an artificial oxide layer, which is considerably thicker than natural oxide layers, but still in the low micron range. Anodizing is used as a pretreatment for corrosion protection systems (e.g. painting) and adhesive bonding. Depending on the anodizing bath, the anodic coating layer can be rather brittle. Cracks can then occur in the anodic surface layer, especially if the load spectrum contains periodic high loads. These cracks can act as a starting point for surface crack nuclei in the base material.

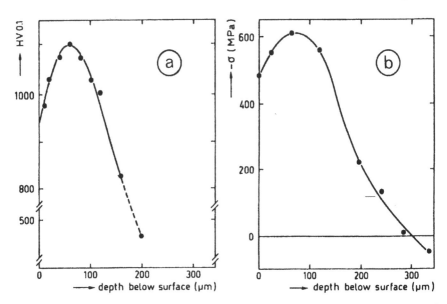

Fig. 14.2 Profiles of Vickers hardness and residual stress in nitrocarbonized surfacel ayer of a CrMo steel (S_U = 1840 MPa) [2].

Surface layers with improved properties are the opposite of the above layers. In contrast to decarburization, a carburizing process is applied to steel to increase the carbon content of the surface layer. It leads to higher fatigue resistance and a much improved wear resistance. Low-alloy high-strength steels (e.g. CrMo steel) are often selected if a high-fatigue strength for a dynamically loaded component is required. Nitriding of these materials can significantly increase the hardness of the surface layer by some precipitation phenomena which also improves the fatigue resistance. Nitriding gives a volume increase and as a result residual compressive stresses in the surface layer. An example of hardness and residual stress profiles is shown in Figure 14.2. Apparently, the depth of the affected surface layers is in the order of a few tenth of a millimeter.

Improvements of the fatigue limit obtained by nitriding are shown in Figure 14.3 with results obtained by Overbeeke and van Lipzig [3]. They performed fatigue tests on rotating beam specimens with shoulder fillet notches. Different K_t-values were obtained by changing the root radius of the fillet. All specimens were heat treated to the same static strength level of $S_U \approx 1000$ MPa. Three different nitriding processes were used. The improvement of the fatigue limit (S_f) is shown as a bar chart representing the ratio of S_f of nitrided specimens and S_f of non-nitrided specimens.

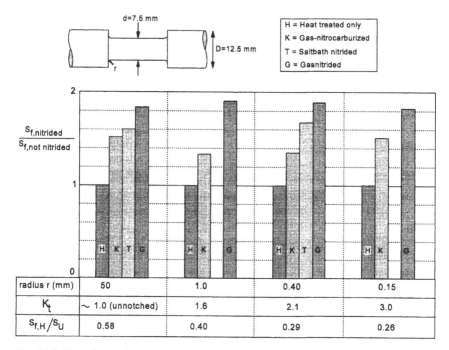

Fig. 14.3 The effect of surface nitriding on the fatigue limit of notched rotating beam specimens of a CrMo-steel (42CrMo4) [3].

Significant improvements were found for all K_t-values. The fatigue limit was almost doubled by gas nitriding (G in Figure 14.3) for all K_t-values. As pointed out in [3], the favorable effect of nitriding can be explained by considering the increased surface hardness only. An increase of the hardness implies an increased strength and a corresponding increased fatigue limit, see Figures 2.11 and 6.8. The compressive residual stresses can be beneficial if microcracks are nucleated at stress amplitudes just above the fatigue limit. The growth of these cracks will be retarded.

Hardening of a surface layer is also done by induction hardening without introducing extra carbon or nitrogen. Results in Figure 14.4a show improvements obtained on a fusee of a truck which in practice gave some fatigue problems at the smaller radius (3 mm). After induction hardening the S-N curve was raised considerably, and fatigue problems did not occur anymore. The etched cross section of the axle in Figure 14.4b reveals the depth of the induction hardened layer which in this case is in the order of a few millimeters.

An interesting aspect of an improved resistance against microcrack nucleation in the surface layer is the occurrence of subsurface cracking.

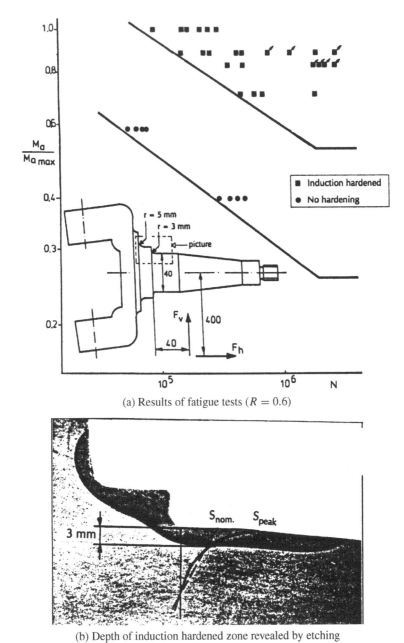

(a) Results of fatigue tests ($R = 0.6$)

(b) Depth of induction hardened zone revealed by etching

Fig. 14.4 The effect of induction hardening on the S-N curve of a car axle of Cr-steel ($S_U = 850$ MPa). (Courtesy: Harry van Lipzig, DAF, Eindhoven)

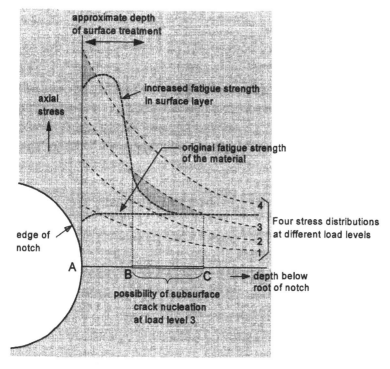

Fig. 14.5 Possibilities for sub-surface cracking due to a surface treatment.

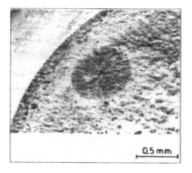

Fig. 14.6 Sub-surface crack nucleation in a notched rotating beam specimen of a CrMo steel with a nitrided surface (curved dark band) [3].

The fatigue resistance of the subsurface core material is not increased, see the schematic picture in Figure 14.5. It then depends on the gradient of the stress distribution in a component whether subsurface material becomes more fatigue critical. Under bending loading, and around notches, a stress gradient is present with the maximum stress at the material surface. Stress distributions as depicted in Figure 14.5 can occur. Without any surface hardening, crack nucleation should occur at the root of the notch, point A in Figure 14.5. However, with surface hardening, the increased fatigue strength is not yet exceeded by the stress distribution of the higher load level 2. An increase of the fatigue limit should be expected. Also, the peak stress of fatigue load level 3 does not yet exceed the increased fatigue strength at point A. However, at some depth below the surface of the notch, the applied stress is larger than the local fatigue strength. Subsurface crack nucleation between B and C may occur. Actually, subsurface crack nucleation below hardened surface layers has been observed at inclusions in the steel matrix, see Figure 14.6 for an example. At a still higher applied fatigue load, level 4 in Figure 14.5, nucleation can occur again at A because the peak stress at the free surface in the root of the notch is exceeding the increased fatigue strength. The explanation of sub-surface crack nucleation in Figure 14.5 is based on simple assumptions. Residual stresses were not considered, but qualitatively the explanation should be correct. It emphasizes the significance of the depth of a surface treatment in relation to the gradient of the stress distribution caused by the applied load. It should be recalled that the stress gradient will be smaller for larger notch root radii (Chapter 3) which are preferred for lower K_t-values.

Surface roughness

Surface roughness is generally associated with the geometric topography of the surface of a material. Roughness will obviously depend very much on the production technique. Pictures of material surfaces obtained by machining, grinding, glass bead peening and steel grid peening of a Ti-alloy are shown in Figure 14.7 [4]. Note the larger magnification in the pictures for machining and grinding, and the much lower magnification for the two peening operations. The first two operations clearly leave traces in a single working direction, but they look quite different. Machining can lead to fairly sharp grooves, depending on the variables of the machining technique and the quality of the cutter. Grinding apparently has led to some smeared traces. The roughness of the peened surfaces is fully different, apparently somewhat similar to orange peel.

The surface roughness can be measured quantitatively with apparatus developed for obtaining a line scan of the surface profile (Talysurf, e.g.). This procedure is used in the workshop for characterizing the surface finish obtained. Two definitions of surface roughness are based on such profiles: (i) the difference between the highest peak and the lowest trough

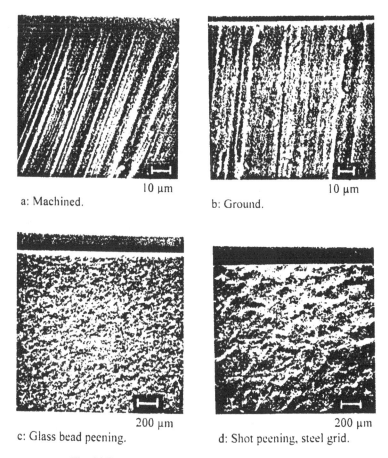

10 μm
a: Machined.

10 μm
b: Ground.

200 μm
c: Glass bead peening.

200 μm
d: Shot peening, steel grid.

Fig. 14.7 Different material surfaces of a Ti-alloy [4].

(R_{max}), and (ii) the average of the absolute value of the amplitude of the profile (R_a). Such values can characterize different surface roughness, but the significance for fatigue performance is limited. Both roughness values do not indicate how sharp the deepest groove will be, while it should be important for the micro-notch effect on fatigue. Moreover, for surfaces as shown in Figures 14.7a and 14.7b, the roughness is different for scanning perpendicular and parallel to the grooves. Even more important, a unique correlation between surface roughness and fatigue properties is disturbed due to work hardening and residual stresses introduced in the material surface layer by the production technique. This is obvious for shotpeening. The material surface after shotpeening can be fairly rough, but significant improvements of the fatigue properties are still achieved. Machining also introduce residual stresses, although it is not always clear whether it leads to compressive or tensile residual stress at the material surface. Actually, it must be expected that machining can lead to residual stresses in a thin surface layer. Machining implies that material is removed by a cutting process. Some material is torn away from the substrate material which is a small-scale failure phenomenon occurring with some local plastic deformation. Because of this plastic deformation it should leave a residual stress distribution depending on the material and the variables of the machining operation.

Suhr [5] carried out fatigue tests on specimens of a CrNiMo steel with different types of surface finish. After specimen manufacture, residual stresses were removed by an additional heat treatment (4 hrs at 590°C), which was confirmed by X-ray diffraction measurements. Comparative CA fatigue tests on these specimens indicated that a significant surface roughness effect was still present. Fractographic observations revealed that cracks nucleated at the deeper scratches, and the fatigue limit was lower for the higher roughness specimens. Unfortunately, the depth of these scratches was not evident from the surface roughness scanning.

Apparently, it is difficult to correlate the surface roughness with fatigue properties. The practical approach was already discussed in Chapter 7 (Section 7.6). A reduction factor γ was defined as

$$\gamma = \frac{S_{fl,\text{specific surface quality}}}{S_{fl,\text{high quality surface}}} \qquad (7.23)$$

It indicates how much the fatigue limit can be reduced for a specific surface quality if compared to the fatigue limit obtained on specimens with a high-quality surface. Graphs with γ-values were published based on empirical results for different types of production techniques. This was previously illustrated by Figure 7.16 for steel. It shows that the fatigue limit is reduced more for high-strength steel, which are known to be more notch sensitive. The same trend was shown in Figure 7.15 for different degrees of surface roughness. Such graphs are not well documented for other materials, although a similar roughness sensitivity is known for Al-alloys. The increased sensitivity of stronger materials for irregularities of the surface topography suggests that surface roughness is related, at least partly, to notch sensitivity, be it for small notches. This is also confirmed by the effect of the direction of the machining grooves with respect to the fatigue stress. Illustrative evidence was already presented in Chapter 2 with results of De Forest in Figure 2.22. Machining grooves perpendicular to the fatigue stress have a more detrimental effect than grooves parallel to the fatigue stress.

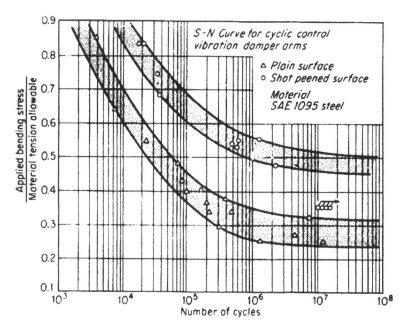

Fig. 14.8 The effect of shot peening on the S-N curve of cantilever springs of a helicopter component (material: high-carbon steel) [6].

Residual stress

The significance of residual stresses for fatigue is very well known. Chapter 4 summarizes various possibilities of introducing residual stress in materials and structures. It includes brief comments on shotpeening and measurements of the peening intensity with the Almen strip (Figure 4.5). A distribution of the residual stress distribution obtained by shotpeening of a high-strength low-alloy steel (SAE 4340) was shown in Figure 4.6. Shotpeening introduces plastic deformation in the surface material which implies stretching of the surface layer. The stretch is restrained by the elastic substrate material, which leads to compressive residual stresses. A most significant improvement of fatigue properties is possible in spite of an obviously rough surface texture after shotpeening. Illustrative results are presented in Figure 14.8 for a cantilever spring of a high-carbon steel. Numerous operational failures occurred in these springs which were not shotpeened. As shown by the fatigue curves in Figure 14.8 shotpeening of the springs significantly increased the fatigue life. The fatigue limit defined as the fatigue strength at $N = 10^7$ cycles was increased by some 70%.

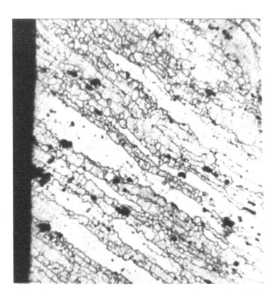

Fig. 14.9 End grain structure of machined forging, material 7075 Al-alloy (width of picture 0.3 mm).

The favorable effect of compressive residual stresses introduced by plastic deformation is generally explained by referring to a reduction of the mean stress. The residual stresses must be added to the cyclic stress caused by the cyclic load on the structure. In Chapter 7, this was already discussed as part of the notch effect problem. Considering the compressive residual stress effect in more detail, it should be recalled that microcracks start at the free surface of the material as a result of cyclic slip caused by cyclic shear stress. The cyclic shear stress is not affected by residual stresses and slip band cracking can still occur. However, compressive residual stresses try to keep such cracks closed, which hampers further growth. Leaf springs of cars are generally shotpeened. Microcracks have been observed in such springs after many years, however, without leading to failure. Apparently, the compressive residual stresses arrested further growth of the microcracks.

A surface zone with compressive residual stresses can be an effective barrier to crack growth. In view of this effect, it is noteworthy that shotpeening is frequently applied to eliminate the detrimental effects of a poor surface quality. This poor quality can be due to a heat treatment at a high temperature, which may cause oxidation and decarburization of steel, or some other unfavorable surface attack. Shotpeening then restores the fatigue resistance to an acceptable level. Another example is shotpeening of machined surfaces of a forging. Machining can imply cutting of the fibrous structure of a forging (end-grain structure). An illustration is given in Figure 14.9 [7]. Inclusions in the fiber direction which are cut by machining will enable crack nucleation. Subsequent crack propagation can be prevented by shotpeening of the material surface. Shotpeening of a material surface with an end-grain structure is also used to reduce the risk of stress corrosion cracking which can occur under residual tensile stresses.

It should be recalled from Chapter 4, that plastic deformation produced by shotpeening can easily lead to some deformation of a component (warpage). Shotpeening is even used to give a specified curvature to plate material, a production process called peen-forming. However, in general warpage is undesirable, and dimensional rectifications may be necessary. Shot-peening is not a panacea for achieving a high-fatigue resistance, but in several cases, it can be a useful tool for preventing fatigue problems.

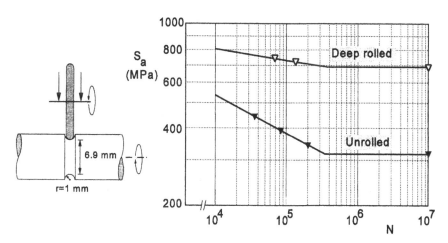

Fig. 14.10 The effect of rolling of the notch root on fatigue under rotating bending. Material 37CrS4 steel (0.4C-1.1Cr-0.2Ni-0.1Mo), $S_U = 1150$ MPa [8].

Surface rolling in another method to introduce compressive residual stresses. In Figure 14.10 a hardened rolling wheel is pressed in the notch of a rotating specimen. As a result, plastic deformation occurs in the material surface layer of the notch root. As shown by the graph in Figure 14.10, the fatigue limit under rotating bending was significantly increased up to more than twice the original value. According to Kloos et al. [8] the improvement is mainly due to high compressive residual stresses and not to work hardening of the surface material. Actually, they have also found small cracks at the root notch which stopped growing due to the compressive residual stresses.

Contrary to shotpeening, surface rolling leaves a smooth surface finish, while initial surface imperfections are flattened.

14.3 Some practical aspects of surface treatments

Designing for fatigue durability includes different problem areas as illustrated in Chapter 1 with the discussion on Figure 1.2. A designer should evaluate several variables in order to achieve a structure with good fatigue properties. Obvious categories of these variables are: (i) geometrical aspects to avoid high stress concentrations, (ii) material selection, and (iii) production techniques, including surface finish and surface treatments.

As emphasized previously, the material surface has a predominant effect on the fatigue limit and the high-cycle fatigue behavior. The material surface properties are less important for low-cycle fatigue. But for high-cycle fatigue problems the material surface quality must be considered. It can be attractive to improve the fatigue performance of a structure by applying a good surface finish or a special surface treatment. However, economic consequences associated with extra production costs will be involved. It must also be recalled that fatigue properties are subject to scatter. As discussed in Chapter 12, the more important source of scatter to be considered by the manufacturer is the variability of his product. The variability depends on the quality of the material, the production technique and the surface treatment. It thus is of great importance to specify the variables in detail, and to carefully consider quality control of these variables. A heat treatments must occur under closely controlled temperatures, atmosphere and durations. Quality control is equally significant for surface treatments.

The beneficial effect of a new surface treatment to be applied by the industry must be verified by experiments. The specimen geometry and size should be similar to the fatigue critical part of the structure. The best solution is to test the real structure itself. Improvements of the fatigue limit and high-cycle fatigue properties can be explored in CA tests. However, if cyclic loads exceeding the fatigue limit occur in service, tests should be carried out under a relevant service-simulation load history. Schütz [9] found substantially different life improvements of shotpeening in CA tests and service-simulation fatigue tests on a high-strength Al-alloy ($S_U = 531$ MPa), a Ti-6Al-6V-2Sn alloy ($S_U = 1250$ MPa) and a high-strength maraging steel ($S_U = 2125$ MPa). The improvements in service-simulation tests were smaller than under CA loading. He emphasized that quantitative fatigue life improvements can not be assessed with CA tests if the component in service is loaded by a VA test spectrum. It should be recognized that a service load spectrum with severe loads may cause plastic deformation which can have some relaxing effect on the residual stress distribution introduced by a surface treatment. Actually, understanding of all variables involved is essential for a realistic scenario of an exploratory test program.

14.4 Summary of major topics of the present chapter

1. Material surface treatments are carried out for various purposes: improvement of fatigue properties, protection against corrosion, improved wear resistance, improving a poor surface quality, and cosmetic reasons. The variety of surface treatments is large.
2. The effect of surface treatments on the fatigue properties of a structure is associated with the fatigue resistance of the surface layer of the material, with residual stresses in this layer, and with surface roughness.
3. Surface treatments can be very effective under high-cycle fatigue. Significant increases of the fatigue limit are possible. Surface treatments are less important for low-cycle fatigue.
4. Shotpeening is used for improving fatigue properties, but also for restoring the fatigue resistance of structural elements with a poor surface finish quality resulting from the production.
5. The effect of surface treatments for a specific application should be verified by experiments on component-type specimens with cyclic loads representative for the application of the structure in service.
6. Quality control of a surface treatment process during production is essential.

References

1. *Metals Handbook*, 9th edn., Vol. 16: Machining, ASM International (1989), and Vol. 5: Surface cleaning, finishing and coating. American Society of Metals (1982).
2. Van Wiggen, P.C., Rozendaal, H.C.F. and Mittemeijer, E.J., *The nitriding behaviour of iron-chromium-carbon alloys*. J. Mater. Sci., Vol. 20 (1985), pp. 4561–4582.
3. Overbeeke, J.L. and van Lipzig, H.T.M., *Nitriding against fatigue*. Fatigue 2000, M.R. Backe et al. (Eds.), EMAS Warley (2000).
4. Franz, H.E., *X-ray measurements of residual stresses after surface machining of TiAl6V4 and TiAl6V6Sn2*. VDI Berichte 313, Verein Deutscher Ingenieure, Düsseldorf (1978), pp. 453–462 [in German].

5. Suhr, R.W., *The effect of surface finish on high cycle fatigue initiation in low alloy steel*. The Behaviour of Short Fatigue Cracks, K.J. Miller and E.R. de los Rios (Eds.), EGF Publication 1, Mechanical Engineering Publications, London (1986), pp. 69–86.

6. Boswell Jr., C.C. and Wagner, R.A., *Fatigue in Rotory-Wing Aircraft*. Metal Fatigue, G. Sines and J.L. Weisman (Eds.), McGraw-Hill (1959), pp. 355–375.

7. Investigation of the National Aerospace Laboratory NLR, Amsterdam (unpublished).

8. Kloos, K.H., Fuchsbauer, B. and Adelmann, J., *Fatigue properties of specimens similar to components deep rolled under optimized conditions*. Int. J. Fatigue, Vol. 9 (1987), pp. 35–42.

9. Schütz, W., *Fatigue life improvement of high-strength materials by shot peening*. First Int. Conf. on Shot Peening. Pergamon Press (1981), pp. 423–433.

Some general references

10. Rice, C.R. (Ed.), *SAE Fatigue Design Handbook*, 3rd edn. AE-22, Society of Automotive Engineers, Warrendale (1997).

11. Zahavi, E. and Torbilo, V., *Fatigue Design. Life Expectancy of Machine Parts*. CRC Press (1996).

12. Leis, B., *Effect of surface condition and processing on fatigue performance*. Fatigue and Fracture, American Society for Materials, Handbook Vol. 19, ASM (1996), pp. 314–320.

13. Harris, J.W. and Syers, G., *Fatigue Evaluation*. Engineering Design Guides 32, Oxford University Press (1979).

14. Frost, N.E., Marsh, K.J. and Pook, L.P., *Metal Fatigue*. Clarendon, Oxford (1974), Section 6.4.

15. Forrest, P.G., *Fatigue of Metals*. Pergamon Press, Oxford (1962), Chapter VI.

Chapter 15
Fretting Corrosion

15.1 Introduction

Fretting corrosion is primarily a surface damage phenomenon occurring as a result of small cyclic movements between two materials caused by cyclic loading. Fretting damage can occur in vacuum, although it then would be better to speak of fretting, rather than fretting corrosion. In normal air, corrosion plays an active role in causing fretting corrosion damage. The other contribution comes from rubbing between two material surfaces. Very small rubbing displacements are sufficient for initiating fretting corrosion damage. Such displacements easily occur in joints; bolted joints, riveted joints, clamped joints, leaf springs, etc. It also occurs inside a bolt hole between the bolt and the wall of the hole. Fretting is even possible between two metallic parts where load transmission does not occur.

Fretting corrosion is a practical problem because it can cause significant reductions of fatigue properties of structural elements. In the present chapter, the fretting corrosion mechanism is discussed in Section 15.2. Important variables on fretting corrosion are discussed in Section 15.3. Methods to avoid fretting corrosion are the subject of Section 15.4. Some practical aspects, which will return in Chapter 18 on joints. Major topics of the present chapter are summarized in Section 15.5.

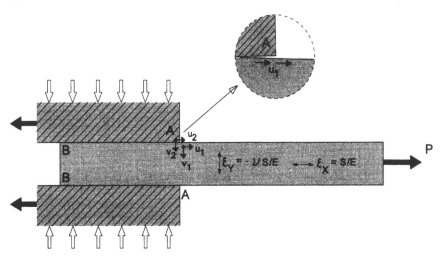

Fig. 15.1 A bar under tension slides out of the clamping area at section AA over a microscopically small distance, which is sufficient for fretting under cyclic load.

15.2 The fretting corrosion mechanism

Fretting can occur in a structure under cyclic load if a piece of material is clamped on an other piece. This will be explained by considering the simple case shown in Figure 15.1. A bar is clamped at one end and loaded in tension at the other end. Load transmission from the bar to the clamping occurs by frictional forces along the two mating surfaces AB. In the free part of the bar, a positive tensile strain occurs, $\varepsilon_x = S/E$, as well as a lateral Poisson contraction, $\varepsilon_y = -\nu S/E$ (ν = Poisson constant). Because of this contraction, the thickness of the free part of the bar including section AA is slightly reduced. Inside the clamping, the tensile stress in the bar (S_x) decreases to zero at the end BB. As a consequence, the lateral thickness constraint varies between A and B, full contraction at A and no contraction at B. The bar then will move slightly out of the clamping at section AA. It might be thought that clamping can prevent this movement. However, that would require equal displacements at point A: $v_2 = v_1$ and $u_2 = u_1$. It implies that the bar and the clamping device would then become an integral part with an extreme stress concentration at point A. However, this is illusory because a physical connection between the material of the bar and the clamping does not exist. Equal lateral displacements of the bar and the edge of the clamping cannot be maintained at point A. Material contact will be lost partly as sketched in the inset figure in Figure 15.1. Under cyclic loading, a cyclic change of the contact between the bar and the clamping occurs

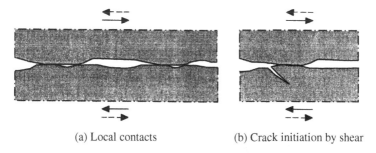

(a) Local contacts   (b) Crack initiation by shear

Fig. 15.2 Cyclic rubbing between two material surfaces.

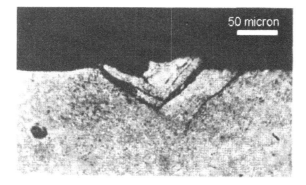

Fig. 15.3 Microcracking leading to loose fretting particles [1].

near point A, and this will cause cyclic slip between the mating surfaces. These slip displacements are very small but on a microscopic level they are substantial and large enough to cause fretting damage by rubbing between the two materials.

It should also be realized that the flat interfaces in the clamping area are not perfectly flat on a microscopic scale. Rubbing contacts are local contacts, see Figure 15.2a. It can lead to high shear stresses which can cause microcracks, see Figure 15.2b. As a results of such microcracks, small fragments of material can be torn out of the material surface, as shown by an example in Figure 15.3. Most technical alloys (ferritic steels, Ti-alloys, Al-alloys) are always covered by an oxide layer. Cracks in this brittle layer enhance the damage process, which in detail may be rather complex. Because of frictional heat development, local welding between the two surfaces is possible, which again is broken up in subsequent cycles. Oxidizing is promoting the damage process, one of the reasons why it is called fretting corrosion. The occurrence of fretting corrosion can result in a severely damaged material surface. A detrimental effect on fatigue should be expected, especially on crack nucleation, and thus on the crack initiation period in high-cycle fatigue and on the fatigue limit.

The above description of the damage process suggests a number of variables which can affect fretting corrosion:

(a) Clamping pressure between the two surfaces.
(b) Amplitude of the rubbing movements.
(c) Materials involved.
(d) Roughness of the rubbing surfaces.
(e) Corrosive contributions of the environments.
(f) Cyclic stress level.
(g) Variable-amplitude loading.

Investigations on the significance of these variables are reported in the literature. In general, fretting is studied by determining the effect on fatigue properties, in particularly the reduction of the fatigue limit and fatigue lives in the high-cycle fatigue regime. The complexity of fretting corrosion does not always allow to draw rigid conclusions, but trends have been recognized. These trends are summarized in the following section.

15.3 Effects on fretting corrosion

Experiments on fretting corrosion are often made by employing unnotched specimens on which one or two blocks are clamped. A test set-up used by Fenner and Field [2] is shown in Figure 15.4. Two pads, both with two small contact areas were clamped on flat surfaces of an aluminium alloy specimen. Under cyclic loading, the pads are causing fretting corrosion damage. The pads and specimens were made of the same Al-alloy (L65 $\approx$ 2014-T3). The clamping pressure was adjusted by two bolts mounted in a calibrated strain-gaged ring around the specimen. Fenner and Field carried out fatigue tests at an amplitude very close to the fatigue limit. Without pads, a very long endurance was obtained, $N = 8000$ kc, see Table 15.1. In another test with the pads on the specimen until failure, the life was most significantly reduced to $N = 100$ kc. Additional tests (see also Table 15.1) were done with the pads on the specimen for a small number of cycles. Removing the pads after 30 kc gave practically the same life reduction as leaving the pads on the specimen. Removing the pads after 12 kc still reduced the fatigue life from 8000 to 1300 kc. Apparently, a relatively small number of cycles under fretting conditions caused significant surface damage, which promoted crack initiation at a relatively early stage of the originally long fatigue life.

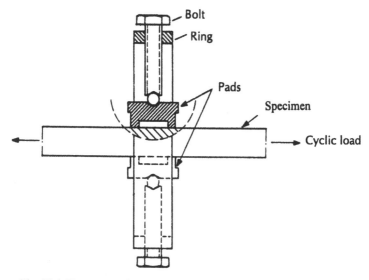

Fig. 15.4 Test set-up of Fenner and Field to study fretting corrosion [2].

Table 15.1 Results of fretting corrosion tests of Fenner and Field [2].

Fatigue tests on unnotched specimens of an Al-Cu alloy $S_m = 193$ MPa, $S_a = 134$ MPa, clamping pressure 4 MPa	Fatigue life (kc)
No pads	8000
Pads until failure	100
Pads removed after 12 kc (=12% of N with pads)	1300
Pads removed after 30 kc (=30% of N with pads)	110

A similar result was found by Endo and Goto [3] for a carbon steel (0.34C, $S_U = 549$ MPa). Obviously, fretting corrosion damage contributes to fatigue crack initiation, and it thus reduces the crack initiation period. Because fretting corrosion damage can occur at very low stress amplitudes, it also can significantly reduce the fatigue limit.

The significance of fretting for the crack growth period should be expected to be negligible because the crack is then growing away from the surface. Some effect might occur if frictional forces at the interface are changing the ΔK-value at the crack tip.

As discussed in Chapter 2, and also in Chapter 14 on surface treatments, surface effects have a large influence in the high-cycle fatigue regime, and a relatively small influence in the low-cycle fatigue regime, see Figure 2.23. This also applies to fretting corrosion as illustrated by S-N curves of a

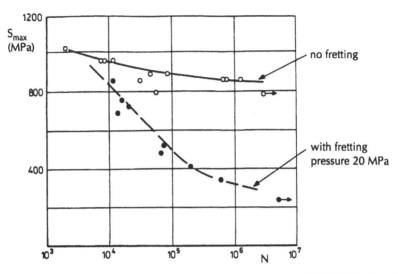

Fig. 15.5 Effect of fretting corrosion on the S-N curve of a Ti-alloy (Ti-6Al-4V) for $R = 0.1$. Results of Hoeppner and Gates [4].

Ti-alloy obtained by Hoeppner and Gates [4], see Figure 15.5. The fatigue limit is reduced by more than a factor of two, while the reduction in the low-cycle regime appears to be small.

Effect of clamping pressure

An increased clamping pressure between the two rubbing surfaces may increase fretting damage. This trend has been observed by visual inspection of surface damage. However, the question is whether it also increases the detrimental effect on fatigue properties. Literature information is not always consistent on the effect of clamping pressure. This might be expected when rubbing is associated with complex tribology aspects. It implies that wear resistance of the material, surface roughness and lubrication are involved. The presence of oxide particles produced by fretting is another complication.

Waterhouse [1] compiled the results shown in Figure 15.6. A fairly drastic reduction of the fatigue strength occurs at a low clamping pressure. Apparently, a small pressure is sufficient for a large effect. A further increase of the pressure does not add much more to a further reduction of the fatigue limit.

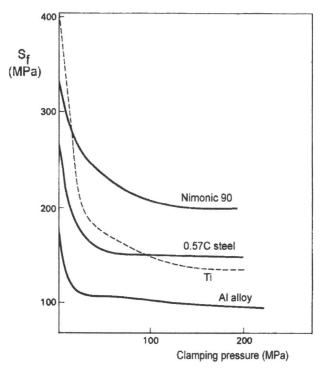

Fig. 15.6 The fatigue limit under fretting corrosion. Effect of the clamping pressure. Results compiled by Waterhouse [1].

Effect of the amplitude of the rubbing movements

Fenner and Field [2], using the test set-up of Figure 15.4, obtained different rubbing amplitudes by changing the distance between the two contact areas of the pads clamped on the specimen. They found indeed a systematically increasing effect for larger rubbing movements until about 8 μm. Still larger rubbing movements did not further decrease the fatigue limit of the Al-alloy. A systematic effect was also found by Funk [5] in fatigue tests on C-steel specimens. The results in Figure 15.7 show that a significant reduction of the fatigue limit was found again, which increased for larger rubbing movements. The difference between the results for 10 and 20 μm is relatively small.

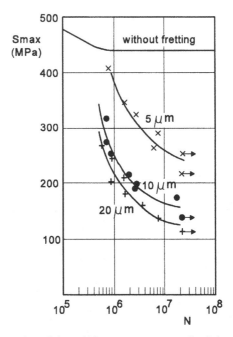

Fig. 15.7 Effect of the size of the rubbing movements on the fatigue curve of an 0.35% C-steel ($S_{min} = 0$). Pressure of fretting pad: 50 MPa. Results of Funk [5].

Fretting corrosion for different materials

The sensitivity for fretting corrosion depends on the type of material. Unfortunately, several high-strength materials used in engineering structures happen to be fretting sensitive. This is illustrated by results in Figure 15.8, based on data of [1], which shows reduction factors, i.e. ratios between the fatigue limit without fretting corrosion and the fatigue limit with fretting corrosion. The results should be considered as indicative; they were obtained in comparative fatigue tests series, each with their own specific test conditions, e.g. clamping pressure. However, some trends have a general meaning. Low ratios are in the order of 1.5, but for the high-strength alloys, reduction factors are in the order of 2 to 4. It is noteworthy that the high-strength low-alloy steels are more sensitive than the low-C steels. The high-strength steels can have a high fatigue limit provided that a high-quality surface is present. But, if damage is introduced by fretting corrosion, it implies a rough surface with micronotches or microcracks. In view of the high notch sensitivity of these materials, a large reduction of the fatigue limit can occur. Another observation is that the reduction factor for 0.7C-steel in the soft annealed condition is lower than in the hard cold drawn condition.

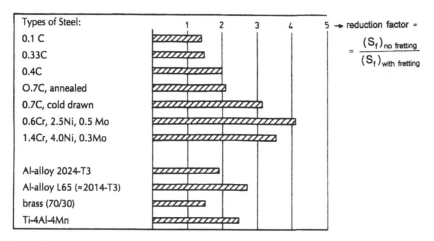

Fig. 15.8 Influence fretting corrosion on the fatigue limit of different materials [1] (specimens and pads of same material).

This difference may also be related to the lower notch sensitivity of soft materials. It does not imply that fretting damage is less in soft materials; it even may be worse and also different. However, softer materials are more tolerant against surface imperfections.

The results of Figure 15.8 apply to fretting corrosion between two surfaces of the same material. Fretting can also occur between dissimilar materials. In general, steel and an Al-alloy are an unfavorable combination. If a hard material is combined with a soft material, fretting corrosion damage of the hard material may be minute because of a superior wear resistance.

Roughness of the material surface

Fretting corrosion of polished surfaces can be more serious than for a rough material. In the latter case, contacts occur mainly at the highest points of the surface profile, which may be damaged and plastically deformed. Fretting particles can be collected in the grooves of the rough surface and thus contribute less to further fretting damage. However, fretting damage of a polished surface creates microcracks which can grow immediately. Actually, roughness effects on fretting corrosion have not been widely investigated.

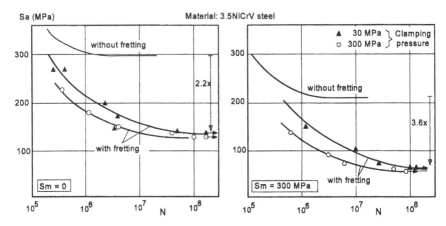

Fig. 15.9 Effect of mean stress on the reduction of the fatigue strength by fretting corrosion. Results of King and Lindley [6].

Contribution of environment on fretting corrosion

As explained earlier, fretting between two surfaces produces oxidized particles, which can enhance further damage of the surface. It then is surprising that fretting in a dry environment may be more severe than in a humid environment. It is possible that local welding between the two surfaces more easily occurs in a dry environment than in a humid environment. This could imply significant damage by tearing the welded asperities in subsequent cycles.

Investigations on fretting corrosion in very aggressive environments are scarce. However, it should be recalled that fatigue properties in such environments are minimal anyway.

Mean stress effect

Constant-amplitude loading is characterized by a stress amplitude (S_a) and a mean stress (S_m). As pointed out earlier, fretting corrosion can have a large detrimental effect on high-cycle fatigue and the fatigue limit, both associated with relatively low stress amplitudes. At high stress amplitudes fretting corrosion can also occur, but its effect on the fatigue life is limited because crack initiation at a high S_a occurs early in the fatigue life, also if fretting does not occur.

With respect to the effect of S_m, it should be recalled that the fatigue limit is a threshold stress level. It is the lowest stress amplitude, which may produce a microcrack that can grow until failure. Cycles with a slightly smaller amplitude cannot overcome barriers for microcrack nucleation or growth of a microcrack to failure. In this context, an increase of the mean stress for the same stress amplitude increases S_{max}. This could assist in overcoming growth barriers for microcracks associated with fretting damage. Furthermore, microcracks are more open at higher S_{max} and can grow more easily. The mean stress effect was confirmed by results of King and Lindley [6] on specimens of a 3.5NiCrMo steel ($S_U = 733$ MPa), as illustrated by the S-N curves in Figure 15.9. The fatigue limit was reduced by a factor of 2.2 times at $S_m = 0$, whereas the reduction factor was 3.6 times at $S_m = 300$ MPa. Fenner and Field [2] for an Al-alloy (L65 $\approx$ 2014-T6) also found a significant mean stress effect on the reduction of the fatigue limit of specimens with fretting corrosion damage.

Fretting corrosion under VA loading

Research on the occurrence of fretting corrosion under VA loading is rarely reported in the literature. However, it must be expected that fretting damage can also be created by load cycles of a VA load history. A detrimental effect was indeed observed in some test programs with random loading carried out on specimens of cast iron and a low-carbon steel [7], and an aircraft aluminium alloy [8]. The fretting damage obtained by clamping pads on the specimens was again observed to occur early in the fatigue life. Although significant fatigue life reductions were found in the random load tests, the impression was that the detrimental effect is not as large as observed in similar tests under CA loading. Furthermore, shot peening [7] and anti-fretting lubrication compounds [8] were beneficial for the fatigue life under fretting corrosion conditions, but the improvement was less under random loading than under CA loading. However, improvements for practical cases in service can hardly be simulated in laboratory test program. Only some qualitative indications can be obtained.

15.4 Methods to avoid fretting corrosion problems

Two different ways can be followed to avoid fretting corrosion problems in structures. The first one is to prevent (or reduce) the occurrence of fretting, which is the more healthy approach (preventing the disease). The second is to accept the occurrence of fretting, but diminish its detrimental effect

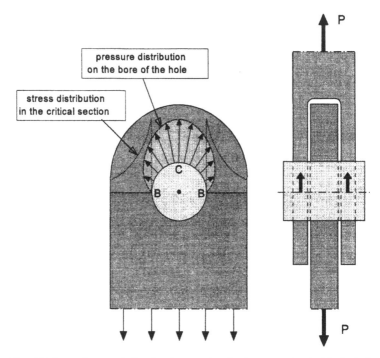

Fig. 15.10 Load transmission in a lug causing fretting corrosion inside the hole.

(controlling symptoms of the disease). It is not always possible to adopt the first method due to practical design restraints of a structure. Illustrative experience is presented below. Some examples are associated with fatigue of lugs because lugs are noteworthy fatigue sensitive structural elements due to fretting corrosion. This will be discussed first.

A lug joint is shown in Figure 15.10. The load in the lug is transmitted by a pin (or a bolt) to a clevis. The load of the pin is applying a pressure distribution on the hole in the lug. The material of the lug between points B and C along the periphery of the hole is carrying a tangential tensile stress, and it thus will be extended. This implies that the material must slide over the pin. As a result, cyclic rubbing between the inside surface of the hole and the outside surface of the pin occurs under fatigue loading. The rubbing displacement is zero at the upper point C for reasons of symmetry, and it increases towards points B. Fretting corrosion damage can occur inside the hole. Although the contact pressure between pin and hole is low at B, the peak value of the stress concentration occurs at this point. Furthermore, the rubbing movements are relatively large near the same point. Indeed, fatigue cracks nucleated at point B or in the neighborhood of this point starting from

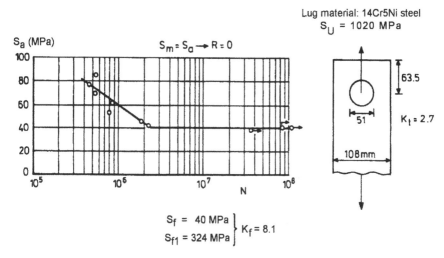

Fig. 15.11 A reduction of the fatigue limit of a lug specimen much larger than predicted by the stress concentration factor. Results of White [9].

fretting corrosion damage. As a result of fretting, the fatigue limit of lugs is exceptionally low. This is well known in the aircraft industry for Al-alloys, but it similarly applies to lugs of steel as illustrated by results of White in Figure 15.11 [9]. The fatigue limit of the lug is 40 MPa. The corresponding fatigue limit at $R = 0$ for the unnotched material is 324 MPa. It implies a notch reduction factor $K_f = 324/40 = 8.1$, which considerably exceeds the stress concentration factor $K_t = 2.7$. The very low-fatigue limit is associated with fretting corrosion inside the hole of the lug. Visual observations in lug holes clearly indicate fretting corrosion debris.

Prevention of metallic contact

The best way to avoid fretting between two rubbing surfaces is to prevent metallic contact between the two surfaces. An unusual way to avoid fretting in a lug hole, based on this principle, is offered by a so-called slotted hole shown in Figure 15.12. A thin layer of the hole edge (0.5 mm) is removed at both sides of the hole by contour milling with a cutter with a small diameter. Metallic contact between the pin and the hole edge is no longer possible between F1 and F2. The net section is slightly reduced by these slots, which gives a small increase of the peak stress at B. However, fretting at B can no longer occur, and the large reduction of the fatigue limit by fretting is eliminated. The fatigue curve of a lug with a slotted hole is compared to

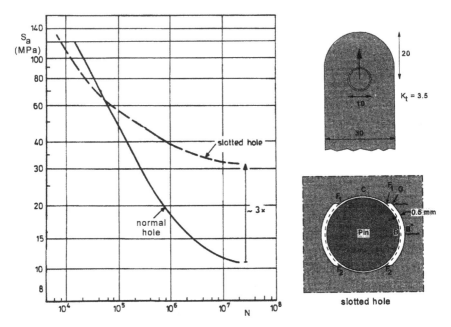

Fig. 15.12 Fretting corrosion effect on the S-N curve of a lug (2024-T3) with a slotted hole to avoid metallic contact between the pin and the edge of the hole.

the S-N curve of a normal lug in Figure 15.12. The curves confirm that an increase of the fatigue limit with a factor of three has been achieved. Fatigue cracks occurred at point B* for high endurances and at point G for low endurances. In both cases the location of a high tangential stress is free from fretting corrosion damage. As a result, a significant increase of the fatigue limit is obtained as shown by the S-N curves in Figure 15.12. Limited fretting corrosion is still possible at point F1, but at this point the tangential stress is zero.

Figure 15.12 also shows that the lug with the slotted hole is slightly inferior at high stress amplitudes where fretting corrosion is less important for fatigue crack initiation. The small reduction of the net section of the lug with a slotted hole then gives a slight increase of the nominal stress level.

A similar example of separation of locations of a maximum peak stress and the location of severe fretting corrosion is a historical one of the 19th century. August Wöhler did his classical research on fatigue failures in railway axles. Wheels were shrink fitted on these axles by a differential heating technique. A very strong joint was obtained, but it was a poor design against fatigue, see Figure 15.13. An extremely high stress concentration occurs in the axle at P, and at the same point fretting corrosion cannot be

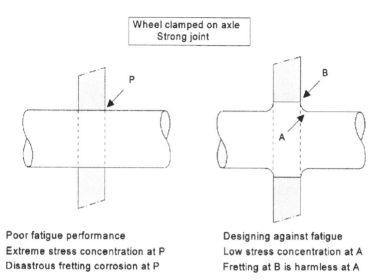

Fig. 15.13 An old improvement of a clamped joint by simple engineering means (Wöhler).

avoided. Fretting corrosion products (iron oxides and hydrates) came out of the joint as a dark-red colored powder, which was labeled as "das Bluten des Eisens" (the bleeding of iron). In the right part of Figure 15.13, the stress concentration is effectively reduced by a shoulder with a generous radius. Moreover, fretting corrosion at the interface between the wheel and the axle now occurs at a location outside the stress concentration area, point B. It was a clever engineering solution, while in the 19th century stress concentration factors and also fundamental knowledge about fretting corrosion were not yet developed.

In Chapter 13 on experiments, it was discussed that a fatigue failure of a specimen can occur at the clamped zone due to fretting corrosion. It was said that these failures could be prevented by inserting a layer of durable paper between the specimen and clamping plates. This is also an example of preventing metallic contact.

Fretting corrosion induced fatigue cracks may also occur if load transmission is not involved. A most simple example is shown in Figure 15.14. A non-loaded angle piece is attached to a cyclically loaded plate. Because the plate is subjected to cyclic strain, rubbing displacements between the plate and the angle piece occur and fretting damage around the bolt hole can initiate fatigue cracks. A simple solution is a thin non-metallic layer between the plate and the angle piece. Adhesive bonding provides a very good layer for preventing metallic contacts. It is one of the major

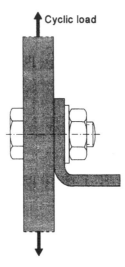

Fig. 15.14 Attachment of angle piece to plate. No load transmission, still fretting.

reasons why adhesive-bonded joints can have significantly better fatigue properties than riveted joints, see Chapter 18.

Alleviation of fretting damage

It is customary to provide lug ends of dynamically loaded connection rods with bushes, see Figure 15.15. Load transmission of the pin to the lug occurs through the bushing. The bushing is pushed with an interference fit into the hole of the lug. As a consequence, residual stresses are introduced. A clamping pressure between the lug and the bushing is present, which can be high enough to prevent rubbing between the bushing and lug. Secondly, residual compressive stresses are introduced in the bushing in the tangential direction. This will be favorable if fretting corrosion occurs between the pin and bushing. Fretting at this interface is also reduced by selecting a wear resistant material for the bushing, e.g. a hard steel or bronze. Furthermore, fretting is reduced by lubrication. Special lubricants have been developed for this purpose, but improvements are uncertain if the lubricant is not periodically replenished.

Fretting can also be controlled by surface treatments. Nitriding and other surface heat treatments reduce the occurrence of fretting damage as well as subsequent microcrack growth, see Chapter 14. Such treatments are often used for gears where sliding contacts between the teeth of the gears may

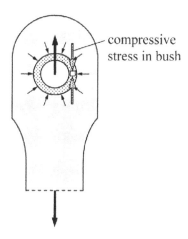

compressive
stress in bush

Fig. 15.15 Severe clamping between the lug and the bush due to an interference fit. As a result, there is compressive stress in the bush.

cause surface damage and fatigue. The motivation is to improve the surface wear resistance.

Shot peening is another way to combat crack growth from fretting induced surface damage. A most effective way to be applied to holes is plastic hole expansion as discussed in Chapter 4, see Figure 4.7. Commercially apparatus is available for this purpose. Residual compressive stresses around a hole can be as high as the compressive yield stress. The fatigue limit of a high-strength Al-alloy (7075-T6) loaded at a mean stress of 123 MPa was raised from 12 MPa (very low-fatigue limit) to 80 MPa (increased 6.7 times). The residual stresses did not prevent fretting between the steel bolt and lug hole. Actually, terrible fretting damage occurred, see Figure 15.16b because of the relatively high S_a and the large number of cycles. However, cracks could not grow through the zone of large compressive residual stress as shown by a microscopic section made afterwards, see Figure 15.16a. Residual stresses do not affect cyclic shear stresses. These shear stresses still caused some microcrack growth along slip bands. However, due to the compressive residual stresses, crack growth occurred in rather erratic directions. In general, fatigue cracks grow perpendicular to the main principle stress (Chapter 2) provided that this stress is a tensile stress which can open the crack tip. In this lug with a plastically expanded hole, a tensile stress did not occur at the edge of the hole and the high compressive residual stresses kept the cracks closed. As a result, strange crack growth directions were followed (Figure 15.16a).

Distance below material surface of hole   View of fretting corrosion damage inside hole
0.61 mm   0.17 mm   of the lug

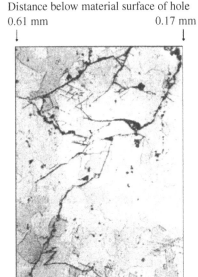

(a) Microscopic cross section of erratic   (b) Lug with plastically expanded hole after $66 \times$
crack growth due to compressive residual 10^6 cycles at $S_a = 78$ MPa ($S_m = 123$ MPa).
stress. Width of picture 0.44 mm   Severe fretting inside hole (diameter 10 mm)

Fig. 15.16 Severe fretting corrosion damage in hole of lug after extensive fatigue loading
without failure [10].

15.5 Topics of the present chapter

1. Fretting corrosion under cyclic load cannot be avoided if metallic
    materials are in contact at mating surfaces. Very small rubbing
    displacements will occur and produce surface damage. The fatigue
    strength in the high-cycle fatigue region is reduced by this surface
    damage. Considerable reductions of the fatigue limit are common,
    especially for high-strength materials, due to the high notch sensitivity
    of these materials.

2. The fretting corrosion mechanism is affected by several variables,
    such as pressure on the interface surface, the amplitude of rubbing
    movements, the materials, surface roughness, environment, and the
    cyclic stress level. Although some systematic trends of the influence
    of these variables are recognized and partly understood in a qualitative
    way, the fretting corrosion mechanism is a complex phenomenon to
    describe in physical detail.

3. Methods to avoid detrimental effects of fretting corrosion fatigue in structures, and particularly in joints, are based on two principles: (i) Avoid metallic contact by structural detail design, or non-metallic interlayers, and (ii) surface treatments to improve surface wear resistance or introduce compressive residual stresses.

References

1. Waterhouse, R.B., *Fretting Corrosion*. Pergamon Press (1972).
2. Fenner, A.J. and Field, J.E., *A study of the onset of fatigue damage due to fretting*. Trans. North East Coast Inst. Engrs. Shipbuilders, Vol. 76 (1960), pp. 183–228.
3. Endo, K. and Goto, H., *Initiation and propagation of fretting fatigue cracks*. Wear, Vol. 38, (1976), pp. 311–324. *Effects of environment on fretting fatigue*. Wear, Vol. 48 (1978), pp. 347–367.
4. Hoeppner, D.W. and Gates, F.L., *Fretting fatigue considerations in engineering design*. Wear, Vol. 70 (1981), pp. 155–164.
5. Funk, W., *Test methods to investigate the influences of fretting corrosion on the endurance*. Materialprüfung, Vol. 11 (1969), pp. 221–227 [in German].
6. King, R.N. and Lindley, T.C., *Fretting fatigue in a $3\frac{1}{2}NiCrMoV$ rotor steel*. Central Electricity Res. Lab., RD/L/N 75/80 (1980).
7. Buxbaum, O. and Fischer, G., *The influence of fretting corrosion on the durability under sinusoidal and random loading*. Konstruktion, Vol. 40 (1988), pp. 333–338 [in German].
8. Edwards, P.R. and Ryman, R.J., *Studies in fretting fatigue under variable amplitude loading*. Royal Aircraft Establishment, Technical Report t TR 75132 (1975).
9. White, D.J., *Fatigue strength of small pinned connections made from alloy steel FV 520 B*. Proc. Inst. Mech. Engrs., Vol. 182 (Part I) (1968), p. 615.
10. Schijve, J., *Fatigue of lugs*. Contributions to the Theory of Aircraft Structures. Nijgh-Wolters Noordhoff University Press (1972), pp. 423–440.

Some general references
11. Lindley, T.C., *Fretting fatigue in engineering alloys*. Int. J. Fatigue, Vol. 19 (1997), pp. 39–49.
12. Shaffer, S.J. and Glaeser, W.A., *Fretting fatigue*. ASM Handbook, Vol. 19, Fatigue and Fracture, ASM International (1996), pp. 321–330.
13. Glaeser, W.A. and Shaffer, S.J., *Contact fatigue*. ASM Handbook, Vol. 19, Fatigue and Fracture, ASM International (1996), pp. 331–336.
14. Waterhouse, R.B. and Lindley, T.C. (Eds.), *Fretting Fatigue*. ESIS 18, Wiley (1994).
15. Attia, M.H. (Ed.), *Standardization of Fretting Fatigue Test Methods and Eequipment*. ASTM STP 1159 (1992).
16. Lindley, T.C. and Nix, K.J., *The role of fretting in the initiation and early growth of fatigue cracks in turbo-generator materials*. Multiaxial Fatigue, K.J. Miller and M.W. Brown (Eds.), ASTM STP 853 (1985), pp. 340–360.
17. Alban, L.E., *Systematic Analysis of Gear Failures*. ASM (1985).
18. Brown, S.R. (Ed.), *Materials Evaluation under Fretting Conditions*. ASTM STP 780 (1982).
19. Waterhouse, R.B., *Fretting Fatigue*. Applied Science Publishers, London (1981).

Chapter 16
Corrosion Fatigue

16.1 Introduction

Corrosion fatigue by definition is fatigue in a corrosive environment. An aggressive environment can be harmful for the fatigue life of a structure, and protection against corrosion is necessary. Designers must consider corrosion in service, not only in view of fatigue. Corrosion is undesirable for reasons related to a safe and economic use of a structure during its service life. Corrosion can also be unacceptable in view of the appearance of a structure, i.e. for cosmetic reasons. Usually, corrosion prevention is considered to be a matter of selecting a corrosion resistant material or applying a suitable surface protection, such as paint or cadmium plating, etc. Unfortunately, these options do not guarantee good fatigue properties. Furthermore, several high-strength materials have a relatively poor corrosion resistance. Disastrous accidents have occurred due to fatigue cracks starting from corrosion damage, in several cases corrosion pits. Whenever corrosion damage can occur to the material surface of a dynamically loaded structure, corrosion fatigue can be a serious problem.

Corrosion fatigue should not be confused with stress corrosion, which is crack initiation and growth under a sustained load or residual stress. Usually, stress corrosion occurs along an intergranular crack growth path, whereas corrosion fatigue in many cases is still a transgranular crack growth phenomenon. Moreover, stress corrosion does not occur in many technical materials, whereas corrosion fatigue can occur in most materials. Corrosion fatigue is also not the same as fatigue of corroded material. Of course, corrosion damage can decrease the fatigue properties because it implies surface damage which will reduce the crack initiation life. The effect of the surface quality was discussed in Chapter 14. The problem considered in the present chapter is technically relevant if a corrosive environment is present during the entire life time of a structure. It implies that the crack initiation period and the crack growth period can be affected both.

As shown by the literature, corrosion fatigue was investigated in various experimental programs, primarily as a material problem. Constant-amplitude (CA) fatigue tests were carried out with closely controlled environments on simple laboratory specimens. The investigations have revealed that damage accumulation during corrosion fatigue is caused by the combined actions of fatigue and corrosion with mutual interactions. The type of material and the environment of the experiments is usually referred to as the material/environment system. Because corrosion is a time-dependent mechanism, it is obvious that the load frequency and the wave shape of the load cycles can be significant for corrosion fatigue.

Corrosion fatigue as a problem of a structure in service is characterized by variables which can be highly different from the material/environment system of laboratory research. The differences were discussed by Schütz in a publication with the title: "Corrosion fatigue. The forgotten factor in assessing durability" [1]. He emphasized the difference between laboratory experience and the exposure of a structure under variable-amplitude (VA) loading in service. Laboratory tests should be completed in an acceptable time period whereas a structure in service is exposed to a non-controlled environment and exposure times of years. The question then is how the basic understanding of corrosion fatigue obtained in fundamental research can be transferred into practical considerations for structural design against fatigue.

In the present chapter, results of laboratory investigations are discussed first (Section 16.2). It is a selection of observations which have contributed to reveal corrosion effects on fatigue, including frequency and wave shape effects. Illustrative results are presented, but it is not an extensive survey of the numerous investigations published in the literature. Practical aspects about corrosion fatigue problems are considered in Section 16.3. A case history is discussed in Section 16.4. The main topics of the present chapter are listed in Section 16.5.

16.2 Aspects of corrosion fatigue

In the previous chapter on fretting corrosion, it was said that fretting primarily affects the crack initiation period and not the crack growth period. However, as already discussed in Chapter 2 (Section 2.5.7), in a corrosive environment, both crack initiation and crack growth are affected by the environment. In some materials with a good corrosion resistance, it is possible that the effect of the environment on crack initiation is insignificant, while fatigue crack growth is accelerated. In such a case, it is just the other way round compared to fretting corrosion. This exceptional behavior has been observed for some Ti-alloys. However, in general, corrosion has an unfavorable influence on both crack initiation and crack growth. The contribution to crack initiation is largely a pure corrosion mechanism causing surface damage. The acceleration of crack growth by a corrosive environment is caused by some interaction between a corrosive mechanism, cyclic slip at the crack tip and a rupture mechanism (decohesion), which leads to an enhanced crack extension.

The large effect of a corrosive environment on an S-N curve was discussed in Section 2.5.7 and illustrated in Figure 2.29 by results for mild steel tested in air, water and salt water. A significant decrease of fatigue lives and fatigue strength is observed in this figure. The large reduction of the fatigue limit is most noteworthy. It implies that crack initiation is possible at low stress amplitudes, which would not have occurred without the corrosive environment. After crack initiation, the environment can enter the crack, reach the crack tip, and enhance the crack growth rate. The corrosion processes involved are chemical, electrochemical in liquid environments, or physical.

Corrosion is a time dependent process. As a consequence, corrosion fatigue should depend on the time scale of the load history. During fatigue at a low frequency, much more time is available for a corrosion mechanism than during fatigue at a high frequency. The frequency effect was also illustrated by the results of Figure 2.29.

Two types of environments should be considered:

1. *Gaseous environments*
   Some gaseous environments do not interact with the material during the fatigue process. The environment is considered to be inert. In this respect, the most pure inert environment is vacuum, which is technically significant for space applications. It is also used in research investigations to study fatigue without any interference from an

environment. The experimental problem is to obtain a very low vacuum pressure in order to be sure that effects on the fatigue mechanism are excluded. Fatigue tests have been carried out with a vacuum pressure as low as 0.1 Pa.[20] An alternative is to use an inert gas, e.g. argon or nitrogen. The gas then should be highly purified, which also implies that it should be an extremely dry gas without water vapor.

Fatigue tests in a laboratory are usually carried out in air. It is more carefully specified as "laboratory air" with data on the temperature (often labeled as room temperature, RT) and the relative humidity (RH). Results of fatigue tests in air are frequently used as a reference for comparison to fatigue in an aggressive environment. Such comparisons might suggest that air is a non-corrosive environment. However, this is not correct. Water vapor and oxygen are active agents during fatigue of several materials. Furthermore, decomposition of water vapor implies that hydrogen can also play some role in fatigue by entering the material.

2. *Liquid environments*

The most well-known liquid environment used in test programs is salt water. The detrimental effect of salt water was known long ago from industrial applications. Experiments on the effect of salt water were already carried out in the early 1930s by Gough and co-workers [2]. Results of this work are shown in Figure 16.1. The large reduction of the fatigue limit is most obvious. The technical relevance of liquid environments is easily understood. Many structures become wet, e.g. structures in the sea, structures outdoors by rain, otherwise by condensation of water vapor, etc. Near coast lines, water will contain salt. In many cases, the presence of water cannot be avoided, while water can contain several impurities which can affect corrosion fatigue. The corrosion mechanism in a liquid is electro-chemical. At a material surface, corrosion can contribute to crack nucleation to create the very first microcrack. As soon as a crack is present, the liquid environment enters the crack. Under cyclic loading, the crack acts as a pump due to cyclic opening and closing of the crack. The environment is drawn into the crack.

Corrosion fatigue has extensively been studied in various research programs reported in the literature. This has significantly contributed to the present

---

[20] 0.1 Pa = 0.1 N/m^2 = 10^{-6} bar $\approx 10^{-6}$ atmosphere. Another frequently used unit is torr, called after Torricelli of the mercury (Hg) barometer. The conversion is: 1 torr (= 1 mm Hg) = 133.3 Pa.

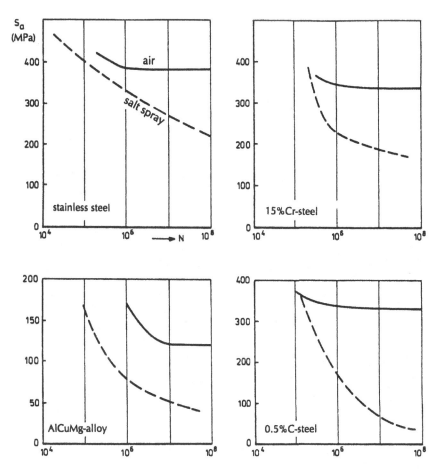

Fig. 16.1 Effect of salt water on S-N curves. Note the large reduction of fatigue limits. Results of Gough and Sopwith [2].

understanding of corrosion fatigue. It has also shown that corrosion fatigue is a rather complex phenomenon. Investigations on the fatigue mechanism under corrosive conditions are difficult. The number of variables involved is large, both with respect to materials and environments. The process zone is extremely small, e.g. crack initiation in a slip band, or crack extension at a crack tip. Dimensions of the crack extension in one cycle are in the μm range, or even in the sub-micron range. It is hard to make in situ observations other than by fractography in the electron microscope. However, the consequences of corrosion fatigue in terms of fatigue lives, fatigue limit, and crack growth rates can be measured. The discussion in this section is restricted to some simplified mechanistic aspects and trends shown in experiments. Aspects of the engineering significance are covered in Section 16.3.

16.2.1 Corrosion fatigue in gaseous environments

Most investigations on corrosion fatigue in gaseous environments were made on fatigue crack growth. Crack initiation, starting in slip bands at a free material surface, has been discussed in Chapter 2. It was pointed out that cyclic slip is causing intrusions (Figure 2.2), which create fresh metal surfaces exposed to the environment. For most technical materials, it implies that oxidation will occur immediately, which will interfere with subsequent cyclic slip. This very first microcrack growth should depend on the presence of oxygen in the environment. It is also possible that water vapor plays a role in this environmental process.

Fatigue tests in vacuum on pure metals have shown that cyclic slip is a more distributed phenomenon occurring at several places along the free surface. It is not hampered by an outside oxide layer on the metal. However, technical materials are always covered by some oxide layer, which usually is rather brittle. Cyclic slip during fatigue in air is more concentrated in a small number of slip bands. It still remains questionable whether gaseous environments can reduce the fatigue limit. However, as soon as microcracks are present, a gaseous environment enters the fatigue crack and an environmental effect on microcrack growth can occur. The effect was confirmed in experimental investigations on macrocracks. It has been shown that water vapor is of essential importance for fatigue crack growth in several materials, including steels and Al-alloys. Figure 16.2 shows results of Pao et al. [3] for fatigue crack growth in AISI 4340 steel ($S_U = 2082$ MPa) in a water vapor environment of a low pressure (585 Pa = 4.4 torr). A significant and systematic frequency effect occurred in the frequency range of 0.1 to 10 Hz. Apparently, a time-dependent mechanism is active during fatigue crack extension. It is generally associated with some hydrogen mechanism. Dissociation of water molecules produces OH^- and H^+ which can affect the cohesion strength at the crack tip. It is referred to as an embrittlement effect during fatigue crack growth. However, it remains a difficult problem to define in detail how this occurs. Moreover, it is not necessarily the same mechanism for different materials. The effect should not be considered to be the same as hydrogen embrittlement under sustained loading, which usually leads to an intergranular failure.

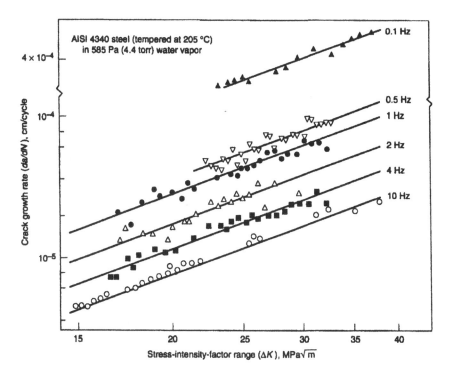

Fig. 16.2 Fatigue crack growth in a high-strength low-alloy steel in water vapor. Effect of frequency. Results of Pao et al. [3].

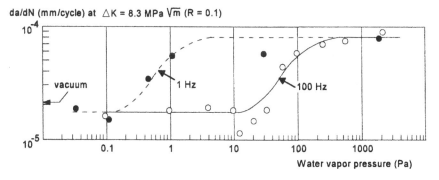

Fig. 16.3 Fatigue crack growth in an Al-alloy (RR58 ≈ 2618). Effect of water vapor pressure. Results of Bradshaw and Wheeler [4].

Similar indications on water vapor effects were observed in fatigue crack growth experiments on Al-alloys. Bradshaw and Wheeler carried out tests in water vapor of different pressure levels [4]. Crack growth rates obtained at a ΔK-level of 8.5 MPa$\sqrt{m}$ are shown in Figure 16.3. At a low water vapor pressure level, the crack growth rate was 0.16 μm/cycle, almost the same as the growth rate in vacuum at the same ΔK. With so little water vapor, it appears to be an inert environment. However, at higher water vapor pressures, the crack growth rate goes to a constant maximum level of about 0.8 μm/cycle, which is five times larger than at the low water vapor pressure level. The transition from the low to the high growth rate occurs in a similar S-curve for 1 and 100 Hz, but at 1 Hz at a 100 times lower pressure range than at 100 Hz. According to Bradshaw and Wheeler, it supports a dynamic adsorption model. More support for the significance of water vapor is coming from the observation that crack growth rates in Al-alloys are higher in humid inert gases than in dry air or dry oxygen.

An interesting observation was made by Broek [5], who found that the crack growth rate in Al-alloy sheet specimens (7075-T6) tested in air at −75°C was significantly lower than in air at RT. At such a low temperature, the water vapor pressure is very low, even at 100% RH, and this leads apparently to a lower crack growth rate. Although the lower ductility of this material at such a low temperature might enhance crack growth, the water vapor effect was predominant in reducing the growth rate.

16.2.2 Corrosion fatigue in liquid environments

Environmental effects

The most noteworthy effect of a liquid environment is the large reduction of the fatigue limit as illustrated by Figure 16.1. Under corrosive conditions fatigue cracks can be nucleated at very low amplitudes, whereas this does not occur if corrosion damage is not present. The initial contribution from corrosion is in creating surface damage. The corrosion pit shown in Figure 16.4 is severe surface damage. Dissolution of material has occurred. The picture also indicates an intergranular corrosive attack. Actually, it is a horrible notch, which should have a large effect on the fatigue limit. The effect of corrosion pits on the S-N curve was illustrated in Chapter 2 with Figure 2.28.

Corrosion damage is generated in the first part of the fatigue life. Surface corrosion damage implies that micronotches are present. Subsequent fatigue crack growth is again assisted by corrosion. Because this can occur at stress amplitudes far below the original fatigue limit, it still can take a large number of cycles until complete failure.

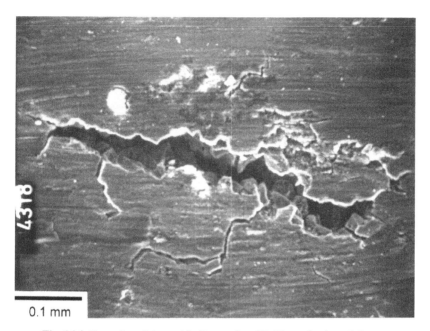

0.1 mm

Fig. 16.4 Corrosion pit in an Al-alloy surface [6]. Tip to tip about 0.5 mm.

Investigations of macrocrack growth have indicated that corrosion in an aqueous environment significantly contributes to fatigue crack growth. The involvement of corrosion variables was borne out by effects on the fatigue crack growth rate when electrochemical conditions were modified. An obvious approach is to carry out comparative tests in different electrolytes. This was done by Duquette and Uhlig using deaerated water in comparison to non-deaerated water in tests on low-carbon steel specimens [7]. The results in Figure 16.5 show that deaerated water gave a better S-N curve than aerated water. The graph indicates that the same trend was also found for tests in a 3% NaCl solution. The S-N curve for fatigue in air in Figure 16.5 suggests that deaeration inhibits early corrosion damage of the material surface. This follows from the reduction of the fatigue limit, which is smaller in deaerated salt water than in aerated water. In deaerated water without salt, the reduction is even absent. But the test results also raise a question; why does deaerated water give a better fatigue life than air?

The corrosion aspect of fatigue crack growth in aqueous solutions is further confirmed by experiments with liquid corrosion inhibitors added to salt water. Crack growth rates in several high-strength steels were significantly reduced by the application of inhibitors as shown by results of Lynch et al. [8]. The corrosion activity is indeed dependent on the

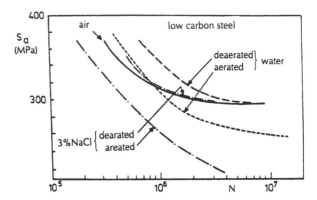

Fig. 16.5 S-N curves of low-carbon steel in water. Effect of deaerating water. Results of Duquette and Uhlig [7].

composition of the electrolyte at the crack tip. Pollution of water can also affect the corrosion fatigue behavior.

Furthermore, electrolytic corrosion is affected by application of an electrical potential on a specimen. Actually, this technique is used as a practical means for corrosion protection of structures operating in water or underground. The application of a potential voltage modifies the anodic or cathodic reaction. In laboratory fatigue experiments, it has been shown that an electrical potential imposed on a specimen can reduce surface corrosion damage, crack initiation and fatigue crack growth.

With respect to explaining faster fatigue crack growth in a liquid solution as compared to fatigue in air, different suggestions are proposed in the literature. Arguments adopted are associated with anodic dissolution phenomena at the crack tip, or again some hydrogen embrittlement mechanism. Plastic deformation at the crack tip can make the material of the crack tip zone more anodic than away from the tip. The crack tip zone then becomes more corrosion sensitive. Moreover, in salt water Cl-ions may weaken the cohesive strength of the material. The weakening can also be due to absorbed hydrogen produced as a result of the local corrosion activity.

Frequency and wave shape effects

Interesting results were published by Barsom [9], see Figure 16.6. He carried out crack growth tests on a maraging steel (12Ni-5Cr-3Mo, $S_U = 1290$ MPa) using different frequencies and wave shapes (positive and negative saw tooth wave shapes, see Figure 2.30). Tests in air did not reveal an influence of

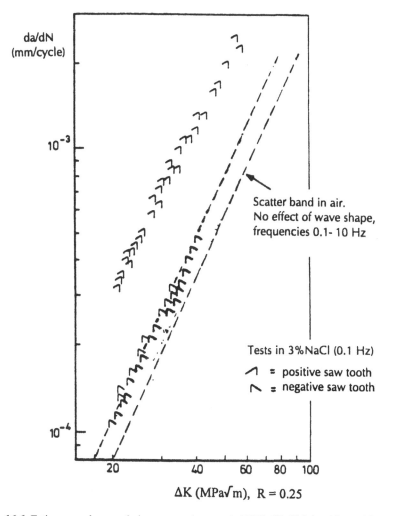

Fig. 16.6 Fatigue crack growth in a maraging steel (12Ni-5Cr-3Mo) with positive and negative saw tooth wave shapes. Results of Barsom [9].

these variables. In salt water, a significant frequency effect was found for sinusoidal load cycles with frequencies of 0.1, 1 and 10 Hz, respectively. Interesting results were obtained with positive and negative saw tooth cycles. The results for negative saw tooth cycles in salt water are practically within the scatter band of the test results in air (Figure 16.6). It suggests that corrosion did not accelerate crack growth. This was explained by the high loading rate when increasing the load from S_{min} to S_{max} in a very short time. In this part of the load cycle, crack extension must occur. In spite of the low overall frequency of 0.1 Hz, the loading part of the cycle is too short for a significant contribution of salt water to crack extension. However, in the tests with the positive saw tooth cycles, the load is increased from S_{min} to S_{max} very slowly; for 0.1 Hz in 10 seconds. There is ample time for the corrosive environment to interact with the crack extension mechanism during the increasing stress intensity at the crack tip. The crack growth rate was about three times faster than for the negative saw tooth cycles.

Another interesting question was addressed by Atkinson and Lindley [10]. They considered slow load cycles with hold times at the maximum load, see the wave shapes in Figure 16.7. The technical relevance is related to structures, for which the working load is maintained for a considerable period of time. Pressure vessels are a good example. The hold times at S_{max} in the test program were 0, 1 and 10 minutes respectively. Two loading rates were applied, characterized by the rise time and fall time, both being 100 seconds (wave shapes A-C) or 10 seconds (wave shapes D-F). Crack growth rates were measured at a constant ΔK-value of 50 MPa$\sqrt{m}$. The results in

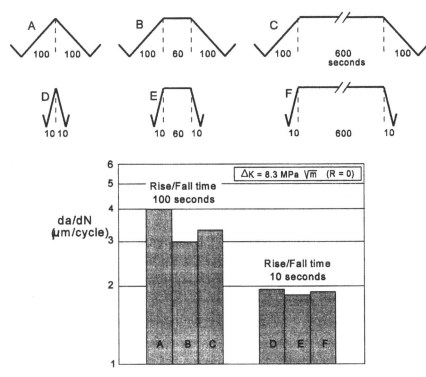

Fig. 16.7 Effect of hold times at maximum load on fatigue crack growth in 0.22C-stetel in lake water. Results of Atkinson and Lindley [10].

Figure 16.7 indicate a systematic influence of the rise time. The crack growth rate for increasing the load from minimum to maximum in 100 seconds was about two times larger than for a rise time of 10 seconds. However, the results for hold times at K_{max} of 1 or 10 minutes, or no hold time at all, did not indicate a systematic difference. It thus should be concluded that crack extension occurred during raising the load from minimum to maximum, and not during the hold time at K_{max}. If some crack extension might occur during the hold time period, it should be associated with stress corrosion. It then may be recalled that stress corrosion in many materials occurs along the grain boundaries (intergranular), whereas fatigue crack growth occurs through the grains (transgranural). It thus is not obvious that crack extension of corrosion fatigue and stress corrosion could be additive.

Comparative fatigue crack growth tests were carried out in Delft [11] on Al-alloy specimens (7075-T6) in salt water. The wave shapes are shown in Figure 16.8. Different rise times (time for raising the load from S_{min} to S_{max}) were used. Hold times were added at either S_{max} or S_{min}, see the inset figures in Figure 16.8. The crack growth tests were carried out at $S_{max} = 90$ MPa and $S_{min} = 30$ MPa ($R = 1/3$). The crack growth rates in Figure 16.8 are da/dN-values at $\Delta K = 10$ MPa$\sqrt{m}$ ($2a \approx 17.0$ mm

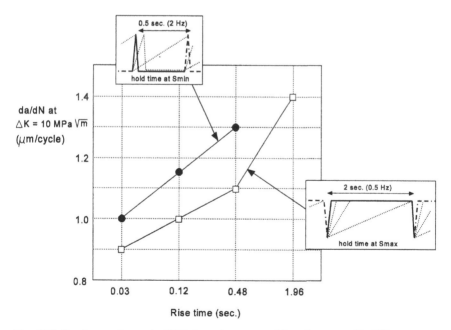

Fig. 16.8 Crack growth rate in 7075-6 specimens tested in salt water with different wave shapes. Effects of rise time and holding at S_{max} or S_{min} [11].

in a 100 mm wide specimen, average results of three similar tests). The crack growth rate increases for a longer rise time, i.e. a slower load increase. This trend agrees with the observation of Barsom and Atkinson and Lindley (Figures 16.6 and 16.7, respectively). However, the crack growth rate was lower for holding times at S_{max} as compared to the results for holding times at S_{min}, perhaps an unexpected trend. It was tried to clarify the trend by fractographic observations in the SEM. It turned out that holding at S_{max} caused secondary cracking in grain boundaries perpendicular to the crack front. These cracks are associated with a stress component in the thickness direction. It was generally observed that the fracture surface for holding at S_{max} had a more complex topography which was responsible for the slower crack growth.

Some additional comments on the crack growth mechanism

Perhaps unexpected, it was found by Komai that fatigue crack growth in low-carbon steel could be slower in salt water than in air [12]. It turned out that corrosion products filled the crack, which prevented a full crack closure at S_{min} and probably caused a smaller effective ΔK. Wanhill and Schra observed that corrosion could increase the threshold ΔK_{th} obtained in air for two Al-alloys (2024-T3 and 7475-T761), even more by sump water than by salt water [13]. The increase of ΔK_{th} was associated with jamming of the crack tip by corrosion products. Remarkably enough, Wanhill and Schra found that sump water considerably accelerated crack growth at higher ΔK-values (tests at 13 Hz). It should be expected that an increased mean stress will enhance the corrosion effect because the crack will be more open. This promotes the accessibility of the environment to the tip of the fatigue crack.

An increased temperature should also enhance the corrosion fatigue mechanism. In the temperature range were most liquid environments can exist, it is expected that activation of the corrosion mechanism could become more significant at an increased temperature, The information in the literature is limited, although increased crack growth rates in the temperature range of 10 to 60°C are occasionally reported, a trend which should be expected. This was confirmed by results of Atkinson and Lindley [14] obtained in fatigue crack growth tests on a carbon steel specimens loaded at $\Delta K = 50$ MPa$\sqrt{m}$ ($R = 0.05$). Crack growth rates for temperatures of 0, 25, 50, 75 and 100°C were 1.0, 2.0, 2.6, 3.6 and 4.6 μm/cycle respectively.

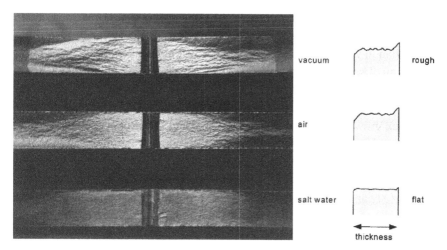

Fig. 16.9 Effect of environment on the fatigue fracture surfaces of fatigue cracks in center-cracked specimens of Al-alloy plate specimens, 7075-T6, thickness 6 mm [16]. $S_{max} = 98$ MPa, $R = 0.1$.

The above examples of corrosion fatigue test results illustrate that corrosion fatigue crack growth is a complex phenomenon, which can be quite different for different material/environment systems. In addition to various aspects of corrosion reactions, the role played by the configuration of the fracture surface is somewhat underexposed in the literature. Different corrosion fatigue mechanisms can cause different fatigue fracture surfaces, both microscopically and macroscopically. On a microscopic level, Stubbington and Forsyth [14] refer to two types of striations in Al-alloys, brittle striations, which are rather flat, and ductile striations, which are more undulated. Brittle striations occurred in an aggressive environment and ductile striations in air as an environment.

Different environments can also lead to macroscopically visible differences of the fracture surface. A noteworthy example is offered by the occurrence of shear lips, described in Section 2.6, see Figure 2.38. Shear lips at a free material surface have been observed on fatigue crack fractures in several steels and Al-alloys. Horibe et al. [15] carried out fatigue crack growth tests on low-carbon specimens in air and in salt water. The fracture surface had a more brittle appearance in salt water, while the transition to shear lips during crack growth was less evident than in air. In an aggressive environment, shear lip formation is suppressed. Apparently, an aggressive environment is promoting tensile decohesion, rather that shear decohesion occurring in the shear lips. A similar corrosion effect was observed in comparative tests on center-cracked Al-alloy specimens tested in vacuum, laboratory air, and salt water [16]. Crack growth in air was about two times faster than in vacuum, while in salt water it was again about three times faster than in air. The faster crack growth in the more aggressive environment could not be correlated with a larger ΔK_{eff} because crack closure measurements indicated the same S_{op} for the three environments.[21] The difference between the crack growth rates should thus be associated with environmental effects. However, the fracture surfaces obtained in the three environments were also different, see Figure 16.9. The fracture surface in vacuum showed the largest shear lips and a more undulated fracture surface over the full thickness. The smallest shear lips and the most flat fracture surface occurred in salt water. The fracture mechanisms are apparently different in the three environments, and a different crack growth resistance should thus be expected. In addition, a more flat fracture surface implies a straight crack front, whereas a corrugated fracture surface is associated with a tortuous crack front. The crack driving force (dU/da) per unit length of the crack front will be smaller for an irregular fracture surface than for a flat fracture surface. In other words, the environmental effect on crack growth is not solely the result of a corrosion mechanism; the fatigue fracture characteristics are also involved.

16.3 Practical aspects of corrosion fatigue

In Section 16.2, several aspects of corrosion fatigue have been discussed and illustrated by experimental data. Some evident trends of these data are: (i) a large reduction of the fatigue limit can occur in liquid environments, (ii) crack growth rates may be significantly increased by a corrosive environment, and (iii) the load frequency and wave shape can have a significant effect on crack growth. It was also pointed out that the detrimental effects could be understood in terms of surface corrosion damage and enhanced crack growth rates by corrosion assisted crack extension during cyclic loading. These effects are dependent on the type of material and environment. The understanding was largely obtained in investigations in the laboratory, usually in CA tests on unnotched specimens or on crack growth specimens (MT and CT). The question remains how this understanding should be translated into design and production practices, and prevention

---

[21] Theoretically, an effect of the environment on the plastic zone size should not be expected. As a consequence, a similar crack opening stress level, and thus a similar ΔK_{eff}, appears to be logical.

of corrosion fatigue in service. As an example of this problem setting, the practical relevance of fatigue tests on aircraft Al-alloys specimens submersed in salt water is not directly obvious. After all, an aircraft is not a submarine. But some lessons can be learned from understanding the corrosion fatigue problem.

Corrosion can cause surface damage, which should be considered to be notches. In spite of a limited depth (a few tenths of millimeters), corrosion pits must be associated with high stress concentrations. Especially for high-strength materials, the corrosion damage can be disastrous. For instance, fatal accidents have occurred due to fatigue failures starting from a corrosion pit in the rotor system or the blades of helicopters. The characteristic feature was corrosion damage at the material surface of a notch which was open to the environment. Corrosion protection had been insufficient. The notch should not be accessible for the environment, or surface treatments should prevent crack growth from corrosion surface damage, e.g. shot peening or other surface treatments (Chapter 14).

It can be tried to make a prediction of the remaining fatigue life if a corrosion pit is present. The corrosion pit should then be represented by an equivalent initial flaw, and the life prediction is a fatigue crack growth prediction. Problems involved are associated with assumptions to be made. The initial flaw size is an important parameter. The crack growth rate is relatively low for a small crack as a result of the low stress intensity factor. As a consequence, assuming an initial crack depth of either 0.2 or 0.4 mm (0.008 or 0.016 inch) can imply a large difference for the predicted crack growth life. Another problem is the crack growth prediction. Which crack growth data will be used? The environment causing the surface damage can also affect subsequent crack growth. Furthermore, is the load spectrum sufficiently known? At best, an estimate of the order of magnitude of the remaining fatigue life is obtained from a prediction, although the estimate can be instructive for considering safety and economic aspects.

In Section 16.2, the discussion of corrosion fatigue was mainly concerned with CA tests on unnotched specimens or crack growth specimens. The results of such tests can still be relevant to notches in a structure if they are directly exposed to the environment, e.g. open holes and fillets. A good environmental accessibility also exists for welded structures in the sea. In other situations, the most fatigue critical locations are not in contact with a liquid environment, for instance inside closed box type structures, or in bolted and riveted joints depending on how these joints are designed and produced. However, it should not be overlooked that air also contains water vapor. It implies that water by condensation can be present at notches which can be reached by air. Furthermore, if plates are joined by bolts or rivets, crevices between the plates may not be fully airtight, and air can penetrate between the plates. Water can then be present also, and cyclic loading will help further penetration between the plates. It can lead to the well-known crevice corrosion. Industries have learned to recognize such problems by service experience. Because corrosion is generally undesirable, also if fatigue is not a problem, various ways have been developed to protect structures against different types of corrosion. At the same time, this is a protection against corrosion fatigue. Actually, it is commonly believed in the engineering world that the corrosion fatigue problem should primarily be handled as a corrosion problem. It thus implies avoidance of corrosion by proper design and material selection, supplemented by suitable surface treatments.

Sometimes, incidental fatigue cracks cannot be avoided. It then may be important to know if a corrosive environment will accelerate fatigue crack growth in service. If fatigue cracks are present, corrosion protection techniques applied to the material surface may be of little help because fatigue cracks can disrupt surface coatings. It could be tried to make a crack growth prediction, but the result should be considered with caution as said earlier. Safety factors are necessary, and the choice of these factors is a matter of judgment. An alternative is to carry out realistic service-simulation fatigue tests. The tests should be realistic, or at least conservative, with respect to the load history to be applied and the environment to be used. The specimen geometry should be representative for the structure under consideration. A realistic simulation of the load-time history can imply that experiments will require long testing times. In view of differences between service-simulation fatigue tests and the real service life conditions, a safety factor should still be considered. According to Schütz [1], a factor 2 may then be sufficient, but larger factors might be considered, depending on safety and economy arguments.

One variable mentioned in the previous section is the temperature which can affect corrosion reactions and increase the crack growth rate. Limited experimental evidence is available on this issue and it should be realized again that this evidence is coming from laboratory tests carried out at relatively high test frequencies in a short period. Engineering judgment as well as qualitative understanding of corrosion fatigue is essential to make decisions in such cases.

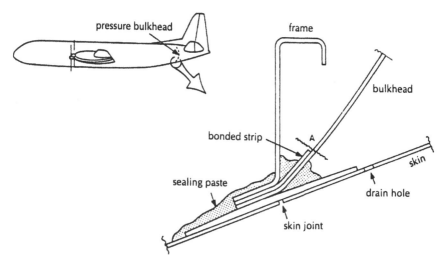

Fig. 16.10 Corrosion fatigue caused a catastrophic failure in a pressure bulkhead of a transport aircraft [17]. The failure occurred along the edge of the bonded strip at location A. Rivets are not shown in this figure.

16.4 A case history

In 1968 an aircraft crashed as a results of a pressure cabin failure occurring at cruising altitude. The accident investigation showed that the rear pressure bulkhead was blown out, which damaged the control system of the empennage rudder and elevators [17]. Nobody survived the accident, which resulted in 85 fatalities. Figure 16.10 shows the position of the rear pressure bulkhead in the aircraft, and a more detailed cross section of the structure at the fatigue critical location. The hemispherical dome of the pressure bulkhead is connected to the fuselage skin which is supported by a frame at the same location. The bulkhead is loaded by the cabin pressure which introduces a tensile membrane stress in the sheet material of the bulkhead. In general, each flight implies one load cycle on the joint of the pressure bulkhead to the fuselage skin. The joint is reinforced with a bonded strip along the edge of the pressure bulkhead. The designer was aware of the corrosive environment. Sump water can be collected in the lower part of the joint, i.e. between the bulkhead and the fuselage skin, and between the bulkhead and the frame.

Corrosion prevention is taken care of by a drain hole in the fuselage skin and by applying a sealing paste at the locations where crevice corrosion might occur. Corrosion fatigue cracks started at A and grew along the edge of the bonded strip. The environment inside the pressure cabin is humid air at room temperature. On the outside of the pressure bulkhead, the temperature at cruising altitude is very low; in the order of $-50°C$. As a result of the temperature difference between both sides of the pressure bulkhead, condensation of water vapor occurs at the inside of the bulkhead. Due to gravity, the condensed water flows down and is trapped in small quantities at the edge of the bonded strip, point A in Figure 16.10. This caused local corrosion damage which initiated fatigue cracks. After some circumferential crack growth, a catastrophic decompression failure occurred. The failure mode was corrosion fatigue.

Typical fatigue and corrosion prevention aspects were involved:

1. The material selection was not optimal. The crack growth properties and corrosion resistance of the Al-alloy involved were relatively poor.
2. From a fatigue point of view, the bonded strip is certainly a better solution then a riveted strip, because the notch effect is smaller. However, it was overlooked that the thin edge of the strip could trap water causing corrosion damage. It may be noted that the bonded strip is positioned at the inside of the pressure bulkhead. If the strip would have been bonded at the outside, it could not have trapped water.
3. The aircraft manufacturer applied sealing paste to the joint to avoid corrosion. If the sealing had included the edge of the bonded strip, corrosion fatigue should have been avoided.
4. Inspection of the location where corrosion fatigue started is difficult.

The case history illustrates that material selection, corrosion prevention and detail design can be important to avoid corrosion fatigue in service. Sump water is a well-known cause for corrosion problems in structures of steel and Al-alloys. Well-known examples are pockets in steel structures which collect stagnant water. The same can occur in the belly of a transport aircraft. Such circumstances should be avoided by a more clever design and corrosion protection.

16.5 Topics of the present chapter

1. Corrosion fatigue is a complex phenomenon due to the corrosion aspects involved. Detrimental effects of corrosion on crack initiation and crack growth under fatigue loading depend on the type of material and environment. Systematic effects have been recognized, but

these effects are not generally applicable to all material/environment combinations.

2. Corrosion fatigue effects are usually associated with material dissolution at the material surface and at the crack tip. At the material surface it will shorten the crack initiation period and substantially reduce the fatigue limit. At the crack tip it will accelerate crack growth which in addition may be increased by some weakening of the cohesive strength of the material at the crack tip.

3. Corrosion damage of the material surface and noteworthy corrosion pits, can lead to a large reduction of the fatigue limit.

4. Some typical aspects of corrosion fatigue are: (i) Damage contribution of corrosion to fatigue crack growth primarily occurs during crack extension in the load-increasing part of the load cycles. The loading rate in this part of the cycle is important. As a consequence, the load frequency can also be important. (ii) Holding times at maximum load are not necessarily contributing to crack extension. (iii) The water vapor pressure is an important variable of gaseous environments.

5. In most investigations, corrosion fatigue experiments were carried out as constant-amplitude tests on unnotched specimens or crack growth specimens under closely controlled environmental conditions. However, the conditions for corrosion fatigue of a structure in service are significantly different, especially with respect to the load-time history, the variable environment and long exposure times (years). These differences should be considered in practical problems.

6. Design considerations on corrosion fatigue are frequently based on the prevention of corrosion. It can be done by prevention of the access of the environment to fatigue critical locations of a structure, selection of a suitable corrosion resistant material, and application of material surface treatments.

References

1. Schütz, W., *Corrosion fatigue. The forgotten factor in assessing durability.* Estimation, Enhancement and Control of Aircraft Fatigue Performance, J.M. Grandage and G.S. Jost (Eds.), EMAS (1995), pp. 1–51.

2. Gough, H.J. and Sopwith, D.G., *Some comparative corrosion fatigue tests employing two types of stressing action.* J. Iron Steel Inst., Vol. 127 (1933), pp. 301–332.

3. Pao, P.S., Wei, W. and Wei, R.P., *Effect of frequency on fatigue crack growth response of AISI 4340 steel in water vapor.* Environment-Sensitive Fracture of Engineering

Materials, Z.A. Foroulis (Ed.), The Metallurgical Society of AIME, Warrendale, USA (1977), pp. 565–580.

4. Bradshaw, F.J. and Wheeler, C., *The effect of gaseous environment and fatigue frequency on the growth of fatigue cracks in some aluminium alloys.* Int. J. Fracture Mech., Vol. 6 (1969), pp. 255–268.

5. Broek. D., *Fatigue crack growth and residual strength of aluminium sheet at low temperatures.* Nat. Aerospace Lab. NLR, Report TR 72096, Amsterdam (1972).

6. 't Hart, W.G.J., Nederveen, A. Nasette, J.H. and Van Wijk, A., *Influence of corrosion damage on fatigue crack initiation.* Nat. Aerospace Lab. NLR, Report TR 75080, Amsterdam (1975).

7. Duquette, D.J. and Uhlig, H.H., *Effect of dissolved oxygen and NaCl on corrosion fatigue of 0.18% carbon steel.* Trans. ASM, Vol. 61 (1968), pp. 449–456.

8. Lynch, C.T., Vahldiek, F.W., Bhansali, K.J. and Summitt, R., *Inhibition of environmentally enhanced crack growth rates in high strength steels.* Environment-Sensitive Fracture of Engineering Materials, Z.A. Foroulis (Ed.), The Metallurgical Society of AIME, Warrendale, USA (1977), pp. 639–658.

9. Barsom, J.M., *Effect of cyclic stress form on corrosion fatigue crack propagation below K_{Iscc} in a high yield strength steel.* Corrosion Fatigue: Chemistry, Mechanics and Microstructure, O.F. Devereux, A.J. McEvily and R.W. Staehle (Eds.), Vol. NACE-2, National Association of Corrosion Engineers, Houston (1972), pp. 424–436.

10. Atkinson, J.D. and Lindley, T.C., *Effect of stress waveform and hold-time on environmentally assisted fatigue crack propagation in a C-Mn structural steel.* Metal Science, Vol. 13 (1979), pp. 444–448.

11. Schijve, J., *The significance of fracture mechanisms for the application of fracture mechanics to fatigue crack growth in Al-alloy structures and materials.* Proc. of the USAF Aircraft Structural Integrity Program Conference, San Antonio (1999).

12. Komai, K., *Corrosion fatigue crack retardation and enhancement and fracture surface reconstruction technique.* Environment Assisted Fatigue, P. Scott and R.A. Cottis (Eds.), EGF7, Mechanical Engineering Publications, London (1990), pp. 189–204.

13. Wanhill, R.J.H. and Schra, L., *Corrosion fatigue crack arrest in aluminium alloys.* ASTM STP 1085 (1990), pp. 144–165.

14. Forsyth, P.J.E., *The Physical Basis of Metal Fatigue.* Blackie and Son, London (1969).

15. Horibe, S., Nakamura, M. and Sumita, M., *The effect of seawater on fracture mode transition in fatigue.* Int. J. Fatigue, Vol. 7 (1985) pp. 224–227.

16. Schijve. J. and Arkema, W.J., *Crack closure and the environmental effect on fatigue crack growth.* Fac. Aerospace Eng., Report VTH-217, Delft (1976).

17. *Review of accident investigation in Report MV-73-03.* Ministère des Communications Administration de l'Aéronautique, Brussels (1973).

Some general references

18. Barsom, J.M. and Rolfe, S., *Fracture and Fatigue Control in Structures. Applications of Fracture Mechanics*, 3rd edn. Butterworth-Heinemann (1999).

19. Pao, P.S., *Mechanisms of corrosion fatigue.* Fatigue and Fracture, American Society for Materials, Handbook Vol. 19, ASM (1996), pp. 185–192.

20. Andresen, P.L., *Corrosion fatigue testing.* Fatigue and Fracture, American Society for Materials, Handbook Vol. 19, ASM (1996), pp. 193–209.

21. Dover, W.D., Dharmavasan, S., Brennan, F.P. and Marsh, K.J. (Eds), *Fatigue Crack Growth in Offshore Structures.* EMAS, Solihull, UK (1995).

22. Carpinteri, A., *Handbook of Fatigue Cracking – Propagation in Metallic Structures.* Elsevier, Amsterdam (1994).
23. Lynch, S.P., *Failures of structures and components by environmentally assisted cracking.* Engineering Failure Analysis, Vol. 1 (1994), pp. 77–90.
24. Gangloff, R.P., *Corrosion fatigue crack propagation in metals.* NASA CR 4301 (1990).
25. Baloun, C.H. (Ed.), *Corrosion in Natural Waters.* ASTM STP 1086 (1990).
26. Scott, P. and Cottis, R.A. (Eds.), *Environment Assisted Fatigue.* EGF Publication 7, Mechanical Engineering Publications, London (1990).
27. Lisagor, W.B., Crooker, T.W. and Leis, B.N. (Eds.), *Environmentally Assisted Cracking. Science and Engineering.* ASTM STP 1049 (1990).
28. Sudarshan, T.S., Srivatsan, T.S. and Harvey II, D.P., *Fatigue processes in metals – Role of aqueous environments.* Engrg. Fracture Mech., Vol. 36 (1990), pp. 827–852.
29. *Standard specification for substitute ocean water.* ASTM-D-01141-90 (1990), American Society for Testing and Materials.
30. Jones, W.J.D. and Blackie, A.P., *Effect of stress ratio on the cyclic tension corrosion fatigue life of notched steel BS970:976M33 in sea water with cathodic protection.* Int. J. Fatigue, Vol. 11 (1989), pp. 417–422.
31. *Corrosion fatigue.* AGARD-CP-316 (1981).
32. Forrest, P.G., *Fatigue of Metals.* Pergamon Press, Oxford (1962).

Chapter 17
High-Temperature and Low-Temperature Fatigue

17.1 Introduction

Material properties are dependent on the temperature. The tensile strength, yield strength and modules of elasticity decrease with increasing temperature. It should be expected that fatigue properties are also affected by the temperature. The effect of a high temperature on mechanical properties can be associated with transformations of the material structure due to diffusion processes, aging, dislocation restructuring (softening), and recrystallization. In general, such processes imply that plastic deformation can occur more easily at an elevated temperature. This can lead to the well-known creep phenomenon defined as continued plastic deformation under sustained load. With respect to fatigue, it can imply that more plastic deformation and creep occur in the plastic zone of a fatigue crack which may apply to both microcracks and macrocracks. As a result, fatigue damage accumulation might be enhanced. Furthermore, other failure mechanisms are possible. During creep under sustained load, creep failures occur by grain boundary sliding, void formation (also often at grain boundaries), void growth and coalescence. In general, fatigue is not an intergranular but a transgranular failure mechanism. It thus is not obvious that fatigue and creep damage are simply additive. Actually, the different failure mechanisms of creep and fatigue suggest that a simple addition of damage contributions is physically not realistic. Furthermore, the combined action of cyclic load and an increased temperature should be expected to be different for different materials and different temperature ranges.

High-temperature fatigue combined with time-dependent temperature variations applies to specific structures. As an example, turbine blades are exposed to high combustion temperatures, high centrifugal forces and vibratory bending loads. In certain cases, the severe conditions of high-temperature fatigue have necessitated the development of new materials. Another aspect of the high-temperature fatigue problem is that the temperature will not be constant. In general, the temperature varies between a high operating temperature and a low non-operational ambient temperature. As a result of the temperature profile, cyclic thermal stresses can be introduced. High-temperature fatigue conditions imply that the fatigue load and temperature vary both as a function of time. In addition to the cyclic load, two more variables are time (t) and temperature (T). It then is easily recognized that the complexity of the problem scenario in practice can be considerable.

In previous chapters, verification tests have been frequently recommended. This is also very much true for high-temperature fatigue problems, but tests with both cyclic loads and cyclic temperature profiles are far from easy. Moreover, realistic high-temperature fatigue tests require that a realistic time scale is adopted which will imply lengthy experiments. Furthermore, also the temperature must be accurately controlled since it has a large effect on creep.

In view of the complexity of the high-temperature fatigue problem and the special materials developed for high-temperature application, the problem is a very special one. Knowledge and experience of the more general problems of fatigue of structures at ambient temperatures have a limited meaning for high-temperature fatigue. Because of the particular nature of the high-temperature fatigue problem, it is briefly covered in this chapter. Two selected case histories are discussed (Section 17.2) followed by some illustrations of temperature effects on fatigue properties in Section 17.3.

Fatigue is also possible in structures operating at temperatures significantly below room temperature, e.g. structures in the open air or sea water. The problem of fatigue at low temperatures is entirely different from fatigue at high temperatures. Trends of low-temperature fatigue are summarized in Section 17.4. Some general comments on temperature effects are made in Section 17.5.

17.2 Two examples of high-temperature fatigue

Turbine blades

The turbine blades of a turbofan engine are exposed to an extreme temperature, at the present time well above 1000°C, and in the order of 70 to 80% of the melting temperature (in °K). High-temperature materials have been developed because a high operating temperature is most important for increasing the efficiency of gas turbine engines.

The blades are cyclically exposed to high centrifugal forces with one cycle corresponding to a single engine run-up and shutdown, which is a low-cycle fatigue condition. Because of the sustained centrifugal force and high working temperature, creep may occur. This obviously occurred to the turbine blade in Figure 17.1a. Furthermore, blade vibrations at a high frequency induce high-cycle fatigue conditions. Fatigue fractures due to these vibratory loading are shown in Figure 17.1b. A third type of load is coming from thermal stresses. During engine run-up, the blade temperature is increased very rapidly which leads to a transient inhomogeneous temperature distribution in the blade. The temperature of the thin trailing edge of the blade is increased much faster than in the bulk of the blade. This also applies to the leading edge, although to a lesser extent. During engine shutdown the cooling of the trailing and leading edge will be faster. The inhomogeneous temperature in the blade and the related thermal expansion and contraction are causing a thermal stress cycle. It has resulted in a fatigue crack in the trailing edge of the turbine blade shown in Figure 17.2 (low-cycle fatigue). Crack growth occurred along the grain boundaries. Plastic deformation in the grain boundaries is very difficult at room temperature, but at a high temperature the grain boundaries become weaker than the matrix of the grains (grain boundary sliding and cavity formation) with intergranular cracking as a result.

The material of the blades shown in Figures 17.1 and 17.2 is a nickel alloy (Nimonic) which was developed as a high-temperature material for turbine blades in view of the high working temperature. The failures shown in these figures occurred in the 1960s. However, a further increase of the working temperature was required for improving the efficiency of turbofan engines. This has lead to the development of new materials which also required new blade production techniques. Transverse grain boundaries were eliminated by producing turbine blades as monocrystals with a unidirectionally solidified eutectic composition [1, 2] which greatly improved the creep resistance. The eutectic structure consists of a matrix

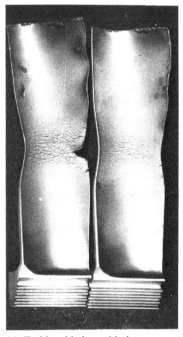

(a) Turbine blades with large creep deformation and cracking due to overheating. Blade length 11 cm

(b) Fatigue failures in turbine blades starting at the trailing edge (blade chord 25 mm)

Fig. 17.1 Turbine blade failures.

as one phase of the eutecticum embedding long fibers as the second phase. Obviously, the directionality of the eutectic composite should be in the length direction of the blade. This is an example of designing a material and a production process for a special application. Special coatings are also applied to turbine blades in order to resist surface oxidation. Furthermore, new blade cooling systems have been introduced. As a consequence, turbine blades are highly complex and expensive [3]. The major problem is not designing against fatigue, but developing materials and blades for extremely high-temperature application. Of course, stress distributions in the blade and the joint between the blades and the turbine disk must still be considered in view of fatigue (see the wiffle tree in Figure 17.1a).

 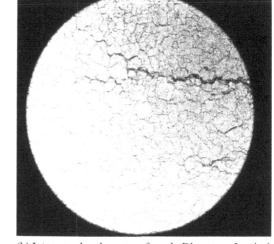

(a) Trailing edge with cracks. Length of lower crack 4 mm

(b) Intergranular character of crack. Diameter of optical field 2.5 mm

Fig. 17.2 Intergranular fatigue crack in trailing edge of a turbine blade due to thermal fatigue.

Supersonic transport aircraft: The Concorde [4]

In the 1960s, the supersonic passenger aircraft, the Concorde, was developed jointly by the French and English aircraft industries. The aircraft had to fly at twice the speed of sound (Ma-2) which implies that aerodynamic heating of the aircraft skin occurs up to a temperature of slightly above 100°C (212°F). The design team expected that aluminium alloys could still be used, but it was recognized that the age-hardened condition of these alloys could be affected by overaging at the elevated temperature. Some softening of the material might decrease the mechanical material properties and creep and fatigue were earmarked as potential problems. Extensive experimental programs were carried out. The Al-alloy selected for the structure was RR58[22] (composition 2.4Cu, 1.6Mg, 1.1Ni, 1.65Fe; ASM equivalent: 2618) which was artificially aged at 190°C (374°F). Because the aging temperature is significantly above the maximum in-flight temperature, it might be expected that a stable material structure is maintained in service. However, it does not guarantee that some softening does not occur and cyclic slip in the material might occur more easily. Fatigue tests were carried out on various types of specimens simulating notched elements of the aircraft

---

[22] RR58 was an existing Rolls Royce alloy used for compressor blades subjected to a moderately elevated temperature.

Fatigue limit (MPa)

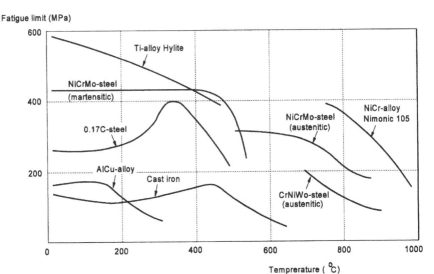

Fig. 17.3 The influence of the temperature on the fatigue limit ($N = 10^7$) of various materials. Data collected by Forrest [6] (graph after Radaj [7]).

structure. Load histories and temperature cycles relevant to supersonic flights of the Concorde were used. Significant thermal stresses are introduced in supersonic flight because aerodynamic heating starts at the outer skin of the structure while the temperature of the internal structure is lagging behind. Thermal stresses in the reversed direction occur after returning to subsonic flight. Finally, a full-scale flight-simulation test on the complete aircraft structure was considered to be necessary in order to gain sufficient confidence with respect to fatigue and creep. A complex test rig for the full-scale test included equipment for heat cycles with periodic heating and cooling of the outside of the structure by air in a kind of tunnel around the aircraft. Due to the complexity of the test and the need for periodic inspection, the test could not be run fast. Acceleration of a test with heat cycles is anyway a problem because heat effects on the material behavior are time dependent and should in principle be simulated on a real time scale. A compromise was obtained by using a slightly higher maximum in the temperature cycle (maximum 130°C) in order to accelerate the thermal damage contribution, but the test still lasted for about seven years. The Concorde fleet has accumulated some 25 years of operation in service without significant thermal fatigue problems. The Concorde was retired from service in 2003 for economic reasons.

17.3 Fatigue properties at high temperatures

Test results of the fatigue limit of various materials as affected by temperature were collected by Forrest [5], see Figure 17.3. The data were obtained under cyclic bending at frequencies in the order of 40 to 50 Hz. For several materials, the fatigue limit above a certain temperature is considerably reduced, which indicates that the material can no longer be used at higher temperatures. A comparison between different materials in Figure 17.3 indicates a relatively high-fatigue strength of the Ti-alloy at temperatures up to about 400°C. Titanium alloys are still used for several high-temperature applications, particularly in several components of turbofan engines. Figure 17.3 also shows that the austenitic steels are superior to the martensitic steels. The superior material in Figure 17.3 is the nickel alloy Nimonic 105.

A remarkable behavior is exhibited by the 0.17C steel (mild steel). The fatigue limit apparently increases at elevated temperatures until a maximum of about 350°C. According to Forrest [5], this should be associated with strain aging due to enhanced diffusion of carbon atoms to dislocations which then are restricted in slip movements.

The reduction of the fatigue limit of the aluminium alloy (Al-Cu) at elevated temperature is a result of overaging of the precipitation hardened material structure. Also, the martensitic structure of the low-alloy high-strength steel is not stable at high temperatures. If the material structure is not stable at a high temperature, time-dependent phenomena can occur in the material, and an effect on fatigue properties should occur. Effects of the frequency and wave shape have been observed. Creep, either on a microscale or a macrolevel, will contribute to the failure mechanism. Creep resistance is essential for the fatigue strength at high temperatures which requires a stable material structure.

The fatigue notch effect at high temperatures can be significantly different from the notch effect at room temperature. If creep is possible, the peak stress at the root of the notch may be relaxed as result of plastic creep deformation. Obviously, life predictions accounting for creep deformation will be complex because the temperature-time history is an additional and relevant variable to be considered. Moreover, retrieving of material data to account for the time-dependent material behavior is not a simple task. Actually, design efforts for a structure operating in a high-temperature environment should be supplemented by experiments which faithfully include all relevant conditions.

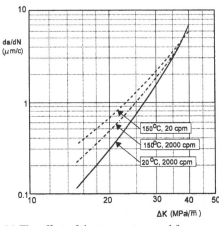

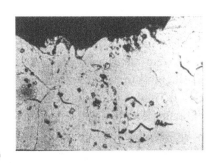

(a) The effect of the temperature and frequency on the fatigue crack growth rate

(b) Intergranular crack growth and grain boundary voids during creep at $S = 200$ MPa and $T = 150°C$ (width of picture 200 μm).

Fig. 17.4 Fatigue and creep crack growth in an Al-alloy (2024-T3) [8].

Most basic material data on the temperature effect were obtained as low-cycle fatigue results of tests on unnotched specimens. Such tests can indicate a safe upper temperature limit which precludes creep deformation, in most cases a criterion to be satisfied. Fatigue crack growth at an elevated but still acceptable temperature is then of interest. Fatigue crack growth data for elevated temperatures were essential for the development of the Concorde. Some results of an old elementary test program on an aluminium alloy are presented in Figure 17.4. Tests were carried out on center cracked specimens at room temperature and 150°C. For the elevated temperature tests, frequencies of 2000 and 20 cycles per minute were used. Crack growth was faster at the elevated temperature, and at this temperature, it was again faster for the lower frequency. Crack growth under all three conditions was still transgranular. Tests at 150°C on precracked specimens ($a_0 = 12$ mm) were also carried out under sustained load. Crack growth by creep until $a = 20$ mm occurred in 32 hours, while crack growth now was intergranular and grain boundary voids were also observed in the wake of the crack, see the illustration in Figure 17.4b. The difference between the fracture mechanisms of fatigue and creep is confirmed.

James [8] reported on fatigue crack growth at 538°C (1000°F) in an austenitic stainless steel (type AISI 304), which was a potential material for a breeder reactor. He carried out crack growth tests at several load

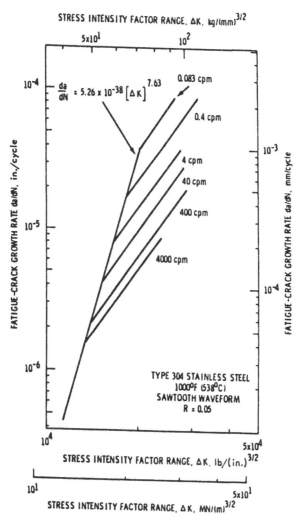

Fig. 17.5 The effect of the frequency on the crack growth rate in annealed stainless steel (AISI 304) [9].

frequencies varying from fast (4000 cpm = 67 Hz) to rather slow (0.083 cpm or 1 cycle in 12 minutes), see Figure 17.5. At low ΔK-values, a frequency effect was not yet observed, but at higher ΔK-values a systematic effect occurred with higher crack growth rates at a lower frequency. Crack growth in most of these tests was still transgranular. James also carried out tests with two different wave shapes: (1) a positive saw tooth, and (2) a trapezoid wave shape with holding at S_{max} for the major part of the load cycle, both at three frequencies of 4, 0.333 and 0.083 cpm respectively. At the

first two frequencies, a systematic wave shape effect was not evident, but
at the lowest frequency of 0.083 cpm, the crack growth rate was lower
for the trapezoid wave shape with holding at S_{max}. Moreover, the fracture
mechanism being predominantly transgranular for the saw tooth wave shape
changed to predominantly intergranular crack growth. In the latter case, the
crack growth path was more irregular while also more secondary cracks were
observed. Recall a similar observation in Chapter 16 (see Figure 16.8) where
a more complex cracking mechanism under corrosion fatigue due to holding
at S_{max} also led to a lower da/dN.

The results of James cannot simply be transferred to other materials.
However, it should be recognized that a high-temperature fatigue crack
growth problem requires similar research to explore the effect of
temperature, loading rates and hold times corresponding to load and
temperature profiles occurring under service conditions.

17.4 Fatigue at low temperatures

Low temperatures can change the material fatigue behavior for two reasons.
First, the mechanical response of the material is different. In general, the
yield strength and tensile strength are higher than at room temperature. This
trend is associated with an increasing resistance against plastic deformation
(lower mobility of dislocations). Second, environmental effects on fatigue
are reduced at a low temperature because reaction rates of chemical
processes and diffusion are lower. Forrest [5] collected fatigue strength data
from the literature and averaged the results of various sources for different
groups of materials. He presented results for three low temperatures as the
ratio between the fatigue strength ($N = 10^6$) at a low temperature ($S_{N,T}$)
and the fatigue strength at room temperature ($S_{N,RT}$) see Figure 17.6. The
ratios in Figure 17.6 are all larger than 1. Apparently, the fatigue strength at
low temperatures is larger than at room temperature, the more so for a lower
temperature. The results in Figure 17.6 indicate that the fatigue strength of
notched specimens is also increased by a lower temperature.

Effects of low temperatures on fatigue crack growth are also reported
in the literature, see e.g. [9]. In general, fatigue crack growth occurs
more slowly at low temperatures if small to moderate ΔK-values are
applicable, whereas faster crack growth has been observed for larger
ΔK-values with K_{max} close to K_c or K_{Ic}. The increased crack growth rate
for large ΔK-values can be understood because of the reduced ductility

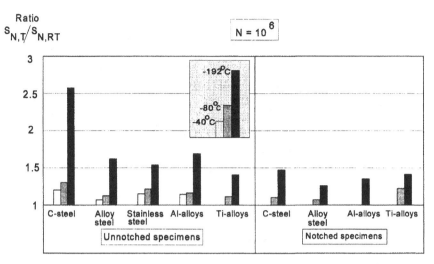

Fig. 17.6 The low-temperature fatigue strength ($S_{N,T}$) divided by the fatigue strength at room temperature ($S_{N,RT}$) for unnotched and notched specimens of different material groups. Data compiled by Forrest [6].

at lower temperatures. However, why should crack growth be slower at low ΔK-values, i.e. in the threshold and Paris region; regions I and II in Figure 8.6. This could be due to reduced cyclic plasticity at the crack tip, but also to some environmental contribution which may be less detrimental at a lower temperature. As discussed in Chapter 16, the water vapor contents of air can affect the crack growth rate. A lower crack growth rate has been observed for a lower water vapor pressure, and the pressure is reduced by a low temperature. Broek carried out crack growth tests on two aluminium alloys (2024-T3 and 7075-T6) at room temperature and at 0, −25, −50 and −75°C respectively [10]. The results in Figure 17.7 confirm that crack growth is slower at lower temperatures. Broek explained this trend as a result of the lower water vapor content of the air at low temperatures. He confirmed the favorable effect of a low water vapor pressure by supplemental tests at room temperature in very dry air, see the crack growth curve for these conditions in Figure 17.7.

A reduced environmental effect on fatigue crack growth at low temperatures may also apply to several materials for crack growth in a liquid environment. If low-temperature conditions should be considered for structures in service, it is advisable to carry out exploratory tests. This requires a careful simulation of the service conditions also with respect to the load frequency.

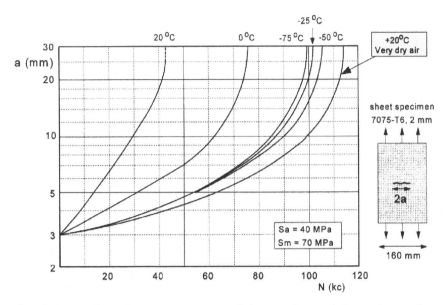

Fig. 17.7 The effect of low temperatures on fatigue crack growth in specimens of an aluminium alloy [11].

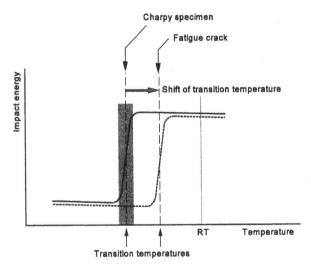

Fig. 17.8 The transition temperature revealed by impact tests on Charpy V-notched specimens of low-carbon steel. A higher transition temperature for fatigue cracks.

Transition from ductile failures to brittle failures

In general, less plastic deformation occurs during static failures at low temperatures. The material ductility is reduced and this is manifest

Fig. 17.9 Two Charpy V-notch specimens, thickness 10 mm. Brittle failure in the front specimen tested below the transition temperature, and ductile failure in the rear specimen tested above the transition temperature.

during fatigue crack growth under severe load cycles with a high K_{max}. Fractographic observations have shown that ductile striations may disappear at low temperatures, while indications of crack extension by a cleavage mechanism have been found depending on the type of material [9]. However, an exceptional transition from ductile to brittle failure is exhibited by low-carbon steels (mild steel). This phenomenon is usually studied by impact tests on Charpy V-notch specimens. The tests are carried out at different temperatures and the impact energy for breaking the specimen is measured. If the temperature is decreased, the impact energy suddenly drops to a substantially lower level within a fairly narrow temperature range, see Figure 17.8. The range is characterized by the transition temperature, T_{trans}. For $T > T_{trans}$, the failure of the Charpy specimen is a ductile failure with much plastic deformation and without separating the specimen in two pieces, see Figure 17.9. However, for $T < T_{trans}$, a brittle failure occurs without apparent plastic deformation also shown in Figure 17.9. Microscopic investigation have revealed that the failure for $T > T_{trans}$ occurs as a quasi-static type of failure by void formation and coalescence. For $T < T_{trans}$, a cleavage type of failure occurs. Although the Charpy test is useful to indicate whether a material is sensitive to cold-brittleness, it should be understood that the transition temperature is not a material constant. In general, T_{trans} will move to a higher temperature if plastic deformation at the tip of the notch of the Charpy specimen is more restrained. A smaller plastic zone and a higher peak stress in this zone are then obtained (see Section 5.8).

This will promote the brittle type of fracture. Plane strain conditions with a triaxial stress distribution at the root of the notch enhance the restraint on plastic deformation. The plane strain character is further promoted by a larger thickness of the specimen, but also by a sharper notch and even more by a fatigue crack. As a consequence, the transition temperature can be higher for a fatigue crack than indicated by tests on Charpy V-notch specimens, see Figure 17.8. Because of the restraint on plastic deformation, the transition temperature is also increased by a higher yield stress, which implies that the risk of brittle failures in structures of mild steel is larger if the hardness of the material is higher. The increased hardness can be due to a higher carbon content or the heat treatment of the steel. A most dramatic example of brittle failures occurred during World War II and also afterwards, when welded Liberty ships in cold water broke in two parts by brittle failures in welded joints [11].

17.5 Some general comments

High-temperature fatigue is in the first place a problem associated with the stability of the material structure. Essential information is related to the load- and temperature-time histories. Where ambient temperature fatigue problems can considerably benefit from stress analysis calculations, material research is more essential for high-temperature fatigue. Temperature limits beyond which the material stability deteriorates should be determined, and fundamental understanding of the material behavior is important. As an example, it has occasionally been overlooked that the creep resistance can significantly depend on the strain-hardening (dislocation structure) of the original material. High-temperature fatigue problems require experimental research while knowledge of material science is indispensable for planning the research.

Low-temperature fatigue is a fully different problem. Plastic deformation at low temperatures is more difficult than at room temperature which has consequences with respect to fracture toughness and fatigue crack growth at high ΔK-values. These aspects should be recognized if the operational environment of the structure includes low temperatures. Designing for fatigue durability in terms of avoiding stress concentrations and limitations on allowable stress levels still remain relevant. The transition from ductile failures to brittle failures at low temperature, especially of low-carbon steels, is a special issue to be considered for (welded) structures of these materials.

References

1. Hertzberg, R.W., *Deformation and Fracture Mechanics of Engineering Materials*, 4th edn. John Wiley & Sons, New York (1996).
2. Ashby, M.F. and Jones, D.R.H., *Engineering Materials*. Pergamon Press, Oxford (1980).
3. Kool, G.A., *Current and future matrials in advanced gas turbine engines*. 39th ASME International Gas Turbine and Aeroengine Congress and Exposition, The Hague (1994).
4. Ripley, E.L., *The philosophy which underlies the structural tests of a supersonic transport aircraft with particular attention to the thermal cycle*. Advanced Approaches to Fatigue Evaluation, Proc. 6th ICAF Symposium. NASA SP-309 (1972), pp. 5–91.
5. Forrest, P.G., *Fatigue of Metals*. Pergamon Press, Oxford (1962).
6. Radaj, D., *Fatigue Strength*. Springer-Verlag, Berlin (1995) [in German].
7. Schijve, J. and De Rijk, P., *The effect of temperature and frequency on the fatigue crack propagation in 2024-T3 Alclad sheet material*. Nat. Aerospace Lab. NLR, Report TR M.2138, Amsterdam (1965).
8. James, L.A., *Hold-time effects on the elevated temperature fatigue-crack propagation of type 304 stainless steel*. Nuclear Technol., Vol. 16 (1972), pp. 521–530.
9. Stephens, R.I. (Ed.), *Fatigue at Low Temperatures*. ASTM STP 857 (1985).
10. Broek, D., *Fatigue crack growth and residual strength of aluminium sheet at low temperatures*. Nat. Aerospace Lab. NLR, Report TR 72096, Amsterdam (1972).
11. Barsom, J.M. and Rolfe, S., *Fracture and Fatigue Control in Structures. Applications of Fracture Mechanics*, 3rd edn. Butterworth-Heinemann (1999).

Some general references

12. Davis, J.R. (Ed.), *Heat Resistant Materials*. ASM Speciality Handbook. ASM International (1997).
13. Piascik, R.S., Gangloff, R.P. and Saxena, A. (Eds.), *Elevated Temperature Effects on Fatigue and Fracture*. ASTM STP 1297, Am. Soc. for Testing and Materials (1997).
14. *Fatigue and Fracture*, American Society for Materials, Handbook Vol. 19, ASM (1996).
15. Webster, G.A. and Ainsworth, R.A., *High Temperature Component Life Assessment*. Chapman and Hall (1994).
16. Larsson, L.H., *High Temperature Structural Design*. Mechanical Engineering Publications, London (1992).
17. Skelton, R.P., *High Temperature Fatigue: Properties and Prediction*. Elsevier Applied Sciences, London (1987).

Part V
Fatigue of Joints and Structures

Chapter 18
Fatigue of Joints

... fatigue is the art of joints ...

18.1 Introduction

Fatigue failures in structures frequently occur in joints. Various catastrophic accidents due to fatigue have been reported in the literature. As a consequence joints are a major issue for designing against fatigue. The prime purpose of a joint is to transmit loads from one element of the structure to an other element. The variety of different joint configurations is very large. Some elementary types of joints are shown in Figure 18.1.

A significant difference between two categories of joints is associated with the question whether it should be possible to disassemble and reassemble a joint, or whether that is not necessary. The lug type joint and the bolted joint are in the first category. However, riveted lap joints, bonded lap joints and welded joint, including spot welded joints, are supposed to remain in the as-produced condition. Another noteworthy aspect of joints is associated with eccentricities in the joint. The bolted joint in Figure 18.1 is fully symmetric with respect to the line of the applied load. But that is not true for the two non-symmetric lap joints in Figure 18.1. As a result of the eccentricity in these joint, bending is introduced by the applied tensile

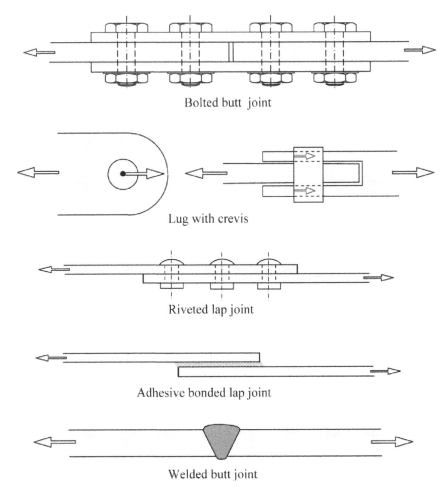

Bolted butt joint

Lug with crevis

Riveted lap joint

Adhesive bonded lap joint

Welded butt joint

Fig. 18.1 Symmetric butt joints and lap joints with eccentricities.

load. The additional bending, referred to as secondary bending, will increase notch effects in these joints. Furthermore, fasteners in a simple lap joints are loaded in single shear whereas bolts in the symmetric joint are loaded in double shear which appears to be a more favorable situation. Still another complexity is the occurrence of fretting in lugs and bolted and riveted joints. Once again, the variety of joints is just large. Moreover, the number of design parameters for each type of joint also contributes to a broad spectrum of design questions. Variables included are associated with joint dimensions, production variables and selected materials.

In view of the above diversity, several types of joints are discussed in separate sections. Fatigue of lugs is covered in Section 18.2. A lug is the most simple and elementary joint. Symmetric joints with more fasteners loaded in shear are discussed in Section 18.3. Bolts loaded in tension offer different problems dealt with in Section 18.4, including pre-tensioning. Characteristic aspects of riveted joints are associated with the riveting process, fretting corrosion and secondary bending. These aspects are discussed in Section 18.5. Adhesive bonded lap joints as an alternative to riveted lap joints are briefly addressed in Section 18.6. Fatigue of welded joints is covered in a separate chapter (Chapter 19) because characteristic aspects of welded joints are entirely different from problems associated with mechanical joints. Actually, welded joints are supposed to be an integral connection rather than joining different structural elements. Because joints offer specific prediction problems, some general comments are presented in Section 18.7. Major aspects of the present chapter are summarized in Section 18.8

18.2 Fatigue of lugs

The lug connection is an elementary joint with a single pin or bolt as described earlier in Chapter 3, see Figure 3.17. The stress concentration factor (K_t) of a lug is relatively high, see Figure 3.19. Attractive features of a lug connection are: (1) assembling and disassembling is relatively easy, and (2) rotation between the two connected parts is possible which may be necessary in view of specific functions, e.g. in push/pull mechanisms. In cases of statistically indeterminate structures, it may be desirable to have a zero moment at a pin connection which can be accomplished by a small rotation in a lug connection.

Observations on fatigue of lugs were already discussed in Chapter 15 on fretting corrosion. It was shown that the fatigue limit could be very low, see Figures 15.11 and 15.12. The fatigue strength reduction factor, K_f, can be significantly larger than the stress concentration factor K_t. The low fatigue limit is caused by fretting corrosion inside the hole, which initiates microcracks at low stress amplitudes. At higher stress amplitudes with lower endurances, the detrimental effect of fretting corrosion is much less because microcracks are then initiated early in the fatigue life anyway, also if fretting does not occur.

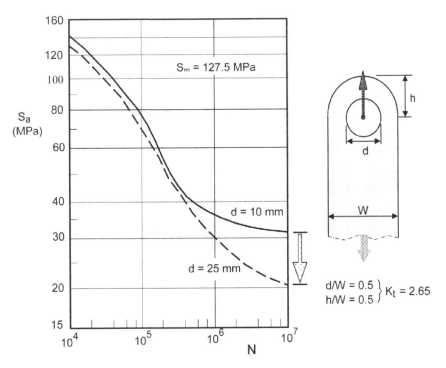

Fig. 18.2 Size effect on the S-N curve of a lug. Material: Al-alloy 2024-T3 [1].

As discussed in Chapter 15, fretting corrosion is depending on the amplitude of the fretting movements, see Figure 15.7. The fretting movement amplitude inside the hole of the lug is larger for a larger lug because the amplitude is proportional to the size of the lug. As a consequence, it should be expected that the fatigue limit of a lug depends on the size of the lug. This has been confirmed by experiments. The results for lugs of an Al-alloy in Figure 18.2 show that the fatigue limit of a large lug is 35% lower than for a small lug although the shapes are similar, and thus the same K_t is applicable. This size effect is much larger than predicted by the size effect equations discussed in Section 7.2 for notches where fretting is not involved. Figure 18.2 also shows that the size effect due to fretting disappears at lower endurances in agreement with the much smaller effect of fretting during low-cycle fatigue.

Larsson [2] has analyzed a large amount of S-N data of lugs of two Al-alloys (2024-T3 and 7075-T6), and later the same was done for Ti-alloys, high-strength steels and Mg-alloys in the aircraft industry [3]. Larsson arrived at an empirical equation to predict the fatigue strength of a lug. A

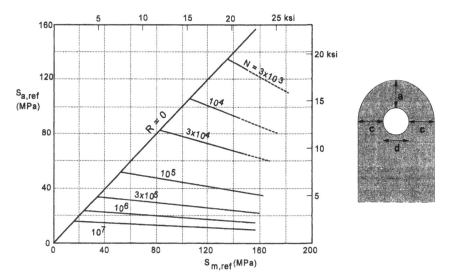

Fig. 18.3 Fatigue diagram for a reference lug with dimensions $a = c = d = 10$ mm (0.4") according to Larsson [2]. Material: Al-alloy 2024-T3.

reference lug was defined by the dimensions $a = c = d = 10$ mm (0.4"), see Figure 18.3. The analysis of many data led to the fatigue diagram for the reference lug, which is presented in Figure 18.3 for the Al-alloy 2024-T3. Larsson derived a second diagram for the Al-alloy 7075-T6. Symbols of the stress amplitude and mean stress for the reference lug are $S_{a,\text{ref}}$ and $S_{m,\text{ref}}$. The fatigue strength for another lug, defined by S_a and S_m, is now obtained by the following equations (stress calculated on the net section):

$$\frac{S_a}{S_{a,\text{ref}}} = \frac{S_m}{S_{m,\text{ref}}} = 1 + \theta(k_1 k_2 - 1)$$

$$\text{with} \quad k_1 = \left(\frac{ad}{c^2}\right)^{0.5}, \quad k_2 = \left(\frac{10}{d}\right)^{0.2} \quad (d \text{ in mm}) \tag{18.1}$$

$\theta = 0.25 \log N - 0.5$ for $N = 10^3$ to 10^6 and $\theta = 1$ for $N \geq 10^6$

As shown in this equation, k_1 accounts for the shape of the lug, k_2 for the size effect, and θ for the decreasing effect of fretting corrosion at lower endurances. It should be emphasized that Larsson's equation is empirical. The equation was obtained by a multi-variable regression analysis of many test data. It gives approximate indications of the fatigue strength. Fatigue data of lugs are also presented in [4]. More accurate information would

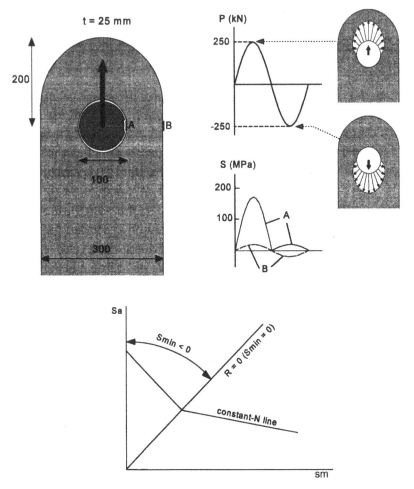

Fig. 18.4 Unusual constant-N lines in the fatigue diagram of lugs due to reversion of the load direction [1].

require fatigue tests on the lug itself. Furthermore, Eq. (18.1) is applicable to positive stresses only; $S_{min} \geq 0$, or $R \geq 0$.

For negative R-values ($S_{min} < 0$), lines for constant N-values in the fatigue diagram are abruptly going upward, see the trend in the lower graph of Figure 18.4. This unusual feature is typical for lugs, and it can easily be understood. If the load on a structure is reversed from tension to compression, all stresses are also reversed (Hooke's law). However, this does not apply to lugs as illustrated by Figure 18.4. If a tensile load P is applied to a lug it causes a high tensile peak stress at point A. However, if the load P is reversed from tension to compression, the bearing pressure on the upper

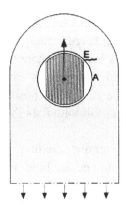

Fig. 18.5 Pin with flats to prevent fretting corrosion at location A of the peak stress [5].

side of the hole disappears instead of being reversed. A bearing pressure is now applied to the lower side of the hole. As a result, the minimum section A-B is hardly loaded. Strain gage measurements have shown that it causes a relatively low tensile stress at A instead of a high compression stress [1]. As a consequence, the peak stress range is not $2K_t S_a$, but only half that value. The nominal fatigue strength increases considerably.

Improvement of the fatigue limit of a lug

The fatigue limit of a lug is very low, see Figures 15.11, 15.12, 18.2 and 18.3. In addition to a relatively high-stress concentration factor, the low fatigue limit is primarily due to fretting corrosion in the hole. The fatigue limit of a lug can be improved in two different ways:

(i)   Prevention of fretting corrosion.
(ii) Reduction of the detrimental effect of fretting corrosion damage by introducing compressive residual stresses.

Fretting corrosion is most effectively eliminated by preventing contact between the pin or bolt and the bore of the hole. This can be achieved by slotted holes as discussed in Chapter 15. The results in Figure 15.12 show that a most impressive improvement of the fatigue limit was obtained. Unfortunately, slotted holes are "expensive" holes. Clarke proposed to avoid metallic contact at points A in Figure 18.5 by using pins with flat sides [5]. A worthwhile increase in life was obtained for lugs of an Al-alloy (L65 $\approx$ 2014). A significant increase of the fatigue limit was also reported in [6] for another Al-alloy (7075-T6), while similar improvements were reported

for lugs of some steels [4]. It was observed that cracks in lugs with pins
with flat sides started at point E (Figure 18.5). The cracks were still initiated
by fretting corrosion, but at a location where the tangential stress is lower
than at point A, and the fretting movements are smaller. Obviously, if a pin
with flat sides in mounted in a 90° rotated position, the fatigue limit will be
reduced rather than improved. Pins with flats should not be considered to be
a practical proposition.

Another method for preventing fretting corrosion is by using an
interference fit between the pin and lug hole. It was shown by fatigue tests
[7] that a hard driven fit increases the fatigue limit, but this is also not a
practical solution. Assembling and disassembling of such a joint is no longer
feasible. However, a bushing pressed into a lug hole is a good alternative, as
already discussed in Chapter 15 (Section 15.4, Figure 15.15). This should
be done with a high interference fit in order to prevent fretting movements
between the bushing and bore of the hole. The interference fit introduces a
high tangential compression stress in the bushing, which suppresses fatigue
in the bushing itself. Moreover, the hole in the lug is protected against
handling damage caused by mounting and dismounting of bolts or pins.

The fatigue properties of lugs can be much improved by plastic hole
expansion, a possibility already discussed in Chapter 4 (Figure 4.7). Plastic
hole expansion results in compressive residual stresses in the tangential
direction around the hole. The large favorable effect on the fatigue limit
of lugs was emphasized in Chapter 15. Fretting damage occurs, but
microcracks starting from this damage can be arrested also at relatively
high stress amplitudes. The fatigue limit is increased considerably. Plastic
hole expansion occurs by pulling an oversized mandrel through the hole,
see Figure 4.7. This requires a fairly high pulling force on the mandrel.
Moreover, in many practical applications only one side of the component
is accessible indicating an apparent need for an expansion technique which
can be applied from one side. Two methods were developed for this purpose.
The first one is the split-sleeve expansion method developed in the 1970s
[8]. The method is illustrated by Figure 18.6. A mandrel is moved through
the hole, and then the split sleeve (a bushing with a single split) is moved
along the mandrel into the hole. Retraction of the mandrel through the
hole and the sleeve in the hole requires a radial expansion of the sleeve
and the hole together. As a result, the diameter of the hole is increased by
plastic deformation around the hole, and compressive residual stresses in
the tangential direction will remain afterwards. A large force is necessary
for retraction of the mandrel, but the sleeve has a split and thus the sleeve

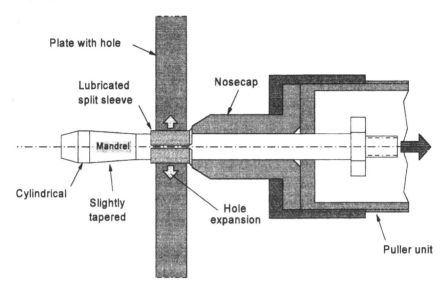

Fig. 18.6 Plastic hole expansion by the split-sleeve method. (Courtesy Len Reid, Fatigue Technology Inc.)

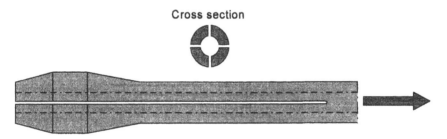

Fig. 18.7 Split mandrel for plastic hole expansion [9].

can easily expand by a small opening of the slit. Furthermore, the friction between the mandrel and the sleeve is reduced by a surface coating baked on the inside of the sleeve. After the mandrel has been retracted, the hole is reamed to the final size and a correct cylindrical shape. The amount of material removed by reaming is small compared to the depth of the plastic zone around the hole. Plastic hole expansion starts from a slightly undersized hole in view of the expansion and subsequent reaming.

The second technique, developed in the 1990s is the split-mandrel method [9]. A mandrel is used with two mutually perpendicular slits along a large part of the mandrel, see Figure 18.7. The maximum diameter is larger than the hole diameter, but the mandrel can still pass through the hole because the slits can be closed by some elastic bending of the four quadrants of the mandrel. Before retracting the mandrel, a pin is inserted in the center of the mandrel. This prevents closure of the slits during the retraction. The frictional forces between the mandrel and the bore of the hole are reduced by a lubricating oil.

Plastic hole expansion has been successfully applied to many aircraft structures. It is also an attractive method for repair if small cracks occur in service at bolt holes. The bolt is removed, the hole is cleaned and reamed, the hole is plastically expanded and a new bolt with a slightly larger diameter is installed. Successful experience with this rejuvenating repair has been claimed. It is also applied to new aircraft if fatigue problems might be expected. An expansion by 2 to 3% is typical which is sufficient to obtain a significant improvement of the fatigue strength. Larger percentages cause too much lateral deformations and may also introduce some rupture in a material with a fibrous structure. In general, designers are not in favor of plastic hole expansion because of extra costs of manufacturing and quality control.

Plastic hole expansion has also been applied to components of high-strength steels and the Ti-6Al-4V alloy, but the pulling loads are high. Residual compressive stresses were introduced inside large holes in a high-strength steel part of a helicopter by roll-peening. Rollers are forcefully rolled on the material surface inside the bore of the hole which introduces plastic deformation in a way somewhat similar to shot peening. Significant improvements were reported by Waters [10], who found that the fatigue limit for a 4130 lug joint was increased by about 100%.

18.3 Symmetric butt joints with rows of bolts or rivets

A simple double strap joint with two rows of bolts at each side of the joint is shown in Figure 18.8. It appears to be a good design from a static point of view if both rows carry the same load because then $P_A = P_B$ which implies the same bearing pressure on the hole and the same shear stress in the bolts. This will be achieved if the sum of the thicknesses of the two straps is equal to the thickness of the plates: $2t_2 = t_1$. However, from a fatigue point of view, the two rows are in a different position. The holes in row B of the plate are loaded by P_B only. In row A, the same load P_A is present on the holes, but the load already introduced in row B is also passing these holes. The latter load is called a bypass load. With S as the nominal net section stress in the plate, the peak stress at the hole in row A can be written as:

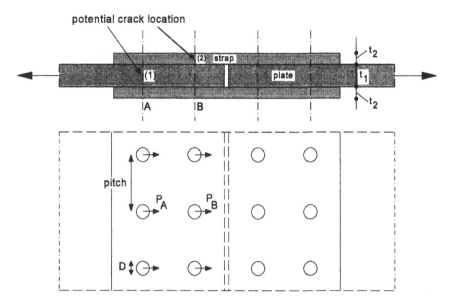

Fig. 18.8 Double strap joint with two rows of bolts (or rivets) at both sides of the joint.

$$(S_{\text{peak}})_{\text{row A}} = \frac{1}{2}S(K_t)_{\text{pin loaded hole}} + \frac{1}{2}S(K_t)_{\text{unloaded hole}} \qquad (18.2)$$

The second term is from the bypass load. For row B the peak stress is:

$$(S_{\text{peak}})_{\text{row B}} = \frac{1}{2}S(K_t)_{\text{pin loaded hole}} \qquad (18.3)$$

For a ratio of hole diameter to hole pitch of 1/5, the K_t-values of a pin loaded hole and an unloaded hole are 5.2 and 2.5 respectively. Substitution in the two equations gives

$$(S_{\text{peak}})_{\text{row A}} = 3.85S \quad \text{and} \quad (S_{\text{peak}})_{\text{row B}} = 2.6S$$

The results are an approximation because for K_t of the bypass load (unloaded hole) it has been assumed that the hole is an open hole whereas a bolt is present in the hole. Also, the K_t-values do not consider fretting corrosion in the hole. Nevertheless, it is still correct to say that the holes in the plate in row A are more severely loaded than the holes in row B and thus will be more fatigue critical. For the straps, the reverse is true for the same reasons; row B is more fatigue critical than row A. Fatigue cracks occur at either location (1) or (2) in Figure 18.8. Cracks in the plate (location 1) are invisible. It may cause an inspection problem.

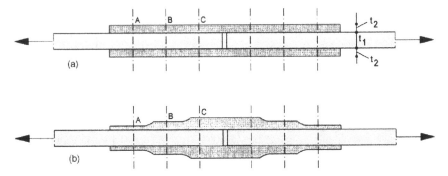

Fig. 18.9 Two double strap joints with a constant and a staggered strap thickness.

Two butt joints with three rows of bolts are shown in Figure 18.9, one joint with a constant thickness of the strap plates, and a second one having strap plates with a staggered thickness. Now, three rows of bolts contribute to load transmission from the plates to the straps. The distribution of the load transmission between the three rows can be obtained by a simple elastic calculation considering equal displacements of bolt holes in the plate and the straps (displacement compatibility). A surprising result is obtained for the joint with the constant-thickness straps in Figure 18.9a again with $t_2 = t_1/2$. The result of the simple displacement compatibility analysis is that $P_A = P_C$ and $P_B = 0$; in other words, the middle row of bolts does not contribute to load transmission. This can easily be understood by considering that the strain between A and C is constant and equal in the plate and the two straps. It implies that the displacements of the holes at B in the plate and in the straps are equal and thus the bolts in row B remain unloaded. A more elaborate FE calculation will indicate that the bolts of the second row carry some load, but still much less than the bolts in the other two rows.[23]

A better distribution of load transmission between the three rows is obtained by staggering the thickness of the straps, see Figure 18.9b. This can be done in such a way that $P_A = P_B = P_C$. However, it is true again that the bolt holes of row A in the plate are the more fatigue critical holes because of the maximum bypass load in addition to the pin-loading. This is called the end-row effect because row A is the last one in the plate. The end-row effect on fatigue can be reduced by designing the staggered strap for a lower load transmission than 1/3. Such improvements are obtained by local variations of the stiffness of the joint.

---

[23] Under static load until failure, the bolts in row B will significantly contribute to load transmission because plastic deformation around the holes will lead to a more homogeneous load distribution between the three rows.

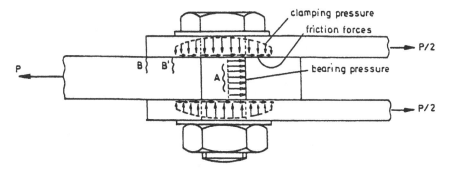

Fig. 18.10 As a result of clamping, load transmission also occurs by friction forces. Shift of crack location is possible.

Clamping bolts of a joint

Bolts can be pre-tensioned by controlled torque with a torque wrench. This can have a large impact on the fatigue life. Load transfer from one element to another one now occurs partly by frictional forces. Moreover, the coefficient of friction, which statically may be in the order of 0.3 for a dry assembling, can increase considerably under cyclic load as a result of fretting between the clamped plates. If clamping is absent, fatigue cracks are initiated at the edge of the hole in the net section, see location A in Figure 18.10. However, after sufficient tightening of the bolt, fretting movements inside the hole are suppressed. Crack initiation shifts to locations B or B'. Clamping will increase the fatigue strength until this shift of the crack location has occurred. More clamping should not give a further improvement.

18.4 Bolts loaded in tension

Bolts loaded in tension are frequently used in structures because it allows easy assembling and disassembling. If a tension bolt is subjected to cyclic loading it should be admitted that the shape of a bolt is poor from a fatigue point of view. Significant stress concentrations occur in a bolt at three locations, see Figure 18.11:

1. At the transition from the bolt head to the shank of the bolt.
2. In the groove between shank and screw thread (not always present).
3. In the screw thread.

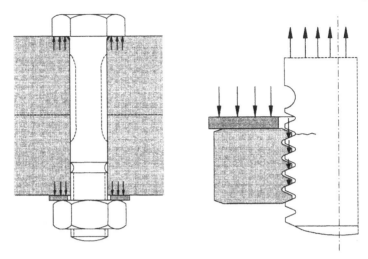

Fig. 18.11 Tension bolt with loads on bolt head, washer and screw threads between nut and bolt.

The bolt head is loaded in an unfavorable way because the load is applied close to the root of the notch. A similar case was discussed in Chapter 3 for a flat anchor bolt (T-head), see Figure 3.18 with high K_t-values. The circular cross section of a normal bolt will give a somewhat lower stress concentration. Unfortunately, the radius between the bolt head and shank cannot be made large in view of interference with the hole edges. Values of K_t of the order of three and even larger are possible. The bottom plane of the bolt head and the plate to be clamped should be perfectly parallel because otherwise the bolt head is also loaded in bending. This can be disastrous for fatigue.

In general, the groove between the shank and screw threads will not be critical because the radius is larger than the root radius in the screw thread itself, see the cross section in Figure 18.11. The root radii of the screw threads are standardized, unfortunately at low values leading to a significant stress concentration. Furthermore, the load transmission from the bolt to the nut is not homogeneously distributed over the screw threads. As indicated in Figure 18.11 by the arrows on the screw threads, the largest contribution to the load transmission occurs by the first full screw thread of the nut. As a result, fatigue cracks initiate at the corresponding screw thread of the bolt, the more so because the loads of the other screw threads also pass through this cross section (bypass load). Most fatigue failures occur at this location of the bolt.

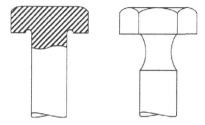

Fig. 18.12 Modified bolt heads to reduce the stress concentration between bolt head and shank.

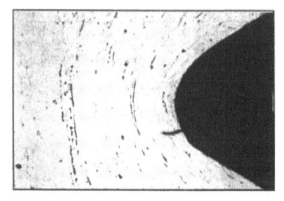

Fig. 18.13 A small fatigue crack in the screw thread of a bolt. Crack length 40 µm. Note the structure of the rolled screw thread as revealed by the curved pearlite strings [14].

Several possibilities have been proposed to improve the bolt design, see e.g. [11–13]. Two possibilities to reduce the root radius between the bolt head and shank are shown in Figure 18.12. The local K_t-value can thus be reduced. Improved screw thread concepts for bolts and nuts were also proposed. However, special bolts and nuts are not generally appreciated to avoid fatigue problems. Designers prefer to rely on high-quality bolts and pre-tensioning the bolt. High quality bolts are made of low-alloy steel which allows a high pre-tension in the bolt.

The screw thread of a bolt is made either by cutting or rolling. It should be recommended to use rolled thread for fatigue loaded bolts. A cutting operation cuts through the fibrous structure of the material, whereas rolling will force the fibrous structure to follow the thread profile, see Figure 18.13. Moreover, rolling leaves a better surface quality and can also introduce residual compressive stresses at the root of the screw thread. High-quality bolts always have rolled threads. The fatigue strength, and especially the fatigue limit are much better than for bolts with cut screw thread.

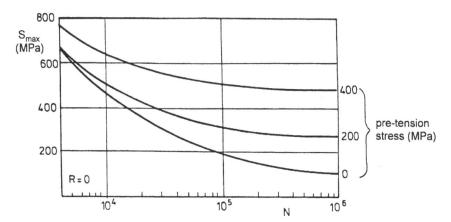

Fig. 18.14 Effect of pre-tension on the S-N curve of a steel bolt loaded in tension (diameter 10 mm) [15].

Pre-tension in bolts

During assembling of a bolted structure, it is usual to tighten a bolt by applying a prescribed torque moment to the nut. This introduces a pre-tension load in the bolt. If a cyclic load is applied to the structure, the pre-tension increases the mean stress in the bolt, but reduces the stress amplitude. In view of the predominant effect of the stress amplitude, a significant gain of the fatigue strength can be obtained. This is illustrated by the S-N curves in Figure 18.14.

The mechanism of pre-tension in the bolt can be explained with reference to Figure 18.15. In this figure two parts A and B are clamped by a pre-tensioned bolt. The pre-tension load in the bolt is causing a compressive load on the contact area between the two parts ($P_{contact}$). In the unloaded joint $P_{contact} = -P_0$, with P_0 as the pre-tension load of the bolt. If a load P is applied on the joint, the load in the bolt (P_{bolt}) will increase, and $P_{contact}$ will become less compressive. However, the joint is still reacting as an integral part as long as contact between the two parts does exist. It implies that the load transmission in the joint occurs partly as load flow through the bolt and for another part via the contact area. The load increment in the bolt will thus be smaller than the increment of the applied load. The simple equation with load increments is:

$$P = \Delta P_{bolt} + \Delta P_{contact} \tag{18.4a}$$

or

$$\Delta P_{bolt} = P - \Delta P_{contact} \tag{18.4b}$$

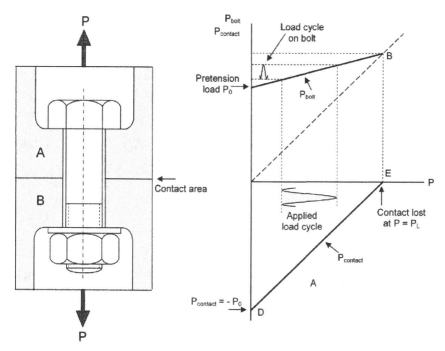

Fig. 18.15 Increased mean load and reduced load amplitude by pre-tension of bolt.

During an increasing load P the contact load will become smaller, which means that $P_{contact}$ becomes less negative, and thus $\Delta P_{contact}$ is positive. Thus,

$$\Delta P_{bolt} < P \qquad (18.5)$$

The effect of pre-tension on the reduction of the incremental load in the bolt is further illustrated in Figure 18.15. Both P_{bolt} and $P_{contact}$ are plotted along the vertical axis as a function of the load P applied on the joint along the horizontal axis. If the load P is increased, the contact load starting at $P_{contact} = -P_0$ will become smaller until $P_{contact} = 0$ at point E in Figure 18.15. The corresponding load is denoted as P_L. For $P > P_L$ contact between parts A and B is lost. During the same increasing load from $P = 0$ to $P = P_L$, the load in the bolt starts at the pre-tension load P_0 and increases to $P = P_L$. It turns out that the load variation in the bolt is significantly smaller than the load applied on the joint. The consequence for a cyclic load is illustrated in Figure 18.15 by a specific applied load cycle and the corresponding load cycle in the bolt. As said before, pre-tensioning of the bolt increases S_m in the bolt but a significant reduction of S_a is obtained which has a predominated effect and the fatigue strength as illustrated by the results in Figure 18.14. However, if $P > P_L$ then the contact between the two connected parts is lost, and P_{bolt} will be equal to the applied load.

The pre-tension effect can be enhanced by reducing the stiffness of the bolt. If the stiffness coefficients of the bolt and the compressed material of the joint are denoted by C_{bolt} and C_C respectively, and the elongation of the bolt by $\Delta \ell$, then

$$\Delta P_{bolt} = C_{bolt}\Delta \ell, \quad \Delta P_C = C_C \Delta \ell \qquad (18.6)$$

Substitution in Eq.(18.4) gives:

$$\frac{\Delta P_{bolt}}{P} = \frac{1}{1 + C_C / C_{bolt}} \qquad (18.7)$$

According to this equation, a lower C_{bolt} will reduce ΔP_{bolt}, and the favorable effect of pre-tensioning increases. A lower C_{bolt} can be obtained by using a waisted bolt, see the dotted lines in Figure 18.11. Another option is to use a bolt of a Ti-alloy which has a lower E-modulus than steel. These bolts are used in aircraft, also because of weight-saving.

Pre-tensioning is effective as long as separation at the contact area of the joint does not occur. For this reason, high-strength steel bolts are pre-tensioned to a high stress level up to 70% of $S_{0.2}$ (P_0 calculated for the minimum diameter of the screw thread). This also helps to come to a more uniform load transfer along the screw thread of the bolt. Usually, the pre-tension is controlled by tightening of the nut with a calibrated torque wrench. However, the relation between the torque moment and the pre-tension load in the bolt is strongly dependent of the friction between the nut and washer and between the screw threads of nut and bolt during tightening of the bolt. Lubrication should be considered because it gives a larger pre-tension in the bolt for the same torque moment. For critical tension bolts, special means are available to assure that the required pre-tension has been installed. This can be done by measuring the length increment of the bolt, but also by using special washers.

It is important to emphasize that the favorable effect of pre-tensioning is fundamentally different from the favorable effect of compressive residual stresses obtained by shot peening or plastic hole expansion. Residual stresses reduce S_m and do not affect S_a, whereas pre-tensioning increases S_m and reduces S_a.

18.5 Riveted and bolted joints with eccentricities

Single lap joints between sheets are used in various structures. Overlapping sheets are joined by a number of fasteners (rivets or bolts). The variety of such joints is large; the material, material thickness, type of fastener, pattern of fastener locations (how many rows, and how many fasteners in a row), hole diameters, distances between fasteners, joining techniques, etc. Some essential aspects with respect to fasteners should be mentioned.

Bolts usually have a clearance fit in the hole, while clamping between the sheets can be controlled by tightening of the bolts. Riveting occurs by a deformation process on the rivet to fill up the hole and to produce the closing head (also called the driven head). Solid rivets are usually of a similar alloy as the sheets or plates. Riveted joints were applied already long ago in steel structures (e.g. bridges, cranes, ships, etc.) and also in aluminium structures, in particular aircraft structures. Riveting occurs by hammering or squeezing. Squeezing is not noisy, in contrast to hammering. Moreover, squeezing can be done in an automated production process which gives a more uniform production quality.

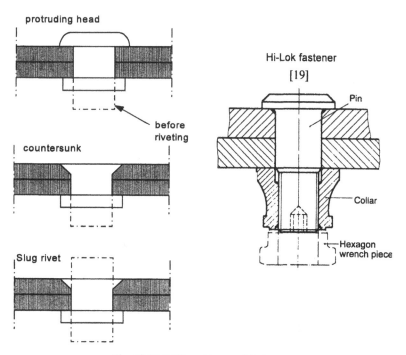

Fig. 18.16 Different types of fasteners.

A few fastener types are shown in Figure 18.16. The protruding head rivet is the older one, which is also often used with a spherical head. The countersunk rivet was introduced to obtain a flat surface at one side of the joint which can be desirable for aerodynamic or hydrodynamic reasons. Due to the countersunk hole, the stress concentration at the hole is larger than for the cylindrical hole of a protruding head rivet [16]. During riveting, the force applied on the rivet is pressing material of the rivet shank into the rivet hole. Hole filling depends on the rivet force, which is also called the squeeze force. If the force is low, the hole is just filled up, but if a larger force is used, the hole is plastically expanded and a much better contact between the entire rivet and the hole is obtained. This leads to better fatigue properties as discussed later.

Several types of fasteners were developed for aircraft applications for special reasons, e.g. easy and fast assembling, good fatigue properties, and high static shear strength [17]. An example is the Hi-Lok fastener also shown in Figure 18.16. It is usually installed with a slight interference fit for good hole filling. The fastener is made from a high-strength steel or Ti-alloy to obtain a large static shear strength. The installation of the nut on the bolt is done from one side. The nut consists of two parts, a collar which is the real nut, and a hexagon wrench piece, see Figure 18.16. During installation, the hexagon piece is tightening the nut until this piece is sheared off from the nut. As a result, a well-controlled high torque is applied which ensures a significant clamping of the sheets by the Hi-Lok fastener, and also a high strength and good fatigue properties. Of course such a fastener is more expensive than a conventional rivet or bolt. A variety of high-tech fasteners is commercially available [18, 19].

Some simple types of riveted joints with eccentricities are schematically indicated in Figure 18.17; three lap joints with 1, 2 and 3 rows of fasteners respectively, and a single-strap butt joint with four rows of fasteners. Actually, the latter joint can also be identified as two lap joints in series with two rows of fasteners each. The eccentricities of the joints imply that the neutral line is not a straight line but contains one or more eccentricities, see Figures 18.18 and 18.19 to be discussed later. Symmetric double strap joints shown earlier (Figures 18.1 and 18.9) do not have such eccentricities. Characteristic differences occur between joints with and without eccentricities:

1. The fasteners in a single lap joint are single-shear fasteners loaded in shear in one cross section of the fastener only. The fasteners in a double strap joint are double shear fasteners with two cross sections equally

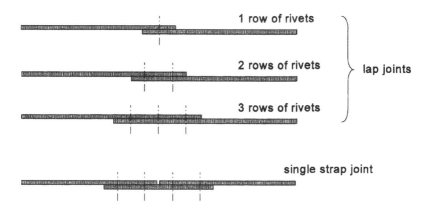

Fig. 18.17 Different types of simple riveted joints with eccentricities at the rivet rows.

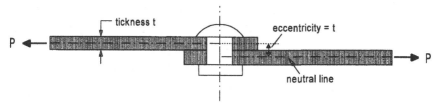

Fig. 18.18 Lap joint with one row of rivets. Large eccentricity at rivet row.

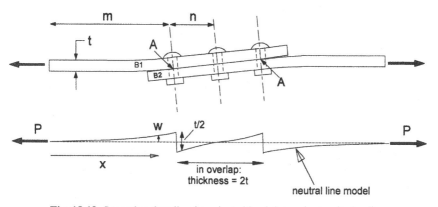

Fig. 18.19 Secondary bending in a riveted lap joint under tensile loading.

loaded in shear. As a result, the shear strength of a double strap joint is larger.

2. In a lap joint, only one contact surface is present between the two plates or sheets of the joint. Load transmission by frictional forces can occur in one mating surface only. In a double strap joint, two such mating planes are available which is favorable for the load transmission.

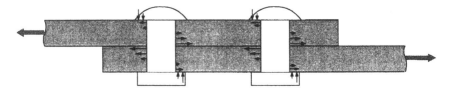

Fig. 18.20 Asymmetric loading on rivets as a result of the eccentricity in a lap joint.

3. As a result of the eccentricities in a lap joint, a tensile load on the joint causes bending of the plates, see Figure 18.19. Maximum bending stresses occur at the eccentricities, i.e. at the fastener rows (point A in Figure 18.19). This will cause an extra stress concentration at the holes of the fasteners. Bending caused by the tensile load is referred to as *secondary bending*. It is a by-product of the tension load.
4. The fastener in a lap joint is asymmetrically loaded, see Figure 18.20, and the fastener will tilt in the hole if the hole filling by the fastener is poor. This gives an inhomogeneous bearing pressure along the hole.

All aspects (1) to (4) are unfavorable for the joint with eccentricities in comparison to symmetric joints without eccentricities. An obvious eccentricity occurs in a riveted lap joint with a single row of rivets. The eccentricity is equal to the sheet thickness, see Figure 18.18.

The bending moment at the rivet row in Figure 18.18 is equal to $Pt/2$, which leads to the following bending stress:

$$S_{\text{bending}} = \frac{\frac{1}{2}Pt}{\frac{1}{6}Wt^2} = 3\frac{P}{Wt} = 3S_{\text{tension}} \qquad (18.8)$$

(W = width of joint). Both S_{bending} and S_{tension} are nominal stress levels, disregarding the presence of the holes filled with rivets. The bending factor, k, is defined as the ratio of the bending stress and the tensile stress:

$$k = \frac{S_{\text{bending}}}{S_{\text{tension}}} \qquad (18.9)$$

For the single rivet-row lap joint, $k = 3$ according to Eq. (18.8). The bending stress is three times the tension stress, which is a significant increase of the stress level at the critical location of the joint. The fatigue strength of riveted lap joints with a single rivet row is poor indeed. The fatigue strength of riveted lap joints with two or more rivet rows is significantly better. Bending is less, and furthermore, the load transmission occurs in more than one rivet row.

The secondary bending caused by the tension load can be calculated by considering the out-of-plane displacements of the neutral line of the joint, denoted by w, see Figure 18.19. It leads to a differential equation for the bending moment M_x:

$$M_x = Pw = E^* I \frac{d^2 w}{dx^2} \tag{18.10}$$

$E^* = E/(1 - v^2)$, E is Young's modulus, v is Poisson's ratio,[24] I is the inertia moment of the cross section of the sheet. The solution of the equation, using boundary conditions and assuming that the length m in Figure 18.19 is much larger than n, leads to Eq. (18.11) for the bending factor. The derivation is given in [20].

$$k = \frac{3}{1 + 2\sqrt{2}\,T_2} \text{ with } T_2 = \tanh\left(\frac{n}{t}\sqrt{1.5\frac{S}{E^*}}\right) \tag{18.11}$$

(t is the sheet thickness, $2n$ is the row distance for a specimen with two rivet rows, and n is the row distance if three rivet rows are present; tanh is the hyperbolic tangent function). It should be noted that the bending factor depends on the applied tensile stress S. The relation between $S_{bending}$ and S is non-linear, which is illustrated by the calculated results in Figure 18.21 for a lap joint with three rows of rivets. The bending factor in this case is $k = 1.36$ at $S = 100$ MPa which is considerably lower than the extreme $k = 3$ for the single-row lap joint. For a three-row lap joint with the same row distance, the calculated bending factor, again at $S = 100$ MPa, is reduced to $k = 0.99$. This reduction is not a consequence of having three rows of rivets, but it is due to a two times longer overlap. Fatigue tests on riveted lap joints with two rows of rivets have shown that the fatigue life is improved by increasing the distance between the two rows [21].

A comparison of S-N curves of three different types of riveted joints is made in Figure 18.22. Obviously, the symmetric butt joint without any secondary bending is superior to the other two types of joints. The bending factor k for the lap joint and the single-strap butt joint are $k = 1.78$ and 2.40 respectively (calculation data $t = 1.2$ mm, $n/t = 5.42$, $E = 72000$ MPa for the 2024-T3 Al-alloy, and $S = 100$ MPa). Equation (18.11) is not valid for the single strap joint, but again a similar k equation was derived with

---

[24] In the previous edition of this book prevention of anti-clastic bending was ignored, and thus E^* was simply assumed to be equal to E. In general, some prevention of anticlastic bending will be present in thin sheets which will cause a small increase of the bending stiffness. Secondary bending calculations with the neutral line model but ignoring anticlastic bending may yield a slightly larger bending stress, in the order of 10% larger.

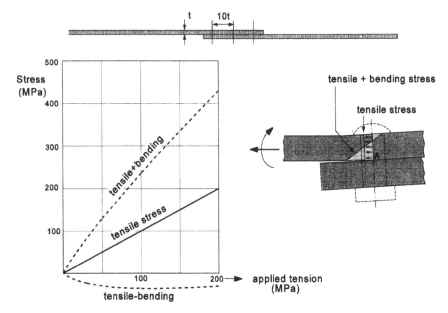

Fig. 18.21 Non-linear secondary bending stresses in a riveted lap joint calculated with Eq. (18.14) for Al-alloy sheet material.

Eq. (18.10). The higher k factor of the single strap joint has indeed led to a lower S-N curve.

The most critical points for crack initiation in a riveted lap joint are points A of the first and last row of the joint indicated in Figure 18.19. To a certain extent, this is the end-row effect discussed before for symmetric bolted joints. However, the effect is still aggravated in a lap joint because of the additional bending stress caused by secondary bending. The maximum bending stress occurs at the same end rows, see point A in Figure 18.21. Furthermore, fretting corrosion between the two sheets cannot be avoided. In Figure 18.19, part B1 of the left sheet carries the full load on the joint, whereas part B2 of the other sheet is unloaded. Fretting between the two sheets occurs, especially around the rivet holes of the end rows. The superposition of stress concentration, secondary bending and fretting corrosion explains why the fatigue limits of the S-N curves in Figure 18.22 are low.

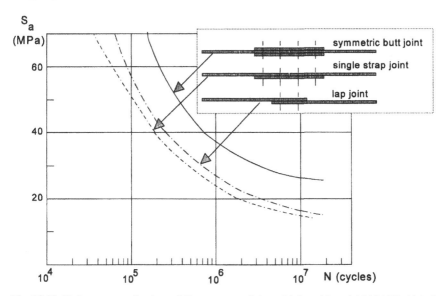

Fig. 18.22 Fatigue curves for three different types of riveted joints. Material 2024-T3 Alclad ($t = 1.2$ mm), $S_m = 69$ MPa [21].

Effect of the rivet squeeze force

As emphasized earlier, the load transmission in a riveted joint is a complex phenomenon. Hole filling has been mentioned as a significant aspect. The shank of a fastener should fully fill up the hole, preferably with an interference fit. This will prevent rivet tilting and the related unfavorable pressure distribution of the rivet on the hole. The shear deformation of the rivet and the surrounding material (rivet flexibility) will be reduced by an improved hole filling. Furthermore, the fatigue sensitivity for the surface finish of the hole will be less.

Hole filling in a riveted joint depends on the plastic deformation of the rivet during the riveting process. It can be improved by increasing the squeeze force on the rivet during riveting. If this force is sufficiently large, the rivet hole will be expanded and an interference fit is obtained. The shank of the rivet is then surrounded by pre-tensioned sheet material. Measurements of the deformation of the driven rivet head as a function of the squeeze force are given in Figure 18.23. The height of the driven head is decreased and the diameter is increased by a larger squeeze force, as should be expected. At the same time, hole expansion is introduced up to a few percent. The favorable effect of a high squeeze force on the fatigue life is shown by Figure 18.24.

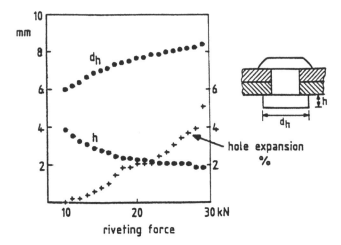

Fig. 18.23 Deformation of the driven rivet head and hole expansion as a function of the rivet squeeze force [22].

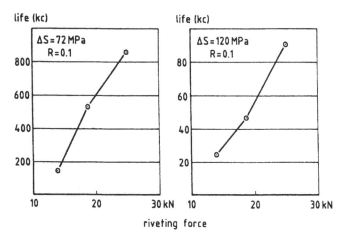

Fig. 18.24 The effect of the rivet squeeze force on the fatigue life of a riveted lap joint. Material: Al-alloy 2024-T3, $t = 2$ mm, rivet diameter 4.8 mm, $R = 0$ [22].

Apparently, a significant increase of the fatigue life can thus be achieved. More results are presented in [23].

An increased squeezing force will also increase the clamping force of the rivets on the sheets. Clamping of the rivets affects the location of the fatigue crack. If the clamping is poor, cracks start at the rivet holes in the minimum section of the most critical end row, i.e. at point A in Figure 18.25. Such a crack is shown in Figure 18.26a. Fretting at that location causes crack initiation. However, if an improved clamping is obtained, fretting movements

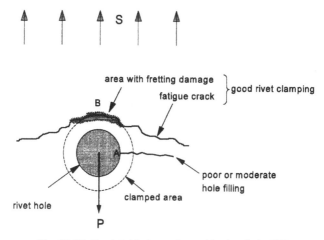

Fig. 18.25 Crack initiation at B outside rivet hole [22].

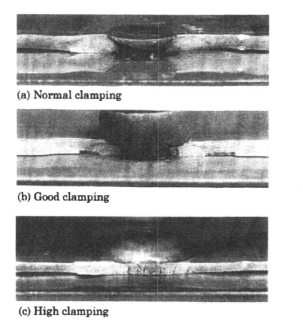

(a) **Normal clamping**

(b) **Good clamping**

(c) **High clamping**

Fig. 18.26 Effect of clamping on fatigue crack at rivet holes [22]. (a) Corner cracks at hole edge. (b) Semi-elliptical cracks slightly away from hole. (c) Crack growing around hole.

are no longer possible near the hole. Crack nucleation starts still in the minimum section by fretting between the two mating surfaces, but it occurs slightly outside the hole, see the semi-elliptical cracks in Figure 18.26b. With a further increase of the squeezing force the clamping of the two sheets by the rivet becomes considerable. Fretting around the hole is highly restrained, but it cannot be prevented at the edge of the clamped area near point B in Figure 18.25 where secondary bending is large which implies that some separation between the two sheets must occur. Several crack nuclei are created simultaneously which are linked up to a single crack with a ragged appearance, see Figure 18.26c. This crack no longer grows through the hole, but around the rivet hole. The phenomenon is somewhat similar to the effect of bolt clamping as described earlier, see Figure 18.10, although secondary bending is not involved in the symmetric joint.

Predictions on the fatigue life of riveted lap joints

Predictions of the fatigue life of a riveted lap joint should be expected to be a complex problem in view of secondary bending, fastener type, fretting corrosion and rivet hole filling; all having a significant effect of the fatigue life. These aspects are not easily accounted for by analytical equations. Moreover, geometric variables of the joint must also be considered. Homan and Jongebreur [24] suggested a prediction method for riveted lap joints in sheet material of aircraft fuselage structures loaded under constant-amplitude loading. The model starts from an S-N curve of a reference lap joint for $R = 0$ for which fatigue data are available. Predictions are extrapolated from this curve by accounting for three contributions to the stress concentration at the rivet holes of the critical end row. The contributions are associated with (i) load transmission by the rivets (pin loading on the hole), (ii) bypass loading of the other rivet rows, and (iii) increased stress by secondary bending. The equation used is the following:

$$K_t = \gamma K_{t,\text{pin}} + (1 - \gamma)K_{t,\text{hole,tension}} + k K_{t,\text{hole,bending}} \qquad (18.12)$$

In this equation, γ is the percentage of the load transmitted to the other sheet in the critical row. Then, $(1 - \gamma)$ is the percentage of the bypass load. The factor k is the secondary bending factor as defined in Eq. (18.9). The three stress concentration factors, $K_{t,\text{pin}}$, $K_{t,\text{hole,tension}}$ and $K_{t,\text{hole,bending}}$, depend on the joint geometry (rivet diameter/rivet pitch). It may be noted that Eq. (18.12) is partly similar to Eq. (18.2) for a symmetric butt joint with two rows of fasteners. The secondary bending term does not occur in Eq. (18.2), and γ of Eq. (18.12) is equal to 0.5 in Eq. (18.2).

The prediction method of Homan and Jongebreur is based on the peak stress calculated with Eq. (18.12). The peak stress is calculated for the reference joint and for the actual joint for which life predictions should be made. The similarity principle is adopted, which implies that similar peak stresses in the reference joint and the actual joint should give similar fatigue lives. Obviously, the fastener type, fretting corrosion and rivet hole filling are not accounted for by Eq. (18.12). As pointed out by Homan and Jongebreur, these conditions should be similar for the reference joint and the actual joint. If this similarity is questionable, more relevant reference data should be used. It might well imply that fatigue tests have to be made on the actual joint.

18.6 Adhesive-bonded joints

Adhesive bonding is widely used for various technical applications, including metal-to-metal bonding and bonding of a metallic material to a non-metallic material. If bonding of metals is done under closely controlled conditions, high-quality and durable joints can be made. Structural applications with a primary load on the adhesive bond line are still largely restricted to aircraft structures. Tensile loads on an adhesive metal-to-metal bond line are generally avoided because of the possibility of peeling failures.[25] However, the static strength and fatigue strength under shear loading are satisfactory.

In principle, adhesive bonding of Al-alloy sheet material in a lap joint should be attractive if compared to a riveted lap joint. In order to see the advantages of adhesive bonding with respect to fatigue, two fundamental differences between the two types of lap joints are important. First, in a riveted lap joint, the overlapping sheets are attached to one another at discreet points only, i.e. by the fasteners. Obviously, severe stress concentrations should occur. However, if the attachment is made continuously in the full overlapping area by adhesive bonding, these stress concentrations do not occur. Secondly, metallic contact between the two sheets is absent in the adhesively bonded joints, and thus fretting between the mating sheets is also eliminated. From a fatigue point of view, adhesive bonding is the preferred jointing method. Fatigue tests have confirmed this conclusion as shown by the S-N curves in Figure 18.27. It should also be noted that the nominal overlap of the two types of specimens in Figure 18.27 are the same, but the effective overlap is larger in the bonded lap joint (60 mm) than in the riveted lap joint (40 mm = distance between two outer rows). As previously mentioned, the larger overlap reduces secondary bending.

---

[25] In the literature, a peeling failure is often associated with the mode I opening as defined in Figure 5.2. The peeling resistance can still be satisfactory if a good adhesive is used in accordance with the prescribed bonding operation.

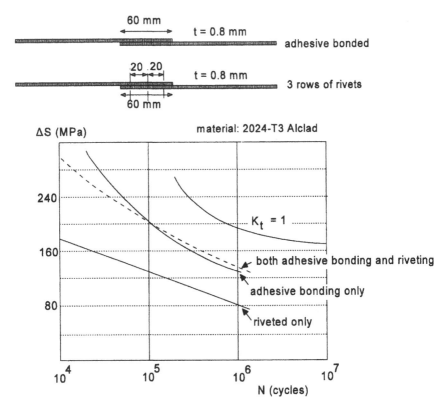

Fig. 18.27 S-N curves of bonded lap joints ($R = 0.1$). Comparison to the S-N curve of a riveted lap joint with a similar geometry, and to the S-N curve of unnotched material [25].

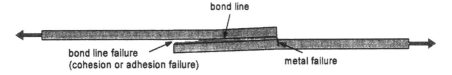

Fig. 18.28 Two different failure modes in an adhesive-bonded joint.

Two fatigue failure modes of an adhesive bonded lap joint are shown in Figure 18.28. Under high-load amplitudes and with a short overlap (e.g. 10 times the sheet thickness or less), failures can occur in the bond line, either in the adhesive itself (cohesion failure) or at the interface between the adhesive and metal (adhesion failure). This will depend on the quality of the pre-treatment of the metal surfaces, adhesive and curing cycle. The quality of the adhesives and the bonding techniques have been much improved, and short overlaps are not used. As a result, sheet metal failure is the predominant fatigue failure mode. The sheet metal failure is the result of the stress concentration at the end the overlap and the occurrence of secondary bending with the maximum bending factor at the same location. An abrupt thickness variation is present at the end of the overlap. A significant stress concentration might be expected, but it is less serious because of the low elastic modulus of the adhesive. Young's modulus is of the order of 3000 MPa compared to 72000 MPa for the aluminium alloy sheet material. The bond line thickness is usually in the range of 0.1 to 0.2 mm. Contrary to intuition, the bond line thickness seems to have a minor influence on the fatigue strength.

18.7 General discussion on predictions of fatigue properties of joints

Fatigue of a variety of joints has been discussed in the previous sections. Apparently, each type of joint has its own specific features depending on how the load transmission occurs in the joint. Several phenomena were revealed, such as fretting corrosion, clamping, secondary bending, bypass loads, which all can affect the fatigue properties. Unfortunately, it turned out to be difficult to account quantitatively in a fully rational way for these influences on the fatigue life. Joints do not allow such simple comparisons as discussed in Chapter 7 on elementary notch geometries fully characterized by K_t, the size of the notch root radius and surface roughness. In the latter case, the fatigue properties were predicted with the fatigue data of unnotched specimens as basic reference data. Although these predictions also had limitations, the similarity concept could still be used. The logic of the similarity concept was: similar conditions in notched components and unnotched specimens should lead to the same fatigue properties. Such a simple similarity cannot be defined for joints due to the more complex conditions of joints. The prediction techniques discussed in the previous sections all start from available fatigue data for joints. These data are extrapolated to other joints. The extrapolation equations are based on a similarity between joints of the same type. The extrapolation is based on empirical evidence. If sufficient data are available, it can imply that the extrapolation is in fact an interpolation. This applies to Larsson's equations for lugs.

The prediction methods for riveted and bolted joints were based on calculated peak stress values. Similar values were supposed to lead to the same fatigue life. The equations to calculate the peak stress values are based on reasonable arguments. However, the representation of the load transmission in the joint is not covering all aspects which are known to affect the fatigue properties. A more refined procedure was proposed by Jarfall [26, 27], who introduced the stress severity factor (SSF). This factor for riveted joints also accounted for such aspects as hole preparation, residual stress from cold working, interference and fastener flexibility. Empirical data on the influence of these aspects are required to use the SSF. According to Jarfall, life predictions with the SSF are still difficult, but the SSF can be used in design studies. Calculations on load transfer by the fasteners should be made. The SSF then gives comparative indications on the stress severity of the fasteners and design modifications can be considered to reduce the severity of the most critical ones. This approach in essence is designing against fatigue by considering suitable field parameters characterizing the fatigue severity of the joint.

Prediction of crack growth life instead of fatigue life until failure can be an interesting option if fatigue crack growth starts almost immediately. However, it is questionable whether it is useful for bolted and riveted joints. The situation for these joints is complex, particularly for small cracks. Much life is spent by crack initiation and initial growth of small cracks. It implies that the finite element calculation of K-values requires a realistic modeling of a joint to calculate K-values of part through cracks. In view of the complex load transmission in riveted and bolted joints, the FE modeling is a problematic issue. The limitations of predictions are not set by the capacity of computers, neither by available calculation programs, but rather by modeling of the joint details and the mechanisms of fatigue crack initiation and early propagation of small cracks. Consequently, a more efficient solution to obtain fatigue life indications of complex joints should be to collect fatigue test results of similar joints. These results should then be translated to the geometrical conditions of the structure by adopting some suitable field parameter. It could be desirable to carry out some exploratory fatigue tests for this purpose. As emphasized previously, such fatigue test should be carried out on test articles which are representative for the geometry with a load spectrum relevant for the structure.

18.8 Major topics of the present chapter

1. The load transmission in joints is essentially different for lugs, joints with bolts in tension, bolted and riveted joints with shear loaded fasteners and adhesive bonded joints.
2. The fatigue limit of joints can be very low due to severe stress concentrations, fretting corrosion and secondary bending. In spite of a high static strength of a joint, the fatigue limit can be low.
3. Prediction of the fatigue life, fatigue strength and fatigue limit is a complex problem for joints because the crack initiation and initial growth of small cracks cannot easily be compared to a similar behavior in unnotched specimens. This excludes predictions based on basic material fatigue properties. Fatigue properties of joints should be derived from fatigue properties of similar joints for which data are available.
4. The size effect on the fatigue limit of lugs is large. The low fatigue limit of lugs can be significantly improved by plastic hole expansion.
5. Bolts loaded in cyclic tension have a relatively low fatigue strength which can be substantially increased by pre-tensioning.
6. The fatigue strength of symmetric butt joints (riveted or bolted) is superior to the fatigue strength of lap joints. The former joints have no eccentricities, whereas the eccentricity in lap joint causes unfavorable secondary bending and a more complex loading of the fastener on the hole.
7. Hole filling is of great importance to riveted joints. A high rivet squeeze force leads to significant life improvements due to plastic hole expansion and a better clamping between the sheets.
8. Fretting corrosion and local load transmission by fasteners are eliminated in adhesive bonded lap joints. It results in a larger fatigue strength in comparison to similar riveted lap joints. But due attention must be paid to the quality and durability of the bonded joint.

References

1. Schijve, J., *Fatigue of Lugs. Contributions to the Theory of Aircraft Structures.* Nijgh-Wolters Noordhoff Universiy Press (1972), pp. 423–440.
2. Larsson, S.E., *The development of a calculation method for the fatigue strength of lugs and a study of test results for lugs of aluminium alloys.* Fatigue Design Procedures, 4th ICAF Symposium. Pergamon Press (1969), pp. 309–339.

3. Unpublished calculation methods in Technical Handbook No. 3. of Fokker Aircraft Factories, TH3.411.1 (1980).
4. *Fatigue-Endurance Data, Vol. 8.* ESDU Engineering Science Data (1990).
5. Clarke, B.C., *The use of pins with flats to increase the fatigue life of aluminium alloy lugs.* Royal Aircraft Establishment, Farnborough, UK, Tech. Report 66015 (1966).
6. Schijve, J., Broek, D. and Jacobs, F.A., *Fatigue tests on aluminium alloy lugs with special reference to fretting.* Nat. Aerospace Lab. NLR, Report TR M.2103, Amsterdam (1962).
7. Hartman, A. and Jacobs, F.A., *The effect of various fits on the fatigue strength of pin-hole joints.* Nat. Aerospace Lab. NLR, Amsterdam, Report M1946 (1954).
8. Champoux, R.L., *An overview of cold expansion methods.* Fatigue Prevention and Design, J.T. Barney (Ed.), EMAS, Warley (1986), pp. 35–52.
9. Leon, A., *Benefits of split mandrel cold working.* Int. J. Fatigue, Vol. 20 (1998), pp. 1–8.
10. Waters, K.T., *Production methods of cold working joints subjected to fretting for improvement of fatigue strength.* Fatigue of Aircraft Structures. ASTM STP 274 (1959) pp. 99–111.
11. Heywood, R.B., *Designing against Fatigue.* Chapman and Hall, London (1962).
12. Peterson, R.E., *Stress Concentration Factors.* John Wiley & Sons, New York (1974).
13. Shin-ichi Nishida, *Failure Analysis in Engineering Applications.* Butterworth-Heinemann, Oxford (1992).
14. Bonnee, W.J.A., *Investigation of horizontal stabilizer attachment bolts of Astir-gliders.* Nat. Aerospace Lab. NLR, Report TR 80019, Amsterdam (1980).
15. Hertel, H., *Fatigue Strength of Structures.* Springer-Verlag, Berlin (1969) [in German].
16. Shivakumar, K.N. and Newman, J.C., *Stress concentrations for straight-shank and countersunk holes in plates subjected to tension, bending, and pin loading.* NASA-TP-003192 (1992).
17. Niu, M.C., *Airframe Structural Design.* Conmilit Press (1988).
18. Hoffer, K., *Permanent Fasteners for Light-Weight Structures.* Aluminium-Verlag, Düsseldorf (1984).
19. Barrett, R.T., *Fastener Design Manual.* NASA Reference Publication 1228 (1990).
20. Schijve, J., *Some elementary calculations on secondary bending in simple lap joints.* National Aerospace Lab. NLR, Report TR 72036, Amsterdam (1972).
21. Hartman, A. and Schijve, J., *The effects of secondary bending on the fatigue strength of 2024-T3 Alclad riveted joints.* National Aerospace Lab. NLR, Report TR 69116, Amsterdam (1969).
22. Schijve, J., *Multiple-site-damage fatigue of riveted joints.* Proc. Int. Workshop on Structural Integrity of Aging Airplanes. Atlanta Technical Publications (1992), pp. 2–27.
23. Müller, R.P.G., *An experimental and analytical investigation on the fatigue behaviour of fuselage riveted lap joints. The significance of the rivet squeeze force, and a comparison of 2024-T3 and Glare 3.* Doctor thesis, Delft University of Technology (1995).
24. Homan, J. and Jongebreur, A.A., *Calculation method for predicting the fatigue life of riveted joints.* Durability and Structural Integrity of Airframes. Proc. 17th ICAF Symposium, A.F. Blom (Ed.). EMAS (1993) pp. 175–190.
25. Hartman, A., *Fatigue tests on single lap joints in clad 2024-T3 aluminium alloy manufactured by a combination of riveting and adhesive bonding.* National Aerospace Lab. NLR, Report M.2170, Amsterdam (1966).
26. Jarfall, L., *Shear loaded fastener installations.* Int. J. Vehicle Design, Vol. 7 (1986), pp. 337–380.
27. Jarfall, L., *Optimum design of joints: The stress severity factor concept.* Aircraft fatigue. Design, Operational and Economic Aspects, J.Y. Mann and I.S. Milligan (Eds.), Pergamon, Australia (1972), pp. 49–63.

Some general references (see also [12, 14, 19, 20])

28. de Rijck, J.J.M., Homan, J.J., Schijve, J. and Benedictus, R., *The driven rivet head dimensions as an indication of the fatigue performance of aircraft lap joints.* Int. J. Fatigue, Vol. 29 (2007), pp. 2208–2218.

29. de Rijck, J.J.M., *Stress analysis of fatigue cracks in mechanically fastened joints. An analytical and experimental investigation.* Doctor thesis, Delft University of Technology, Delft University Press (2005).

30. Segerfrojd, G., Wang, G.S., Palmberg, B. and Blom, A.F., *Fatigue behaviour of mechanical joints: Critical experiments and statistical analysis.* Fatigue in New and Ageing Aircraft, Proc. 19th ICAF Symposium, R. Cook and P. Poole (Eds.), EMAS, Solihull, UK (1997), pp. 575–598.

31. Toor, P.M. (Ed.), *Structural Integrity of Fasteners.* ASTM STP 1236 (1995).

32. Matek, W., Muhs, D., Wittel, H. and Becker, M., *Roloff/Matek Machinenelementen,* 12th edn. Vieweg & Sohn, Braunschweig (1992) [in German].

33. Bickford, J.H., *An Introduction to the Design and Behaviour of Bolted Joints,* 2nd edn. Marcel Dekker, New York (1990).

34. Jensen, W.J., *Failures of mechanical fasteners.* Failure Analysis and Prevention, Metals Handbook, Vol. 11, (1986), pp. 529–549.

35. Potter, J.M., *Fatigue in mechanically fastened composite and metallic joints.* ASTM STP 927 (1986).

Chapter 19
Fatigue of Welded Joints

19.1 Introduction

Welding of metals is applied on a very wide scale, especially for building up structures by welding of steel plates and girders of different cross sections (I-beams, U-beams, angle beams). Welding provides many structural design options which cannot be simply realized with other production techniques. Major applications are found in bridges, cranes, ships, offshore structures, pressure vessels, buildings and various types of spatial frames.

Welding as a production technique is associated with various problems, which are characteristic for welding only. As a result, the subject "welding" became practically a discipline on its own as illustrated by the existence of welding institutes and organizations, standards and design codes, journals, and an extensive literature. Within the welding discipline, much attention has been paid to problems related to different welding techniques known under general names as: arc welding, gas welding, electron beam welding, laser welding, resistance spot welding, friction welding, and more recently stir friction welding. Welded joint designs and notch effects of welds are typical for welded structures. Welded joints are also known for a number of characteristic weld defects. These defects have created new issues for non-destructive inspections (NDI), which have stimulated developments of X-ray and ultrasonic equipment. Moreover, fatigue properties of welded joints can exhibit considerable scatter because of a variety of imperfections of these joints. As a consequence, fatigue of welded joints has always been a matter of concern, but good welding practice can be specified for fatigue critical structures including non-destructive inspections of all welds.

The present section does not give a full survey of fatigue of welded joints, but instructive books have been published, e.g. by Gurney [1], Madox [2], Radai [3] and Lancaster [4]. The present text is a brief and elementary account of fatigue problems of welded joints which are related to aspects of geometries of welded joints and fatigue lives. Welding techniques and metallurgy problems are not covered. It should be understood that welding is not attractive for materials with a high static strength obtained by a heat treatment. The welding process will then destroy the heat treatment. Also, welding of thin sheet material is not popular in view of geometric distortions due to the heat flux. However, spot welding of thin sheets can be an attractive production technique for the industry.

19.2 Some general aspects

Figure 19.1 shows a sketch of a welded butt joint between two plates. The joint is welded from one side only. A few terms are recalled in this figure, and some defects are indicated. Under a cyclic tension load, the root failure (lack of penetration) is a most serious one. It can occur over a considerable distance and the defect is similar to a surface crack. The undercut at the weld toe may be serious if the profile is sharp at the bottom of the undercut. If an undercut is not present, the transition of the excess weld material[26] to the base material still gives a stress concentration at the weld toe. Slag inclusions can be serious defects for fatigue crack initiation, more than porosity due the shape of these defects. Weld defects determine the weld quality. In this respect, significant differences can exist between manually made welds and those made by automated production. The quality of a manually welding is dependent on the competence of the welding operator. It requires training, practice, and skill to make good welds. Automated welding processes have been developed specifically for fast production of long weld seams and to eliminate part of the human factor. A more homogeneous weld quality is

---

[26] The excess weld material is also called the reinforcement which is a strange term.

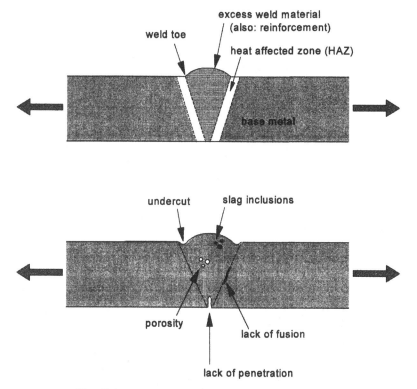

Fig. 19.1 Some terms and defects of a butt welded joint.

obtained. In any case, fatigue critical welds should always be inspected by suitable NDI techniques.

Another aspect to be mentioned here is related to thermal stresses. During welding, the weld material cools down from the melting temperature to room temperature. The weld material will contract, but this is restrained by the cooler plates. As a result, residual stresses are introduced with residual tensile stresses in the weld direction, see Figure 19.2. If such a weld is loaded in this direction, fatigue crack initiation at the ripples of the weld and related defects may be promoted. Unfortunately, residual stresses perpendicular to the weld can also be introduced by the welding process, commonly with residual tensile stress at the material surface and residual compressive stress at mid-thickness of the plate. These stresses are relevant if the fatigue load is perpendicular to the weld seam. The residual stresses introduced by welding depend on the welding technique and the design of the structure.

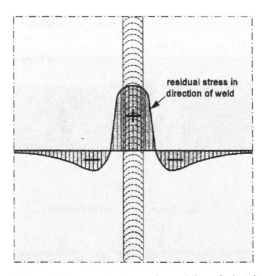

Fig. 19.2 Residual stress distribution after welding of a butt joint [2].

19.3 Geometry aspects of welds

The geometry of welded structures covers two aspects: (i) the layout of
the structure, and (ii) the local geometry of the weld joint. The layout of
a structure is a design problem which allows a large variety of solutions. As
an illustration, Figure 19.3 shows different design options for a corner joint
between two I-section girders of a frame, a plate structure of a bulge corner
of the bottom of a ship, and a nozzle of a pressure vessel [3]. This figure
illustrates that a variety of different solutions is available to the designer. The
choice will depend on considerations like ease of production, quality to be
obtained, etc. In general terms, this is a problem of production costs versus
quality of the product obtained, including durability and safety. Qualitative
fatigue life considerations on the corner joints between the I-section girders
in Figure 19.3 suggest that solution (a) is inferior to the other three options.
For the bulge corner of the bottom of a ship, the better solution should be
expected to be option (b) because the location of a high stress concentration
(arrow in the figure) is then separated from the location of complex weld
geometries. This also applies to option (d) for the nozzle of a pressure
vessel. Of course, the designer can think of more solutions than shown in
Figure 19.3. It requires judgement to design against fatigue. Unfortunately,
the best solution for fatigue is usually not the most profitable one for
production costs.

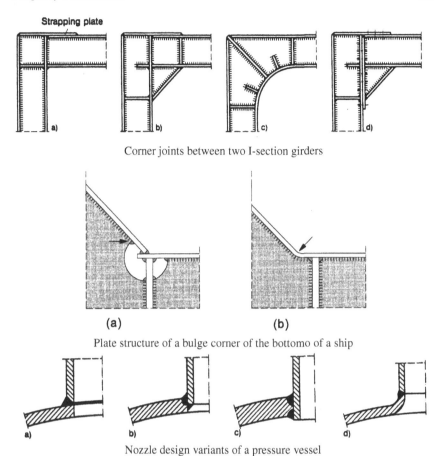

Corner joints between two I-section girders

Plate structure of a bulge corner of the bottomo of a ship

Nozzle design variants of a pressure vessel

Fig. 19.3 Different designs of welded structures (figures from [3].)

The layout of the structure determines the nominal stress level of the welds. The fatigue resistance of a structure is then dependent on the geometric details of the various types of welds. The latter problem has been studied in numerous experimental research programs by fatigue tests on a variety of specimens. These specimens should simulate characteristic features of welds in structures. Most tests were carried out on steel specimens under CA loading, but later tests were also done with VA loading and other materials, notably weldable Al-alloys.

Several types of significantly different specimens are shown in Figure 19.4. Specimens (a) to (c) are simple specimens without any stress raising elements except for the weld itself. It is of some interest to consider the reduction of the fatigue strength of these specimens in comparison to the

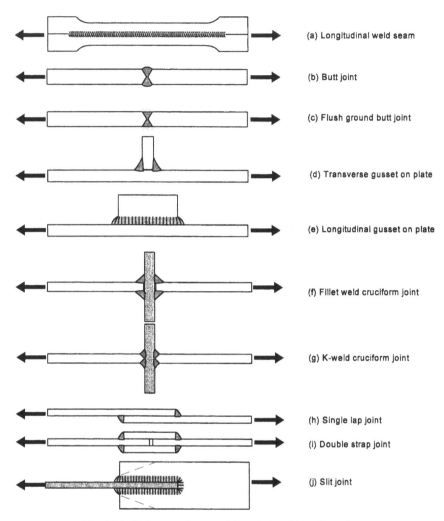

Fig. 19.4 Various specimens with typical weld configurations.

fatigue strength of the base material. The reduction of the fatigue strength is relatively small for specimen (a) with the weld seam in the loading direction. But ripples of the weld surface or weld defects may cause some reduction of the fatigue limit.

Larger reductions have been observed for specimen (b) depending on the profile of the weld reinforcement. This is illustrated by the results in Figure 19.5 which shows the fatigue strength as a function of the reinforcement angle θ defined in the graph. For $\theta = 120°$, the weld is raising fairly abruptly away form the plate surface. According to Figure 19.5, the

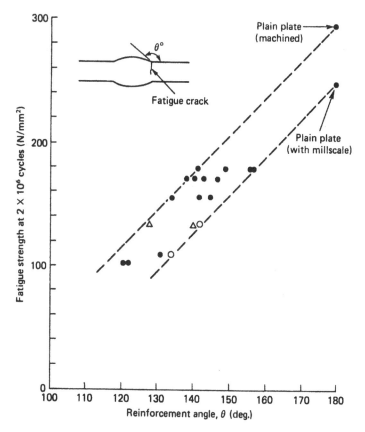

Fig. 19.5 The effect of the reinforcement angle on the fatigue strength of butt joints in steel plate [1].

fatigue strength is more than halved compared to the fatigue limit of the base material. Obviously, flush grinding of the reinforcement, specimen (c), should then improve the fatigue strength. The stress concentration of the reinforcement is eliminated. In general, a significant improvement of the fatigue strength by flush grinding is possible, but the improvement depends on possible defects in the weld material itself. Flush grinding should be more beneficial for good quality welds. In poor quality welds, crack initiation still occurs at weld defects in spite of flush grinding.

Specimens (d) and (e) of Figure 19.4 are plate specimens with a transverse and a longitudinal gusset respectively. These geometries are characteristic for several built-up structures. The fatigue strength with the transverse gusset can be moderate, although it depends on the weld toe geometry. Improvements can be obtained by removing some material at the weld

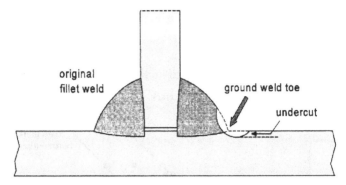

Fig. 19.6 Increased root radius of the weld toe obtained by grinding.

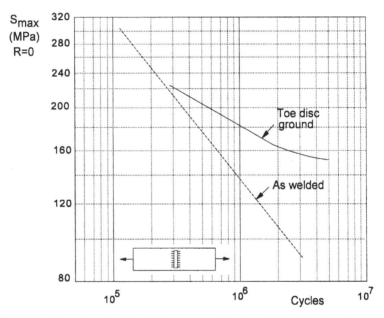

Fig. 19.7 The effect of a reduction of the toe root radius by grinding. Steel specimen with a transverse gusset [1].

toe by grinding to obtain a larger radius, see Figure 19.6. In spite of a slight undercut, the fatigue strength is increased significantly, see the results in Figure 19.7. It should be noted that the improvement is large at high endurances, whereas the improvement vanishes at low endurances. A similar trend was discussed in previous chapters (e.g. Chapter 14). The fatigue crack initiation period is relatively short at high stress amplitudes with low endurances. As a result, the major part of the fatigue life is covered by the crack growth period. Because grinding of the weld toe does not have an

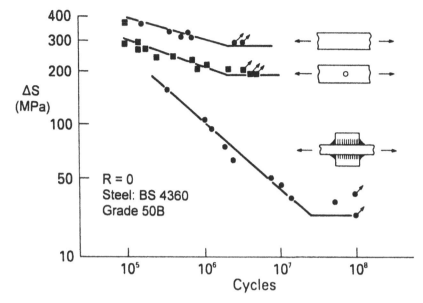

Fig. 19.8 Comparison between S-N curves of a welded specimen with a longitudinal gusset, a hole notched specimen and an unnotched specimen [2].

important influence on the crack growth period, its effect is relatively small at high stress amplitudes. But the effect can be large at low stress amplitudes with a significant crack initiation period.

An increased weld toe radius has also been obtained by so-called TIG dressing. In this process, the root of the weld is remelted to a shallow depth with a TIG welding torch (Tungsten Inert Gas), which leads to a smooth transition between the plate surface and fillet weld. According to [1], a most significant improvement of the fatigue limit is possible.

The specimen with the longitudinal gusset (specimen (e) in Figure 19.4) represents an unfavorable geometry for fatigue. The vertical gusset is causing a large discontinuity of the stiffness at the ends of the gusset. The gusset is attracting load which must be transmitted again to the plate at the ends of the gusset. A severe stress concentration can occur at this location. The effect is illustrated by Figure 19.8 by a comparison between S-N curves for an unnotched specimen, a specimen with a central hole, and a specimen with a vertical gusset. The comparison between the two upper curves indicates a moderate notch sensitivity of the base material. The fatigue limit of the hole-notched specimen is only 1.5 times lower than S_f of the base material. However, for the gusset specimen, the fatigue limit is about eight times lower, which illustrates that the fatigue limit is very low. It should also be noted that the knee in the S-N curves of the unnotched and the hole-notched specimen occurs approximately at $N = 2 \times 10^6$ cycles. However, for the gusset specimen, the knee is found at a significantly higher fatigue life, about 2×10^7 cycles. Apparently, crack initiation can occur in the gusset specimen at very low stress amplitudes, but due to slow crack growth this requires a high number of cycles before failure occurs. The same gusset specimen was also tested after a stress relieving heat treatment to eliminate residual stresses in the weld [1]. The fatigue strength at $N = 2 \times 10^7$ was raised from 37 to 55 MPa, relatively a significant improvement, but still a low fatigue limit. Residual tensile stresses in the non-stress relieved specimens have contributed to the extremely low fatigue strength at high endurances.

The two cruciform specimens, (f) and (g) in Figure 19.4, are significantly different. In specimen (f), the fillet welds leave an internal separation between the two longitudinal plates and the transverse plate. This is equivalent to having two internal cracks from which cracks can nucleate. In specimen (g), the K-weld of the two double-bevel butt ends does eliminate the internal separation. Better fatigue strength properties were reported for the latter specimen. Radai [3], citing Kaufmann, mentions a reduction of 30% of the fatigue strength if compared to the base material, whereas this percentage was 60% for the other specimen with the internal plate separation, specimen (f).

Figure 19.4 shows a single lap joint and a double strap joint to connect two plates, specimens (h) and (i) respectively. The double strap joint is free from bending and it should be expected to have a larger fatigue strength. However, a significant stress concentration still occurs at the edges of the joint. The fatigue strength is still moderate, but inferior properties should be expected for the lap joint of specimen (h).

Finally, specimen (j) in Figure 19.4 connects two plate elements which are mutually perpendicular to one another. The stress concentration is somewhat similar to the situation of specimen (e) with the longitudinal gusset. The fatigue strength should be expected to be poor. It might be improved by tapering the ends of the two plate element, see the dashed lines in Figure 19.4, which reduces the abrupt change of the stiffness.

A geometric mistake not covered by the specimens in Figure 19.4 is misalignment of welded plates. Misalignment occurs in butt joints if the central lines of the two plates are not fully parallel (small-angle misalignment), or if a small shift between these lines is present. A tension load on the joint then introduces plate bending at the weld which can significantly impair the fatigue strength. Misalignments should not occur in fatigue critical structures.

19.4 Fatigue life considerations for CA loading

The previous discussion has shown that the fatigue strength of a welded structure depends on the layout of the structure (Figure 19.3) and characteristic details of the weld illustrated by the specimens in Figure 19.4. These two aspects are interrelated because the layout may require specific types of welds. Welded structures offer their own specific design problems. Citing Maddox [2]:

> The avoidance of fatigue failure (in welded structures) is very much the province of the design engineer, including the wise choice of weld details to optimize fatigue strength, recognition of potential fatigue problems associated with welded joints and full appreciation of the fatigue loading to be experienced by the structure.

Estimations of the fatigue properties of welded structures require two inputs: (i) information on the nominal stress levels at the weld, and (ii) fatigue data of relevant welded specimens. A comparison can then be made between the structure and the welded specimen. It is assumed that similar stress levels in the structure and in the specimen will give similar fatigue lives. In essence, this is the similarity approach discussed in Chapter 7 on fatigue predictions of notched elements. This concept can be useful if the fatigue life is mainly covered by the crack initiation period. Recall that the crack initiation period includes microcrack nucleation and growth of very small cracks as long as the growth of these cracks should be considered to be a surface phenomenon (Chapter 2). The crack initiation period is then followed by the crack growth period which covers fatigue crack growth away from the material surface controlled by the crack growth resistance of the material as a bulk property independent of the material surface.

Unfortunately, fatigue cracks in a welded joint frequently start from some weld defect early in the fatigue life. For that reason, it is often thought that the crack initiation period as defined above is negligible. The fatigue life is supposed to be largely covered by crack growth only. This can be applicable if crack initiation starts at either a weld material defect or a weld geometry defect. It then is meaningful to predict the fatigue life as being a crack growth period. The crack growth life can be estimated by integration of the Paris equation: $da/dN = C \cdot \Delta K^m$. The result of the integration was given previously (Eq. 8.22) which can be written as

$$\Delta S^m \cdot N = \int_{a_0}^{a_f} \frac{da}{(\beta \sqrt{\pi a})^m} = \text{constant} = C \qquad (19.1)$$

This equation is similar to the Basquin relation: $S_a^k \cdot N = \text{constant}$ (Eq. 6.2) (note that $\Delta S = 2S_a$). The equation implies a linear relation between $\log(S)$ and $\log(N)$ with the slope equal to $-1/k$. According to Eq. (19.1), k should be equal to the exponent m in the Paris equation. Analysis of large numbers of test series on welded specimens has confirmed that k is approximately equal to 3.0 [1], while values close to $m = 3$ are observed in crack growth tests on steel specimens. This result is interesting, but it should be understood that the crack growth life in Eq. (19.1) depends of the constant C. This constant is a function of a_0, the starting crack length at the beginning of the crack growth life, and a_f, the crack length at failure. As discussed in Section 8.6.1, the effect of a_f on the crack growth life is relatively small because at a large crack length the crack growth rate is high. However, the influence of the initial crack length, a_0 is large due to the low crack growth rates for a small crack length. Although the size of a_0 should be associated with the size of a weld defect, the choice of a_0 is not obvious. This question is addressed later in the discussion on a worst case analysis. Furthermore, predictions based only on fatigue crack growth can raise questions if the predicted life turns out to be very long for low stress amplitudes. In such a case, it must be expected that a substantial part of the fatigue life is spent in the crack initiation period to overcome initial thresholds for the growth of very small cracks. Unfortunately, predictions on the crack initiation period are problematic in view of the variability of the geometry of the weld toe profile, residual stresses and basic material reference fatigue data relevant for the heat affected zone (HAZ) next to the weld [5]. The definition of the crack size at the transition from the initiation period to the crack growth period is another complication.

A practical approach to the estimation of fatigue lives of welded structures is based on the development of design codes. As a result of national and international efforts, welded joint configurations have been grouped in different classes. These classes are characterized by different severities of weld configurations. As a result of analyzing large numbers of fatigue test series, an average S-N curve is adopted for each class. A survey of the codes of different countries and related literature is given by Radai in [3] and Chapter 7. The set of S-N curves given by the British Standard Institution [6] are discussed in [1, 2]. These curves shown in Figure 19.9 are discussed below. The curves are marked by capital letters as a reference to the class involved. In Class B the fatigue strength is high, whereas in Class G the fatigue strength is poor. Figure 19.9 illustrates that a large variation in the fatigue strength of welded joints exists depending on the geometry of the welded joint. The classification is based on the configuration of the joint and

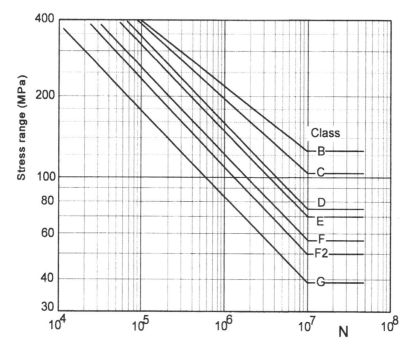

Fig. 19.9 The S-N curves for the various classes of welded joints [1, 2].

the type of the weld, provided that a good welding practice applies. A flush ground butt joint (type (c) in Figure 19.4) is suggested to be in Class B, and an as-welded butt joint (type (b) in Figure 19.4) is in Class C, the two upper curves in Figure 19.9. A transverse gusset joint (type (e) in Figure 19.4) can be in Class F, and a longitudinal gusset joint (type (f) in Figure 19.4) in Class F2, two lower curves in Figure 19.9. Joints with a rather primitive design geometry are found in class G.

Various types of joint geometries are described in the classification codes. These geometries should cover welded joint configurations as they occur in structures. The joints have a more complex geometry than shown for the specimens in Figure 19.4, which is easily recognized in Figure 19.3. As said previously, the classification does not refer to the weld quality, but to the design features of the weld geometry. However, some aspects of the weld quality are still incorporated in the classification. As an example, a butt welded joint is referred to Class C, D or E depending on the overfill profile (angle θ in Figure 19.5). Another restriction used in the classification is that a class is justified provided the weld is made by automatic welding without making intermittent stops during the welding operation. Such stops can cause weld defects on resuming the welding process. Also, NDI can be required for the high fatigue strength classes to be sure that the classification is not impaired by flaws or other weld defects. Furthermore, reference is sometimes made to the plate thickness. Experimental evidence [7] has shown that similar welded joints in thicker plates can have a lower fatigue strength. This is associated with a geometric thickness influence on stress concentrations (K_t) and stress intensity factors (K) which in the BS code [6] is called a plate thickness design penalty. As a matter of fact, it is not easy to allocate a specific class to a welded joint configuration which does not occur in the listed examples. It requires experience and engineering judgement to decide whether the S-N curves give a reasonable estimate of the fatigue strength. In addition to this conclusion, some more comments should be made on the S-N curves.

First, the S-N curves in Figure 19.9 are supposed to be valid for various structural steels with an ultimate strength in the range of 400 up to 690 MPa. Apparently, a higher static strength does not mean that the fatigue strength is increased also. This observation should probably be associated with a larger notch sensitivity and hardly any improvement of the crack growth resistance if S_U and $S_{0.2}$ are increased by modifications of the chemical composition or heat treatment.

Second, the S-N curves are based on fatigue tests carried out under a fatigue load with $S_{min} = 0$ ($R = 0$). It is generally thought that the mean stress effect on the fatigue strength of welded structures is relatively small. This should imply that residual stresses in welded joints must also have a minor effect. However, it has been shown that the mean stress can still have a systematic effect, the more so for welded joints with a high fatigue strength. If a mean stress effect is present, it should be expected that residual stresses can also be significant. As discussed before, the fatigue strength for high endurances could be improved by a heat treatment which relieves tensile residual stresses in the welded joint. Unfortunately, residual stress distributions cannot easily be forecast for welded structures of various complexities. As a consequence, general recommendations for stress relieving treatments are questionable. However, a favorable effect of residual compressive stress was confirmed by improvements of the fatigue strength after shot peening of the toe of fillet welds.

Third, the S-N curves in Figure 19.9 are average curves, representing the mean value of scatter bands. In the BS joint classification code [6], standard deviations are suggested for each class, varying from $\sigma_{\log N} = 0.18$ to 0.25. Furthermore, it is proposed to decrease the N-values of the curves with two standard deviations, which implies fatigue life reduction factors of 2.3 and 3.2 respectively. These reductions should account for scatter of the fatigue

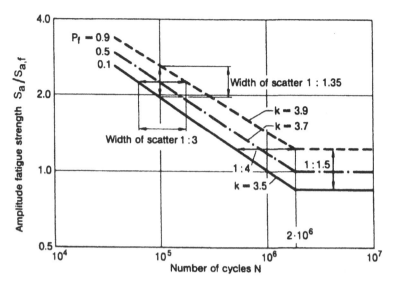

Fig. 19.10 Normalized S-N curves for welds in structural steels according to Haibach [8].

life, i.e. scatter which is considered to be representative and acceptable for variations of a normal weld quality.

An interesting graph of Haibach [8] is reproduced from [3] in Figure 19.10. Haibach proposed a normalized S-N curve obtained by dividing the stress amplitude (S_a) by the fatigue limit ($S_{a,f}$), actually the fatigue strength at $N = 2 \times 10^6$ cycles. This normalizing procedure has led to an $S_a/S_{a,f} - N$ curve, which should be valid for carbon steels with 0.12% to 0.20%C, and for different types of welded joints and R ratios. Haibach proposed scatter bands around the average curve (probability of failure $P_f = 0.50$) for probabilities of failure of 10% and 90% ($P_f = 0.10$ and 0.90 respectively). The Basquin relation was adopted, $S_a^k \cdot N = $ constant with a k-value of 3.7 for the average curve ($P_f = 0.50$). The width of the scatter band is slightly narrower at lower endurances which agrees with general experience. As a consequence, the slope factor for $P_f = 10\%$ is lower; $k = 3.5$. This value may be compared to $k = 4.0$ for Class B, $k = 3.5$ for Class C, and $k = 3.0$ for the other classes of the joint classification in Figure 19.9. The width of the scatter band, defined by N for $P_f = 0.10$ and N for $P_f = 0.90$, is given in the Haibach graph as a life ratio. The ratio is 1 : 3 at $N = 10^5$ and 1 : 4 at $N = 10^6$. The width between the two P_f-values corresponds to 2.58 standard deviations (normal distribution assumed). It implies that $\sigma_{\log N} = 0.186$ and 0.235 for the two ratios respectively. These values are again close to the standard deviations adopted in the joint classification code cited previously, which are in the range from 0.18 to 0.25. Apparently, a good deal of agreement is found between Haibach's analysis and the S-N curves of the BS-joint classification code.

It is also of some interest to consider scatter of the fatigue strength. The scatter band of the fatigue strength at 2×10^6 cycles in Figure 19.10 is accounted for by a ratio 1 : 1.5. This corresponds to a standard deviation $\sigma_{\log(S_a)}$ of $0.069 \approx 7\%$.

Some further comments on fatigue life estimates of welded joints should still be made. According to the previous discussion, a fatigue life estimate for a welded joint starts with considering the joint classification. After such a choice has been made, the S-N curve follows e.g. from Figure 19.9. This curve should be reduced with two standard deviations to account for scatter. A fatigue life of the welded structures is then obtained by reading the N-value at the nominal stress level of the welded structure. Two problems are easily recognized; first the selection of the joint classification, and second, the assessment of the nominal stress level of the structure. Furthermore, corrections could be considered for the effects of mean stress and plate thickness. The result of all these steps is affected by uncertainties. With some judgement about these issues and conservative assumptions, it is possible that the estimated fatigue property is satisfactory in comparison to the design goal. This could imply that an acceptable margin of safety is still left. However, if the result does not give sufficient confidence in comparison to the design goal, exploratory fatigue tests should be considered. It then is necessary to simulate all characteristic details of the welded structure in the specimen to be tested. The fatigue life in the test provides a measure of the fatigue quality of the weld design. From this result, an S-N curve is obtained by adopting the Basquin relation using an assumed k-value, e.g. $k = 3.0$.

Of course, the best solution is to carry out a fatigue test on the structure itself with a representative service load-time history. However, FE calculations should also be considered to explore the stress distribution at critical locations in a welded structure. Sometimes strain gage measurements may be instructive. Strain gages then should be located at critical points where crack initiation may occur. However, at the root of a weld it is difficult to apply strain gages due to the irregular profile of the weld surface. Strain gages can then be applied at a small distance away from the root of a weld which has been labeled as the hot spot stress location. The indicative significance of such measurements requires a good understanding of local stress gradients.

19.5 Fatigue endurances of welded joints under VA loading

Another problem arises if the fatigue load in service is associated with VA loading. The Miner rule is generally considered to be the only calculation rule available for welded structures, although it is also claimed to be unconservative because $\sum n/N < 1$ results are found. Limitations of the Miner rule were previously discussed in Chapter 10. It was pointed out that load cycles with amplitudes below the fatigue limit can still contribute to fatigue damage. Because fatigue crack growth is an important part of finite lives of welded structures, small cycles with amplitudes below the fatigue limit can contribute to the growth of cracks initiated by cycles with amplitudes exceeding the fatigue limit. As discussed in Chapter 10, extrapolation of S-N curves below the fatigue limit must be advised for life calculations with the Miner rule. The extrapolation was shown in Figure 10.11 as line B. It implies that the Basquin relation ($S_a^k \cdot N$ = constant) is assumed to be also applicable to cycles with stress amplitudes below the fatigue limit. This does not mean that the Miner prediction becomes accurate. However, it is more realistic and more conservative to account for fatigue damage contributions from small cycles.

An other proposition was made by Haibach [8], line H in Figure 10.11, with a slope factor $2k - 1$ (Basquin relation: $S_a^{2k-1} \cdot N$ = constant). Later, in the welding code [6] a slope factor of $k + 2$ was proposed with the knee in the S-N curve at 10^7 cycles. Because the value of k for welded specimens is in the order of 3, the two factors are practically equal ($2k - 1 = k + 2$ for $k = 3$). Of course, a prediction with such modified S-N curves is more conservative than the original Miner rule prediction.

Gurney [9] analyzed $\sum n/N$-values obtained in a large number of VA test series on welded specimens. The $\sum n/N$-values were obtained with S-N curves extrapolated below the fatigue limit in accordance with the Basquin equation. He found as an average value ($\sum n/N)_{average} = 1.2$, while for 99% of the data $\sum n/N$ was larger than 0.35. It suggests that a rough estimate of the fatigue life can be obtained. It should still be recalled from the discussion in Chapter 10 that the Miner rule prediction does not account for any interaction effect. Miner prediction results must be considered with caution, and safety factors on life should be considered.

19.6 Two special cases

Two special cases of welded structures are related to pressure vessels and large tubular offshore structures. They are briefly addressed below.

Pressure vessel

A pressure vessel is special for the following reasons: (i) The load spectrum applicable to many pressure vessels is rather flat and can be approximated by constant-amplitude loading with a zero minimum load ($R = 0$). (ii) The required fatigue life in numbers of pressurization cycles is not excessive. The design goal may be in the order of 10^5 cycles. (iii) Safety is an important issue in view of explosive failures. Inspection of the weld quality is obligatory.

Fatigue life estimates can be made with S-N curves which are assumed to be relevant to the critical welds of the pressure vessel. These curves can give indications of the safety margins of the structure. However, in view of the limited fatigue life and safety considerations, it is desirable to know how fast a crack will grow if it is present. A fracture mechanics problem must then be considered. A "worst case" analysis should be recommended. It implies that an initial defect has to be assumed with a size depending on the limitations of NDI techniques used. The size must be related to the minimum size that will not escape detection by NDI. Crack growth should then be predicted by using relevant K solutions and crack growth data of the material involved. The crack growth life is continued until a critical K-value is reached causing failure ($K_{\max} = K_c$), or until a "leak before break" situation arises. A prediction for a semi-elliptical surface crack was discussed in Section 8.6.3, where it was assumed that a full through crack was immediately present after the crack penetrated through the full plate thickness, see Figure 8.24. In general, the geometry of the structure and an initial flaw will be more complex. As a result, it may be necessary to make more elaborate FE calculations to obtain relevant K-values. Another problem involved is associated with the leak-before-break condition. Usually, pressure vessels are made from ductile steel. At the moment of break through of the crack, considerable plastic deformation may occur. Elasto-plastic fracture mechanics should then be applied, which is not really a simple procedure. Furthermore, if the pressure vessel is filled with a fluid, the pressure will rapidly decrease after some leakage. The crack driving force decreases simultaneously and a catastrophic failure need not occur. However, this is

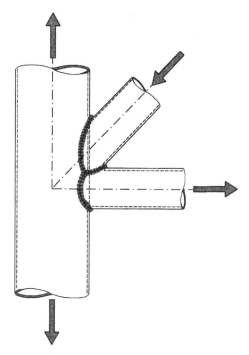

Fig. 19.11 Example of tubular welded joints.

not necessarily true for a gas filled pressure vessel. For thick-walled pressure vessels, leak-before-break is obviously preferable to a complete failure in view of safety issues, but if the thickness is large it may be difficult to satisfy the leak-before-break criterion. Of course, crack detection before a break through occurs is the better solution. Such a situation might be monitored by combining crack growth predictions, NDI and safety factors for quantitatively unknown influences.

Tubular offshore structures

An entirely different problem setting is applicable to fatigue cracks in welded tubular frames used in offshore structures. Complex nodes are present where several tubes meet. An example is schematically shown in Figure 19.11 with two side braces connected to a main pillar. The size of these tubes is immense. Diameters in the order of 2 m (6 ft) and larger are used, and wall thicknesses can be as high as 5 cm (2 inch). As a result, weld seams are long and do not occur along straight lines. Furthermore, the welded structure operates in salt water which is an aggressive environment, while the load spectrum depends on the sea waves and weather conditions. The load spectrum contains many small cycles, but in stormy weather cycles with large amplitudes occur. The fatigue life design goal may be 40 years or more, which implies a large number of cycles. Fatigue cracks are highly undesirable in view of difficult inspections and repairs. As part of the design analysis, much effort is spent on fatigue problems. Extensive FE calculations are made to obtain detailed pictures of the stress distribution in the joints. The results should indicate where the most fatigue critical areas of the welds are located. Local stress levels at these critical areas can be used for preliminary fatigue life estimates, but the similarity between the tubular nodes and simple specimens is a problematic issue. Moreover, in view of the VA load spectrum, a Miner rule calculation must be made, which introduces other uncertainties.

In some laboratories, full-scale tests are carried out on very large specimens representing a typical node joint. These tests are also carried out to study crack growth along the welds of the joints. Obviously, a scenario is also required for the very expensive offshore structures. In comparison to the pressure vessel case, the problem is more complex because of the geometries involved, the occurrence of variable-amplitude loading, and the salt water environment. Furthermore, in a welded joint of a large structure, several fatigue crack nuclei are initiated simultaneously along the weld. Initially, these cracks grow independently until they coalesce and then grow as a single crack with a large length along the weld toe. Predictions on this type of crack growth with fracture mechanics require considerable efforts.

The influence of salt water is confusing. It is well known that the fatigue limit of unnotched specimens in a salty environment is very low and almost non-existent, see Figures 2.29 and 16.1. The effect of salt water on fatigue crack growth is generally detrimental, but less disastrous than the effect on the fatigue limit. As discussed in Chapter 16, a corrosive environment can enable fatigue crack initiation at very low stress amplitudes by a surface corrosion process. Probably, this aspect is less important for fatigue of welded joints in a worst case analysis because a small crack is supposed to be present at the beginning of the fatigue life. During fatigue crack growth, the corrosion effect on the growth rate depends on the accessibility of the environment to the crack tip and the corrosion products left inside the crack. It is difficult to analyze this phenomenon in a realistic model. Comparative fatigue tests on welded steel specimens in air and in salt water under variable-amplitude loading have indicated lower endurances in salt water. In the high-cycle region, the effect was found to be a life reduction by a factor in the order of three to four. It should be understood that the magnitude of this factor is based on empirical evidence from laboratory experiments, and not from experiments in the open sea. Actually, safety factors to account for the corrosive influence cannot be chosen on rational arguments only. Engineering judgement based on understanding of possible influences, experience and economic and safety consequences of fatigue failures should lead to reasonable decisions, also with respect to corrosion protection and inspections.

19.7 Spot welded joints

Spot welded joints are entirely different from welded joints with continuous weld seams. Spot welding is a local attachment between sheets, plates or sections. As a design option, spot welded joints are more comparable to riveted or bolted joints. Spot welding is usually restricted to structural configurations with low material thicknesses which may be of the order of 1 to 2 mm (0.04 to 0.08 inches). However, spot welding can also be applied to thicker material, and fatigue evaluations of such joints between steel plates were carried out by Overbeeke and Draisma [10, 11]. Spot welding is frequently used in the automotive industry for assembling preformed sheet metal parts. The attractive feature is that spot welding allows high production rates by full automation of the production process. In the past, spot welding was also used in aircraft structures, particularly for attaching stiffeners to sheet metal skins, both made from Al-alloys. Load transfer in such joints can be negligible and the joint should not be fatigue critical. However, such joints should be carefully sealed to prevent moisture penetration and corrosion in the joint which can activate fatigue crack nucleation.

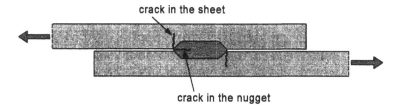

Fig. 19.12 Cracks in a spot welded joint.s

The fatigue strength of a spot welded lap joint is very low. The geometry of a spot welded joint is somewhat similar to the geometry of a riveted joint. However, contrary to a riveted joint, holes and rivets are absent in a spot welded joint. The weld nugget and the sheets are an integral part. As a consequence, the stress concentration at the edge of the nugget is high. Cracks are nucleated at this critical location, see Figure 19.12. The predominant failure mode is cracking in the sheets, but cracks in the nugget have been observed. From a fatigue point of view, spot welded joints should be considered with caution and full-scale fatigue tests are recommended. An interesting method for the evaluation of the severity of spot welds in parts of a motor car has been proposed by Rupp et al. [12].

19.8 Major topics of the present chapter

1. The fatigue behavior of welded joints is entirely different from the behavior of joints with fasteners. The fatigue critical locations of welded structures occur at the welds, while the nominal stress level at these welds depends on the layout of the welded structure. Furthermore, the variety of welding processes is large and several geometric imperfections and defects in the weld itself can occur. The S-N curve of a welded structure depends very much on the design of the joints and the quality of the welding. Preliminary information on S-N curves is given in the Welding Codes.

2. Estimates of the fatigue life of welded structures loaded by a variable-amplitude load history can be obtained with a Miner calculation. But the S-N curves should then be extended to high N-values for damage contributions of fatigue cycles with amplitudes below the fatigue limit.

3. Environmental and load frequency effects for welded structures in sea water should be accounted for by safety factors while periodic inspections are desirable.

4. A worst case analysis must be considered for welded structures if serious safety or economic problems are a relevant issue if fatigue cracks can occur. The fatigue life must be assumed to be fully covered by fatigue crack growth starting from a possible initial defect. The life prediction is then replaced by a crack growth prediction. The result is significantly depending on the assumed size of the initial defect.

References

1. Gurney, T.R., *Fatigue of Welded Structures*, 2nd edn. Cambridge University Press, Cambridge, UK (1979).
2. Maddox, S.J., *Fatigue Strength of Welded Structures*, 2nd edn. Abington Publishing, Cambridge, UK (1991).
3. Radaj, D., *Design and Analysis of Fatigue Resistant Welded Structures*. Abington Publishing, Cambridge UK (1990).
4. Lancaster, J., *Metallurgy of Welding*, 6th edn. Abington Publishing (1999).
5. Skorupa, M., *Fatigue life prediction of cruciform joints failing at the weld toe*. Welding Research Supplement, Welding Journal, Aug. (1992), pp. 269S–275S.
6. *Code of practice for fatigue design and assessment of steel structures*. BS 7608: 1993. British Standards Institution, London (1993). See [3] for more information on other codes.
7. Noordhoek, C. and de Back, J. (Eds.), *Steel in Marine Structures*, Proc. 3rd Int. ESCS Offshore Conference on Steel in Marine Structures (SIMS '87). Elsevier, Amsterdam (1987).
8. Haibach, E., *Fatigue Strength in Service*. VDI-Verlag GmbH, Düsseldorf (1989) [in German].
9. Gurney, T.R., *Cumulative damage of welded joints: II. Test results*. Joining Mater., Vol. 2, (1989), pp. 390–395.
10. Overbeeke, J.L., *The fatigue behaviour of heavy-duty spot welded lap joints under random loading conditions*. Welding Res. Int., Vol. 7 (1977), pp. 254–275.
11. Overbeeke, J.L. and Draisma, J., *Fatigue characteristic of heavy-duty spot-welded lap joints*. Metal Constr. British Welding J., Vol. 6 (1974), pp. 213–219.
12. Rupp, A., Störzel, K. and Grubisic, V., *Computer aided dimensioning of spot-welded automotive structures*. SAE Paper 950711 (1995).

Some general references (see also [1–4])

13. Sonsino, C.M., *Course of SN-curves especially in the high-cycle fatigue regime with regard to component design and safety*. Int. J. Fatigue, Vol. 29 (2007), pp. 2246–2258.
14. Radaj, D., Sonsino, C.M. and Fricke, W. (Eds.), *Fatigue Assessment of Welded Joints by Local Approaches*. Woodhead Publishing (2006).
15. Samuelson, J. (Ed.), *Design and Analysis of Welded High Strength Steel Structures*. EMAS, UK (2002).
16. Savaidis, G. and Vormwald, M., *Hot-spot stress evaluation of fatigue in welded structural connections supported by finite element analysis*. Int. J. Fatigue, Vol. 22 (2000), pp. 85–91.
17. Barsom, J.M. and Rolfe, S., *Fracture and Fatigue Control in Structures. Applications of Fracture Mechanics*, 3rd edn. Butterworth-Heinemann (1999).
18. Van der Sluys, W.A., *Effects of the Environment on the Initiation of Crack Growth*. ASTM STP 1298 (1997).
19. Radaj, D., *Review of fatigue strength assessment of non-welded and welded structures based on local parameters*. Int. J. Fatigue, Vol. 18 (1996), pp. 153–170.
20. Lawrence, F.H., Dimitrakis, S.D. and Munse, W.H., *Factors influencing weldment fatigue*. Fatigue and Fracture, American Society for Materials, Handbook Vol. 19, ASM International (1996), pp. 274–286.

21. Tarsem, J., *Fatigue and fracture control of weldments*. Fatigue and Fracture, American Society for Materials, Handbook Vol. 19, ASM International (1996), pp. 434–449.

22. Dover, W.D., Dharmavasan, S., Brennan, F.P. and Marsh, K.J. (Eds.), *Fatigue Crack Growth in Offshore Structures*. EMAS, Solihull, UK (1995).

23. Monahan, C.C., *Early Fatigue Crack Growth at Welds*. Topics in Engineering, WIT Press (1995).

24. Scott, P. and Cottis, R.A. (Eds.), *Environment Assisted Fatigue*. EGF Publication 7, Mechanical Engineering Publications, London (1990) (four papers on welded steel joints in salt water).

25. McHenry, H.I. and Potter, J.M. (Eds.), *Fatigue and Fracture Testing of Weldments*. ASTM STP 1058 (1990).

26. Booth, G.S. (Ed.), *Improving the Fatigue Strength of Welded Joints*. The Welding Institute, Cambridge, UK (1983).

27. *Residual Stresses and Their Effect*. The Welding Institute, Cambridge UK (1981).

28. Hoeppner, D.W. (Ed.), *Fatigue Testing of Weldments*. ASTM STP 648 (1978).

Chapter 20
Designing against Fatigue of Structures

> *... convince your manager about designing against fatigue ... ?*
> *... first convince yourself ...*
> *... next your colleagues ...*
>
> *...*

20.1 Introduction

The present chapter is a kind of a reflection on previous chapters. It starts with a brief survey of different types of structures and related prediction problems in Section 20.2, followed by a repetition of design tools in Section 20.3. Uncertainties of predictions and safety factors are addressed in Section 20.4. Some illustrative case histories of structural fatigue problems are presented in Section 20.5. The chapter is completed with summarizing conclusions.

20.2 Different types of structural fatigue problems

The question about how to define problems of designing a structure against fatigue is obviously associated with the goals to be achieved. In principle it implies that satisfactory fatigue properties of a structure should be obtained, but it depends on the type of structures which fatigue properties should be explored. For the present discussion three categories are considered:

1. *Structures for which fatigue failures are unacceptable.*
2. *Structures in which fatigue cracks may occur after a sufficient lifetime but without the risk of a complete failure.*
3. *Structures for which crack initiation and crack growth until a complete failure are acceptable, but for which a reasonable lifetime is still desirable.*

Rotating blades of turboprop engines, wind turbines and compressors are examples in the first category. Many components of various engines are also in this category with a crankshaft as a well-known case. A fatigue failure in such components would be a kind of a disaster. The fatigue limit of the structure is the important fatigue property and high-cycle fatigue is an important issue. However, fatigue failures may also be unacceptable in pressure vessels for which the number of pressurization cycles is not very large, e.g. not exceeding 10^5. If all cycles have practically the same load range, the relevant fatigue property is the crack initiation life under CA loading.

A variety of structures can also occur in the second category. Obviously the crack initiation life is again of interest, and it should be large enough for a satisfactory lifetime in service. If a complete failure is unacceptable, a reliable inspection procedure is indispensable. This applies to aircraft structures, and it can also be applicable to several welded structures. As a consequence both the crack initiation life and the crack growth life are of interest. Moreover, fatigue under VA amplitude loading may also be a relevant condition.

The third category includes various utilities for which final failure simply implies that it must be replaced by a new one. Various housekeeping articles are in this category, e.g. washing-machines, vacuum cleaners, but not stairs. Bicycles are another typical example in which fatigue failures do occur. The fatigue property is the fatigue life until failure with lifetime as an economical criterion. Data on crack initiation life and crack growth properties are not required, but again both CA and VA load histories can be significant.

The three categories of structure have been defined because within each category similar fatigue properties should be predicted. The literature on fatigue prediction problems is quite diverse. The world of building steel bridges and the world of manufacturing wind turbines are two different cultures, but still with similar fatigue problems. As an example, in both worlds load spectra are consisting of a combination of deterministic loads and random loads.

In practice the designer who is faced with fatigue endurance problems, must also consider other durability issues, such as: maintenance, inspections, repairs, replacements, service conditions with implications for corrosion, wear and tear. They are all a matter of concern dealing with the structure as an object that should be in function for a long time. Anyway, the possibility of fatigue crack initiation is a relevant problem because it can have a large economic impact. Designing against fatigue crack initiation is one of the responsibilities of the designer of the structure. Figure 1.2 of Chapter 1 is reproduced here in a slightly different layout. It shows that predictions require:

(i)   information about the structure,
(ii) analysis and fatigue data, and
(iii) last but not least, the load spectrum.

In the literature it is sometimes suggested that our fatigue problems are solved if an accurate prediction model would be available. This is misleading. The present physical understanding about fatigue damage accumulation is reasonably well developed in a qualitative sense. And just because of this understanding, it must be accepted that accurate quantitative predictions on fatigue lives are illusory.

Problems of fatigue life and crack growth prediction were discussed in Chapters 7, 8, 10 and 11 for notched elements, and in Chapters 18 and 19 for joints. It was indicated how estimates of the fatigue limit of notched elements could be obtained. Unfortunately, similar prediction procedures are not applicable to fatigue of joints. Empirical data of joints must be available to arrive at estimated of the fatigue limit. Predictions on the fatigue life under VA loading is even more complicated. The Miner rule is physically rather primitive. The rule starts from the idea that damage can be characterized by a single damage parameter which essentially disagrees with the present knowledge about fatigue damage accumulation. At best, the Miner rule gives some weighted indication of the load spectrum, but not of the severity of the load spectrum. The Miner rule fully breaks down in comparisons of load spectra severities. When the Miner rule is used to obtain some rough indication of the fatigue life under VA loading, one should realize that the prediction is an extrapolation of S-N data which by itself have already a limited reliability.

The situation appears to be more convenient for predictions of fatigue crack growth. Crack growth prediction for CA loading based on the well-known fracture mechanics methodology can be reasonably reliable. But the situation is less satisfactory for fatigue crack growth under VA loading, see the discussion in Chapter 11. A major problem is to account for interaction effects of cycles with different amplitudes. If the interaction effects are ignored, predictions will probably be conservative, but it can lead to significant under-predictions for crack growth under steep load spectra.

In general terms, it must be accepted that fatigue predictions are speculative in a way that the order of magnitude may be instructive, but the predictions should be evaluated with appreciable judgement. In cases of doubt, the design variables should be reconsidered to see where weak links are present. Estimates of fatigue properties can be improved by experiments. Whether this is really necessary depends on safety margins and costs involved. Detailed stress analysis, fatigue experiments and load spectrum measurements can improve the significance of predictions. It is possible that a simple fatigue analysis shows that the occurrence of a fatigue failure problem is very unlikely, and no further design improvements are necessary. It is also possible that fatigue failures in service are acceptable because a simple replacement of the failed element is not expensive and safety is not involved. In such cases, a cost-benefit analysis can show that efforts to improve the fatigue prediction are not really worthwhile. But it is also possible that a simple prediction indicates that structural improvements must be considered, i.e. designing against fatigue. It then is useful to have some idea about the accuracy of preliminary fatigue life predictions. Several sources of uncertainties in the prediction technology should be considered, including the strategy of applying safety factors.

20.3 Designing against fatigue

The fatigue life as discussed in Chapter 2 is divided into two periods with final failure at the end of the life:

crack initiation period → *crack growth life* → final failure

The practical significance of recognizing the differences between the two periods has been emphasized in several chapters. *A designer should know*

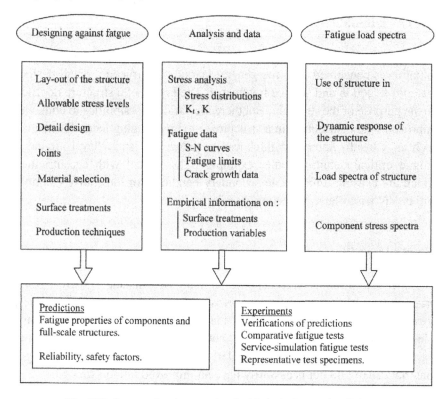

Fig. 20.1 Survey of topics associated with designing against fatigue.

whether he is designing against crack initiation, or for an acceptable crack growth behavior, or for both. Moreover, he also should be aware of the question whether his problem is associated with high-cycle fatigue or low-cycle fatigue.

The initiation period is basically a material surface phenomenon, whereas crack growth is a matter of crack growth resistance of the material as a bulk property. As a consequence, fatigue related influences are essentially different for the two periods. The crack initiation period and the fatigue limit are heavily depending on material surface conditions, whereas most of these conditions are practically irrelevant for the crack growth period. Understanding of the effects of these variables is essential for designing against fatigue.

The crack initiation aspect

It is easily understood from Figure 20.1 that designing against fatigue crack initiation is concerned with the general layout of a structure, detail design, material selection and surface treatments. The layout of a structure depends on the purpose of the structure. But there are various possibilities to obtain an improved load distribution in a structure, e.g. by changing local dimensions such as a locally increased thickness to reduce the stress level around a fatigue critical detail. Another example is associated with eccentricities which are causing unfavorable secondary bending. An illustrative example will be discussed later in Section 20.5.2.

Material selection

The selection of the material depends on many circumstances, such as static properties, workshop properties, corrosion resistance, thermal properties, costs, etc. It may be recalled that a material with a higher $S_{0.2}$ may have a higher S_f for unnotched specimens, but also an increased notch sensitivity, see the discussion in Chapter 7. Similarly, welded joints of a higher strength material usually do not necessarily have an improved fatigue strength.

If a new material is considered for a structural application, it should be supported by results of service-simulation fatigue tests on specimens which are representative for fatigue critical details of the structure under consideration. A different question arises when advanced fiber-metal laminates and composites materials are considered as an alternative for the more classical materials. Especially for the black composites it implies an entirely different design and production discipline, and thus another technological culture. Fatigue aspects of fiber-metal laminates are briefly discussed in Chapter 21.

Surface treatments

The designer can specify the quality of the material surface, and also certain surface treatments. Several options for surface improvements and preventing unfavorable surface effects were discussed in Chapters 14 to 16. Some typical examples of surface treatments are: fine machining, nitriding of steel, shot peening, surface rolling, prevention of fretting and corrosion protection. It has been pointed out that surface treatments are carried out for various purposes: improvement of fatigue properties, protection against corrosion, improved wear resistance, restoring poor surface quality, and cosmetic reasons. Surface treatments can increase the hardness of a surface layer, and thus hamper cyclic microplasticity. At the same time, residual compressive stresses restrain the opening of microcracks in the surface layer and thus will reduce or even arrest the growth of these cracks. As a result, the major benefit of surface treatments is on the crack initiation period. They are important for high-cycle fatigue, and in particular for the fatigue limit.

Detail design for an improved stress distribution

An essentially different approach is associated with the reduction of stress concentrations. For fatigue critical notches it generally boils down to increasing root radii or applying stress relieving grooves if that is possible. Non-circular fillets are rarely considered. Recently K_t-values were calculated for elliptical fillets which are not covered in the book by Peterson [1]. Results have shown significant reductions of the K_t-value to be discussed in Section 20.5.1. It may be repeated here that various K_t-values in the book by Peterson are instructive but not always very accurate as a result of older techniques used to determine the various graphs in the book. By now FE analysis can produce more accurate K_t-values as well as stress gradients.

Large-scale design issues

Detail design as discussed previously, is associated with dimensions which are significant for local stress concentration, e.g. a hole diameter or notch root radii. On a larger scale, the designer is considering the general concept of the structure. As an example, although perhaps a somewhat curious one, rather different concepts can be contemplated for designing a bridge. Another noteworthy example, in various structures joints are present, but the variety of different joint concepts is also large. Decisions to be made on the type of structure are generally depending on experience of the industry, and in the industry on economic implications.

20.4 Uncertainties, scatter and safety margins

The purpose of designing against fatigue is to prevent disasters, and also to avoid non-fatal incidents in view of unwanted economic consequences. Unfortunately, uncertainties about the fatigue performance of a structure cannot be solved by accurate and rational arguments. It implies that some philosophy about safety factors or other measures should be considered. A solid rational frame work to arrive at safety factors cannot be formulated. Statistical distribution function are unknown. Information about scatter of fatigue properties is largely coming from laboratory test series (see Chapter 12), and not from service experience. The choice of reasonable safety factors is a matter of experience and engineering judgement. Both economic and safety consequences of the occurrence of a premature fatigue failure must be considered. In view of limited accuracies of quantitative fatigue predictions, it must be asked how this situation should be carried on. The variety of sources for uncertainties is fairly large. They will be briefly discussed.

20.4.1 Uncertainties

Three reasons for uncertainties about the prediction of the fatigue performance of a structure are easily recognized:

(i)   Uncertainties about the load spectrum.
(ii)   Uncertainties about the fatigue properties of the structure.
(iii) Uncertainties about the reliability of predictions.

Not all these uncertainties can be associated with scatter of some properties. Variations of the load spectrum were discussed in Chapter 9 including differences between deterministic loads (applied by the operator) and stochastic loads (random type loads depending on environmental conditions). With respect to the deterministic loads, all structures of the same type are not used in exactly the same way. The designer must consider the variability of loads which should be taken into account (functional loads, maneuvers). But another part of the variability of the load spectrum does not depend so much on the operator. Stochastic loads are relevant to structures operating under a variety of weather conditions (aircraft, boats, drilling platforms) or moving over various roads (passenger cars, trucks, coaches). Statistical distribution functions and power density spectra can be involved for air turbulence, sea-waves and road roughness. Several types of structures will see a combination of deterministic and stochastic loads. Cranes, bridges and buildings offer interesting combinations of both types of loads.

The second reason for uncertainties is associated with fatigue properties of the structure. These uncertainties are of an entirely different nature. Statistical variations are related with material properties and production quality. The fatigue properties of a material with a standardized composition may be obtained from data banks, but it cannot be guaranteed that these properties are always the same. Scatter may occur as a result of batch to batch differences, but even in a single plate statistical variations are possible. Moreover, the crack initiation fatigue life is depending on the quality of the production of components. There are sufficient reasons why components produced during a number of years cannot be considered to be samples of the same statistical population.

Finally, the third source of uncertainties is associated with the reliability and accuracy of a prediction model. As discussed in several chapters, estimates can be obtained for S-N curves, fatigue limits and crack growth. But is was also explained why accurate predictions are problematic, especially for the crack initiation life, while VA loading is an additional problem.

20.4.2 Scatter and safety factors

The fatigue limit and the safety factor

As discussed earlier, an important category of problems of designing against fatigue is associated with high-cycle fatigue and a flat load spectrum. If fatigue failures are unacceptable, the criterion is that all load cycles should be below the fatigue limit of the structure. The variables involved are the maximum load cycles occurring in service load spectra and the fatigue limit of the structure. They are both affected by uncertainties. It can be tried to obtain an estimated value of the fatigue limit of the structure, but even if the analysis would be supported by experiments, some unknown scatter must be expected. With respect to the load spectrum uncertainties are involved, not so much as a consequence of scatter, but due to different utilizations. Under these twofold conditions, a safety factor cannot be defined with rational arguments. If a fatigue failure of the structure would cause a fatal accident, relevant experimental efforts should be considered. A full-scale fatigue test

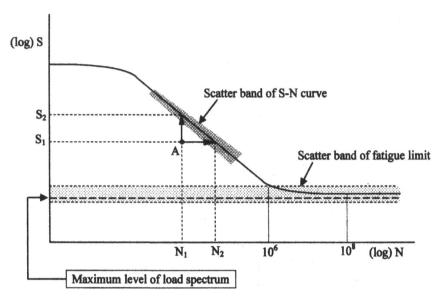

Fig. 20.2 Safety margin on load level S_1 for required life time N_1. A similar margin for the fatigue limit is unrealistic.

on a representative part of the structure with the step by step increasing load (see Figure 13.2) can give useful information about the fatigue limit.

A full-scale CA load tests at the estimated load level of the load spectrum cannot be recommended. The test should be continued to a very high number of cycles, say $> 10^8$. However, if the load level is just below the unknown fatigue limit, see Figure 20.2, then failure will not occur. In view of the scatter band of the fatigue limit, an other similar structure can fail at a fatigue life between 10^6 and 10^7 cycles, just above the average fatigue limit. It implies that information about the safety level remains unknown. In the high-cycle fatigue regime and for the fatigue limit, scatter of fatigue lives is not the relevant issue. Scatter of the fatigue strength, and in this case of the fatigue limit, is crucial. For this reason the step by step increasing test of Figure 13.2 should be preferred. Of course the number of cycles in each step (ΔN) should be large enough in order to be in the high-cycle fatigue regime, for instance $\Delta N = 10^6$ or 2×10^6 cycles.

The fatigue limit S_f obtained with the step-by-step method and also the load spectrum in service are not free from uncertainties. A safety factor should be adopted. Since quantitative indications on scatter are lacking, an intelligent guess must be made. Possible consequences of fatigue failures in service have to be considered. It is believed that a safety factor of 1.5 can be sufficient in many cases. However, if more confidence is desirable, more fatigue tests should be carried out. Another approach is to carry out load history measurements in service to have more information about the load spectrum.

Safety factors for finite fatigue life problems under CA loading

Crack initiation cannot be avoided if stress amplitudes above the fatigue limit occur in the service load spectrum. As a consequence, fatigue crack initiation is possible and a finite life should be considered. A typical example is represented by a pressure vessel. A safe approximation of the load spectrum is that the pressure vessel is always loaded to the same maximum operational pressure. Load spectra of other structures with a flat load spectrum can be approximated in the same way. A safety factor can now be defined in two different ways. The factor can be applied to the fatigue life or to the fatigue strength. If a finite life is envisaged, the natural approach is to think in terms of endurances which guarantee a sufficient lifetime. If N_1 is the required lifetime and N_2 the estimated fatigue life, see Figure 20.2, then the safety factor is $f_N = N_2/N_1$. However, in terms of the fatigue strength, if S_1 is the required fatigue strength and S_2 is the estimated fatigue strength, then the safety factor $f_S = S_2/S_1$. Adopting the Basquin relation ($S_k \cdot N$ = constant), the relation between the two safety factors is $f_N = (f_S)^k$. If loads exceeding S_1 should not be expected or even be impossible, then the safety factor for the fatigue life should be considered. However, if required lifetimes larger that N_1 are of little interest then the safety factor for the stress level is more appropriate. The size of these safety factors to be adopted depend on the consequences of a fatigue failure. Obviously larger factors are necessary if fatal accidents are possible, say 1.5 on the stress level or 6.0 on lifetimes. In such a case, a realistic experimental verification test must be advised. If the consequence of a final failure are not serious, a smaller safety factor can be adopted, say 1.2 on the stress level, or 2.5 on the fatigue life. If the quality of the stress raisers is poor (e.g. in low-quality welds), larger values may be worthwhile. Engineering judgement and experience from previous structures should be practiced.

Safety factors for finite fatigue life problems under VA loading

The VA load case offers an additional uncertainty if compared to the CA load case. Predictions for a VA load history are affected by the unreliability of the Miner rule. It is difficult to understand how this might be accounted for by a safety factor. As said in Sections 10.4.2, when using the Miner rule, it appears to be wise to extrapolate S-N data below the fatigue limit. In cases of doubt, some exploratory service-simulation fatigue tests are much recommended.

Safety factors and fatigue crack growth

Safety factors on fatigue crack growth have to be considered if the crack growth period covers an essential part of the lifetime in service. This can occur when cracks are initiated at material defects, corrosion pits, or sharp corners with a high stress concentration. It can also start from unintentional surface damage caused in-service (nicks, dents, scratches, impact damage, etc.). In welded structures, crack initiation is possible from weld defects, but also at the edge of the weld toe due to a locally unfavorable profile. All these situations are undesirable, but they cannot always be avoided. In view of safety, it may be necessary to consider fatigue lives with a practically zero crack initiation period. It is kind of a worst case analysis which should be made if complete failure is unacceptable.

Two different cases can be defined:

1. Crack growth is accepted, but the occurrence of a complete failure must be prevented by periodic inspections.
2. The crack growth period until failure should be larger than the design lifetime of the structure because inspections for cracks in service are undesirable or not feasible.

The first case is well-known for aircraft structures for which so-called damage tolerance requirements are laid down in official airworthiness regulations. It can also be applicable to nuclear pressure vessels or other structures if fatigue failures are inadmissable and periodic inspections must be done to detect fatigue cracks before failure occurs. The problem setting is illustrated in Figure 20.3 by a schematic crack growth curve and a corresponding curve of the decreasing static strength of the structure caused by the growing fatigue crack. Failure of the structure is supposed to occur at a critical crack length, a_c. Cracks can be detected at the crack length denoted

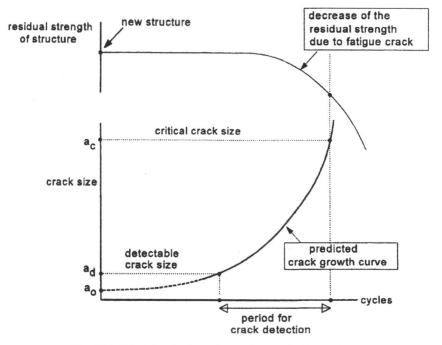

Fig. 20.3 Principle of safe crack growth by period inspections.

as a_d. The period for crack detection covers crack growth from a_d to a_c, see Figure 20.3. The number of uncertainties is fairly large: (i) the initial crack length a_0, (ii) the final crack length a_c, (iii) the crack growth data of the material, (iv) the load spectrum, (v) the crack growth prediction model, and (vi) the probability of detecting a fatigue crack.

All uncertain topics are quite obvious. They are addressed here with a some comments only. Limitations of crack growth predictions have been discussed in Chapters 8 and 11. The probability of crack detection depends on the non-destructive techniques adopted. Questions can be raised whether a surface crack with a length of a few millimeters can be detected. In general, very small cracks, say 1 mm (0.04 inches) cannot be detected reliably. Crack detection of invisible cracks, e.g. in joints, must be done with special inspection techniques.

Secondly, it must also be established how far the crack may grow before the risk of a large failure is present. The crack must be found within the crack growth range between the detectable crack size (a_d) and the critical crack size (a_c), see Figure 20.3. A safety factor should then be applied to this period to assess the inspection period. In the past, a factor 3 has been used for transport aircraft, but more recently, the tendency is to use a factor 2. Obviously, the choice of the safety factor is a matter of judgement, which requires that all sources of uncertainties are recognized and understood. It should also include the human factor of the inspection procedure. If a large number of structures must be inspected, most of which will be free from cracks, an occasionally occurring small crack might escape detection. Situations of finding cracks in order to prevent dangerous situations are not confined to aircraft. It also applies to other types of structures if a fatigue failure cannot be accepted, e.g. for pressure vessels. Operators of large structures try to combine inspections with periodic maintenance for economic reasons. Actually, operators prefer structures which do not require inspections.

The size of the initial crack length (a_0) must be associated with the size of some initial defect. This is a difficult issue because the crack growth rate of initially small cracks is very low. As a consequence, the predicted crack growth life will significantly increase for a smaller value of a_0 as discussed in Chapter 8 (see Table 8.2). It is more conservative to select a larger a_0-value, but which size? The final crack length, a_c, is reached at the moment of failure. It requires that the reduction of the residual strength of the structure is calculated as a function of the increasing crack length, which is not a simple calculation because macroplasticity will occur. However, the crack growth rate in the last part of the crack growth period is relatively high, and assuming a lower a_c will have a small effect on the crack growth period, see again Table 8.2.

The crack growth prediction model is less problematic for a CA load spectrum than for a VA load spectrum. In case of CA loading, predictions may give reasonably reliable results provided that K solutions are available. Quite often, K solutions are not available, even for structural elements with a simple geometry. Small cracks are usually part through cracks at the material surface. If K-values are not available, they can be calculated with FE techniques, but it requires expertise on this topic.

Predictions on crack growth during VA loading offer problems due to interaction effects discussed in Chapter 11. Ignoring these effects should be expected to give a conservative prediction for most load spectra. The basic CA crack growth data used in the prediction are also subjected to uncertainties. Variations can occur between nominally similar materials from different producers. Even differences between batches from the same producer have been found, see Figure 8.16. It may be recalled from Chapter 11 that small cycles with $\Delta K < \Delta K_{th}$ can still contribute to crack growth. It was proposed to extrapolate the $da/dN - \Delta K$ function in the Paris regime to low $\Delta K < \Delta K_{th}$.

Safety aspects associated with a corrosive environment and low frequency fatigue

The previous discussion did not include possible effects of a corrosive environment and load cycles with a very low frequency. The problem of corrosion fatigue was discussed in Chapter 16, where it was pointed out that the effect of corrosion on fatigue depends on the material/environment system. Unfortunately, most types of steel and aluminium alloys are sensitive to corrosion. It can imply that these materials are also sensitive to the frequency and wave shape of the load cycles. Unfortunately, the effect of corrosion fatigue cannot simply be described by a quantitative model. Experience should indicate how to deal with safety issues introduced by a corrosive environment.

In Chapter 16, it was pointed out that corrosion can affect both crack initiation and crack growth. The fatigue limit case was discussed in the present chapter as being relevant to problems where crack initiation is not allowed (flat load spectra with all cycles below the fatigue limit). Obviously, the application of safety factors does not preclude the occurrence of corrosion. Pitting and other local corrosion phenomena can occur in a corrosive environment, and subsequent crack growth will be activated. It might be hoped that cracks should not grow at low stress amplitudes, but it would require a high safety factor (see Figure 2.29 for mild steel). The best solution is to prevent corrosion at the material surface. Sometimes this is done by preventing the access of the aggressive environment to a fatigue critical element of a structure. Corrosion resistant surface layers can be considered also, but experience should indicate whether this will be successful. Another solution is shotpeening of the material surface. This would not prevent corrosion at the material surface, but the residual compressive stresses may prevent crack opening and further crack growth. An example of this application is shotpeening of springs used in cars.

If water is trapped in the structure, the consequences of a stagnant water environment may be disastrous. An example was discussed in Chapter 16.4. Trapping of water should be avoided, either by design or sealing of critical locations.

Corrosion fatigue can be problematic for structures used in the open air or in the sea, e.g. for bridges, cranes, ships, offshore structures, but also for many other structures. In the open air, rain and fog are causing a moist environment of usually polluted water, which is an aggressive environment. After fatigue cracks have been initiated, the corrosive environment can enhance crack growth. As discussed in Section 19.6 on welded joints, accelerated crack growth has been observed in comparative tests carried out in air and salt water. In salt water, crack growth could be about three times faster. A safety factor of three applied on the crack growth life may be reasonable. If fatigue failures in the environment of the structure would have serious consequences, it might be necessary to support the fatigue analysis by relevant experimental work. The problem is how a service-simulation fatigue test should then be carried out in view of corrosion being a time dependent phenomenon. The frequency of the cyclic loads in service may be low and an exact simulation can imply an unacceptably long duration of the test. A compromise should be considered. Certain parts of the load-time history can be simulated faster than the history in service, while the more damaging load cycles can be applied with the loading rates relevant for the service load-history. It then should be recognized that the increasing load part of a cycle is the most important part for fatigue crack increments, see the saw tooth effect discussed in Section 16.2.2.

Another interesting alternative to service-simulation tests is to build a few prototypes of the structure and to test these prototypes in a realistic but severe application in service. This has been done for cars and trucks, which were tested by severe driving along selected tracks with rough road conditions. Actually, such tests are not done for fatigue only. It should show a satisfactory functioning of all parts of a structure under severe conditions. However, it also can reveal insufficient fatigue properties.

20.5 Some case histories

Four case histories are presented in this section as illustrations of design aspects discussed previously. It has been repeated that accurate predictions of fatigue properties are illusory. At best, a reasonable estimate can be made, and it requires a good deal of understanding and experience about fatigue problems in order to evaluate the significance of calculated results. It has been pointed out that a verification with realistic experiments should be considered in cases of doubt. The case histories to discussed are associated with:

1. Improved shoulder fillets.

2. Secondary bending introduced by non-symmetric hole reinforcements.
3. Cracked aircraft wing panel repaired with a poorly designed patch.
4. Online structural health monitoring of the Tsing Ma Bridge.

20.5.1 *Improved shoulder fillets*

Peterson, in his book *Stress Concentration Factors* [1, 2], presents graphs with K_t-values for shoulder fillets, both for flat plates and round bars. Shoulder fillets occur in various structures. They are also important for laboratory specimens in order to avoid fatigue failures near the clamping of the specimens. It is well known that large fillet radii will give lower K_t-values, but in a structure that is not always possible. Quarter circular fillet are still frequently applied. Recently FE calculations were made for circular fillets, and in addition for quarter elliptical fillets [3]. It was expected that elliptical fillets would be superior to circular fillets, but K_t-values for these fillets are not in the book by Peterson. Results are presented in Figure 20.4 for both flat plates and circular bars. Increasing the root radius from $r/t = 1$ (quarter circular) to $r/t = 3$ gives a substantial reduction of the K_t-value. However, with the quarter elliptical fillets another interesting reduction is obtained.

It is of some interest to compare the K_t-values of the circular fillets obtained with the FE calculations with results obtained from graphs in the book by Peterson. For the shoulder fillet in a plate K_t-values for the dimensions in Figure 20.4 are not in his book. But values can be obtained with interpolation between the graphs for a stepped flat bar which may be considered as consisting of two mirrored shoulder fillets. The agreement with the present FE results is fully satisfactory. For the quarter circular shoulder fillets in a round bar Peterson shows K_t-values up to $r/t = 0.3$. With a small extrapolation and interpolation it suggests $K_t = 1.43$ to be compared with $K_t = 1.388$ of the FE calculations. Peterson warns that the data in his book for this case are approximations. In conclusion, with simple FE calculations accurate stress concentration factors can be obtained where as a lot of published data on K_t-values were obtained with photo-elastic techniques or with Neuber interpolation techniques applied to analytical solutions.

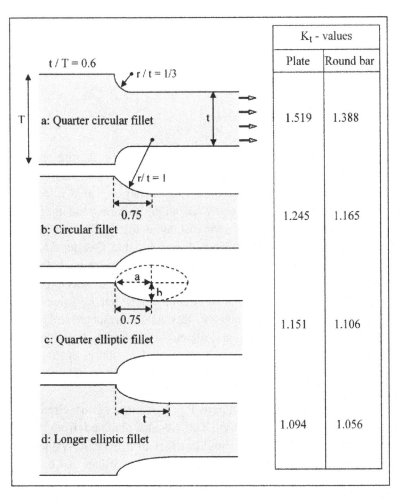

Fig. 20.4 Circular and elliptical shoulder fillets of a flat and a round bar tensile specimen.

20.5.2 *Secondary bending introduced by non-symmetric hole reinforcements*

Open holes in a structure may be fatigue critical because a significant stress concentration can be involved. Without any reinforcement around the hole a stress concentration with a K_t-value slightly below 3.0 will be present. Several decades ago a drain hole was machined in a plate of the tension skin of the wing of a large aircraft. The plate thickness around the edge of the hole was increased as shown in Figure 20.5. The purpose was to reduce the stress concentration. The aluminium alloy plate was produced by

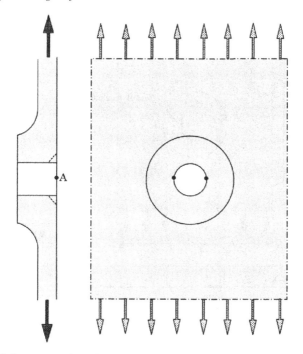

Fig. 20.5 Reinforcement of the hole edge introduces an eccentricity and thus secondary bending. Crack initiation occurred at point A.

computer controlled machining. However, it was overlooked that the thicker hole edge introduces an eccentricity which will cause secondary bending (see Section 18.5). Moreover, the reinforcement is adding locally increased stiffness to the plate around the hole which can attract load to the hole. Because of the additional bending stress, crack nucleation occurred at point A. The crack was found in a full-scale tests. A number of aircraft was already in service. A provisional solution was adopted consisting of tapering the hole (dashed lines) and shot peening of the tapered area.

The stress distribution including the secondary bending can be analyzed with FE calculations. Recently such calculations were made [3]. Selected results are presented in Figure 20.6. The maximum stress for the unreinforced hole agrees with the result obtained with the K_t-value in the book by Peterson. Secondary bending is still avoided if the thickness is increased at both sides of the plate, which is the second case in Figure 20.6. The maximum stress of the unreinforced hole is decreased from 303 to 229 MPa, a reduction with 25%.

In practical situations it is often required that one side of the plate remains flat and the increased thickness occurs at one side of the plate only. It implies

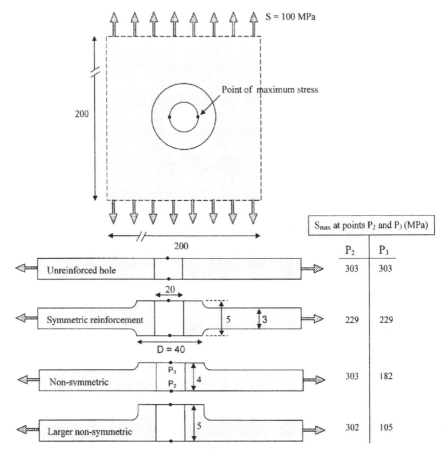

Fig. 20.6 Effect of secondary bending on the maximum stress at the edge of a reinforced hole.

that an eccentricity is introduced because the plate is no longer symmetric around the midthickness plane of the plate. Due to the secondary bending the maximum stress is larger at point P_2 than at point P_3 and even more so for the larger non-symmetry in the last case in Figure 20.6, as should be expected. But it is a striking result that the maximum stress at point P_2 is practically the same as for the unreinforced hole. Another surprising result was obtained by carrying out the same calculations for a larger diameter D of the reinforcing ring around the hole, i.e. for $D = 60$ mm instead of $D = 40$ in Figure 20.6. Actually, this is a rather heavy reinforcement, but it turned out that the maximum stress at point P_2 changed by a few percent only. The lesson to be learned is that adding or removing material around notches in a structure can satisfactorily be investigated with FE calculations. It can show instructive results for decisions on detail design questions.

20.5.3 Cracked aircraft wing panel repaired with a poorly designed patch

A fatigue crack was observed in the lower wing skin of a transport aircraft with four turbo prop engines. The tension skin of the wing box between the front and rear spar consisted of five planks with integral blade stiffeners. Figure 20.7 shows a single plank. A fatigue crack was found at the edge of a fuel access hole in middle plank after about 17000 flying hours [4]. The repair of this crack was done by a fairly large external patch and two angle sections nested inside the wing box against the skin and blade stiffeners. After some 11000 additional flying hours the repaired plank failed, but the failure was stopped at the edges of the adjacent planks which did not fail and the aircraft made a safe landing. Apparently, the fail safety feature of the five parallel planks was effective. It is now of interest to see why the repaired central plank failed. Fatigue cracks were initiated at fasteners A and B below the patch, and these fatigue cracks became unstable during a flight in severe turbulent air. The cracks occurred in the last critical end row of the repair as should be expected, see the discussion in Section 18.5. The patch and two angle sections considerably increase the local stiffness of the plank, which is good for the original fatigue crack but bad for the end row of fasteners A and B. Because of the significantly increased stiffness, load is attracted to the repair which is unfavorable for the four fasteners in the end row. As a result of the end row effect and eccentric loading on the fasteners of the end row, new fatigue cracks could easily be initiated. Note that fatigue cracks at the other end row (at the bottom in Figure 20.7) were also initiated. A better solution would be a lower stiffness repair, and a thickness tapered patch instead of a width tapered patch. A much better repair can be designed with a better understanding of the load transmission in and around the repair. A further optimization can be obtained with FE analysis.

20.5.4 Online structural health monitoring of the Tsing Ma Bridge

The Tsing Ma Bridge in Hong Kong was opened in 1997 (Figure 20.8) [5]. At that moment it was the longest suspension bridge in the world. The span

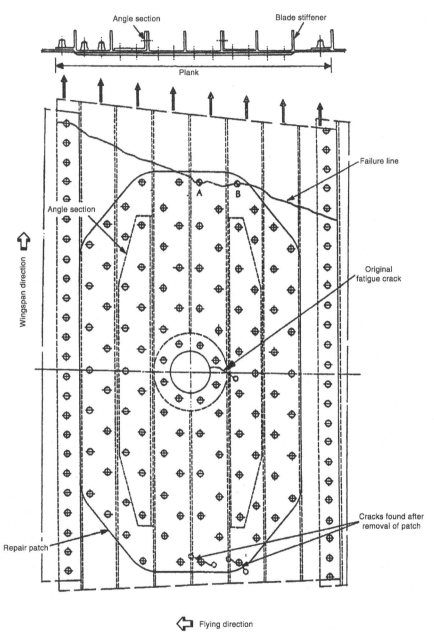

Fig. 20.7 Repair of a fatigue crack in an aircraft tension skin.

of the major part was 1377 meters, see Figure 20.9a. Details of the bridge are
shown in Figures 20.9b and 20.9c. Several carriage ways and two railways

Fig. 20.8 The Tsing Ma suspension bridge in Hong Kong.

are available to take care of busy traffic. The deck section was built up as a welded framework of steel beams. It was realized that fatigue could be one of the major lifetime issues. A large number of strain gages was bonded near fatigue critical locations. A number of these gages are used for a continuous online recording. A record during 24 hours is shown in Figure 20.10a with samples in Figures 20.10b and 20.10c. Similar records were obtained on other days. Some systematic trends are easily observed. The record shows straight lines between successive maxima and minima. Little traffic between 2 and 6 o'clock (night-time), and a systematic mean stress variation with a period of one day (probably a night and day cycle, not discussed in [5]). Furthermore, a random character of the fatigue loads with a few large loads, and non-symmetric loads with respect to the average load.

A computer program was developed for a statistical analysis of the records. The results were then translated into fatigue damage for which a modified Miner rule was adopted. The procedure is known as online structural health monitoring. The approach implies that the load spectrum is continuously obtained and translated in a fatigue damage parameter. Structural health monitoring is not a design tool, but it substantially reduces uncertainties about the load history encountered by the structure in service. Actually, the message is simple. If you do not know what happens to the structure in service, just measure it. Of course, the question remains how much of the lifetime has been consumed by the measured load history.

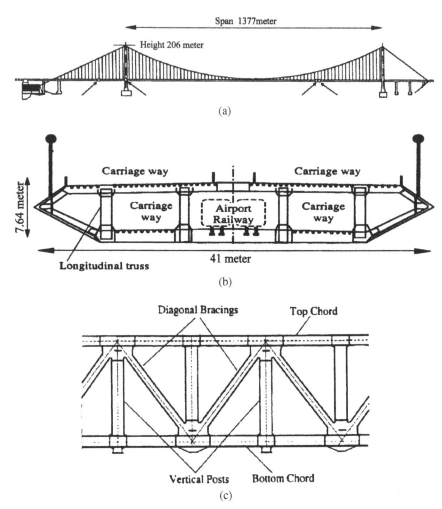

Fig. 20.9 Structure of the Tsing Ma Bridge [5]. (a) Span and height of the bridge. Arrows indicate locations of straingage. (b) Desk section of the bridge. (c) Typical longitudinal truss.

Structural health monitoring is also applied to military aircraft. The advantage then is that the load spectra from different aircraft can be compared. Because structural health monitoring requires a significant investment, it will only be done if thorough and risky conditions are applicable. But under such conditions it may well be recommended.

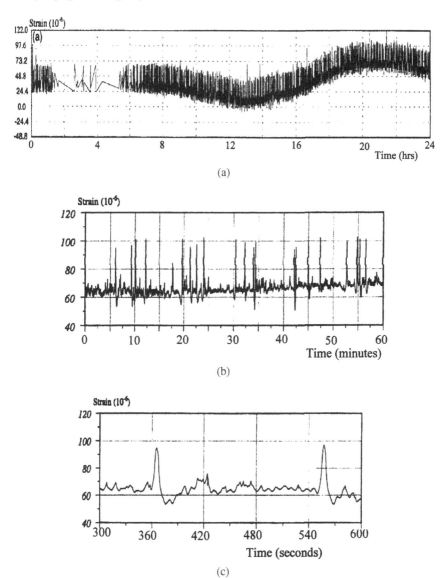

Fig. 20.10 (a) Strain-time history at May 20th, 1999. (b) Record during 1 hour. (c) Record during 5 minutes.

20.6 Summarizing conclusions

1. The present chapter is a collection of reflections on problems encountered when designing against fatigue. Various problem settings

are reviewed and apparently the variety of aspects involved is large. Structural design options are related to the lay-out of the structure, design of fatigue critical notches in the structure, various types of joints, material selection, surface treatments and production variables. Another essential part of the problem is associated with load spectra in service. Load spectra depend on the operator of the structure, but in various cases also on environmental conditions such as air turbulence, sea-waves, road roughness and other usage circumstances. The designer should carefully consider all aspects of dealing with a particular fatigue problem.

2. The purpose of designing against fatigue is to achieve satisfactory fatigue properties, but the definition of this goal can be highly different for different types of structures. Three different categories of structures are considered: (i) structures for which fatigue failures are unacceptable, (ii) structures in which fatigue cracks may occur, but the risk of a complete failure must be maintained at a very low level, and (iii) structures for which crack initiation and growth until failure after a reasonable lifetime are acceptable.

3. Designing against fatigue is more than avoiding high stress concentrations and selecting fatigue resistant materials. It also includes considerations on stiffness variations in the structure, load flow in the structure, avoidance of eccentricities, application of surface treatments, etc. Special problems are associated with joints.

4. The present knowledge about fatigue crack initiation and crack propagation in metallic materials is qualitatively well developed but quantitatively limited, and because of this it must be concluded that accurate predictions are illusory. Methods for qualitative estimates of fatigue properties can be adopted, but in case of doubt about the results, experimental verifications should be considered.

5. An experimental verification of predictions or estimates of fatigue properties of a structure should be obtained in service-simulation fatigue tests. Both fatigue critical details of the structure and the applied load history should be representative for the particular problem.

6. Safety factors can be applied on estimated load spectra, predictions of fatigue lives, fatigue limit and crack growth, design stress levels and stress levels applied in supporting experiments. The choice of safety factors should take into account various conditions and uncertainties, as well as the economic and safety consequences of premature fatigue failures. Here, engineering judgement and experience are essential.

7. The problem of corrosive environments is primarily a problem of corrosion prevention. If this is not feasible, safety factors and realistic experiments should be considered.

8. Nowadays, the tools for dealing with structural fatigue problems are powerful. FE analysis of load and stress distributions in a structure is well developed. Experimental tools for realistic fatigue tests can also meet the most demanding questions. Finally, techniques for load history measurement can provide extensive information about load histories in service. The question is how and when to adopt these tools into the scenarios of current problems of designing against fatigue.

9. Designing against fatigue requires imagination, understanding and experience. It is a real challenge.

References

1. Peterson, R.E., *Stress Concentration Factors*. John Wiley & Sons, New York (1974).
2. Pilkey, W.D. and Pilkey, D.F., *Peterson's Stress Concentration Factors*, 3rd revised edn. John Wiley & Sons (2008).
3. Schijve, J., Campoli, G. and Monaco, A., *Some FE calculations on stress concentration factors*. To be published.
4. Holshouser, W.L. and Mayner, R.D., *Fatigue failure of metal components as a factor in civil aircraft accidents*. Advanced Approaches to Fatigue Evaluation, Proc. 6th ICAF Symposium. NASA SP-309 (1972), pp. 611–630.
5. (a) Li, Z.X., Chan, T.H.T. and Ko, J.M., *Fatigue analysis and life prediction of bridges with structural health monitoring data. Part I: Methodology and strategy*. Int. J. Fatigue, Vol. 23 (2001), pp. 45–53. (b) Chan, T.H.T., Li, Z.X. and Ko, J.M., *Part II: Applications*. Int. J. Fatigue, Vol. 23 (2001), pp. 55–64.

Some general references

6. Grandt, Jr, A.F., *Fundamentals of Structural Integrity. Damage Tolerant Design and Nondestructive Evaluation*. John Wiley & Sons (2003).
7. Miller, K.J., *Structural integrity – Whose responsibility?* The 36th John Player Memorial Lecture. Institution of Mechanical Engineers (2001).
8. Fuentes, M., Elices, M., Martin-Meizoso, A. and Martinez-Esnaola, J.-M., *Fracture Mechanics: Applications and Challenges*. ESIS Publication 26, Elsevier (2000).
9. Barsom, J.M. and Rolfe, S., *Fracture and Fatigue Control in Structures. Applications of Fracture Mechanics*, 3rd edn. Butterworth-Heinemann (1999).
10. Marquis, G. and Solin, J. (Eds.), *Fatigue Design and Reliability*, 3rd Int. Symp. on Fatigue Design 98. ESIS Publication 23. Elsevier, Amsterdam (1999).
11. Marquis, G. and Solin, J. (Eds.), *Fatigue Design of Components*, 2nd Int. Symp. on Fatigue Design 95. ESIS Publication 22, Elsevier, Amsterdam (1997).
12. Farahmand, B., Bockrath, G. and Glassco, J., *Fatigue and Fracture Mechanics of High Risk Parts*. Chapman & Hall, New York (1997).

13. *Materials Selection and Design*. ASM Handbook Volume 20. ASM International (1997).
14. Hudson, C.M. and Rich, T.P. (Eds.), *Case Histories Involving Fatigue and Fracture Mechanics*. ASTM STP 918 (1986).
15. Abelkis, P.R. and Hudson, C.M., *Design of Fatigue and Fracture Resistant Structures*. ASTM STP 761 (1982).

Part VI
Fatigue Resistance of Fiber-Metal Laminates

Chapter 21
Fatigue Resistance of Fiber-Metal Laminates

21.1 Introduction

The history of mankind has been characterized by an interesting development of materials originally used for tools, housing, weapons and other needs. Initially wood and clay were available materials, followed by stone (Stone Age) and much later, but still about 3000 years ago, by iron (Iron Age). Apart from the availability of building materials, the production and working processes were also decisive for the success of a material (which in fact is still true in the present time). In the past, material properties were related to strength, stiffness, and durability. It was recognized that stones could carry high-compression loads but not high-tension loads. Later, the engineering approach to new materials included the development of composite materials with the aim to combine favorable properties of different materials into a single composite material. Reinforced concrete is a well-known example and fiber-reinforced plastics another typical case. Several composite materials were designed for specific applications. Developing materials and designing composite materials for specific purposes is often essential for advanced applications. A noteworthy example is offered by modern ceramics for high-temperature applications with the space shuttle as an outstanding example.

The history of aircraft structures has seen a large variation of different materials. In the very beginning, the time of the Wright brothers, the materials of the aircraft structure were wood, steel and linen. In the first half of the 20th century, design criteria for materials used in aircraft structures were associated with low weight and sufficient strength. The introduction of aluminium alloys in the late 1920s was a kind of a revolution because it drastically changed structural design concepts. The efficiency of aircraft structures was significantly improved. Aluminium alloys also penetrated many other applications because of the low specific weight, e.g. in many household appliances.

In the second half of the 20th century, developments of aircraft materials had to face more criteria than just low weight and sufficient strength. It turned out that civil transport aircraft were going to be used for a service life well over 20 years which poses requirements for durability of the aircraft structure, in particular with respect to fatigue and corrosion. However, the aging aircraft problem also included safety aspects. Damage to aircraft structures can include fatigue cracks, corrosion damage, impact damage and other kinds of incidental damage. The danger is that cracked parts of the structure may no longer have sufficient strength. This has led to so-called damage tolerance requirements. Cracks should not grow too fast in order to detect the damage during periodic inspections of the aircraft, see the discussion on Figure 20.3 in the previous chapter. The designer can introduce structural elements to obtain crack growth retardation or even crack arrest. *But a different approach is to develop a material which has a high crack growth resistance as an inherent material property.* This was the basic idea for the development of the fiber-metal laminates. A number of thin sheets was bonded to a single laminate with long high-strength fibers embedded in the intermediate adhesive layers. The development of fiber-metal laminates started in the late seventies in the laboratory of Structures and Material of the Faculty of Aerospace Engineering of the Delft University of Technology. Originally aramid fibers were adopted and the commercial name of the laminate was Arall (Aramid reinforced aluminium laminates). Some ten years later Glare (Glass reinforced) was introduced in view of certain shortcomings of the aramid fibers. The aramid fibers were replaced by advanced glass fibers. Both Arall and Glare were built up with thin aluminium alloy sheets. The history is described in [1]. More recently a new version of the laminate family was developed for thick plates in aircraft wing structures with the name CentrAl [2].

A different evolution of aircraft materials occurred in the 2nd half of the 20th century, i.e. the development of the composite materials, an epoxy matrix with long fibers, usually carbon fibers with a high strength and stiffness, the so-called black composites, also referred to as advanced composites. The structural application of these composites has encountered entirely different problems because of essentially different design concepts, production techniques and material properties. For these reasons the black composites are not discussed in this textbook.

In the present chapter, fatigue crack growth in laminated sheet material without fibers is discussed in Section 21.2. The major topic is fatigue of fiber-metal laminates covered in Section 21.3. Other properties and production aspects of Glare are briefly addressed in Section 21.4. The chapter is completed with some general remarks.

21.2 Laminated sheet material without fibers

The development of the fiber-metal laminates (Arall and Glare) was preceded by research on fatigue of laminated sheet material without fibers. These investigations were stimulated by the successful application of adhesive bonded aircraft structures designed by the Fokker Aircraft Industries since the 1960s. The research on laminated sheets of aluminum alloys has been instructive for the application of bonded structures in general, but also for the development of the fiber-metal laminates starting around 1980. The experience is briefly summarized in this section.

Various fatigue crack growth tests were carried out on sheet laminate specimens [3]. Illustrative results of crack growth under CA loading are shown in Figure 21.1 by a comparison of the crack growth curves for through cracks in a thick specimen (thickness 5 mm, 0.2 inch), a specimen of laminated material obtained by adhesive bonding of 5 thin sheets (thickness 5×1 mm) and a single sheet specimen (thickness 1 mm). The graph shows that the crack growth life is about 60% larger for the laminated material as compared to the solid material. The longer fatigue life results from a sheet thickness effect. In the laminated material with a through crack, crack growth occurs simultaneously in all thin sheets in a similar way as in the single thin sheet specimen. The intermediate adhesive layer has a very low stiffness, the elastic modulus is more than 25 times lower than for the aluminium alloy. As

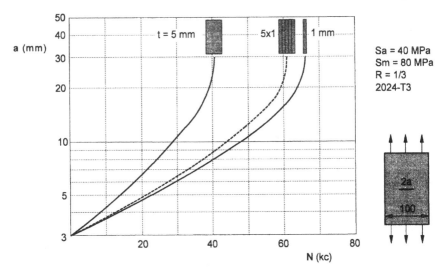

Fig. 21.1 Fatigue crack growth curves of solid and laminated material [3].

a result, fatigue cracks can grow independently in all layers of the laminate without mutual interference. The results confirm the tendency to faster crack growth under a more predominant plane strain situation at the crack tip in thicker material (see the discussion in Section 8.4.2). Thin sheets exhibit a better crack growth resistance.

Completely different results were obtained for part-through surface cracks as illustrated by Figure 21.2. A solid 5 mm thick specimen was tested with a semi-elliptical surface crack, crack depth 1 mm. The crack growth life is about 30% larger than for the through crack because the elliptical crack must first grow through the thickness. However, for the laminated specimen with a crack in the surface layer only, the fatigue life is approximately 10 times larger than the crack growth life obtained for the through crack. The same trend was observed in fatigue tests on lug specimens provided with a 1 mm deep saw cut as an initial through crack. The results in Figure 21.2 indicate a 1.5 times longer crack growth life for the laminated lug as compared to the life for the solid lug; whereas, the crack growth life for the corner cracks is about 5 times longer for the laminated material. Lug specimens were also tested without crack starter notches. The fatigue life then showed a good deal of scatter which is associated with crack initiation by fretting corrosion inside the hole. Cracks started at both sides of the hole, while several crack nuclei were observed at each side of the hole, not exactly in the same cracking plane. As a consequence, radial ridges occurred in the solid material where nuclei overlap. The tortuous crack front contributes to

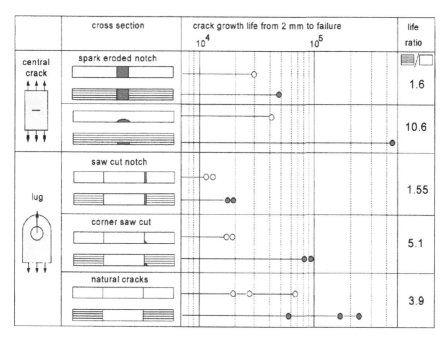

Fig. 21.2 Crack growth lives for solid material ($t = 5$ mm, 0.2 inch) and laminated material (5×1 mm, 5×0.04 inch). Central cracked specimens ($W = 100$ mm, 4 inch) and lug specimens ($W = 60$ mm, 2.4 inch, $D = 25$ mm, 1 inch), material: alluminium alloy 2024-T3 [3].

more scatter. In the laminated lug, cracks did not start simultaneously in all sheets, which again caused significant scatter. Anyway, the average fatigue life of the laminated lug is considerably larger (about four times) than for the lug of the solid material.

The large improvement of crack growth lives for part-through cracks seems to be logical because the adhesive layers might be a barrier for crack growth in the thickness direction. However, this argument does not correctly recognize the crack growth delaying effect. In addition to the visible crack growth in the surface sheet, crack growth in the second and the fourth sheet was measured by a special electrical potential drop method. The results are shown in Figure 21.3. Initially, very slow crack growth occurred in the first sheet only, much slower than compared to crack growth in a single sheet specimen. This should be explained by a significant restraint on crack opening in the first sheet because the other sheets are still uncracked. As a consequence, the stress intensity factor K will decrease as the crack becomes longer. Only after crack initiation in the second sheet, and afterwards in the other sheets, crack growth acceleration occurs. This finding

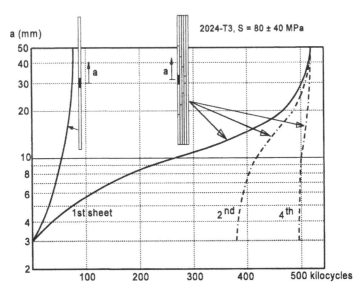

Fig. 21.3 Crack growth from a part through crack in laminated sheet material (5 × 1 mm). Slow crack growth in the first sheet due to crack opening restraint by the other sheets [3].

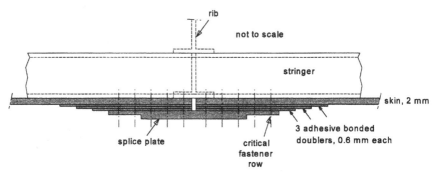

Fig. 21.4 A skin joint with adhesive bonded doublers in an aircraft wing tension skin [4].

on the sheet-metal laminates has been instructive for the development of the fiber-metal laminates.

A similar observation was made on crack growth developments in full-scale fatigue tests on wing tension skin panels (length 8.3 m, 27.3 ft) [4]. Two identical skin joints were present in the panels. As shown by Figure 21.4, the joint consisted of an outside splice plate and continuous inside stringers. The skin joint was locally reinforced with three adhesively bonded and staggered doublers. Invisible crack growth occurred at the rivet holes of the end row of rivets, a classical location for fatigue crack nucleation (Chapter 18). Cracks were observed after 40 to 50% of the total life of the

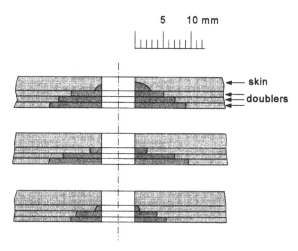

Fig. 21.5 Fatigue crack nuclei at three fastener holes of the joint in Figure 21.4. Small or no crack nuclei in skin.

panel defined by failure at another more critical location. Crack growth was followed by X-ray inspections. The cracks first occurred in the outer doubler, later in the inner doublers, and finally in the skin. This was confirmed by crack nuclei configurations revealed after opening the fracture at the end of the test, see Figure 21.5. Apparently, cracking developed slowly thanks to lamination of the material. Much faster crack growth would have occurred in an integrally machined wing structure.

21.3 Fiber-metal laminates Arall and Glare

The principle ideas of the fiber-metal laminates are discussed first, followed by a section on fatigue crack nucleation and crack growth, and another section on some experience of Glare in full-scale structural components.

21.3.1 The fiber-metal laminate concept

It has been shown in the previous section that crack growth in a single sheet of a laminated material was slowed down after some crack growth because the other sheets restrain crack opening and thus reduce the stress intensity at the crack tip. But this mechanism cannot be active for a through crack when cracks in all sheets of the laminate are growing simultaneously. One of the

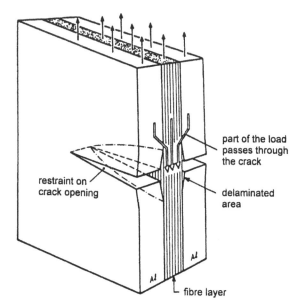

Fig. 21.6 Unbroken fibers in the wake of the crack are restraining crack opening. Part of the load is transmitted by the fibers through the cracked area.

basic ideas of the development of fiber-metal laminates is to restrain crack opening of through cracks by unbroken fibers in the wake of a fatigue crack. This situation is illustrated in Figure 21.6 for a laminate of two sheets with an intermediate adhesive layer as a matrix for uni-directional fibers in the loading direction. If the fibers would not be present, the remote load must be transmitted around the crack. However, as shown in Figure 21.6, the fatigue cracks in the metal layers are bridged by the fibers, which is advantageous: (i) the crack bridging fibers exert a significant restraint on crack tip opening, and (ii) the unbroken fibers in the cracked area imply that part of the load in the aluminium alloy sheets is still transmitted through the crack. As a result, a most significant reduction on the stress intensity factor K will occur.

The load in the crack bridging fibers are causing a shear stress on the interfaces between the fiber layers and the sheets, a problem first analyzed by Marissen [5]. Due to the shear stress some delamination will occur along the crack edges. A second basic idea of the fiber-metal laminates is that the shear stress can be significantly reduced by creating more interfaces. It has been accomplished by implementing a larger number of metal sheets with a low thickness. Typical values are 0.3 and 0.4 mm (0.012 and 0.016 inch). A certain laminate thickness is then obtained by more thin sheets and thus more intermediate fiber layers. A third aspect is that fiber failure in the wake of the crack must be prevented which is achieved by using fibers with a high tensile strength and more fibers in each layer. The latter aspect means that a sufficiently large fiber volume content should be obtained. Experiments have shown that the delamination is limited and acceptable for a well balanced fiber volume ratio and sufficient interfaces. First Arall was developed with aramid fibers. Later advanced glass fibers were adopted in Glare [6].

During fatigue tests on Arall specimens it was observed that failure of the crack bridging aramid fibers could occur under certain fatigue load conditions. In the wake of the crack in the metal sheets, the epoxy matrix is also cracked. The crack bridging fibers are loaded in tension during an increasing load. To some extent, the fibres will be pulled out of the cracked area of the matrix material. This is facilitated because the adhesion between the aramid fibers and the matrix is rather week. This pulling out process is not fully reversible during unloading. As a consequence, fiber buckling can occur during many cycles. The fibers will be damaged and fiber failure at S_{max} is possible. Even then, there are still unbroken fibers closely behind the crack tip which take care of a K reduction and slow crack growth. Fortunately this fiber failure mechanism does not occur in Glare aslo because of a good adherence between the glass fibers and the adhesive matrix. Aspects of the fatigue phenomenon in Arall and Glare were extensively studied by Roebroeks [7]. He showed that fatigue is a fairly complex phenomenon in these hybrid materials, but in a qualitative way the fatigue mechanism is reasonably well understood. By now, Arall is surpassed by Glare. Moreover, significant break-throughs for design and production of Glare structures have been achieved, e.g. splices and CentrAl to be addressed later. But the discussion in this chapter is primarily restricted to Glare.

21.3.2 Fiber-metal laminates as sheet material

Glare sheets are produced with standard bonding technology. The thin aluminium alloy sheets of 2024-T3 or 7475-T76 are pretreated for adhesive bonding. The fiber layers consist of a number of prepregs consisting of unidirectional fibers in a thin adhesive film. The strength of the fibers is 4000 MPa, the stiffness is 88 GPa, and the strain at failure 4.5%. A survey of different Glare grades now used for structural application is presented in Table 21.1 [8]. Prepreg layers can be arranged with fibers in different direction depending on the dominant load on a structural component. If all fibers are in the same direction, e.g. for Glare 1 and Glare 2, the anisotropy

Table 21.1 Standard Glare grades [8].

Glare grade	Sub-grade	Metal sheet thickness (mm) and alloy	Prepreg orientation* in each fibre layer**	Main beneficial characteristics
Glare 1	–	0.3–0.4 7475-T761	0/0	fatigue, strength, yield stress
Glare 2	Glare 2A	0.2–0.5 2024-T3	0/0	fatigue, strength
	Glare 2B	0.2–0.5 2024	90/90	fatigue, strength
Glare 3	–	0.2–0.5 2024	0/90	fatigue, impact
Glare 4	Glare 4A	0.2–0.5 2024	0/90/0	fatigue strength in 0° direction
	Glare 4B	0.2–0.5 2024	90/0/90	fatigue strength in 90° direction
Glare 5	–	0.2–0.5 2024	0/90/90/0	impact
Glare 6	Glare 6A	0.2–0.5 2024	+45/–45	shear, off-axis properties
	Glare 6B	0.2–0.5 2024	–45/+45	shear, off-axis properties

*The rolling direction is defined as 0°.
**The number of orientations in this column is equal to the number of prepregs in each fiber layer.

Table 21.2 Mechanical properties and density of Glare 1 and Glare 3. L = longitudinal direction, T = transverse direction.

	Aluminium alloys				Glare			
Sheet material Fiber orientation	2024-T3 –		7475-T761 –		Glare 1 uni-directional		Glare 3 cross ply	
Properties	L	T	L	T	L	T	L	T
S_U (MPa)	455	448	520	531	1282	352	717	716
$S_{0.2}$ (MPa)	359	324	476	466	545	333	305	283
Elongation (%)	19	19	11	11	4.2	7.7	4.7	4.7
E (GPa)	72	72	69	69	65	50	58	58
Density (g/cm^3)	2.78		2.78		2.52		2.52	

is significant. The anisotropie is illustrated by a survey of static properties of Glare 1 and Glare 3 in Table 21.2. Properties of the aluminium alloys of the two Glare grades are given in the same table. Cross ply introduced in Glare 3 and Glare 4 can be advantageous for biaxial loading, e.g. for the skin of an aircraft fuselage with the hoop stress as the larger stress component. For all Glare grades the number of layers and the thickness of the aluminium alloy sheets must be chosen in accordance with the loading conditions of a specific component. A coding system is used to refer to a specific composition of Glare. As an example, Glare 4B-4/3-0.4 applies to

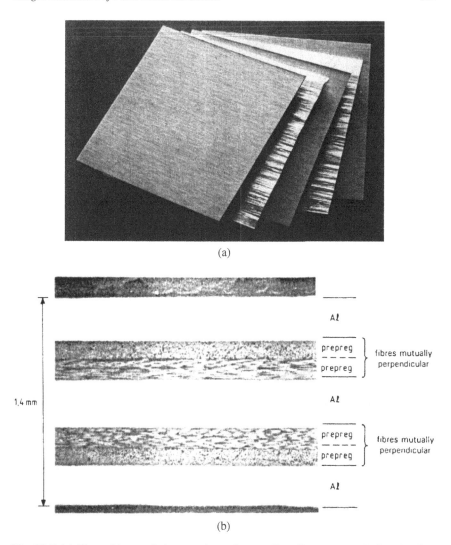

(a)

(b)

Fig. 21.7 (a) Three thin metal sheets and two intermediate fiber prepregs before bonding occurs. (b) Cross section of "biaxial" Glare 3 with three 2024-T3 sheets and uni-directional fiber layers in the L and T directions. (Courtesy G. Roebroeks)

Glare 4B with 4 aluminium alloy sheets and 3 fiber layers (a so-called 4/3 lay up), and thickness of 0.4 mm of the metal sheets. A cross section of Glare 3 is given in Figure 21.7.

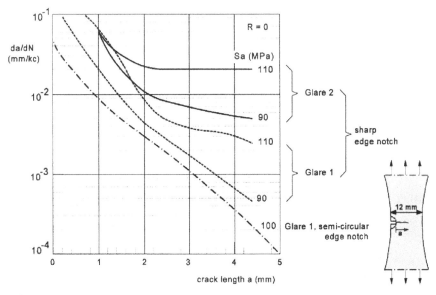

Fig. 21.8 The growth of small cracks in Glare under CA loading [9, 10].

21.3.3 Crack growth in Glare

About crack nucleation in Glare

The growth of fatigue cracks in Glare occurs very slowly as a consequence of the crack bridging effect. However, crack bridging can only occur if a fatigue crack is present. It implies that fatigue crack nucleation might occur earlier in Glare than in solid sheet material because of the slightly lower stiffness of Glare, see Table 21.2. The cyclic strain at the root of a notch will be somewhat larger for the same applied fatigue load. This question was investigated by Marissen for Arall specimens in 1988 [5]. He observed that microcrack growth initially occurred reasonably fast. But already at a crack length equal to the thickness of the aluminium sheets (0.5 mm) it was followed by a significant reduction of the crack growth rate. Later experiments were carried out on Glare specimens by Maria Papakyriacou [9, 10], see the results in Figure 21.8. Specimens with a sharp edge notch (depth 2 mm) and a semi-circular edge notch were tested in a resonance device at a very high frequency (21 kHz). Although very low crack rates occurred, the tests could still be done in a reasonably time. Results for Glare 1 and Glare 2 show a systematically decreasing crack growth rate already for cracks with a length of a few tenth of a mm. Apparently, some favorable

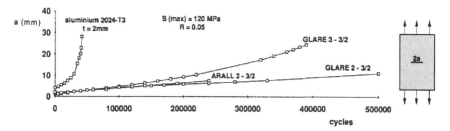

Fig. 21.9 Fatigue crack growth in fiber-metal laminates under CA loading [7].

effect of the fibers is observed as soon as some minute cracking has been initiated. How fast and how far da/dN will decrease will depend on the shape of the specimen and the notch geometry.

Crack growth of macrocracks in Glare

Crack growth results for constant-amplitude (CA) loading are shown in Figure 21.9 with a crack growth curve for 2024-T3 for comparison. The much slower crack growth in Glare as compared to crack growth in solid sheet material is evident. At a crack length $a = 10$ mm (0.4 inch) the crack growth rate in Glare as compared to the growth rate in 2024-T3 is seven times slower in Glare 3 and 20 times slower for Glare 2. Slow crack growth in Glare has been confirmed in several test programs, see e.g. [6, 7].

Fatigue crack growth in Glare has also been studied under VA loading. Simple CA tests with OLs have shown the crack growth retardation in a similar way as discussed in Section 11.2. The delay periods were smaller which should be associated with smaller plastic zones created by the OLs due to restrained crack opening because of crack bridging.

Crack growth in Arall and Glare specimens with 2024-T3 sheet material was also investigated under flight-simulation loading with the miniTwist load history [11]. Specimens with open holes were adopted in order to be informed about the crack initiation period as well. Illustrative results are shown in Figure 21.10. The initial crack growth period until a crack length of 1 to 2 mm (0.04 to 0.08 inch) is in the order of 10000 to 15000 flights. But the remaining crack growth life is very large. Actually, tests were stopped after 270000 flights for three of the four types of fiber-metal laminates because of marginal crack growth. The same tests were also carried out on similarly laminated specimens without fibers [12]. The crack initiation fatigue lives were of the same order of magnitude as in the tests on the fiber-metal

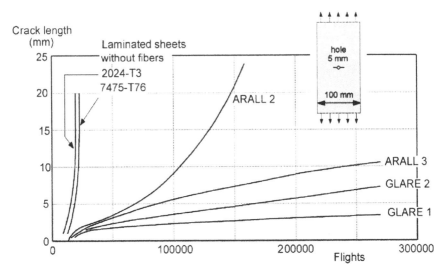

Fig. 21.10 Fatigue crack growth in open-hole specimens of fiber-metal laminates under flight-simulation loading (miniTWIST) [11].

laminated specimens, but the crack growth lives are highly different. The crack growth life until failure in the laminated specimens without fibers is in the order of 10000 kc. However, after 270000 flights the crack length was no more than approximately 3 and 7 mm in the Glare 1 and Glare 2 specimen respectively.

The effect of the truncation level of the gust spectrum was also considered in [11]. The gust spectrum of miniTwist is a steep spectrum, and the truncation level of the larger amplitude cycle has a large effect on fatigue crack growth in 2024-T3 specimen. as discussed in Chapter 11 (see the discussion on Figure 11.16). It turned out that a similar large effect was found for Glare 1 and Glare 2, i.e. slower crack growth for a higher truncation level. It appears that the crack growth behavior in the aluminium sheets of the fiber-metal laminates is similar to the behavior of solid sheet specimens. This observation is important for considering prediction models for fatigue crack growth in fiber-metal laminates. A major problem then is to predict the history of the stress intensity factor in the metal layers of fiber-metal laminates. A more elementary problem is to predict the *K* history in fiber-metal laminates under CA loading. This is the problem which was addressed by Marissen for Arall [5]. He made several simplifying assumptions to define the first prediction model for fatigue crack growth in Arall sheet specimens. A major problem was to predict the delamination zone around the growing fatigue crack in the aluminium sheets, and in

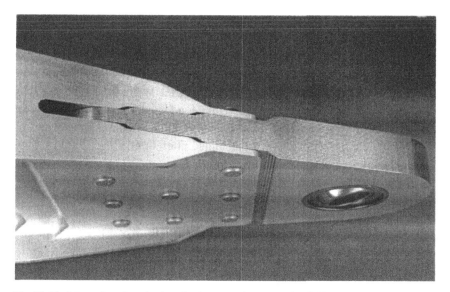

Fig. 21.11 A lug of an aircraft wing-fuselage attachment made of Glare 1 [16]. Lug thickness 19.0 mm (0.75 inch), hole diameter 44.45 mm (1.75 inch).

addition the crack bridging forces which are required for predictions of the K-value at the tip of the crack. Variants on the model of Marissen has been published in the literature. A more realistic prediction model was proposed by Alderliesten in 2005 [13, 14]. It includes the interaction of the increasing delamination zone and fatigue crack growth in the aluminium sheets. In addition to the crack growth resistance of the sheet material, the resistance to the simultaneous delamination is accounted for in order to predict the shape of the delamination zone. Prediction for VA loading are now investigated [15].

21.3.4 Fatigue properties of Glare components

Experience of three cases will be summarized:

- *Wing-fuselage connection by lugs*
  The connection between the wing and the fuselage of the CN-235 aircraft occurs with four lugs, two at the front spar and two at the rear spar. In view of damage tolerance it is unacceptable that a lug should fail by fatigue. A low stress level was adopted in view of the risk of a fatigue failure and the low fatigue limit of lugs due to fretting

corrosion (Chapter 15). An investigation was made to produce the lug as a Glare 1 component with all dimensions similar to a lug of the CN-235 aircraft, which implied a thickness of 19 mm (0.75 inch) (and a hole diameter of 44.45 mm (17.5 inch) [16]. The lug is shown in Figure 21.11. It was built up with 25 metal layers [16]. The lug was tested in a flight-simulation test with a load history consisting of blocks of 1000 flights with 10 different types of flights in a random sequence. No indications of cracks were obtained after simulating 120000 flights at the original design stress level for the lug connection. All stress levels were then increased by a factor of 1.25 and again 120000 flights were simulated without any cracking. This increasing of the load level was done three times, bringing the load level to $1.25^3 = 1.95$ times the original design load level, almost doubling the stress level. Failure of the lug still did not occur, but fracture occurred in the aluminium alloy clamping of the lug after another 92000 flights at the last load level. Small cracks could be observed inside the hole in three of the 25 metal layers which should have an insignificant effect on the static strength of the joint. The test shows a large safety margin and favorable information for long inspection periods.

• *Riveted lap joint*

In view of the application of fiber-metal laminates as a skin material for a pressurized aircraft fuselage, various fatigue tests were carried out on a riveted lap joints. Some results are shown in Figure 21.12 for lap joints with three rivet rows and seven rivets in each row [17]. Specimens were made from Glare 3 (cross ply) and monolithic 2024-T3. The Glare sheets consisted of three aluminium alloy layers with intermediate prepreg layers. The specimens were tested until multiples of 100000 cycles and then pulled to failure to determine the residual strength and to observe the amount of fatigue cracking on the fracture surface. The residual strength is plotted in Figure 21.12. Some small cracks were present in a 2024-T3 specimen after 100000 cycles. Larger cracks were found in other specimens after longer fatigue cycling periods. Finally, a specimen failed at a fatigue life of 470000 cycles and the residual strength is then equal to S_{max} of the fatigue load. Fatigue cracks in the monolithic 2024-T3 sheets can penetrate through the full thickness of the sheet. The increasing fatigue damage as a function of the number of load cycles is shown in Figure 21.12. It is typical "multiple site damage". However, cracking in the Glare 3 specimens occurred for a large number of cycles in one of the three metal layers only, which is the mating surface layer of the lap joint. This layer carries the maximum

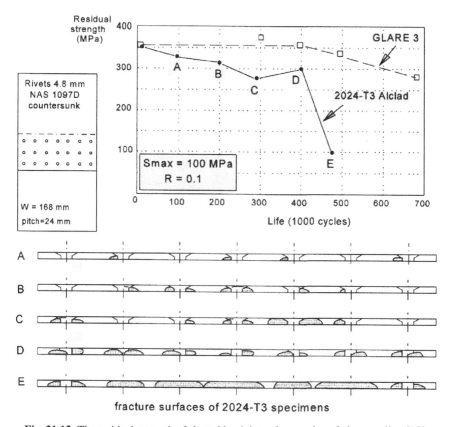

Fig. 21.12 The residual strength of riveted lap joints after previous fatigue cycling [17].

bending stress. However, after 500000 cycles fatigue cracks occurred in the mating surface layer only without any cracks in the other two layers. Only for the last Glare 3 result in Figure 21.12 after fatigue testing to 700000 cycles, cracks were found in the second sheet layer. Apparently, a superior damage tolerance behavior if compared to the full metal lap joint.

• *Crack growth in the Airbus barrel test*

Results of fatigue crack growth in a full-scale test representative for the Airbus A-340 fuselage are shown in Figure 21.13 [18]. Different skin materials were used in the fuselage structure and two different cases were investigated: (i) crack growth under a broken stringer and (ii) crack growth under a broken frame. The cracks started from a saw cut of 75 mm (3 inch). The results show a considerably slower crack growth in Glare 3 (thickness 1.4 mm) than in 2024-T3 (thickness 1.6 mm). This illustrates the potential usefulness of Glare 3, also because of weight

Crack length
(2a, mm)

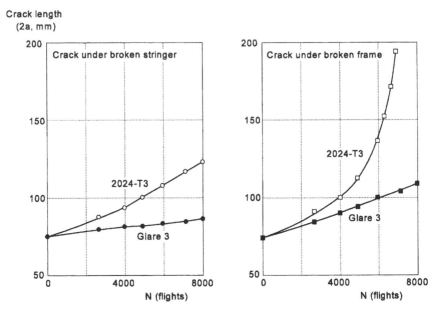

Fig. 21.13 Crack growth results obtained in the Airbus barrel test [18].

savings in view of the lower thickness and in addition the lower specific mass (2.52 g/cm³ for Glare 3 and 2.78 g/cm³ for 2024-T3).

21.4 More about Glare

In general, Glare should be considered as a potential material to be selected for structural elements for which fatigue is a critical design issue. In Section 21.3, the fatigue properties of Glare have been highlighted. However, the introduction and structural application of a new material requires more information than data about the static and fatigue properties only. Various aspects about durability, physical properties, and technological aspects must be explored. A few remarkable aspects of Glare associated with the laminated composition will be summarized in Section 21.4.1. Furthermore, the production techniques for sheet metal products are not completely similar to the standard techniques for aluminium alloy sheet material. Aspects of this topic are briefly covered in Section 21.4.2.

21.4.1 Some typical properties of Glare

The corrosion resistance of the fiber-metal laminates is good, partly because the thin sheets are anodized and coated with a corrosion-inhibiting primer prior to the bonding process. A specific feature is that the depth of corrosion damage, also in severe corrosive environments, is restricted to the outer metal layer. The first fiber-epoxy layer has been shown to be an efficient barrier for a deeper penetration of corrosion damage. The durability of Glare as applied in the fuselage of the Airbus A380 is extensively investigated as reported by Beumler [19]. Various specimens including joints have been exposed in extreme outdours environments, e.g. in Australia, which after several years are tested in the laboratory. Until now (2008) specimens exposed for six years have given satisfactory results.

A similar barrier function also occurs in case of a fire. The extraordinary flame resistance of Glare follows from the ability of the glass fiber epoxy layers to prevent fire penetration. Experiments have shown that Glare can be an attractive material for fire walls.

Another interesting quality of Glare laminates is associated with the impact damage resistance. Both low and high-velocity impact damage of Glare parts do not have significant consequences with respect to durability and aircraft inspection intervals if compared to damage of aircraft structures of aluminium sheet material. To some extent the impact resistance of Glare is comparable to the resistance of the aluminium alloys used in Glare. However, for an aircraft under a severe hailstorm Glare can even perform better. In sheet metal aircraft structures crack can be induced by a hailstorm. However, damage is less severe for Glare skins due to the fact that the fiber layers take care of maintaining the coherence of skin panels.

21.4.2 Production and design aspects of Glare structures

Workshop properties

The production techniques applied to aluminium alloy sheet material are largely adopted in a similar way for Glare. It includes techniques as cutting, contour milling, drilling, sheet metal bending, riveting and bonding. But there are differences which should be recognized and explored in trial experiments. Drilling holes and contour milling implies that new sheet edges

Metal layer   Fiber layer

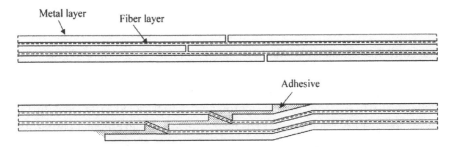

Adhesive

Fig. 21.14 Splicing concepts for obtaining large sheets [8].

are made which may cause delamination damage along the edges. Care must be taken.

Sheet metal bending of uni-directional fiber-metal laminates perpendicular to the fiber direction is possible and adopted for the production of stiffeners of various shapes. Again the occurrence of delamination should be explored.

Curved panels

In view of the application of Glare to fuselage structure, the question of producing single curved and double curved Glare sheet panels is also considered. Both types of panels have been made in the autoclave by using a single or double curved mould [8]. An attractive feature is that it occurs as part of the Glare production operation, in other words, in a single production cycle. The thorough stretching operation for double curved panels of solid metal sheets is thus avoided.

Large sheets

Initially the size of Glare sheets was limited to a width of 1.65 m (65 inch) because this was the maximum width to be obtained in the rolling production of the thin metal sheets. Roebroeks [8] developed a splicing concept to produce large Glare sheets by a bonding operation implemented in the production of Glare. The basic idea is illustrated in Figure 21.14a for a 3/2 lay-up. The interruptions of the three metal layers are located at a staggered position. The interruptions imply a local reduction of the static strength. If this is not acceptable, the overlap splice configuration shown in Figure 21.14b can be adopted. Large panels obtained with splices are

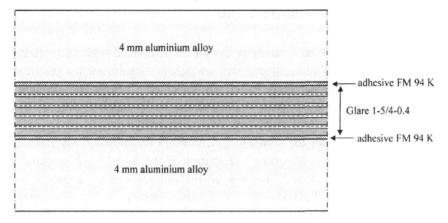

Fig. 21.15 Typical lay-up of CentrAl [2].

now applied in the fuselage of the Airbus A380. In addition to good fatigue properties another advantage is that many riveted joints can be eliminate.

Thick hybrid plates

Glare sheets were primarily developed as sheet material for pressurized transport aircraft. However, the tension skin of a wing structure may also be fatigue critical. The thickness of these skins can be fairly large. This has led to the development of CentrAl as described by Roebroeks [2]. CentrAl combines Glare and solid sheets in a single plate. An illustrative example is shown in Figure 21.15. Glare in the center of the thickness is bonded to the relatively thick outside metal sheets by a special adhesive without fibers in order to reduce possible delamination between Glare and the outside metal layers. CentrAl is a kind of a superposition of a crack growth retarding material and another solid material. Various mixed lay-ups of Glare sheets combined with solid metal sheets are possible. The designer should specify which combination is most useful for his structure. Actually, he should design the constitution of an optimal variant of Central. Roebroeks carried out fatigue tests on several types of CentrAl specimens which showed substantial improvements of the crack growth life. The crack growth retardation may be less than the impressive retardation shown for pure Glare in Figure 21.9, but increasing the crack growth life with a factor of 3 to 10 can already be most attractive for a damage-tolerant structure.

21.5 Some summarizing remarks

The development of new materials cannot be completed without research of a large variety of material properties which are significant for practical application. Much is learned by experiments and service experience of initial application in structures. With respect to the basic understanding of the fiber-metal laminates Arall and Glare, a vast amount of development research was carried out under the stimulating leadership of Professor Vogelesang in the Structures and Materials Laboratory of the Faculty of Aerospace Engineering of the Delft University of Technology. Many students were involved, including PhD students of several countries (USA, Germany, Belgium, China, and the Netherlands). The cooperation with Fokker Aircraft Industries, Airbus, and the National Aerospace Laboratory NLR should also be mentioned.

The original motivation for the development of the fiber-metal laminates was to obtain a material with a high fatigue resistance. Fatigue properties are discussed in the previous section. It is logical that fiber-metal laminates should be considered in particular for fatigue critical components. This can be a small component, e.g. the lug shown in Figure 21.11. For such a small part, the lower specific mass of Glare may not be significant, but the durability, limited inspection effort, and high safety are important improvements. For large parts, e.g. the aircraft fuselage skin, weight saving can certainly be of interest. Good fatigue properties of a fatigue critical component may imply that the design stress level can be increased. The weight saving is then coming from the smaller material volume, and in addition from the lower specific mass of the fiber-metal laminates.

It is remarkable that the fiber-metal laminates also have some favorable properties not related to fatigue but still associated with the laminated stacking of thin layers. Corrosion resistance, fire resistance and low-impact damage sensitivity are mentioned in Section 21.4.1. Several aircraft components made of fiber-metal laminates (Arall and Glare) have been installed in operational aircraft. The fiber-metal laminates were selected for reasons of fatigue, damage tolerance, weight saving and impact resistance. The behavior in service of these components is fully satisfactory. This is also true for components originally made of aluminium alloy material which showed early fatigue problems in service. Replacing the components by components of fiber-metal laminates eliminated the fatigue problems. The most appealing application of Glare is associated with the Airbus A380. Large panels on the fuselage are made from Glare. As said before, designing a structure from Glare requires various experiments to explore the behavior in a specific structural component for load histories applicable to the component. This is extensively reported in the doctor thesis of Beumler [19]. His work includes fatigue crack initiation and propagation in riveted joints under relevant load histories, variables of riveting, effects of temperature and load frequency, and environmental effects. The outdoor exposure program was already mentioned in Section 21.4.1. It may well be emphasized that extensive and diverse investigations are necessary for the development of a new material until it becomes a mature and potentially useful material for structural application. Physical and mechanical understanding of the material is essential. Production techniques are another important topic to be explored. The fiber-metal laminates offer another dimension to designing against fatigue, especially for fatigue critical components in aircraft, but possibly also for other structures if damage tolerance and durability are important issues.

References

1. Vlot, A., *Glare, History of the Development of a New Aircraft Material.* Kluwer Academic Publishers, Dordrecht (2001).
2. Roebroeks, G.H.J.J., Hooijmeijer, P.A., Kroon, E.J. and Heinemann, M.G., *The development of CentrAl.* Paper in [20].
3. Schijve, J., Van Lipzig, H.T.M., Van Gestel, G.F.J.A. and Hoeymakers, A.H.W., *Fatigue properties of adhesive-bonded laminated sheet material of aluminium alloys.* Engrg. Fracture Mech., Vol. 12 (1979), pp. 561–579.
4. Schijve, J., Broek, D., De Rijk, P., Nederveen, A. and Sevenhuysen, P.J., *Fatigue tests with random and programmed load sequences with and without ground-to-air cycles. A comparative study on full-scale wing center sections.* Nat. Aerospace Lab. NLR, Report TRS.613, Amsterdam (1965). Also issued as: Technical Report AFFDL-TR-66-142, Air Force Flight Dymanics Laboratory, Wright Patterson AFB, Ohio (1966).
5. Marissen, R., *Fatigue crack growth in ARALL, a hybrid aluminium-aramid composite material. Crack growth mechanisms and quantitative predictions of the crack growth rates.* Doctor Thesis, Delft University of Technology (1988).
6. Roebroeks, G.H.J.J., *Towards GLARE – The development of a fatigue insensitive and damage tolerant aircraft material.* Doctor Thesis, Delft University of Technology (1991).
7. Roebroeks, G.H.J.J., *Fiber metal laminates. Recent developments and applications.* J. Fatigue, Vol. 16 (1994), pp. 33–42.
8. Roebroeks, G.H.J.J., *Glare features.* Paper in [21], pp. 23–37.
9. Papakyriacou, M., Stanzl-Tschegg, S.E., Mayer, H.R. and Schijve, J., *Fatigue crack growth in Glare, role of glass fibers.* Structural Integrity: Experiments, Models and Application, Proc. ECF 10, K.-H. Schwalbe and C. Berger (Eds.). EMAS, Warley (1994), pp. 1193–1199.

10. Papakyriacou, M., Schijve, J. and Stanzl-Tschegg, S.E., *Fatigue crack growth behaviour of fibre-metal laminate Glare-1 and metal Laminate 7475 with different blunt notches.* Fatigue Fract. Engrg. Mater. Struct., Vol. 20 (1997), pp. 1573–1584.

11. Schijve, J., Wiltink, F.J. and van Bodegom, V.J.W., *Flight-simulation tests on notched specimens of fiber-metal laminates.* Report LRV-10, Delft University of Technology (1994).

12. Schijve, J. and Wiltink, F.J., *Fatigue tests on notched specimens of laminates of very thin Al-alloy sheet materials.* Report LRV-12, Delft University of Technology (1994).

13. Alderliesten, R.C., *Fatigue crack propagation and delamination growth in Glare.* Doctor Thesis, Delft University of Technology (2005).

14. Alderliesten, R.C., *Analytical prediction model for fatigue crack propagation and delamination growth in Glare.* Int. J. Fatigue, Vol. 29 (2007), pp. 628–646.

15. Plokker, H.M., Alderliesten, R.C. and Benedictus, R., *Crack closure in fibre metal laminates.* Fatigue Fract. Engrg. Mater. Struct., Vol. 30 (2007), pp. 608–620.

16. Mattousch, A.C., *Structural application of stacked GLARE – Design, production and testing of the CN-235 forward attachment fitting lug.* Master Thesis, Faculty of Aerospace Engineering, Delft (1992).

17. Soetikno, T.P., *Residual strength of Glare 3 riveted lap joints after fatigue testing.* Master Thesis, Faculty of Aerospace Engineering, Delft (1992). (Results summarized in: LR-Document b2-93-6, Delft, 1993).

18. Vogelesang, L.B., Schijve, J. and Fredell, R., *Fibre-metal laminates: Damage tolerant aerospace materials.* Case Studies in Manufacturing with Advanced Materials, Vol. 2, A. Demaid and J.H.W. de Wit (Eds.). Elsevier (1995), pp. 253–271.

19. Beumler, Th., *Flying GLARE.* Doctor Thesis, Delft University of Technology, Delft University Press (2004).

Some general references

20. Benedictus, R. (Ed.), *Damage tolerance of aircraft structures.* Papers of a Symposium in Delft, September 2007, to be published in Int. J. Fatigue.

21. Vogelesang, L.B., *Fibre MetalLaminates – The development of a new family of hybrid materials.* Plantema Memorial Lecture, Proceedings ICAF Symposium, Lucern. EMAS Publishing, UK (2003).

22. Vlot, A. and Gunnink, J.W., *Fibre Metal Laminates. An Introduction.* Kluwer Academic Publishers (2001).

23. Vlot, A., *Bonded aircraft repairs under variable amplitude fatigue loading and at low temperatures.* Fatigue Fracture Engrg. Mater. Struct., Vol. 23 (1999), pp. 9–18.

24. Vercammen, R.W.A. and Ottens, H.H., *Full-scale Glare fuselage panel tests.* Fatigue in New and Aging Aircraft, Proc. 19th ICAF Symp., R. Cook and P. Poole (Eds.). EMAS, Warley (1997), pp. 1015–1026.

25. de Vries, T.J. and Vlot, A., *Delamination behavior of spliced fiber metal laminates (FML). Part 1: Experimental results (ARALL, GLARE, A330, A3XX).* Composite Struct., Vol. 46 (1999), pp. 131–145.

26. Vlot, A., *Impact loading on fiber-metal laminates.* Int. J. Impact Engrg., Vol. 18 (1996), pp. 291–307.

27. Vogelesang, L.B., Gunnink, J.W., Roebroeks, G.H.J.J. and Müller, R.P.G., *Toward the supportable and durable aircraft fuselage structure.* Estimation, Enhancement and Control of Aircraft fatigue Performance, Proc. 18th ICAF Symposium, J.M. Grandage and G.S. Jost (Eds.). EMAS, Warley (1995), pp. 257–271.

28. Vlot, A., *Impact properties of fibre metal laminates.* Composite Engrg., Vol. 10 (1993), pp. 911–927.
29. Van Veggel, L.H., Jongebreur, A.A. and Gunnink, J.W., *Damage tolerance aspects of an experimental ARALL F-27 lower wing skin panel.* New Materials and Fatigue Resistant Aircraft Design, Proc. 14th ICAF Symposium, D.L. Simpson (Ed.). EMAS, Warley (1987), pp. 465–502.

Index

imagine ...observe ... think .....

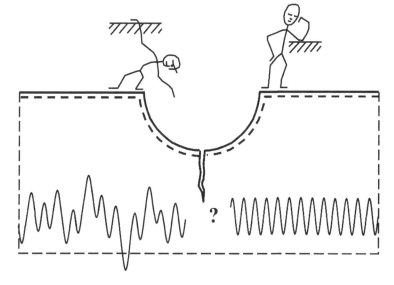